Chromatin

Chromatin

Structure and Function

Third Edition

A. WOLFFE

*National Institutes of Health,
Bethesda, Maryland, USA*

ACADEMIC PRESS

San Diego London Boston
New York Sydney Tokyo Toronto

Academic Press
A Harcourt Science and Technology Company
Harcourt Place, 32 Jamestown Road, London NW1 7BY, UK
http://www.academicpress.com

Academic Press
A Harcourt Science and Technology Company
525 B Street, Suite 1900, San Diego, California 92101-4495, USA
http://www.academicpress.com

ISBN 0-12-761915-1 (Pbk)

A catalogue record for this book is available from the British Library

Printed and bound by CPI Antony Rowe, Eastbourne

Contents

Preface to the First Edition

Research on chromatin structure and function is expanding rapidly. Technical advances allow us to follow the events regulating gene expression in the eukaryotic nucleus in molecular detail. Within the chromosome, alterations in the organization and accessibility of key regulatory DNA sequences can be documented and interpreted. This book is intended to introduce scientists to this exciting field, in the expectation that many more contributions will be required before we understand completely how the nucleus of a eukaryotic cell functions.

The book has five sections. The first section is a brief overview of the issues discussed and an historical account of their development. The second section describes the structure of chromatin and chromosomes as far as it is known. Concepts concerning chromatin structure are already very well developed; indeed, many of the biophysical techniques and paradigms for studying protein–nucleic acid interactions were pioneered using the basic unit of chromatin, the nucleosome, as a model. In contrast, large-scale chromosomal architecture is much less well defined, as is the influence of modifications of structural proteins on chromatin and chromosome organization. How these changes may contribute to the various requirements for correct chromosomal function is a recurring theme.

A complete understanding of the eukaryotic nucleus requires not only that we know how to take it apart, but also that we can assemble it from the various component macromolecules. The third section describes the approaches, results and interpretations of experiments designed to accomplish this task. The biological constraints of assembling a chromosome rapidly are discussed with reference to its final form and properties.

Form and function are intimately related. Once a complete understanding of a process is achieved, it is impossible to separate one from the other. The fourth section describes the multitude of approaches taken towards resolving how DNA can be folded into a chromosome and still remain accessible to the regulatory proteins, and allow processive enzymes to move along the length of the DNA molecules. It is in this field of research that much of the current progress on the interrelationship of chromatin structure and function is taking place. The final section offers a perspective on where prospects for future development might lie.

I would like to thank participants in the NIH chromatin group for sharing their ideas and results, especially Drs Trevor Archer, David Clarke and Sharon Roth. I am indebted to Drs Randall Morse, Geneviève Almouzni, Jeffrey Hayes and my Editor Dr Susan King for their comments on the text. Appreciation and thanks are given to Ms Thuy Vo and Mr William Mapes for preparing the manuscript and figures. Finally I thank my wife Elizabeth for her patience and support during the preparation of this book.

Alan Wolffe

Preface to the Second Edition

The impact of chromatin structure on gene activity and many other nuclear events has become increasingly apparent over the past four years. Tremendous progress has been made concerning the structure and function of the nucleoprotein structures regulating transcription, replication and repair within the eukaryotic chromosome. Important recent advances include the determination of the internal organization of the nucleosome. The histones are found to have unexpected structural similarities to known transcription factors. Similar structures point to similar functions and this emphasizes the importance of considering both the architectural roles of histones and transcription factors in regulatory complexes. Genetic experiments have introduced a whole new significance both to the histones and to other proteins that control long-range chromosomal compaction and regulate differential gene activity. The current text has been extensively modified to incorporate such new discoveries into the framework of established knowledge. The principal aim remains to introduce interested scientists to chromatin.

I would like to thank my colleagues at NIH for sharing their ideas and results. I am indebted to Drs Dmitry Pruss, Horace Drew, Jeffrey Hayes, Stefan Dimitrov, Mary Dasso and Geneviève Almouzni for invaluable discussions. Drs Randall Morse and Jeffrey Hansen read the text for which I am particularly grateful. The interpretation of data and any errors are my own. Appreciation and thanks are given to my Editor Dr Tessa Picknett and to Ms Thuy Vo for help with preparation of the text. Finally, I thank Elizabeth and Max for their patience and support.

Alan Wolffe
February 1995

Preface to the Third Edition

Progress in chromatin research in the past three years has been remarkable. Pre-eminent in recent discoveries is the role of transcriptional coactivators and corepressors as histone modification enzymes. Scientists investigating transcriptional control and signal transduction are now faced with the need to consider chromatin structural modifications as a primary regulatory mechanism. Other advances concerning the nucleosome include the definition of unusual chromatin architecture on human disease genes, the expansion of the families of proteins that resemble chromatin components, and the solution of the crystal structure of the nucleosome core. The nucleus itself is also increasingly recognized as having structural and functional compartmentalization. This organization can contribute to epigenetic effects that have important roles in gene expression and development. The reversibility of such compartmentalization has been dramatically demonstrated through the successful mammalian cloning experiments. New sections and extensive rewriting have integrated these discoveries into the framework of established knowledge. The principal aim remains to introduce interested scientists to chromatin.

I would like to recognize the contributions of my colleagues at NIH and especially the Chromatin Interest Group in sharing their ideas and results. I am indebted to Drs Dmitry Pruss for many of the illustrations, and to Drs Mary Dasso, Jeffrey Hansen, Stefan Kass, Hitoshi Kurumizaka, Nicoletta Landsberger, Guofu Li, John Strouboulis, Alexander Strunnikov, Paul Wade and Jiemin Wong for invaluable discussions. A special thanks to Ms Thuy Vo for the preparation of the text. I thank my wife Elizabeth for her patience and support, Max and Katherine for limiting their destruction of the manuscript.

I also thank Dr Tessa Picknett and Siân Davies at Academic Press, for their assistance in bringing this edition to print. Thanks also to Blackwell Science, the Company of Biologists, Elsevier Science and Oxford University Press for permission to reproduce previously published material.

Alan Wolffe
September 1997

CHAPTER ONE

Overview

1.1 INTRODUCTORY COMMENTS

Our knowledge of how the hereditary information within eukaryotic chromosomes is organized and used by a cell has increased enormously through the application of molecular biology and genetics. Technical advances now allow individual DNA sequences to be isolated and their association with proteins within the cell nucleus to be determined. Experimental progress has led the biologist to explore long-standing questions concerning how a particular cell acquires and maintains its individual identity. Developmental biologists have used new methodologies to investigate at a molecular level how an egg differentiates into different cell types. These questions have led scientists to the realization that growth, development and differentiation are directed by regulated changes in the form and composition of specific complexes of protein and DNA within the nucleus. Understanding how these complexes are assembled and function has become a central theme in modern biology.

Many of the techniques used to probe protein–DNA interactions were developed by researchers interested in the basic structural matrix of chromosomes – chromatin. This complex of DNA, histones and non-histone proteins has been exposed to a multitude of biochemical, biophysical, molecular biological and genetic manipulations. The structure of chromatin is by now well understood, but how it is folded and compacted into a chromosome is not. Knowledge of how chromatin is constructed preceded the development of methods capable of exploring function. The purification and cloning of non-histone proteins required to perform the complex events involved in DNA transcription, replication, recombination and repair is the focus of a continuing and intense research effort. Investigators now make use of their experience with chromatin structure and assembly to examine the function of the structural proteins and enzymes required for the maintenance, expression and duplication of the genome in a true chromosomal environment.

The conclusion from this research effort is that the organization of DNA into chromatin and chromosomes is essential for regulated processes within the nucleus. Histones, nucleosomes and the chromatin structures they assemble function as integral components of the machinery determining transcriptional activity, cellular identity and fate. It might be anticipated that a comparable integration of structure and function will have occurred with the molecular machines controlling replication, recombination and repair.

1.2 DEVELOPMENT OF RESEARCH INTO CHROMATIN STRUCTURE AND FUNCTION

Towards the end of the nineteenth century numerous investigators formulated the theory that chromosomes determined inherited characteristics (see Voeller, 1968). These studies were almost entirely based on cytological observations with the light microscope. Although chromosomes are clearly only present in the nucleus, the influence of components of the cytoplasm on inherited characteristics was examined by forcing embryonic nuclei into regions of the cytoplasm in which they would not normally be found (Wilson, 1925). These experiments and others led Morgan (1934) to propose the theory that differentiation depended on variation in the activity of genes in different cell types. The genes were clearly in the chromosomes, but their biochemical composition remained completely unknown.

The last quarter of the nineteenth century also saw the recognition of RNA (first identified as yeast nucleic acid), DNA (thymus nucleic acid) and the discovery of histones. Albrecht Kossel isolated nuclei from the erythrocytes of geese and examined the basic proteins in his preparations, which he named the histones (reviewed by Kossel, 1928). The apparent biochemical simplicity of DNA and the obvious complexity of protein in chromosomes led investigators mistakenly to regard the latter component as the major constituent of the elusive genes (Stedman and Stedman, 1947). Only the gradual acceptance of experiments on the capacity of DNA alone to change the genetic characteristics of the cell (Avery *et al.*, 1944) led to the recognition of nucleic acid as the key structural component of a gene.

The elucidation of the double helical structure of DNA with its immediate implications for self-duplication, opened up the new approaches of molecular biology to clarifying the nature of genes (Watson and Crick, 1953). Although the double helix was now recognized as containing the requisite information to specify a genetic function, how this information was controlled was not understood. The apparent heterogeneity of the histones due to proteolysis and the various modifications of these proteins suggested that they might be important in regulating genes. Eventually methodological improvements for isolating and resolving the different histones demonstrated that they were highly conserved in eukaryotes and that only a few basic types existed (Fitzsimmons and Wolstenholme, 1976). This lack of variety implied that histones themselves were unlikely to be the determinants of gene specific transcription. However, a key role for histone modification remained central to prevailing ideas of transcriptional regulation (Allfrey *et al.*, 1964).

A major breakthrough came in the 1970s when a combination of methodologies, including nuclease digestion, protein–protein cross-linking, electron microscopy and sedimentation analysis, determined that chromatin consisted of a repetitive fundamental nucleoprotein complex, which came to be called the nucleosome.

Structural studies on the nucleosome continue to the present time. Current and past research reveals the nucleosome to be a remarkably complex structure in which DNA is wrapped around the histones. The integrity of the nucleosome depends on highly specific histone–histone interactions, and the recognition by the histones of DNA structural features as the nucleosome is assembled. The core histones are present as an octamer, consisting of two molecules of H2A, H2B, H3 and H4. Histones H3 and H4 assemble a tetramer (($H3, H4)_2$) that wraps DNA such that two dimers of H2A and H2B can stably associate. Once two turns of DNA are wrapped around the octamer, a fifth linker histone, such as histone H1, can be stably incorporated to complete the assembly process. Although all nucleosomes maintain these architectural features, there are many variations built upon this common theme.

Nucleosomal structures can contain different forms of particular core histones or linker histones. These histone variants are the products of distinct genes which may be differentially expressed during development (Newrock *et al.*, 1977). The histones can also be post-translationally modified to different extents. Early experiments associated different types of histone modification with particular nuclear functions such as transcription (Allfrey *et al.*, 1964). Many early attempts were made to interrelate general differences in the transcriptional activity of genes to the solubility properties of chromatin dependent on histone modification or differences in histone content.

Recombinant DNA methodologies facilitated the isolation and cloning of defined DNA sequences, and DNA sequencing enabled the *cis*-acting elements potentially controlling gene expression to be defined (Brown, 1981). Hybridization analysis allowed the transcriptional activity of specific genes to be related to their accessibility to nucleases such as DNase I (Weintraub and Groudine, 1976). More detailed studies revealed that the regulatory DNA, such as promoter and enhancer sequences, was hypersensitive to DNase I cleavage (Wu *et al.*, 1979). Chromatin was perceived as having a precise organization that was certainly modified by the transcription process. It was even possible to infer that structural features of chromatin might actually determine the potential for transcription to occur. Nevertheless, analysis of the nuclease sensitivity of chromatin was primarily descriptive. The molecules that actually directed the transcription of specific eukaryotic genes could not be determined through these approaches.

The enzymatic activities of the eukaryotic RNA polymerases had been characterized through the early 1970s. An initially disappointing conclusion from these studies was that these polymerases alone did not recognize the regulatory elements of eukaryotic genes with any specificity when the template was naked DNA. Roeder and colleagues (Parker and Roeder, 1977; Jaehning and Roeder, 1977) made the seminal discovery that RNA polymerases would accurately transcribe genes within chromatin, but not as naked DNA. The hunt was now on for the auxiliary proteins that would determine the specific initiation of transcription by RNA polymerase.

The early searches for these transcription factors were dependent on the development of *in vivo* and *in vitro* assays for transcription. Microinjection of purified or cloned genes into the nuclei of eukaryotic cells was an early assay system used to define the *cis*-acting sequences recognized by transcription factors (Brown and Gurdon, 1977). Subsequent assays relied on *in vitro* transcription extracts (Wu, 1978; Weil *et al.*, 1979). These assays led to the purification and characterization of the first gene-specific eukaryotic transcription factor in 1980 (Engelke *et al.*, 1980; Pelham and Brown, 1980).

Much of the research effort on transcriptional regulation during the 1980s focused on the further definition of *cis*-acting elements and *trans*-acting factors involved in the initiation of the transcription process (Johnson and McKnight, 1989). The *in vitro* transcription or transfection assays used to examine the function of transcription factors did not require templates to be within their normal chromosomal environment for transcription to occur. In general these assays examined mechanisms that stimulated gene transcription, but did not examine the repression of transcription or the regulation of transcription in a physiological context.

Although far from the mainstream of research on transcription, the 1980s also saw the discovery of nucleosome positioning around eukaryotic genes (Simpson and Stafford, 1983). Application of genomic footprinting methodologies established that this phenomenon was a feature of several regulatory DNA sequences (Almer *et al.*, 1986; Richard-Foy and Hager, 1987). Histones were increasingly perceived as having the potential for specific effects on the transcription process. Experiments that combined *in vitro* transcription systems with natural chromosomal templates revealed a specific role for histones in transcriptional regulation (Schlissel and Brown, 1984). All of this work relied upon the detailed analysis of particular promoters in individual laboratories. The overall relevance of chromatin structure to the eukaryotic transcription process was difficult to establish from these studies. Nevertheless, they provided the foundation for the interpretation of genetic experiments that did in fact determine the general significance for transcription of assembling DNA into nucleosomes.

In a series of insightful experiments Grunstein, Winston and colleagues (Han *et al.*, 1987, 1988; Han and Grunstein, 1988; Clark-Adams *et al.*, 1988) determined that changes in nucleosomal packaging had pleiotropic effects on gene activity. Subsequent work by these investigators and Mitch Smith established that very specific modifications in histone structure could either activate or repress specific genes (Megee *et al.*, 1990; Durrin *et al.*, 1991; Mann and Grunstein, 1992). This led directly to the resurgence of interest towards understanding gene activity in the natural chromosomal environment that has characterized much of the research effort in eukaryotic transcriptional regulation over the past few years.

The new-found interest in the role of chromatin in transcriptional regulation has been fuelled by progress in two specific areas. Structural studies led to the recognition that the histones were isomorphous with components of the transcriptional machinery (Arents and Moudrianakis, 1993; Clark *et al.*, 1993; Ramakrishnan *et al.*, 1993; Xie *et al.*, 1996; Luger *et al.*, 1997). These observations provided an architectural foundation for examining the specific roles of histones and transcription factors in the assembly and function of regulatory nucleoprotein complexes. Specific modifications to nucleosomal architecture through histone acetylation, removal of histones H2A/H2B or H1 were shown to alleviate the repressive effects of chromatin assembly (Lee *et al.*, 1993; Bouvet *et al.*, 1994; Ura *et al.*, 1995). In certain instances chromatin assembly was also shown to stimulate the transcription process (Schild *et al.*, 1993). Thus the potential roles of nucleosomal proteins in gene control became more interesting (van Holde, 1993). Biochemical purification of histone acetyltransferases and deacetylases (Brownell *et al.*, 1996; Taunton *et al.*, 1996) provided an even closer link between chromatin and the transcriptional machinery. Histone acetyltransferases were discovered to be components of large macromolecular complexes known as coactivators, which are targeted to specific promoters by transcriptional activators. Therefore a direct link was established between histone acetylation and transcriptional activation. Histone deacetylases were found within corepressor complexes that turn genes off. Once again, histone chemistry became an important variable to consider in transcriptional control.

It is now recognized that to understand transcriptional control or any other regulated event in the nucleus it is necessary to define the chromatin structure within which DNA is utilized. Aside from the characterization of specific architecture, we must also determine how structure might change. Chromatin is not static, but dynamic. Targeted histone modifications within regulatory nucleoprotein complexes have emerged as a means of modulating the stability of repressive chromatin structures and the transcription process itself. The observations made using simple model systems are having an impact on our understanding of both development and disease. It is now probable that our increasing knowledge of both chromatin and chromosome structure and function in the nucleus will provide many avenues for future advances in biotechnological and medical fields.

CHAPTER TWO

Chromatin Structure

Chromosomes represent the largest and most visible physical structures involved in the transfer of genetic information. Surprisingly, our understanding of chromosome organization is most complete for the smallest and most fundamental structural units. These units are the nucleosomes which contain both DNA and histones. Long folded arrays of nucleosomes comprise the vast majority of chromatin. In this section I discuss the structural features of DNA and histones, how they assemble into nucleosomes and how nucleosomes fold into chromatin fibres. Finally, I describe what we know about the organization of the chromatin fibre into a chromosome and how this can be modified in various ways.

2.1 DNA AND HISTONES

The most striking property of a chromosome is the length of each molecule of DNA incorporated and folded into it. The human genome of 3×10^9 bp would extend over a metre if unravelled; however, this is compacted into a nucleus of only 10^{-5} m in diameter. It is an astonishing feat of engineering to organize the long linear DNA molecule within ordered structures that can reversibly fold and unfold within the chromosome. Not surprisingly, many aspects of chromosome structure reflect the impediments and constraints imposed by having to bend and distort DNA.

2.1.1 DNA structure

DNA has an elegant and simple structure around which the chromosome is assembled. The DNA molecule exists as a long unbranched double helix consisting of two antiparallel polynucleotide chains. DNA always contains an equivalent amount of the deoxyribonucleotide containing the base adenine (A) to that with the base thymine (T), and likewise of the deoxyribonucleotide containing the base guanine (G) to that with the base cytosine (C) (Fig. 2.1). Each base is linked to the pentose sugar ring (2-deoxyribose) and a phosphate group. The 5′ position of one pentose ring is connected to the 3′ position of the next pentose ring via the phosphate group (a 5′-3′ linkage) to create the polynucleotide chain (Fig. 2.2). The two antiparallel polynucleotide chains are attached to each other by hydrogen bonding between the bases. G is always base paired to C, and A is always base paired to T. In addition to the stability imparted by hydrogen bonding, hydrophobic base stacking interactions occur along the middle of the double helix (Fig. 2.3) (see Calladine and Drew, 1997 or Sinden, 1994 for details).

Physical studies using X-ray diffraction indicate that under conditions of physiological ionic strength, DNA is a regular helix, making a complete turn every 3.4 nm with a diameter of 2 nm. This particular DNA structure is known as B-DNA and has approximately 10.5 bp/turn of the helix. This means that every base pair is rotated

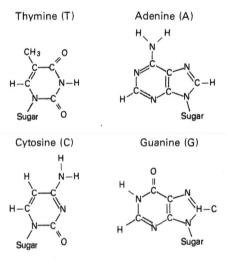

Figure 2.1.   The four bases found in DNA.

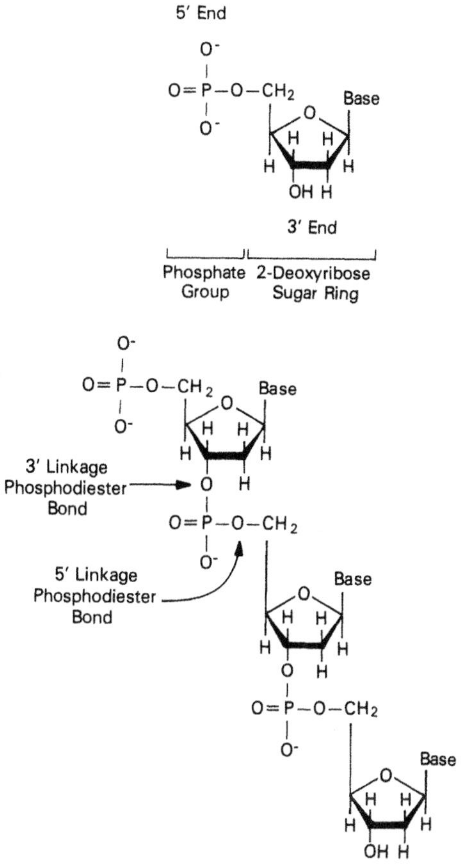

Figure 2.2. A nucleotide and a polynucleotide chain.

approximately 34° around the axis of the helix relative to the next base pair. This results in a twisting of the two polynucleotide strands around each other. A double helix is formed that has a minor groove (approximately 1.2 nm across) and a major groove (approximately 2.2 nm across). The geometry of the major and minor grooves of DNA will be seen later to be crucial in determining the interaction of proteins with the DNA backbone. The double helix is right handed (Fig. 2.4).

Beyond this basic description, DNA structure is exceedingly plastic. Crystallization of various oligonucleotides indicates that a variety of DNA sequences will yield recognizable B-form DNA structures (Privé et al., 1991; Yanagi et al., 1991). More severe alterations in the conditions under which DNA is examined do, however, generate

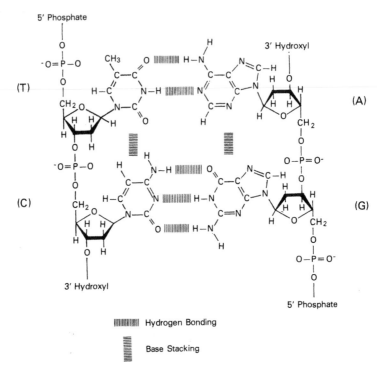

Figure 2.3.   The interactions stabilizing the two antiparallel poly-nucleotide chains in DNA.

distinct conformations. Dehydrating the fibre will cause the double helix to take up a structure known as A-DNA (11 bp/turn); or placing DNA with a defined sequence of alternating G and C bases in solutions of high ionic strength will lead to the formation of a left-handed helix known as Z-DNA (12 bp/turn). The existence of either of these extreme structures in the eukaryotic nucleus under normal physiological conditions is controversial. However, their formation indicates the gross morphological changes that DNA can be forced to undergo (Drew *et al.*, 1988; Calladine and Drew, 1997).

How do we know what structure populations of DNA molecules have in solution? Two experimental methodologies have been commonly used. The first employs DNA cleavage reagents and a flat crystal surface (Rhodes and Klug, 1980). When DNA is absorbed from solution on to a flat calcium phosphate surface and cut with DNase I, the enzyme cuts DNA most readily where it is exposed away from the surface. The average spacing between the sites of cleavage gives the approximate number of base pairs per turn of DNA (Fig. 2.5). This is

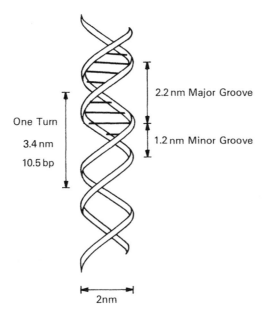

Figure 2.4. The dimensions of DNA.
Base pairs are shown as horizontal lines for one turn of the double helix.

determined by the electrophoresis of denatured molecules through a polyacrylamide gel. A better reagent for this purpose is the hydroxyl radical. Hydroxyl radicals are generated by the Fenton reaction in which an Fe(II) EDTA complex reduces hydrogen peroxide to a hydroxide anion and a hydroxyl radical.

$$[Fe(EDTA)]^{2-} + H_2O_2 \rightarrow [Fe(EDTA)]^{1-} + OH^- + \cdot OH$$

The radical is about the size of a water molecule and has little sequence specificity in cleaving DNA. This it does by breaking the pentose sugar rings of individual deoxyribonucleotides. In contrast, DNase I is a large enzyme which has considerable sequence preferences. In both instances, the number of base pairs per turn of a large population of different DNA sequences bound to a crystal surface is found to be 10.5 (Tullius and Dombroski, 1985). This result is consistent with DNA having a B-form configuration as determined by X-ray studies.

The second method to examine DNA structure in solution reaches the similar conclusion that DNA has a B-form conformation at physiological ionic strength; however, a completely different strategy is used. It is generally found that a population of closed circular DNA

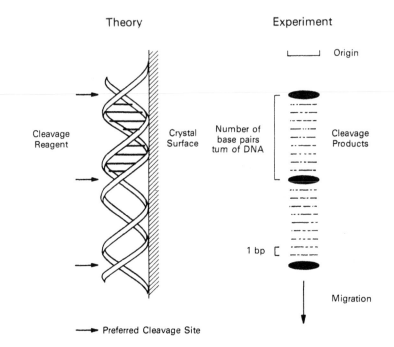

Figure 2.5.   Determining the helical periodicity of DNA in 'solution' through binding to a flat crystal surface and cleavage with an enzyme or a chemical reagent.

In theory the most exposed region of the double helix will be cut preferentially, experimentally this is reflected in a larger population of DNA fragments cut at this site after resolution on a polyacrylamide gel (darker bands). The distance between darker bands in base pairs is the helical periodicity (number of base pairs per turn) of DNA.

molecules, identical in length and sequence, contains different numbers of superhelical turns. Superhelical turns can be simply defined by the following description: a single superhelical turn is introduced into a closed circular DNA molecule if the molecule is broken, one end of the molecule is then fixed, the other is rotated once and the two ends then rejoined. Supercoils can be positive or negative depending on which way the free DNA end is rotated. Closed circular molecules of the same length and sequence with different numbers of superhelical turns are known as topoisomers. Each population of small closed circular DNA molecules that differ in length by a few base pairs will exist as a distribution of topoisomers. These can be resolved by electrophoresis through an agarose gel matrix. A molecule which has a length corresponding to an integral number of helical turns will exist predominantly as a single topoisomer whereas a molecule which deviates from this by half a helical turn will be equally likely to exist with the superhelical turn in a positive or negative sense. The number of DNA molecules with a particular mobility in the agarose gel will be reduced by half since the molecules exist as an equal mixture of topoisomers. Examining the relationship between DNA length and the distribution of topoisomers allows the number of base pairs per turn of DNA to be calculated. The result of 10.5 bp/turn is close to that derived from crystal binding studies (Horowitz and Wang, 1984). Finally, theoretical calculations of the most stable configuration of DNA, which actually preceded much of the experimental work, suggested a value of 10.6 bp/turn (Levitt, 1978). The range of values around 10.5 bp/turn, obtained both experimentally and theoretically, provides a sound basis for considering alterations in this structure based on DNA sequence content and histone–DNA interaction.

Aside from the dramatic changes in DNA structure seen on formation of A- or Z-DNA, local variations in DNA sequence can significantly influence DNA conformation and properties of the helix. Our most extensive knowledge of the local changes in B-form DNA structure due to sequence content comes from studying AT-rich DNAs. For example, oligo(dA).oligo(dT) tracts are found experimentally, using both spectroscopic techniques and DNA cleavage reagents such as the hydroxyl radical, to be straight and rigid with a constant narrow minor groove width (Nelson *et al.*, 1987; Hayes *et al.*, 1991a). This is believed to be a consequence of maximizing the hydrophobic base stacking interactions between adjacent A.T base pairs in the DNA helix (Fig. 2.3). This stabilization process requires the bases to be more twisted relative to each other than would normally be found in typical B-form DNA. Chains of these base pairs have the correct geometry to allow at least two water molecules per base pair to become highly ordered along the DNA backbone. This creates a 'spine of hydration' which contributes to the rigidity of oligo(dA) .oligo(dT) tracts (Berman, 1991). Changes in sequence that affect these structural features lead to widening of the minor groove; for example, a G.C base pair will disrupt the straight path and rigidity of an oligo(dA).oligo(dT) tract. In contrast to oligo(dA).oligo(dT), oligo [d(AT)] tracts are conformationally flexible. This flexibility is a consequence of not being able to achieve efficient hydrophobic base stacking interactions between consecutive T.A and A.T base pairs without severely distorting the DNA helix (Travers and Klug, 1987; Travers, 1989). Finally, short oligo(dA).oligo(dT) tracts (4–6 bp in length) that are phased with a periodicity similar to that of the DNA helix itself will cause the molecule to be curved. This is due to a narrowing of the minor groove every turn of DNA caused by the phased oligo(dA).oligo(dT) tract (Koo *et al.*, 1986). Periodicities that are greater or smaller than 10–11 bp will cause the normally straight DNA to take on a 'corkscrew-like' path. In spite of this wide variation in 'B-form' DNA structure, all of these DNA sequences can be assembled into chromatin (Section 2.2.5).

DNA structure is thought to have an important role in certain human genetic diseases characterized by the presence of repeats of particular trinucleotide sequences. These trinucleotide repeats are found in the gene whose aberrant expression leads to the disease phenotype (Bates and Lehrach, 1994; Sutherland and Richards, 1995). The segments of DNA containing trinucleotide repeats are unstable with the potential to expand from generation to generation. Two trinucleotide repeat sequences are of particular interest: $(CTG)_n$ is associated with many diseases including Huntington's disease, myotonic dystrophy, spinocerebellar ataxia type 1, and hereditary dentatorubral-pallidoluysian atrophy; $(CCG)_n$ is associated with fragile X mental retardation. Normal individuals have relatively few copies of these repeat sequences whereas diseased individuals have many copies (> 50). The number of repeats influences both the expansion process and the disease. How this influence is exerted has been the focus of a great deal of attention (see also Section 2.2.5).

Of all the many potential trinucleotides present in the genome only reiterated CTG and CCG sequences show the special properties of instability and tendency towards expansion (Han *et al.*, 1994). These sequences have the capacity to form stable hairpin structures when they reach a certain threshold length of 40 to 50 repeats (Gacy *et al.*, 1995). It has been suggested that the ability to form stable hairpins might explain both the dependence on particular trinucleotides and the length of sequence for repeat expansion. The favoured model for expansion predicts that DNA polymerase might 'slip' on reiterated sequences during replication leading to small increases in trinucleotide repeat copy number. Once the copy number becomes large enough, the single-stranded DNA at the replication fork might form a stable hairpin looping out intervening DNA and leading a small 'slip' to generate a large expansion of the trinucleotide repeat sequence (Gacy *et al.*, 1995). Under these special circumstances the capacity to form unusual hairpin DNA structures might contribute to the generation of a disease phenotype.

The $(CTG)_n$ and $(CCG)_n$ sequences appear to have no reason to adopt unusual structural features when present as duplex DNA, however there is evidence that these sequences might differ from conventional DNA. Both $(CTG)_n$ and $(CCG)_n$ containing a methylated CpG dinucleotide form very stable complexes with histones (see Section 2.2.5). The location of these sequences within the nucleosome suggests that they might have unusual structural properties (Godde and Wolffe, 1996; Godde *et al.*, 1996). DNA fragments containing CTG repeats have anomalous electrophoretic mobilities suggesting either a reduction in DNA flexibility or the presence of hairpin structures (Chastain *et al.*, 1995; Pearson and Sinden, 1996). This is surprising because B-form flexible DNAs are generally favoured for nucleosome assembly due to requirement for DNA to be wrapped around the histones, and hairpins tend to inhibit nucleosome formation (Satchwell *et al.*, 1986; Nobile *et al.*, 1986; see Section 2.2.5). Clearly there is much more to be understood about the properties of DNA and how they might influence events in the nucleus.

Summary
Under most physiologically relevant conditions DNA is a stable B-form double helix, with 10.5 bp/helical turn, a major and a minor groove. Local variations in sequence content can direct DNA to have intrinsic rigidity, flexibility or curvature. Under special circumstances certain trinucleotide repeats $(CTG)_n$ and $(CGG)_n$ associated with human genetic disease can form non B-form hairpin structures.

2.1.2 The histones

The primary proteins whose properties mediate the folding of DNA into chromatin are the histones. Aside from the compaction of DNA, the histone proteins undertake protein–protein interactions between themselves and other distinct chromosomal proteins. These interactions lead to several constraints on the properties of histones contributing to maintaining their high degree of evolutionary conservation. Not all eukaryotic cells have histones, for example dinoflagellates are reported to package the majority of their DNA with small basic proteins completely unlike histones (Vernet *et al.*, 1990); and in mammalian species the majority of DNA in spermatozoa is compacted through interaction with basic proteins known as protamines (Section 2.5.5).

Each nucleosome contains a core of histones around which DNA is wrapped. This core contains two molecules of each of four different histone proteins: H2A, H2B, H3 and H4. These are known as the core histones. Since histones can be removed from DNA by high salt concentrations, the major interactions between DNA and the core histones appear to be electrostatic in nature. Histones H2A and H2B dissociate first as the salt concentration is raised followed by histones H3 and H4 (see Section 2.2.2). Studies of this type, coupled to chemical cross-linking, demonstrated that histones H2A and H2B form a stable dimer (H2A/H2B), whereas histones H3 and H4 form a stable tetramer ((H3/H4)$_2$) in the absence of DNA (Kornberg, 1974; Kornberg and Thomas, 1974). Many histone proteins have been purified and their amino acid sequences determined. Subsequently, histone genes have been cloned and a very complete picture of core histone sequence properties established.

All core histones are remarkably conserved in length and amino acid sequence through evolution. Histones H3 and H4 are the most highly conserved, for example calf and pea histone H4 differ at only two sites in 102 residues (De Lange *et al.*, 1969a,b). Histones H3 and H4 have a central role both within the nucleosome and in many chromosomal processes (Sections 2.2.2 and 2.2.4). These functional and structural requirements presumably contribute to their remarkable sequence conservation. Histones H2A and H2B are slightly less conserved. All of the core histones are small basic proteins (11 000–16 000 Da molecular weight) containing relatively large amounts of lysine and arginine (more than 20% of the amino acids). Histones H2A and H2B contain more lysine (14 out of 129, and 20 out of 125 amino acids, respectively, in calf), and histones H3 and H4 contain more arginine (18 out of 135, and 14 out of 102 amino acids, respectively, in calf) (van Holde, 1988). All four histones contain an extended histone-fold domain at the carboxyl (C-) terminal end of the protein through which histone–histone and histone–DNA interactions occur, and charged tails at the amino (N-) terminal end which contain the bulk of the lysine residues (Fig. 2.6) (Arents *et al.*, 1991). The C-terminal histone fold domains contain three α-helices. The histone fold domains might be expected to be conserved due to their central structural role in the nucleosome; however, the amino acid sequence of the charged N-terminal tails is also conserved. These charged tails are the sites of many post-translational modifications of the histone proteins (Fig. 2.6, sites of acetylation are indicated, Section 2.5.2). The conservation of the N-terminal tails is now recognized to be a consequence of both the targeting of post-translational modifications to the histone tails by transcriptional regulatory proteins and a key architectural role for the tails through interaction with other structural components of chromatin. These coactivators modify specific amino acids in the N-terminal tails (Kuo *et al.*, 1996). The N-terminal tails are the target

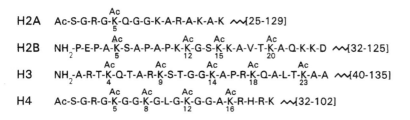

Figure 2.6. The organization of calf thymus histones.
The amino-terminal histone tails are shown. Sites of post-translational modification of lysine residues by acetylation are indicated (Ac). The structured histone fold domain consisting of three α-helices (cylinders) at the carboxyl terminus is shown.

for signal transduction pathways that modify chromatin structure (Section 2.5.4). The N-terminal tails of the core histones also provide contact surfaces for interaction with other proteins that organize higher-order chromatin structures (Hecht *et al.*, 1995; Edmondson *et al.*, 1996). Thus there are considerable constraints on the amino acid sequence of the N-terminal tails. Core histone variants in which the primary amino acid sequence is changed because of expression of different alleles of a histone gene, have important consequences for chromatin structure and function in many contexts, but especially during development (Sections 2.5.1 and 3.4).

Eukaryotic cells contain a fifth histone called the linker histone, of which the most common is called histone H1. In addition, many studies have examined the properties of a specialized linker histone from chicken erythrocytes known as histone H5. Both histone H1 and histone H5 are highly basic, being particularly rich in lysine and are slightly larger than core histones (> 20 000 molecular weight) (Fig. 2.7). Linker histones are the least tightly bound of all histones to DNA, and are readily dissociated by solutions of moderate ionic strength (> 0.35 M NaCl). The metazoan linker histones have a central structured winged-helix domain and highly charged tails at both the N- and C- termini. The central domain contains three α-helices attached to a three stranded β-sheet (Ramakrishnan *et al.*, 1993). The structured domain of the linker histone associates with the nucleosome, stabilizing histone–DNA interactions throughout the nucleosome core (Section 2.2.2). In addition, the linker histone tails interact with the DNA between nucleosomes.

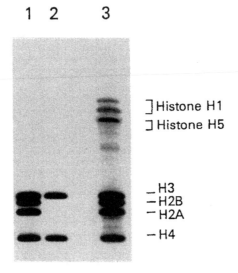

Figure 2.7. The histones are shown resolved on a denaturing polyacrylamide gel, separated by virtue of their size.
Core histones (H3, H2B, H2A and H4) and linker histones (H1, H5) are indicated. Histones were prepared from chicken erythrocytes.

There is considerable diversity in the structure of linker histones. In *Tetrahymena* the linker histone lacks a central structured domain entirely, consisting only of a peptide 163 amino acids long, with a very similar sequence composition to the C-terminal domain of a metazoan linker histone (Wu *et al.*, 1986). In contrast, within the yeast genome there is a gene encoding a histone H1-like protein with two structured winged helix domains (Landsman, 1996). Linker histones are also extensively post-translationally modified both during the cell cycle and during development (Section 2.5.3). These structural modifications have important consequences for the functional properties of the chromatin fibre.

Summary
Two types of histones exist, the highly conserved core histones and the much more variable linker histones. In metazoans, both types have a domain that is inherently structured and both have highly charged basic tails. These tail regions are the site of post-translational covalent modifications.

2.2 THE NUCLEOSOME

The nucleosome is the fundamental repeating unit of chromatin. Many of the techniques used to examine protein–nucleic acid interactions that are in common use today were pioneered on the nucleosome. Outlining how the current model of the nucleosome has been developed introduces the use of nucleases and chemical probes both of DNA structure and protein–DNA interaction (DNA footprinting reagents), non-denaturing gels to study large complexes of protein and DNA in their native state (mobility shift assays), together with various applications of spectroscopic analysis and other biophysical techniques.

2.2.1 The nucleosome hypothesis

The first clear insights into the nucleosomal organization of chromatin came from nuclease experiments (both intended and accidental) in which the DNA in chromatin was found to degrade to a series of discrete fragment sizes separated by multiples of 180–200 bp (Williamson, 1970; Hewish and Burgoyne, 1973). Each step in fragment size is now known to represent the DNA associated with a single nucleosome. Extensive nuclease digestion allowed each DNA fragment to be isolated as a complex with protein (Sahasrabuddhe and van Holde, 1974). These particles were found by sedimentation analysis in the analytical ultracentrifuge to have a mass of around 200 000 Da (176 000 was measured) of which the protein content was close to 110 000 Da (105 000 was measured). This we now know corresponds to the octamer of core histones in a nucleosome (two molecules each of histones H2A/H2B/H3 and H4) plus approximately 146 bp of DNA.

Electron microscopic analysis of chromatin provided further evidence for a structure consisting of discrete complexes of protein and nucleic acid arrayed along the DNA backbone. The pictures of 'beads on a string' were compelling evidence for a repeating particulate structure for chromatin (Woodcock, 1973; Olins and Olins, 1974). Each particle along the DNA backbone was approximately 10 nm in diameter, similar to that of the particles isolated by extensive nuclease digestion. Chemical cross-linking experiments led to the realization that the core histones existed in a precise stoichiometry (Kornberg and Thomas, 1974).

These observations, over a few years in the early 1970s, led to the proposal by Kornberg of the nucleosome model (Kornberg, 1974). This hypothesis suggested that each particle consisted of DNA and histones. DNA was wrapped around an octamer of the core histones, each octamer consisting of a tetramer of histones H3 and H4 ((H3/H4)$_2$) and two dimers of H2A/H2B (H2A/H2B). Initially it was thought possible that only a small fraction of chromatin in the nucleus might be organized in this way. However, the structural significance of the nucleosomal organization of chromatin was made clear by micrococcal nuclease digestion of whole nuclei. These studies revealed that over 80% of DNA in the nucleus was incorporated into nucleosomes (Noll, 1974a). Thus, the general relevance of the nucleosome for the folding of DNA in the eukaryotic nucleus was firmly established. Subsequent studies have considerably refined our understanding of its organization.

Summary

Studies involving nuclease digestion, analytical centrifugation, electron microscopy and chemical cross-linking led to the proposal that a fundamental repeating unit of chromatin existed, consisting of a precise stoichiometry of histones and DNA. This particulate structure became known as the nucleosome. The vast majority of DNA in the cell nucleus is organized into nucleosomes.

2.2.2 The organization of DNA and histones in the nucleosome

Like many scientific fields, the study of chromatin has developed a particular nomenclature: a *nucleosome* consists of one repeating length of DNA in the nucleus, generally determined by very slight micrococcal nuclease cleavage (see below), plus an octamer of core histones and a single molecule of linker histone; a *nucleosome core particle* consists of the octamer of core histones and the length of DNA that resists digestion even after extensive micrococcal nuclease cleavage, this being the 146 bp of DNA that have the strongest contacts with the histone core. The DNA that is in between nucleosome core particles, and that is lost when a nucleosome is trimmed to a core particle, is called *linker* DNA (Fig. 2.8).

Most nucleosomes have been isolated for study by micrococcal nuclease cleavage. Micrococcal nuclease cleaves chromatin in the most accessible DNA (Fig. 2.9). First the linker DNA between nucleosomes is cut, then the nuclease digests the rest of the linker before it cuts DNA directly across both strands within the nucleosome itself. The

Nucleosome Core Particle

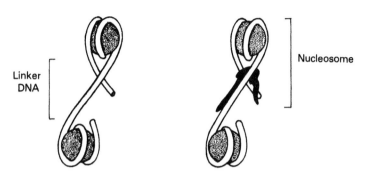

Linker
DNA

Nucleosome

Figure 2.8. The organization of a nucleosome core particle, a nucleosome and the position of linker DNA.
Core histones form the hatched spheres, DNA is the connecting tube and the linker histone is the solid shape. One potential position for the linker histone is shown, others exist (Section 2.3.1).

nucleosome, representing the first product of very slight micrococcal nuclease digestion, will therefore be progressively trimmed to a nucleosome core particle as digestion with micrococcal nuclease proceeds. The nucleosome core particle itself represents only a kinetic intermediate in the digestion of DNA. Eventually micrococcal nuclease can degrade the DNA in this residual structure and the core particle will fall apart. Separation of nucleosomes from nucleosome core particles demonstrated that the initial digestion of the linker DNA led to the loss of linker histone from the nucleosome (Noll and Kornberg, 1977).

The next advance in understanding nucleosomal structure came from the use of non-denaturing gel electrophoresis to examine large complexes of DNA and proteins in their native state. The mobility of free DNA is retarded through association with protein, producing a mobility shift. Following micrococcal nuclease digestion, crude nucleosomal fractions were first resolved on sucrose gradients. The smallest (mononucleosome) fractions were then electrophoresed through a polyacrylamide gel matrix in a mobility shift assay, and two complexes were resolved. The large (slower migrating) complex was found to contain histone H1, whereas the smaller complex did not (Varshavsky *et al.*, 1976). Subsequently, careful gradation in the extent

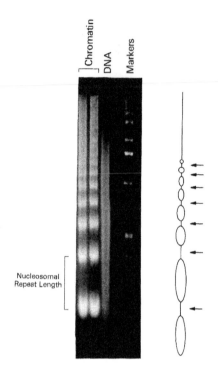

Figure 2.9.   Micrococcal nuclease cleavage of chromatin.
Chromatin and naked DNA are shown after treatment with micrococcal nuclease, removal of protein and resolution on an agarose gel. The nucleosome ladder of bands is clearly seen in the chromatin samples. Cleavage of naked DNA generates a wide distribution of DNA fragments visible as a smear. The visible bands in the chromatin samples are separated from each other by a single nucleosome repeat length of DNA.

of nuclease digestion allowed Simpson (1978) to isolate a discrete particle, called the chromatosome, consisting of an octamer of histones, one molecule of the linker histone H1 and about 160 bp of DNA.

The influence of histone H1 on nucleosome integrity was examined using biophysical techniques. Spectroscopic experiments on chromatosomes examined DNA conformational transitions in these particles dependent on increasing temperature in comparison to particles depleted of histone H1. Thermal denaturation of DNA requires the progressive disruption of both hydrogen bonds and base stacking interactions between the base pairs in DNA, and eventually the separation of the two strands of the DNA helix ('melting'). Histone H1 association was found to significantly stabilize DNA within the nucleosomal core particle (Simpson, 1978). These experiments led to the suggestion that histone H1 influenced not only the organization of linker DNA but also that of DNA wrapped around the histone octamer. These linker-histone induced changes are potentially important both for the folding of nucleosomal arrays into the chromatin fibre and for the interaction of *trans*-acting factors with DNA (Sections 2.3.2 and 4.2.2).

Having defined linker DNA and its interaction with the linker histone H1 in the nucleosome, we can now discuss the organization of the nucleosome core particle itself. Early studies of the organization of DNA in chromatin used nucleosomal core particles obtained by extensive micrococcal nuclease digestion of nuclei. These particles presumably contain representatives of every DNA sequence present in nucleosomes within the nucleus (> 80%, see Section 2.2.1). Using the enzyme DNase I, Noll found that single nicks were made in the DNA of the core particles with a periodicity of 10–11 bp. This periodicity of cleavage reflects the wrapping of DNA around the histone core. Compare these results with those from the nuclease cleavage of DNA when bound to a crystal surface (see Section 2.1.1). This wrapping implies that the minor groove of the double helix, which is recognized and cut by DNase I (Drew, 1984), is exposed only once per turn of the helix (Fig. 2.10). Noll found that the entire length of DNA within the core particle is exposed to nicking by DNase I every 10-11 bp which means that all 146 bp of DNA must lie on the surface of the histone core (Noll, 1974b).

Noll's cleavage data for the organization of DNA in the nucleosome were indirect. The most substantial evidence for DNA being wrapped around a histone core prior to crystallization came from neutron scattering in solution. These experiments relied on the observation that the intensity of neutron scattering by a particular macromolecule depends on the intensity of scattering by the solvent in which the macromolecule is bathed. By adjusting the scattering of the solvent through altering the ratio of heavy water (D_2O) to normal water (H_2O), certain macromolecules in solution could be made invisible. Thus, by looking at the scattering of DNA or protein in the nucleosome in aqueous solutions, it was found that DNA had a larger radius of gyration than the protein component. Therefore DNA was wrapped around the histones (Pardon *et al.*, 1975).

The observation that DNA lies on the surface of the nucleosome was greatly extended in detail by the original crystallization of nucleosome core particles by Klug and colleagues (Finch *et al.*, 1977; Richmond *et al.*, 1984). Analysis of the crystals to 7 Å resolution revealed the nucleosome core to have the disc-like shape dimensions

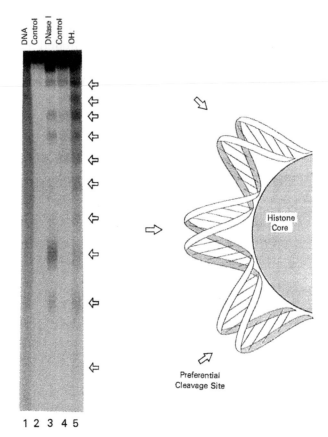

1 2 3 4 5

Figure 2.10.   DNase I and hydroxyl radical cleavage of DNA in the nucleosome.

DNA is made radioactive at one end, by the introduction of a phosphate group containing [32]P. Hydroxyl radical cleavage of naked DNA is shown (DNA). Control lanes are nucleosome core particles that do not have a cleavage reagent added to them. In the DNase I and OH• lanes the indicated cleavage reagent is added. The DNA fragments are denatured to single-strands and are then resolved on a denaturing polyacrylamide gel on the basis of size. Where DNA is exposed away from the histone core it is cut as indicated in the line drawing.

predicted from electron microscopy, sedimentation analysis and neutron scattering. The disc was 11 nm in diameter and 5.6 nm in height. DNA was wrapped in 1.75 turns of a left-handed double helix around the histone core (Fig. 2.11). However, the bending of the DNA around the histone core was not uniform. There were very sharp bends approximately one and four helical turns to either side away from the centre of the nucleosomal DNA. The existence of these

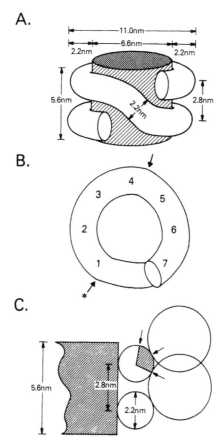

Figure 2.11. The organization of DNA in the nucleosome.
A. The dimensions of the nucleosome: the histone core is represented by the hatched cylinder, DNA by the open tube. This view overlooks the centre of nucleosomal DNA, at this point the DNA superhelix rises very steeply. B. The path of one turn of DNA. Numbers refer to turns of the DNA helix away from the dyad axis. The arrows represent the positions of more severe bending in DNA seen in the crystal structure. The asterisk marks the approximate position of a structural discontinuity detected by singlet oxygen or hydroxyl radical cleavage. C. The problem of transcription factor access to DNA in the nucleosome. A cross-section of one side of the nucleosome is shown. Transcription factor (large circle) access to DNA (small circle) is restricted by the histone core (hatched box) and the adjacent turn of DNA. The only freely accessible region of the DNA helix is the hatched segment – marked by arrows.

distortions in the path of nucleosomal DNA has since been confirmed through cleavage studies with chemical and enzymatic reagents sensitive to DNA structure (Hogan *et al.*, 1987; Pruss *et al.*, 1994; see Section 2.2.3). The nucleosome core appears almost symmetric, hence the centre of nucleosomal DNA also represents the dyad axis of the nucleosome core. It should also be recognized that because of the variation in DNA sequence the nucleosome core can never be truly symmetric unless it contains a 73 bp inverted repeat centered at the dyad axis. The overall bending of DNA and potential perturbations of the path of the double helix from a uniform curve have important implications for the phenomenon of nucleosome positioning (Section 2.2.5). The organization of the individual core histones within the nucleosomal core particle was not determined from the initial low resolution crystal structure of the histones with DNA (Richmond *et al.*, 1984). However, subsequent work on the crystal structure of the histone octamer in isolation has substantially clarified the internal structure of the nucleosome (Arents *et al.*, 1991; Arents and Moudrianakis, 1993) (Section 2.2.4). These predictions have been dramatically extended by the crytallization of a synthetic nucleosome core particle whose structure has been solved to less than 3 Å (Luger *et al.*, 1997).

Summary
Linker histones, such as histone H1, bind to the linker DNA and alter the stability with which DNA within the nucleosome core is associated with the histone octamer. DNA (146 bp) is wrapped in 1.75 left-handed superhelical turns around an octamer of the core histones.

2.2.3 The structure of DNA in a nucleosome

The structure of DNA in a nucleosome was initially refined from the 7 Å crystal structure using chemical probes. Solution studies are important because of possible distortion of DNA during crystal-lization (see Richmond *et al.*, 1984, Luger *et al.*, 1997). Independent evidence for a non-uniform deformation of the DNA double helix came from using highly reactive singlet oxygen. This reagent was found to react preferentially with DNA about 1.5 turns either side of the centre of the nucleosomal DNA in solution (Hogan *et al.*, 1987). Singlet oxygen is a sensitive probe for structural deformations in DNA that are associated with an unstacking of adjacent base pairs. DNA that is more easily distorted may prefer to occur 1.5 helical turns from the dyad axis and unstable TA steps are preferentially found at these sites within the nucleosome core (Satchwell *et al.*, 1986). Other evidence for a deformation of DNA at this position came from hydroxyl radical cleavage of DNA in nucleosomal core particles.

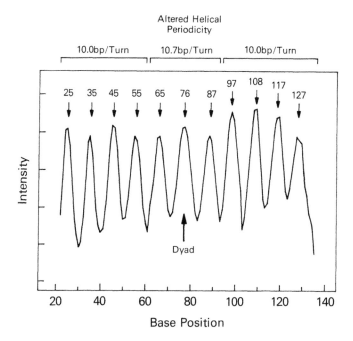

Figure 2.12. DNA has two different structures in the nucleosome. The three turns of DNA around the dyad axis have a different helical periodicity to those in the remainder of the nucleosome. A plot of hydroxyl radical cleavage frequency at each nucleotide of the DNA within nucleosome core particles is shown. Approximate positions of maximum cleavage frequency are indicated. Numbers indicate the absolute length of the single-stranded DNA fragments.

Single base pair resolution analysis of hydroxyl radical cleavage frequencies in nucleosomal DNA (Fig. 2.12) reveals that the central three turns of DNA in the nucleosome have a different number of base pairs per turn (10.7) than the remainder of the structure (10.0) (Fig. 2.12). When these two different structures juxtapose, DNA is likely to be distorted (Hayes *et al.*, 1990, 1991b; Puhl and Behe, 1993).

These sites of DNA distortion are important biologically since they are where the integrase encoded by the human immunodeficiency virus (HIV) prefers to direct integration of the viral genome within the nucleosome (Pruss *et al.*, 1994). This enzyme binds to DNA through the major groove and distorts DNA as a component of its reaction mechanism, hence the preference for DNA that has a wide major groove or that is already severely distorted in the nucleosome (Fig. 2.13). The HIV integrase is a useful example of how the association of DNA with histones can actually facilitate a process requiring DNA as a substrate.

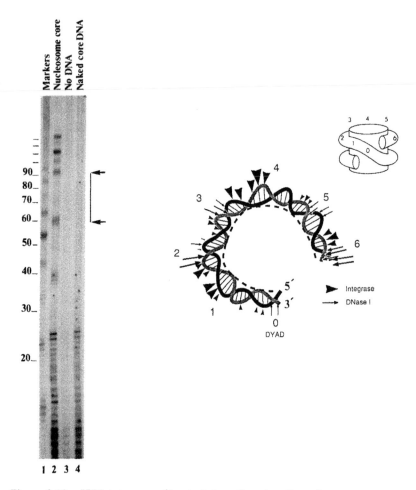

Figure 2.13. HIV integrase directs integration to sites of severe DNA distortion within the nucleosome core.

A. Integration products made with or without a heterologous target (nucleosome cores or naked core DNA) were resolved in a sequencing gel. Lane 1 shows markers of DNA fragments end-labelled with 1 $[\gamma-^{32}p]ATP$ and T4 polynucleotide kinase following DNase I digestion of chicken erythrocyte chromatin. Lane 2 integration into nucleosome cores, lane 3 self-integration products in the absence of target DNA and lane 4 integration into naked DNA extracted from nucleosome cores. The numbers represent actual distance between the integration site and the DNA 3'-terminus. The arrows indicate the sites of preferred integration 1.5 turns to either side of the dyad axis of the nucleosome core at which integration is favoured, the line between them indicates the 32 bp of DNA that includes the dyad axis at which integration occurs less frequently. B. Sites of integration (arrowheads) are shown for a single superhelical turn of DNA within the nucleosome. Arrows indicate sites of DNase I cleavage. Numbers are in Fig. 1B. The inset shows the position of the numbers on the surface of the nucleosome (see Pruss *et al.*, 1994 for details).

The hydroxyl radical cleavage pattern of DNA in the nucleosome core also demonstrates that the structure of DNA changes on association with the histones from that in solution (10.5 bp/turn) (Hayes *et al.*, 1990, 1991b). This has important consequences for the association of sequence-specific DNA binding proteins (Sections 4.2.2 and 4.2.3). As 146 bp of DNA are present in the nucleosome core particle, a simple calculation based on the hydroxyl radical cleavage results predicts that DNA wrapped around the histone core will have an average number of base pairs per helical turn of 10.2. This is in substantial agreement with several other independent measurements of this important value. Synthetic DNA curves designed to have different helical periodicities (see Section 2.1.1) reveal a preference of the histone core for a 10.2 bp/turn structure overall (Shrader and Crothers, 1990). DNase I cleavage analysis of DNA in the nucleosome revealed an average cleavage frequency of 10.4 bp/turn; however, the periodicity of DNase I cleavage sites at the two ends of DNA in nucleosomes was close to 10.0 bp, while the spacing of cleavage sites towards the centre was closer to 10.6 bp/turn (Lutter, 1978). Interpretation of these results as directly reflecting the helical periodicity of DNA in a nucleosome core particle is not possible. This is due to the large enzyme, DNase I, being sterically hindered from having equivalent access to DNA all around the histone core by the turn of DNA adjacent to the one it is cutting (Klug and Lutter, 1981). However, the use of other methodologies, such as examination of the frequency of thymidine dimer formation (photofootprinting) in nucleosomal DNA, reveals modulation with a periodicity of 10.0 bp/ turn on either side of a central region, which has a periodicity of about 10.5 bp/turn (Gale and Smerdon, 1988a) (Section 4.4). The most gruelling approach to this problem has required sequence analysis of DNA within 177 nucleosome core particles. Quantitation of the sequence data reveals periodic modulations in the frequency with which certain runs of three base pairs are found. The phases of these periodic modulations are offset by 2–3 bp across the central 2–3 turns of the nucleosome (Satchwell *et al.*, 1986). All of these experiments strongly suggest a change in helical periodicity for the central region of nucleosomal DNA.

What is precisely responsible for the distortion of the central portion of DNA in the nucleosome has been suggested from the high resolution structure of the histone octamer (Arents and Moudrianakis, 1993). Histones H3 and H4 assemble a surface at the centre of the nucleosome core along which arginine residues are spaced such that they will interact preferentially with DNA having a helical periodicity of 10.7 bp/turn (Section 2.2.4). The spacing of basic amino acids along the DNA binding surface of the histone octamer away from the central region (Section 2.2.4) favours DNA with a helical periodicity of 10.0 bp/turn. Theoretical calculations by Levitt had earlier predicted a 10.6 bp/turn structure for DNA free in solution and a 10.0 bp/turn structure for DNA constrained into an 80 bp circle as found in the nucleosome (Levitt, 1978).

The change in DNA structure on incorporation into a nucleosome from 10.5 to an average of 10.2 bp/turn partially resolves a long-standing problem that has come to be known as the 'linking number paradox' (Klug and Lutter, 1981). This follows from the fact that the nucleosome has at least 1.75 superhelical turns of DNA around the histone core, yet the measured number of superhelical turns introduced into a relaxed DNA molecule (in the presence of topoisomerases) is 1. This discrepancy is explained partially by the change in helical periodicity of DNA upon incorporation into a nucleosome, resulting in a decrease in the average number of base pairs/turn. This overwinding of DNA can account for the disappearance of 0.4 superhelical turns, but does not completely explain the 'paradox' (Hayes *et al.*, 1990). A more contorted path than the wrapping of DNA around a simple cylinder could account for the small differences remaining (White *et al.*, 1988).

Summary

DNA in the nucleosome is severely distorted into a superhelix containing two circles, each approximately 80 bp in length. The helical periodicity of this DNA changes from that in solution. Two regions of 10.0 bp/turn flank the three central turns of nucleosomal DNA which have a periodicity of 10.7 bp/turn (average for the nucleosomal core particle equals 10.2 bp/turn). The junctions between these regions of different DNA structure are the sites of the most severe DNA distortion in the nucleosome. These sites of distortion are recognized by HIV integrase. The activity of this enzyme is a useful example of how the assembly of chromatin can facilitate a biological process.

2.2.4 The position of the core histones in the nucleosome

Where are the individual core histones along the DNA backbone in the nucleosome core particle? Our understanding of their position has been greatly advanced through the solution of the crystal structure of the histone octamer and of the nucleosome core (Arents *et al.*, 1991;

Arents and Moudrianakis, 1993; Luger *et al.*, 1997). The high resolution structure of histones bound to DNA in a synthetic nucleosome core is a major advance, however we still need to determine if the structure of DNA and histones found in the crystal are the same in solution. In particular it is interesting to examine the potential for conformational changes in histone–DNA interactions dependent on the inclusion of proteins such as histone H1 and the HMGs or following post-translational modification of the histone tail domains. Early work indicated that the histone H3/H4 tetramer (H3/H4)$_2$, could organize DNA into nucleosome-like particles (Camerini-Otero *et al.*, 1976). Hydroxyl radical cleavage patterns of DNA in such particles demonstrate that the tetramer organizes the central 120 bp of DNA of a nucleosome identically to that of the nucleosomal core particle (Hayes *et al.*, 1991b; Fig. 2.14). Inclusion of the histone dimers H2A/H2B, protects DNA to each side of the 120 bp segment. This type of footprinting analysis of subnucleosomal particles agrees well with chemical cross-linking experiments which physically map histone–DNA contacts along nucleosomal DNA. Most of the chemistry for this

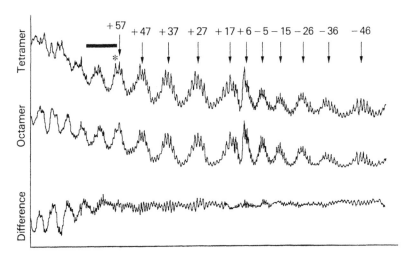

Figure 2.14. A comparison of DNA interactions with a complete octamer of core histones to those with the histone tetramer (H3/H4)$_2$.
A nucleosome structure including the 5S RNA gene is used. Densitometric analyses of hydroxyl radical cleavage patterns of the complexes of the tetramer and octamer are shown, together with a difference plot. The positions of the peaks in hydroxyl radical cleavage with respect to position +1 of the 5S RNA gene (numbers) are shown. The asterisk indicates the position of the first reproducible difference between the two patterns. The horizontal bar indicates the region where the octamer pattern diverges from that of the tetramer (see Hayes *et al.*, 1991b for details).

type of approach has been developed by Mirzabekov (Mirzabekov *et al.*, 1977, 1982; Shick *et al.*, 1980; Pruss and Bavykin, 1997).

Chemical cross-linking of histones to DNA is generally performed after the 5′ end of the DNA within nucleosome core particles has been labelled with ^{32}P so that the position of histone–DNA contacts relative to the end of the molecule (as a reference point) can be determined. This is followed by reaction with dimethyl sulphate which methylates purine bases. The methylated product is then depurinated to an aldehyde. A Schiff base is formed between the modified DNA backbone and available lysine amino acids in the histones, which can be further stabilized by reduction with sodium borohydride (Fig. 2.15). Very low levels of reaction are allowed so that a wide distribution of DNA–histone cross-links is generated. The DNA–histone complex is denatured and the histone–DNA adducts are resolved from each other by electrophoresis. The mobility of the DNA molecule

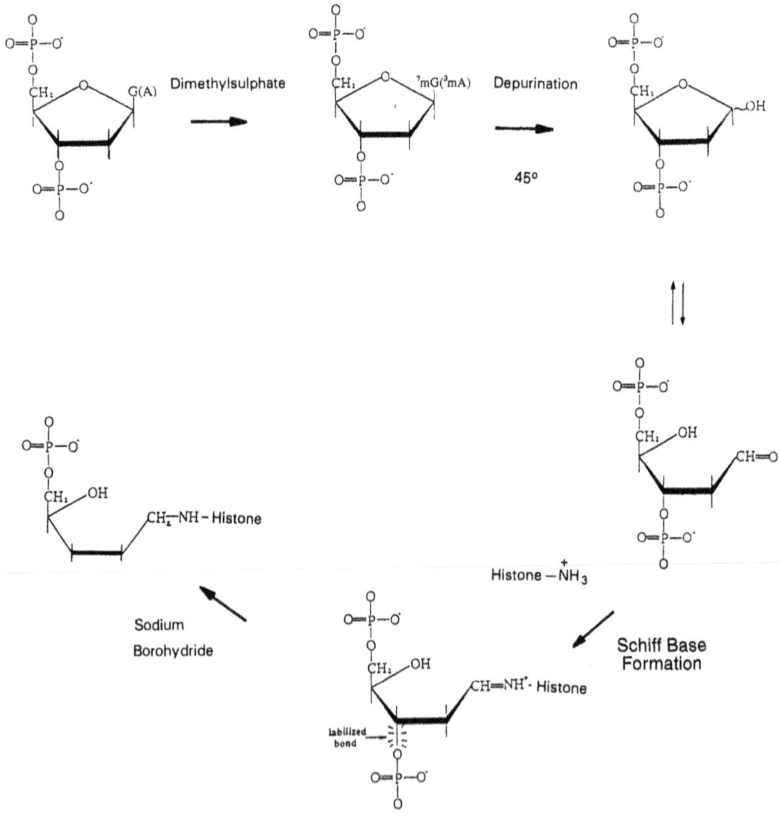

Figure 2.15.   Chemistry of cross-linking histones to DNA.

is reduced by the cross-linked protein. Resolution in a second dimension, after removal of histones using a protease, allows sizing of DNA pieces. The relative mobility of protein-bound and free DNA permits the organization of the histones relative to the end of DNA in the nucleosomal core particle to be determined (Fig. 2.16).

Several conclusions emerge from this analysis. A major result was the realization that the histones bind to nucleosomal DNA in a symmetrical linear array. Histone H3 is found to have weak

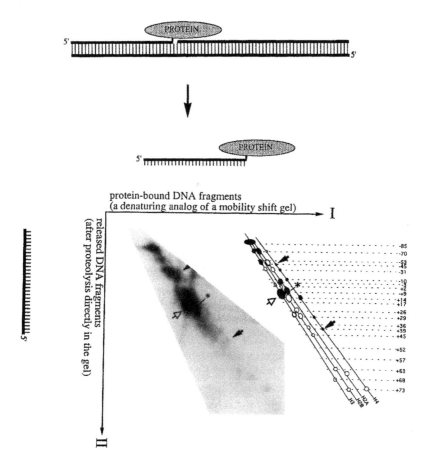

Figure 2.16. Two-dimensional mapping of histone–DNA cross-links.
A covalent histone–DNA complex is shown generated using the chemistry of Fig. 2.15. The histone is cross-linked next to a break in the phosphodiester backbone of DNA. The first dimension (I) of electrophoresis is a denaturing nucleoprotein gel, the second dimension (II) involves proteolytic removal of protein and denaturing resolution of DNA fragments. A typical result is shown for the 5S nucleosome (see Pruss and Wolffe, 1993 for details).

interactions with DNA where it enters and leaves the core particle. We see later that the amino acids involved in these contacts may be modified with important consequences (Section 2.5.4). Moving away from the ends of nucleosomal DNA, histones H2A/H2B bind at the periphery and histones H3/H4 bind towards the centre. Particularly strong protein–DNA interactions occur where histones H3 and H4 organize the path of the central turns of DNA in the nucleosome. We see that these contacts account for the change in DNA helical periodicity within this region (Arents and Moudrianakis, 1993) (Section 2.2.3). These observations are completely consistent with all of the biophysical results (see Section 2.2.2).

The crystal structure of the histone octamer and nucleosome core confirmed and considerably extended existing information about the core histones themselves (Arents *et al.*, 1991; Richmond *et al.*, 1993; Luger *et al.*, 1997). The four core histones were known to have very selective interactions with each other. Histone H2A forms a hetero-dimer with H2B, and H3 forms a heterodimer with H4. The C-terminal histone fold domain of each of the core histones is very similar. It is predominantly α-helical with a long central helix bordered on each side by a loop segment and a shorter helix (Fig. 2.17). Each of the loop segments has some β-strand structure. The long central helix acts as a dimerization interface. The interface between the histones is described as a 'handshake' motif (Arents *et al.*, 1991) (Fig. 2.17). The area of the interface between the two (H3, H4) heterodimers is less extensive than that between the (H3, H4) and (H2A, H2B) heterodimers. However the interface between the (H3, H4) and (H2A, H2B) heterodimers is more accessible to solvent and is consequently less stable (Eickbusch and Moudrianakis, 1978; Karantza *et al.*, 1996). Mutagenesis experiments indicate that tyrosines in the C-terminal α helix of histone H4 have a role in stabilizing the contacts between the (H2A, H2B) heterodimer and the (H3, H4) heterodimer (Santisteban *et al.*, 1997; Zweidler, 1992). Surprisingly the N-terminal tail of histone H4, including all sites of acetylation and the C-terminal α-helix of the H4 histone fold domain are dispensable for assembly into chromatin *in vivo* (Figs. 2.18 and 2.19; Freeman *et al.*, 1996). The N-terminal tail and N-terminal α-helix of H3 are also dispensable for chromatin assembly. However the remainder of the H3 and H4 histone folds are essential for the incorporation of these proteins in chromatin.

The shape of the histone octamer is that of a wedge in which a central V-shaped tetramer ((H3, H4)$_2$) is bordered by two flattened spheres of (H2A, H2B) dimers. The octamer structure has several grooves and ridges on its surface. These make a left-handed helical ramp on to which DNA will wrap. Within this ramp are eight histone

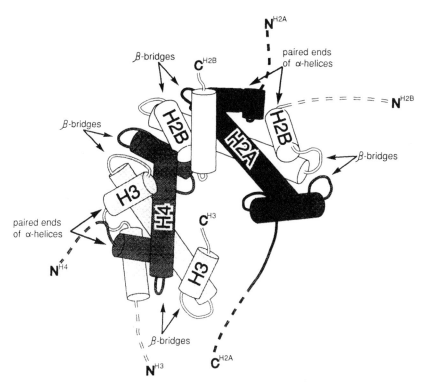

Figure 2.17. Heterodimers of H2A, H2B and H3, H4.
The sites of primary interaction of the histone fold domains with DNA: the
paired ends of helices and β-bridge motifs are indicated.

fold motifs containing 16 loop segments. Owing to the dimerization of
the histones, loop segments from each half of the dimer are paired to
form eight parallel β-bridge segments, two of which are found within
each of the histone dimers (H3, H4) and (H2A, H2B) (Fig. 2.17). Each
β-bridge segment is associated with at least two positively charged
amino acids which are available to make contact with DNA on the
surface of the histone octamer (Arents and Moudrianakis, 1993). The
second repeating motif within the nucleosome is assembled from the
pairing of the N-terminal ends of the first helical domain of each of the
histones in the heterodimers (Fig. 2.17). These four 'paired ends of
helices' motifs also appear to contact DNA. Thus each of the four
heterodimers within the core can make at least three pseudosymmet-
rical contiguous contacts with three consecutive inward-facing minor
grooves of DNA (Fig. 2.20). The eight parallel β-bridges and the four
paired end of helices motifs provide contact sites for 12 out of 14
helical turns of DNA in the nucleosome core; the remaining two turns

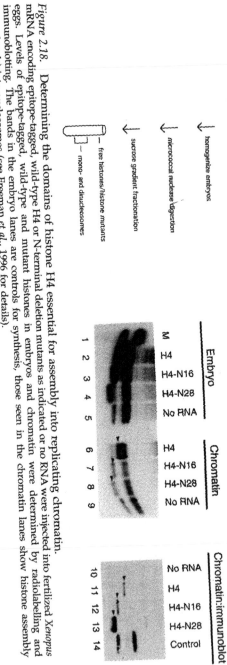
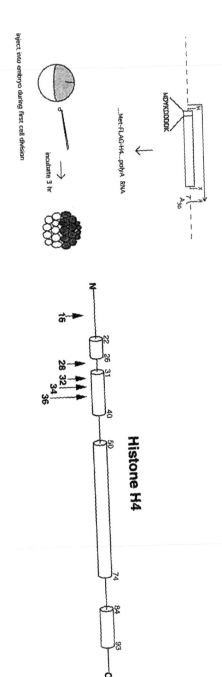

Figure 2.18. **Determining the domains of histone H4 essential for assembly into replicating chromatin.** mRNA encoding epitope-tagged, wild-type H4 or N-terminal deletion mutants as indicated or no RNA were injected into fertilized *Xenopus* eggs. Levels of epitope-tagged, wild-type and mutant histones in embryos and chromatin were determined by radiolabelling and immunoblotting. The bands in the embryo lanes are controls for synthesis, those seen in the chromatin lanes show histone assembly (arrowheads) into nucleosomes (see Freeman *et al.*, 1996 for details).

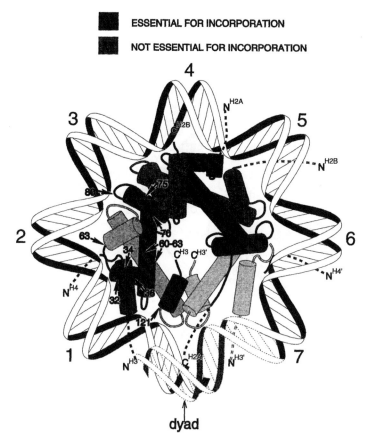

Figure 2.19. Summary of the role of domains of H4 in nucleosome assembly.
Mutated or deleted regions are as indicated. The view shown is down the superhelical axis of the DNA in the nucleosome. For simplicity, the DNA is shown as a uniform superhelix. The helical turns are numbered relative to the dyad axis (0). Only one complete heterotypic tetramer of H2A, H2B, H3 and H4 is shown. N and C termini of the histones are indicated; the dashed lines indicate the N-terminal tails, the exact path of which is not known at this time. The shading (see boxes) indicates which regions of the histone protein are essential for assembly into the nucleosome (see Freeman *et al.*, 1996 for details).

of DNA may be bound by additional helix-loop segments of histone H3 on the flanks of this superhelical ramp (Pruss *et al.*, 1995).

Considerable evidence supports electrostatic interactions of the phosphodiester backbone of DNA with arginine residues present in the histone fold domains of the octamer as being most important for organizing DNA in the nucleosome. The accessibility of arginine and

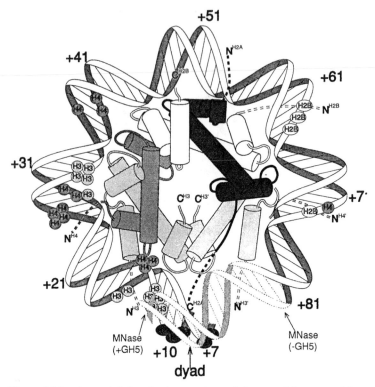

Figure 2.20.   A model for the interaction of the core histones with DNA in the nucleosome.
This view is of one turn of DNA. For clarity only one molecule of H2A, H2B and H4 is shown. Two molecules of H3 are shown to indicate the interface between the two (H3, H4) heterodimers. The numbers show base pairs relative to *X. borealis* somatic 5S ribosomal RNA gene (Pruss *et al.*, 1995).

lysine residues to low molecular weight chemicals (2,4,6-trinitroben-zoic acid and 2,3-butanediol, respectively) in various subnucleosomal particles reveals that only 14 arginine residues are required within the histone-fold domains to maintain the wrapping of DNA in the nucleosome core particle. This major role for arginine may be due to the capacity of this residue to form both hydrogen-bonding and electrostatic interactions with phosphate residues along the DNA backbone (Ichimura *et al.*, 1982). These arginine residues are found in the β-bridges and paired end of helices motifs of the histone octamer (Arents and Moudrianakis, 1993).

Solution data are consistent with the path of DNA on the surface of the histone octamer seen in the crystals (Arents and Moudrianakis,

1993; Luger *et al.*, 1997). Aside from the results already discussed, the N-terminal tail of histone H4 can be cross-linked to DNA at 1.5 turns to either side of the dyad axis (Ebralidse *et al.*, 1988; Pruss and Wolffe, 1993). These contacts with histone H4 are coincident with the sites of strong DNA deformation (Ebralidse *et al.*, 1988). A second feature of histone–DNA interaction around the dyad axis of the nucleosome core is strong histone H3 cross-linking (Bavykin *et al.*, 1990). Towards the periphery of the nucleosome, histone H4 contacts DNA over 60 bp from the dyad axis (Pruss and Wolffe, 1993), consistent with the potential interaction of 120 bp of DNA with the $(H3/H4)_2$ tetramer. The (H2A, H2B) dimer interacts with DNA both around the dyad axis and at the periphery of the nucleosome. Histone H2A is unique among the core histones in having both an N- and C-terminal basic tail. The C-terminal tail binds to DNA around the dyad axis (Gushchin *et al.*, 1991; Usachenko *et al.*, 1994). The N-terminal tail of H2B and H2A contact DNA towards the periphery of the nucleosome (Pruss and Wolffe, 1993; Lee and Hayes, 1997). These results are in excellent agreement with the proximity of the N- or C-terminal helices of the individual core histones to DNA determined through the modelling studies (Fig. 2.20).

Several studies indicate that the highly charged N-terminal tails of the histones do not themselves contribute significantly to the primary wrapping of DNA in the nucleosome. It is possible to remove the positively charged histone tails from nucleosome core particles without altering the accessibility of DNA to DNase I or the hydroxyl radical (Fig. 2.21, Hayes *et al.*, 1991b). The helical periodicity of DNA on the histone surface is not changed by removal of the tails. The stability of the trypsinized nucleosome core particles to physical perturbations such as increased temperature or high salt concentrations reveals that the proteolysed particle undergoes no major changes in stability in comparison to intact particles. This suggests that the histone tails have no essential role in maintaining the integrity of the nucleosome (Ausio *et al.*, 1989). Nevertheless, we see later that post-transcriptional modification of the histone tails may alter more subtle aspects of nucleosome conformation and the quality of histone–DNA interactions (Section 2.5.2).

At first sight it might be thought that the histone tails would stably associate with the major groove of DNA wrapped around the nucleosome core particle. Early experimental evidence indicated otherwise. Dimethyl sulphate has been used to probe the accessibility of bases within the major groove of DNA in the nucleosome. Results indicate that there is almost uniform reactivity of DNA with this small molecule suggesting that the tails do not bind stably in the major groove (McGhee and Felsenfeld, 1979). Moreover, DNA that has glucose groups within the major groove still forms apparently normal

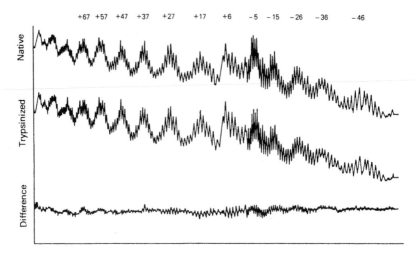

Figure 2.21. A comparison of DNA interactions with an intact octamer of core histones to those with an octamer from which the histone tails have been removed with trypsin.

A nucleosome structure including the 5S RNA gene is used. Densitometric analyses of hydroxyl radical cleavage patterns of the complexes with the intact and trypsinized octamer are shown, together with a difference plot. The position of the peaks in hydroxyl radical cleavage with respect to position +1 of the 5S RNA gene (numbers) are shown (see Hayes *et al.*, 1991b for details).

nucleosomes, even though access by proteins to base pairs would be severely hindered (McGhee and Felsenfeld, 1982). Thomas and colleagues found very little protection of lysines in the N-terminal tails within the nucleosome, suggesting that the tail domains were frequently exposed to solution (Thomas, 1989). These experiments led to the suggestion that the histone tails do not have very stable interactions with DNA through the major groove in the nucleosome. This view of the N-terminal histone tails as relatively unstructured and unstable contributors to nucleosome architecture has been recently challenged by the use of a site-specific histone–DNA cross-linking method (Lee and Hayes, 1997). In these experiments a photoactivatable cross-linking reagent is attached to a particular site within the N-terminal tail of the histone. Upon UV irradiation a covalent bond is formed between DNA and histone. Results for the N-terminal tail of H2A reveal very specific crosslinks about 40 bp from the nucleosomal dyad. This location of the H2A N-terminus agrees well with the contacts anticipated from the fitting of DNA to the surface of the histone octamer (Arents and Moudrianakis, 1993). The surprise is in the very localized distribution of cross-links, which suggests that the N-terminal tail can only adopt a very limited number of conformations (Lee and Hayes, 1997). In addition to any specific role in nucleosome structure itself, the N-terminal tails have a distinct and equally important role in protein–protein interactions outside of the nucleosome (Section 2.5.2).

The apparent simplicity of the histone–DNA interactions maintaining the integrity of the nucleosome core particle has some interesting consequences. It has long been known that nucleosome core particles can be dissociated into their DNA and histone components by elevating ionic strength. This confirms that the primary interactions responsible for the stability of the particle are electrostatic. Histones H2A/H2B dissociate first, at salt concentrations above 0.8 M NaCl, whereas histones H3/H4 only dissociate from DNA when the concentration rises above 1.2 M NaCl (Ohlenbusch *et al.*, 1967). An important aspect of this process is that it is reversible. This means that on dilution from these high ionic strength salt solutions, core histones will reassociate with any available DNA to reassemble nucleosomal structures. In a strict nomenclature, these structures are not nucleosomes, because they have no histone H1, nor are they nucleosome core particles because they contain more than 145 bp of DNA. I have chosen to call them *nucleosome structures*. Much of our current understanding of nucleosomal organization derives from analysis of these synthetic nucleosome structures (Section 2.2.5). In fact it is such a synthetic nucleosome structure that has yielded the most definitive structural information (Luger *et al.*, 1997).

Summary
Biophysical analysis, chemical cross-linking, molecular modelling and a crystal structure reveal that the tetramer of histones H3/H4 has the central role in organizing the nucleosome. Histones H2A/H2B interact to either side of the tetramer and consequently with DNA towards the ends of the molecule as it wraps around the histone core. The histone fold domains of the core histones are responsible for organizing DNA around the histone core, primarily through electrostatic interactions between arginine residues and the phosphodiester backbone.

2.2.5 DNA sequence-directed positioning of nucleosomes

Most current research into the organization of DNA in the nucleosome relies on methodologies to reassemble nucleosomal structures using defined DNA templates (Simpson and Stafford, 1983; Ramsay *et al.*, 1984). Early experiments of this type demonstrated that histone cores cannot interact efficiently with double-stranded RNA or RNA–DNA heteroduplexes to form nucleosomal structures. Extensive homopoly-meric stretches (> 60 bp) of rigid oligo(dA).oligo(dT) or of the left-handed Z-DNA duplex are also not favoured for incorporation into nucleosomes (Prunell, 1982; Garner and Felsenfeld, 1987) (see Section 2.1.2). Cruciform and hairpin DNA structures also exclude nucleo-somes (Nobile *et al.*, 1986). All of these results reflect the energetic cost of deforming the unusual nucleic acid structures (A-form, rigid DNA, Z-form or hairpins) into the bent 10.0 bp/turn structure of nucleo-somal DNA. These represent extreme circumstances, since most naturally occurring DNA sequences including the AT-rich sequences discussed earlier (Section 2.1.1) are incorporated into a nucleosome at little energetic cost (Hayes *et al.*, 1991a; Puhl and Behe, 1995). However, the local influences of DNA rigidity and curvature will affect the precise translational and rotational position the double helix adopts with respect to the histone core (Sivolob and Khrapunov, 1995). This phenomenon is known as nucleosome positioning and has important biological consequences (Section 4.2). As we will see most nucleosome positioning that has been defined occurs in the vicinity of regulatory DNA. There is no need to package the vast bulk of the genome into highly organized chromatin structures and experimental evidence shows that the ubiquitous positioning of nucleosomes does not occur (Lowary and Widom, 1997).

The translational position of a nucleosome refers to where the histone octamer begins and finishes being associated with DNA. The rotational position refers to which face of the double helix is in contact with, or exposed away from, the histone octamer. Nucleases and chemical probes have been very useful in determining both of these parameters. Considerable evidence exists in support of sequence-directed positioning of nucleosomes on DNA. For example, chicken and frog core histones *in vitro* and yeast histones *in vivo* all recognize the same structural features of a 5S ribosomal RNA gene which direct nucleosome position (Fig. 2.22; Simpson, 1991). Mutagenesis experi-ments in which DNA sequences in and around the 5S RNA gene were perturbed indicate that a region comprising 20–30 bp to either side of the centre of nucleosomal DNA contains the elements necessary for positioning (FitzGerald and Simpson, 1985). This region contains the sharp bends and structural discontinuities in DNA structure observed within the nucleosome core (Section 2.2.3).

The nucleosome positioning sequence defined for the 5S ribosomal RNA gene has a DNA structure with a periodic modulation in minor

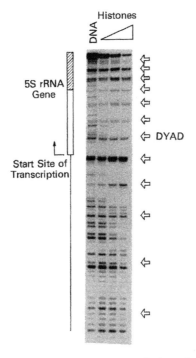

Figure 2.22. A positioned nucleosome includes the site of transcription initiation for the *Xenopus* 5S RNA gene.

The gene is shown as an open box, key regulatory elements inside the gene as a hatched box, and the start of transcription as a hooked arrow. In this experiment DNA is radiolabelled at one end, it is then cleaved with DNase I either as naked DNA lane 1 or as a nucleosome (lanes 2–4). DNA is then denatured, single strands separated with respect to size on an acrylamide gel and autoradiographed. Major cleavage sites and the dyad axis of the nucleosome are indicated.

groove width. If such modulations occur every helical turn of DNA, as they do for the 5S nucleosome positioning element, this directs the DNA molecule to have an intrinsic curvature (Hayes *et al.*, 1990) (Fig. 2.23). It is energetically favourable to incorporate a DNA sequence that is already curved into a nucleosome, as DNA has to be bent around the histone core anyway (Drew and Travers, 1985; Shrader and Crothers, 1989). Narrowed minor grooves will face into the nucleosome and wide minor grooves will face out. In the region about 10–15 bp from the midpoint of the nucleosome core particle DNA, the minor groove varies from 0.7 nm on the inner face of the helix to 1.3 nm on the outside (Morse and Simpson, 1988). A consequence of this is that synthetic DNA curves have proven very useful for directing nucleosome position both translationally and rotationally (Wolffe and Drew, 1989).

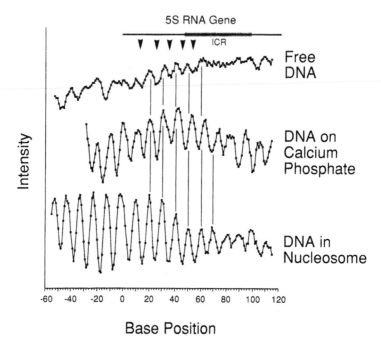

Figure 2.23.   Structure of DNA in a nucleosome, on a crystal surface (calcium phosphate) and free in solution as determined by hydroxyl radical cleavage.

Densitometric scans are shown for 5S DNA cleaved with hydroxyl radical under the conditions indicated. Vertical lines indicate common structural features over 5S DNA (the 5S RNA gene is shown schematically at the top of the figure). Arrowheads indicate regions of narrow minor groove width detected in free DNA that have a periodicity of 10–11 bp indicative of intrinsic curvature. This structural feature is maintained and exaggerated in nucleosomal DNA and when DNA is bound to calcium phosphate.

An innovative recent approach to the identification of natural determinants of nucleosome positioning was to select those DNA sequences that form the most stable nucleosome cores from an entire genome (Widlund *et al.*, 1997). Nucleosome cores were prepared from the entire mouse genome by controlled digestion using micrococcal nuclease. The DNA fragments (146 bp long) were isolated from the nucleosome cores and short DNA sequences were added to their ends for amplification using the polymerase chain reaction (PCR). PCR is a convenient technique for amplifying individual DNA sequences (Fig. 2.24). Those DNA fragments that formed the most stable nucleosome cores were selected under competitive conditions in which histones were limiting for nucleosome assembly (see Chapter 3). The DNA fragments that won the competition were amplified by PCR and the

Bulk chromatin

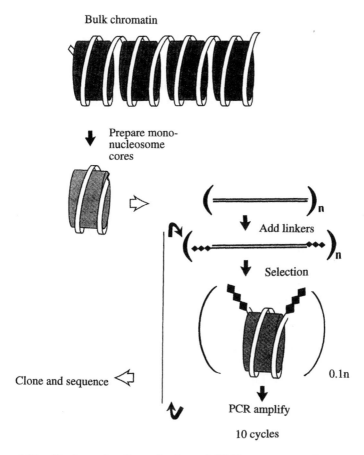

Figure 2.24. Strategy for the selection of DNA sequences that preferentially interact with histone octamers.
Bulk chromatin is prepared from mouse cells. Micrococcal nuclease digestion is used to prepare nucleosome core particles. Core particle DNA is purified and linkers added (diamonds). Primers that hybridize with the linkers are then used in the polymerase chain reaction (PCR) to amplify DNA sequences for use in a selection experiment. In the selection the amplified DNA is reconstituted with histone octamers under competitive conditions such that only 10% of input DNA is assembled into a nucleosome. This 'nucleosomal' DNA is recovered and the procedure repeated through ten cycles. Finally the selected DNA is cloned and sequenced (see Widlund *et al.*, 1997 for details).

process repeated for 10 cycles. The DNA sequences that were recovered were then determined.

As expected several of the DNA sequences contained intrinsically curved DNA with oligo(dA).oligo(dT) tracts (4–6 bp) in length that are phased with a periodicity similar to that of the DNA helix itself

(Section 2.1.1). However, other favoured sequences for nucleosome formation included TATA boxes that were repeated every 10 bp, reiterated CA dinucleotides and reiterated CTG trinucleotides. The favoured assembly of phased TATA boxes into nucleosomes is interesting because such sequences are a common feature of many transcriptional promoter elements (Chapter 4). Phased TATA boxes are known to precisely position nucleosomes on promoters *in vivo* (Li *et al.*, 1998; Fig. 2.25). It is important to prevent inappropriate gene activity and the assembly of these elements into stable nucleosomes may contribute to the repression of genes.

CTG trinucleotide repeats are found in human disease genes (Section 2.1.1). Griffith and coworkers discovered that these sequences would direct the translational positioning of nucleosomes at sites of reiteration in the myotonic dystrophy gene (Wang *et al.*, 1994). Subsequent work *in vitro* found that the CTG repeats were positioned right at the dyad axis of the nucleosome (Godde and Wolffe, 1996) and had exceptionally strong interactions with the histones in the nucleosome (Wang and Griffith, 1995; Wang *et al.*, 1995; Godde and Wolffe, 1996). The significance of nucleosome positioning directed by $(CTG)_n$ is that it provides a mechanism whereby chromatin structure might be altered dependent on the expansion of the repeated sequence. These chromatin alterations occur in individuals with the disease (Otten and Tapscott, 1995), and may contribute to reducing gene expression and consequently the disease phenotype (Thornton *et al.*, 1997; Klesert *et al.*, 1997).

The other trinucleotide repeat associated with human disease genes is $(CCG)_n$. When short segments (13 repeats) of $(CCG)_n$ contain methyl CpG dinucleotides this creates a very favoured binding site for the histones just like $(CTG)_n$ (Godde *et al.*, 1996). However, long segments (76 repeats) repel histones when the CpG dinucleotides are methylated (Wang and Griffith, 1996). The molecular mechanisms that favour nucleosome assembly on short $(CTG)_n$ or methylated $(CCG)_n$ sequences are unknown. Potential mechanisms include sequence-specific interactions of the core histones with these sequences. Precedent for this is the strong preference of the histone H3 variant CENP-A for α-satellite DNA sequences (Sullivan *et al.*, 1994; Sections 2.4.2 and 2.5.1). Alternatively, these particular trinucleotide repeats might be especially deformable to accommodate constraints on the path of DNA at the nucleosomal dyad (Section 2.2.3).

Nucleosome positioning signals in the 5S RNA gene are recognized primarily by the $(H3, H4)_2$ tetramer (Hayes *et al.*, 1991b; Dong and van Holde, 1991), the (H2A, H2B) dimer or the amino terminal tails do not

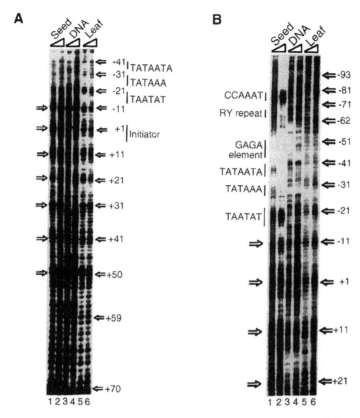

Figure 2.25. DNase I footprinting *in vivo* of the top strand of the proximal region of the phaseolin gene promoter.

For *in vivo* DNase I footprinting, nuclei were isolated from seed or leaf tissue, treated with various levels of DNase I, then DNA was extracted and subjected to ligation mediated PCR. A. DNase I footprinting of +70 to –40 region of the *phas* promoter from seed nuclei (lane 1, 2) and leaf nuclei (lane 5, 6) with naked DNA as a reference (lane 3, 4), using primers annealed to the +90 region. The DNase I concentrations used were: lane 1, 6 units; lane 2, 12 units; lane 3, 1 unit; lane 4, 2 units; lane 5, 6 units; lane 6, 12 units. B. DNase I footprinting of the +21 to –91 region of the *phas* promoter in seed nuclei (lane 1) and leaf nuclei (lane 4) with naked DNA as a control, using primers annealed to the +45 region. The DNase I concentrations used were: lane 1, 6 units; lane 2, 1 unit; lane 3, 2 units. For both A and B, open arrows show a 10 bp cleavage pattern characteristic of DNA wrapped around a nucleosome or TFIID. Base positions are numbered relative to the transcription start site and potentially important *cis* elements are labelled (see Li *et al.*, 1998).

recognize positioning signals (Dong *et al.*, 1989; Hayes *et al.*, 1991b). Linker histones can also influence the exact translational position of the nucleosome (Meersseman *et al.*, 1991; Chipev and Wolffe, 1992; Ura *et al.*, 1995).

Histone H1 has a causal role in repressing *Xenopus* 5S rRNA genes during development (Bouvet *et al.*, 1994; Kandolf, 1994). The exact molecular determinants of this selective gene repression are still to be defined; however, reconstruction experiments using purified histones provide useful information suggesting that nucleosome positioning may have a role. Histone octamers are relatively mobile with respect to DNA sequence; they can occupy a number of distinct translational positions along the 5S rRNA gene that are in continual equilibrium (Meersseman *et al.*, 1991) (Fig. 2.26). The addition of linker histones severely restricts nucleosome mobility (Pennings *et al.*, 1994) and directs the exact translational position (Ura *et al.*, 1995) (Fig. 2.27). Mobile histone octamers on the 5S rRNA genes are accessible to the pol III transcriptional machinery, but in the presence of linker histone H1 this accessibility is severely restricted due to nucleosome positioning and transcription is repressed (Ura *et al.*, 1995). The

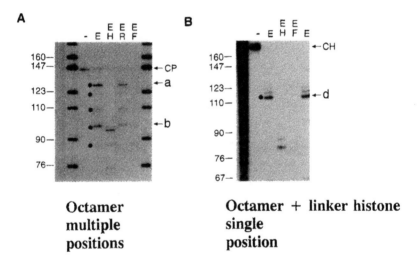

**Octamer
multiple
positions**

**Octamer + linker histone
single
position**

Figure 2.26. Micrococcal nuclease mapping of core particle and chromatosome positions on dinucleosome complexes.
DNA from the nucleosome core particle (CP) and chromatosome (CH) was recovered from an acrylamide gel and digested with *Eco*RV (E) and additional restriction enzymes, *Eco*RI (R) *Hgi*AI (H) and *Fnu*4HI (F), to determine the positions of the boundaries of histone–DNA complexes. A. Micrococcal nuclease mapping of positions of the core particle and B the chromatosomes containing 167 bp DNA. DNA fragments of core particle and chromatosome were digested with restriction enzymes, as indicated. Predominant products of *Eco*RV digestion are labelled a, b, and d, which indicate the 5' boundaries of DNA contacts made by the histones. Fragment lengths were determined using MspI-digested pBR322 size markers and DNA fragments from a hydroxyl radical cleavage reaction (see Ura *et al.*, 1995 for details).

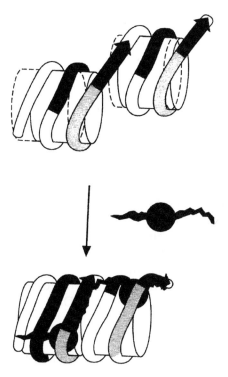

Figure 2.27. Model for structural changes in the dinucleosomal template following the inclusion of linker histone.
Upper panel: a dinucleosome showing the 5S rRNA genes (dark grey) and internal control regions (light grey). Areas covered transiently by mobile nucleosomes are indicated by dotted lines. Average nucleosomal dyads are indicated by dots. Lower panel: addition of linker histone (black) immobilizes nucleosomes, fixes position and constrains linker DNA.

sequence-selective interaction of the globular domain of H1 with nucleosomal DNA has an important role in stabilizing the 5S nucleosome. In this particular case, the globular domain binds inside the DNA turns, making contacts with DNA in the major groove (Hayes, 1996; Pruss *et al.*, 1996; see Section 2.3.1). The capacity of histone H1 to bind to nucleosomal DNA in a sequence-selective manner explains the positioning of nucleosomes on the *Xenopus* oocyte-type 5S rRNA gene *in vivo* (Chipev and Wolffe, 1992; Sera and Wolffe, 1998). Linker histones may also contribute to nucleosome phasing on the chicken β globin gene (Liu *et al.*, 1993; Davey *et al.*, 1995). A similar role has been suggested for HNF3 in the assembly of a phased nucleosomal array on the mouse serum albumin enhancer (McPherson *et al.*, 1993). The globular domain of histone H1 is isomorphous with the winged helix domain of HNF3 (Clark *et al.*, 1993b) and HNF3 appears to replace H1 and direct the position of histone octamers with respect to DNA sequence on the albumin enhancer (Section 4.2).

Summary
Owing to the structural variations in DNA within the nucleosome, rigid or curved DNA sequences will influence the position of the histone octamer along the DNA backbone (translational nucleosome positioning). This will lead to the double helix having a particular face orientated towards the histone core (rotational nucleosome positioning). Trinucleotide repeat sequences $(CTG)_n$ and $(CCG)_n$ found in human disease genes also position nucleosomes through unknown mechanisms. The positioning of the histone octamer appears to be primarily determined by the wrapping of DNA around the $(H3, H4)_2$ tetramer. However, linker histones can also undergo sequence selective interactions with DNA and influence the translational position of core histones with respect to DNA sequence.

2.3 THE ORGANIZATION OF NUCLEOSOMES INTO THE CHROMATIN FIBRE

Although it is relatively easy to visualize the array of nucleosomes along the DNA molecule, this represents only the first level of compaction of DNA in the nucleus. Our understanding of the compaction of DNA into higher order structures than nucleosomal arrays is much weaker than our knowledge of the nucleosome. This probably reflects not only unavoidable heterogeneity in the naturally occurring structures, but also the difficulty in isolating and studying large macromolecular complexes by conventional techniques.

2.3.1 Histone H1 and the compaction of nucleosomal arrays

A key molecule in directing the formation of higher order structure in a nucleosomal array is the linker histone H1. However, histone H1 is not essential for chromatin folding. The folding of histone H1-deficient chromatin provides us with important insights into how and why nucleosomal arrays are compacted. Several studies using 'natural' chromatin depleted of histone H1 and synthetic chromatin, in which nucleosome structures are positioned at varying distances apart, have clearly demonstrated that nucleosomal arrays can be compacted simply by varying the concentration of mono- and divalent cations in solution (Hansen *et al.*, 1989; Clark and Kimura, 1990; Hansen and Wolffe, 1992; Garcia-Ramirez *et al.*, 1992). Under appropriate conditions it is possible to compact chromatin to the level approximately the same as that observed *in vivo* (Schwarz and Hansen, 1994). These results indicate that a limiting factor in chromatin compaction is the degree of shielding of charge along the phosphodiester backbone of DNA. An important question is whether this shielding involves all of the DNA in the nucleosome or just the linker DNA between nucleosomes. Comparison of the salt-induced folding of chromatin in the presence or absence of histone H1 suggests that interactions of both the core histone histone N-terminal tail domains and H1 with the linker DNA facilitate the compaction of nucleosome arrays (Clark and Kimura, 1990; Garcia-Ramirez *et al.*, 1992, 1995; Fletcher and Hansen, 1995; Schwarz *et al.*, 1996). This is a simple and important result that suggests that charge neutralization probably underlies most of the folding of nucleosomes into higher order structures. Nevertheless between different types of linker histone there is considerable specificity in their interaction with DNA in the nucleosome and important consequences for the folding of nucleosomal arrays. It is also probable that mechanisms will exist for interactions between nucleosomal arrays that exist on distinct DNA molecules (Fletcher and Hansen, 1996). These interactions may contribute substantially to the chromatin environment found in the nucleus.

The first experiments to examine the role of histone H1 in chromatin structure concerned its disappearance from the nucleosome as linker DNA is digested away by micrococcal nuclease, and the isolation under controlled digestion conditions of the chromatosome containing the histone octamer, about 160 bp of DNA and a single molecule of histone H1 (Section 2.2.2). The symmetry of the nucleosome core particle suggested that this single molecule of histone H1 would interact with an extra 10 bp to either side of the central 146 bp of the nucleosomal core particle. However, this has yet to be rigorously established. Histone H1 is an asymmetric molecule, which on interaction with a nucleosomal core particle will create an asymmetric nucleosome. Experiments with proteolytic fragments of histone H1 and nucleosomes containing mixtures of all the DNA sequences in the genome suggest that the structured domain of histone H1 binds where DNA enters and exits the nucleosome and across the few central turns of DNA in the structure (Fig. 2.28; Allan *et al.*, 1980, 1981, 1986; Staynov and Crane-Robinson, 1988). This

Figure 2.28.   A model (top left) for the association of the structured (globular) domain of histone H1 (black ball) with DNA in the nucleosome (Allan *et al.*, 1980).
In this diagram the structured domain binds where DNA enters and exits wrapping around the histone octamer and across the few central turns of DNA in the structure. At bottom right is shown a model for the chromatin fibre (Thoma *et al.*, 1979) in which histone H1 interactions at the centre of the fibre facilitate the folding of nucleosomal arrays.

interaction at the periphery of the particle has been proposed to seal two turns of DNA around the histone octamer, leading to further stabilization of the interaction of DNA with the octamer (Simpson, 1978). The extended carboxyl-terminal tail of the linker histone is then proposed to have extensive interaction with the linker DNA between nucleosomes and may adopt an α-helical conformation (Clark *et al.*, 1988).

Linker histones confer particular properties on chromatin which are consistent with their having direct contacts with DNA at the boundaries of the nucleosome. Extended fibres of chromatin containing histone H1 have a zig-zag structure when visualized in the electron microscope consistent with the points of entry and exit of DNA from individual nucleosomes being close together (Worcel, 1978). In histone-H1 depleted chromatin, the fibres are more extended, which is consistent with the entry and exit points of DNA wrapping around the octamer being further apart (Thoma *et al.*, 1979). It has been suggested that the structured domain of histone H1 might constrain the path of DNA at the entry and exit points of the nucleosome. However even chromatin containing specialized histone H1 molecules that lack the structured domain can have a zig-zag appearance in the microscope (Martin *et al.*, 1995). This would emphasize the need for caution in interpreting low resolution images. The structured domain of histone H1 exhibits considerable selectivity for binding to super-coiled rather than linear DNA (Singer and Singer, 1976). In supercoiled DNA, the double helix crosses over itself more frequently than seen with relaxed circular or linear molecules. Histone H1 is predicted to interact preferentially with these cross-over points. This selectivity for supercoiled DNA is consistent with the structured domain of the linker histone having at least two DNA-binding surfaces and potentially interacting with the ends of DNA wrapped around the nucleosome core particle (Goytisolo *et al.*, 1996). The selective association of linker histones with four-way junction DNA in which two double helices are juxtaposed at a relatively sharp angle might therefore reflect a preference of the linker histone for DNA cross-overs (Varga-Weisz *et al.*, 1994). An alternate explanation for these observa-tions is that the structured domain of the linker histone might prefer to bind to DNA that is curved or bent. Supercoiled and four-way junction DNA are more distorted or constrained than linear B-form DNA. Consistent with this possibility is the observation that it is possible to have selective association of the linker histone with nucleosomal DNA when only a single DNA binding site is present in the structured domain (Hayes *et al.*, 1996).

The structured globular domains of the linker histones are known to bind to *naked* DNA molecules co-operatively by constraining two double helices next to each other and stacking linker histones between them (Draves *et al.*, 1992; Thomas *et al.*, 1992). This result is consistent with each globular domain having two binding sites for DNA, however it remains unclear whether it is a single linker histone molecule or a dimer which tries the two DNA molecules together. In crystals the globular domain of H5 dimerizes and exposes its putative DNA binding domains towards the outside of the dimer structure. The spacing between these domains is compatible with that between the two DNA molecules in a binary complex of DNA and the globular domain (Ramakrishnan *et al.*, 1993). The possibility of further interaction of intact linker histones with other linker histones within adjacent nucleosomes, or with the DNA in adjacent nucleosomes, led to the hypothesis that nucleosomal arrays might fold into highly ordered chromatin fibres with linker histones positioned down the centre of the fibre (Thoma *et al.*, 1979; Fig. 2.28) (Section 2.3.2). Consistent with this hypothesis, neutron scattering studies demon-strate that histone H1 is located towards the interior of chromatin fibre (Graziano *et al.*, 1994). However, more recent systematic analysis of higher order chromatin structure in the presence or absence of linker histones demonstrates that the folding of nucleosomal arrays is not particularly regular (see Section 2.3.2).

Recently, the long-established paradigm described in the preceding paragraphs for how linker histones associate with DNA in the nucleosome and the consequences for assembly of the chromatin fibre has been questioned (Travers, 1994; Pruss *et al.*, 1995). Problems with the paradigm have arisen following the solution of the crystal structure of part of the linker histone H5 (Ramakrishnan *et al.*, 1993) (Fig. 2.29). The central structured domain of the linker histone H5 shows remarkable similarity to that of the DNA-binding domain of transcription factor HNF3 (Ramakrishnan *et al.*, 1993; Clark *et al.*, 1993b). Both consist of a bundle of three α-helices attached to a three-stranded antiparallel β-sheet (see Fig. 4.10). HNF3 binds across the major groove of DNA (Section 4.2) as a monomer suggesting that the structured domain of H5 will contact nucleosomal DNA in the same way. Since the structured domain of H5 interacts with at least one other DNA molecule, a second DNA binding site can be proposed to lie on the opposite side of the protein, separated by about 3 nm from the α-helix lying in the major groove. These interactions would be consistent with the known contacts of histone H5 with DNA in the nucleosome determined by chemical cross-linking (Mirzabekov *et al.*, 1990). This information creates several challenges for the long-standing view of how linker histones are bound in the nucleosome

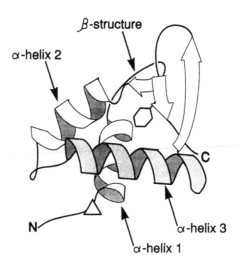

Figure 2.29. A model for the structure of histone H5.
The three α helices and β ribbon structure are shown. Helix 3 is proposed to interact with DNA in the major groove.

(Simpson, 1978; Allan *et al.*, 1980) (Fig. 2.28). None of the new structural data suggests a strong interaction of linker histones with the minor groove of DNA, such as might be envisioned if the linker histone were bound directly over the nucleosomal dyad (Richmond *et al.*, 1984). Nevertheless, interaction of the structured domain of the linker histone with nucleosomal DNA could still potentially occur away from the central superhelical turn. A symmetrical interaction with DNA entering and exiting the nucleosome away from the central turn of DNA at the dyad axis could still protect 20 bp of DNA from micrococcal nuclease digestion (Fig. 2.30).

However, other information suggests that a distant interaction of the structured domain of the linker histone with DNA away from the surface of the histone octamer is unlikely. Neutron scattering studies of chromatosomes place the linker histone near the core histone surface (Lambert *et al.*, 1991). Moreover, protein–protein cross-linking data imply contacts between the C-terminal histone fold domain of histone H2A and the structured domain of histone H1 (Boulikas *et al.*, 1980). This is consistent with the fact that the histone dimer (H2A, H2B) needs to be assembled into the nucleosome before linker histones can bind (Hayes *et al.*, 1994).

An alternative hypothesis for linker histone association is that the single molecule of linker histone in the nucleosome binds at an asymmetrically located position (Hayes and Wolffe, 1993). Nucleosomes

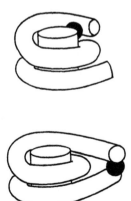

Figure 2.30. Two models for the association of the structured (globular) domain of histone H1 (black ball) with DNA in the nucleosome based on recent data (Pruss *et al.*, 1995).
Top, an asymmetric association of the structured domain inside the superhelical turns of DNA, with the linker histone in close contact with the core histones. Bottom, a symmetric interaction with DNA entering and exiting the nucleosome away from the central turn of DNA at the dyad axis.

incorporating specific DNA sequences have proven useful in establishing this model; however, it should be emphasized that any results derived from these specific structures may not be generally applicable due to the potential for sequence selective association of the linker histone with nucleosomal DNA. Linker histones bind preferentially to a *Xenopus* 5S RNA gene associated with a histone octamer rather than to naked DNA. This preferential binding requires free linker DNA and results in the protection of an additional 20 bp from micrococcal nuclease digestion. Importantly, this additional linker DNA is asymmetrically distributed to either side of the nucleosome core. Formally, asymmetric protection of linker DNA could occur through interaction of the linker histone near the nucleosomal dyad. However, incorporation of linker histones causes no change to the cleavage of DNA in the 5S nucleosome as detected by hydroxyl radical or DNase I cleavage. These results are difficult to reconcile with the binding of the linker histone being across the dyad axis (Fig. 2.28). Histone–DNA cross-linking and the use of site directed cleavage by linker histone Fe(II)EDTA conjugates on the 5S nucleosome reveals a single major contact between the structured domain of histone H5 and DNA over 65 bp from the nucleosomal dyad (Hayes *et al.*, 1994; Pruss *et al.*, 1996; Hayes, 1996). This binding site could either be on the outside of the nucleosome where the major groove of DNA faces out towards solution or it could be inside the turns of DNA in the nucleosome where the major groove of DNA faces in toward the core histones (Fig. 2.31). The available evidence suggests an internal location (Hayes, 1996; Pruss *et al.*, 1996). Similar asymmetric binding of linker histones has now been observed on five other nucleosomes (Weng *et al.*, 1997b; An *et al.*, 1998; Guschin *et al.*, 1998; Sera and Wolffe, 1998).

The asymmetric model would be consistent with the protein–protein interactions of linker histones with H2A, the nuclease protection data and the known dimensions and putative interaction of the structured domain of H5 with DNA (Pruss *et al.*, 1995). Asymmetric binding would presumably be directed by local DNA sequence preferences of the structured domain (Satchwell and Travers, 1989). This type of association would also explain why up to two molecules of a linker histone can associate with a single nucleosome core (Nelson *et al.*, 1979; Bates and Thomas, 1981; Nightingale *et al.*, 1996). However, the asymmetric model does not offer a simple explanation for the protection of an additional 20 bp of linker DNA including regions at both ends of the nucleosome or for the protection at the dyad axis from DNase I cleavage observed in the footprinting of dinucleosomes (Staynov and Crane-Robinson, 1988). Recent experimental results do, however, suggest potential explanations for these apparent discrepancies.

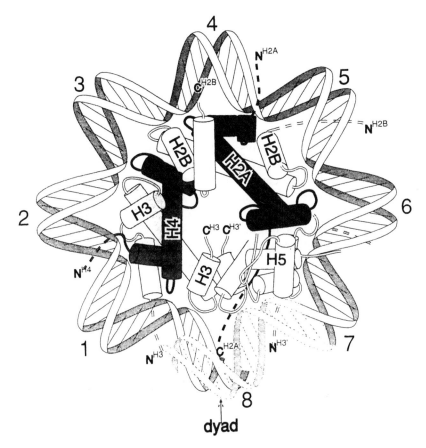

Figure 2.31. A model with the structured domain of the linker histone (H5) associating with nucleosomal DNA inside the super-helical turns of DNA around the histone octamer.

Protein–DNA cross-linking of the core histones within nucleosomes in the presence or absence of the structured domain of linker histone reveals major changes in the contacts of histone H3 and H2A with DNA (Hayes *et al.*, 1994; Lee and Hayes, 1988; Guschin *et al.*, 1998). This suggests that the association of the linker histone may cause an allosteric change in the folding of the histone octamer that could lead to stabilization of core histone–DNA interactions (Simpson, 1978; Usa-chenko *et al.*, 1996). An explanation for the protection from nuclease cleavage by linker histones within dinucleosomes comes from experiments examining the compaction of linker DNA within dinucleosomes (Yao *et al.*, 1990, 1991). These results indicate that dinucleosomes containing linker histones can fold together more readily than linker histone-deficient particles. Thus the adjacent nucleosome could occlude access to the dyad axis by DNase I in the presence of histone H1. The use of model dinucleosomes in which linker histones can be reconstituted should allow this proposal to be tested experimentally (Ura *et al.*, 1995).

Summary

Although nucleosomal arrays will fold in its absence, histone H1 binding to linker DNA will facilitate the process. Several models exist for how histone H1 association might lead to constraint of DNA where it enters and leaves wrapping around the histone core. Histone H1 also makes direct protein–protein contacts with a core histone (histone H2A). These contacts might allow histone H1 to direct allosteric changes in the nucleosome core.

2.3.2 The chromatin fibre

Much information concerning the higher-order organization of nucleosomal arrays has followed visualization of chromatin fragments prepared under various conditions in the electron microscope (Thoma *et al.*, 1979). These observations show that at very low salt (0.2 mM EDTA, 1 mM triethanolamine chloride) chromatin appears as a zigzag fibre of nucleosomes (see also Woodcock *et al.*, 1993; Leuba *et al.*, 1994a; Bednar *et al.*, 1995). At slightly higher salt (0.2 mM EDTA, 5 mM triethanolamine chloride) a flat ribbon forms about 25 nm wide, whereas at moderate ionic strengths (100 mM NaCl, approaching physiological relevance) chromatin condenses to form irregular rod-like structures with a diameter of about 30 nm (Fig. 2.32). This level of chromatin folding can influence the transcription process. The 30-nm fibre structures are similar to those that have been observed in nuclei prepared by a variety of techniques (see Section 2.4.1), hence the interchangeable description of the chromatin fibre as the 30-nm fibre. Much of the chromatin in the cell nucleus of higher eukaryotes has been believed to exist in this configuration. Nevertheless, important concerns remain regarding the regularity, the assembly and even the existence of a stable chromatin fibre. Almost as many models have been proposed as experiments have been carried out (van Holde and Zlatonova, 1995a, 1996; Zlatanova and van Holde, 1996). This in itself points to the instability of the chromatin fibre and to the heterogeneity of its structure. Every model will not be reviewed in detail, but the

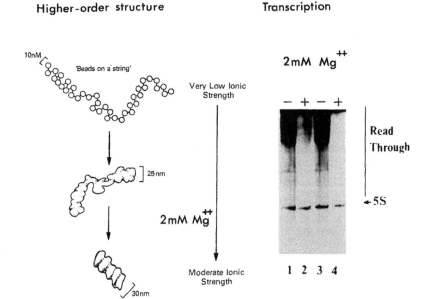

Figure 2.32. Folding of chromatin as visualized in the electron microscope.

As ionic strength increases chromatin changes from a beads-on-a-string configuration to a flat fibre, 30 nm in diameter. It is possible to correlate the Mg^{2+}-dependent folding of chromatin as determined in the analytical ultracentrifuge with the transcriptional properties of chromatin. Reproduced, with permission, from Hansen, J.C. and Wolffe, A.P. (1994) *Proc. Natl Acad. Sci. USA* **91**, 2339. Copyright 1994 by the National Academy of Sciences.

focus will be on the proposals for which the most extensive experimental support exists.

Klug, Koller and colleagues used the electron microscope to examine the unfolding and folding of long chromatin fragments containing several kilobase pairs of DNA, dependent on ionic conditions (Thoma *et al.*, 1979). Based on their observations they proposed a model for the structure of the 30-nm fibre in which the chain of nucleosomes is compacted by winding into a simple solenoidal structure. In this model there would be approximately six nucleosomes per helical turn and the pitch of each turn would be about 11 nm. The pitch was decided on the basis of visible 11 nm striations in the rod-like 30 nm wide structures seen in electron micrographs (Fig. 2.32). Each turn of the solenoid might therefore correspond to the nucleosomes stacking with their long axes parallel to the fibre, since the nucleosomes are individually shaped like 11 nm wide discs (Thoma *et al.*, 1979). An implicit part of this model is that the folding of nucleosomal arrays might be promoted by interaction between the linker histones down the axis of the fibre. Histone H1 is known to bind co-operatively to naked DNA and it has been presumed that comparable interactions exist in chromatin (Clark and Thomas, 1986; Thomas *et al.*, 1992, but see below). These contacts might be dependent on the asymmetry of the histone H1 molecule, which will impart a directionality to the fibre axis. Alternatively if the linker histone associates preferentially with one end of the nucleosome (Hayes and Wolffe, 1993), an asymmetrical nucleosome might also impart a directionality to the chromatin fibre. This directionality might be propagated over extensive regions of chromatin (> 1 kb) since it is possible to detect a polar, head to tail arrangement of histone H5 molecules within chromatin fragments of this size (Lennard and Thomas, 1985).

Whether the folding of nucleosomal arrays is mediated through interactions between linker histone molecules and linker DNA in a local process, or whether co-operative interactions occur down the central axis of the chromatin fibre, remains a subject of controversy. Neutron scattering studies clearly indicate that histone H1 is located towards the interior of the chromatin fibre (Graziano *et al.*, 1994). Experiments examining the accessibility of linker histones to antibodies and proteases demonstrate a reduced accessibility of the linker histones when an array of nucleosomes is compacted (Losa *et al.*, 1984; Dimitrov *et al.*, 1987). This has been interpreted as evidence for the location of linker histones at the centre of the chromatin fibre. However, the decrease in accessibility of linker histones might occur in ways other than sequestration within a defined central axis. Neutron scattering, antibody and protease protection studies do not discriminate between these possibilities. For example, it has been suggested that the local compaction of linker DNA between adjacent nucleosomes which occurs as the ionic strength is increased (Yao *et al.*, 1990, 1991) might account for the decrease in accessibility of linker histones in the absence of the assembly of higher-order structures (Leuba *et al.*, 1993, 1994b). This interpretation is supported by the observation that short chromatin fragments of 10 nucleosomes in length assemble into coiled structures with a constant diameter of 5–7 nucleosomes per turn (Bartolomé *et al.*, 1994).

Chemical cross-linking studies have clearly demonstrated that significant changes in the interaction of linker histones with DNA occur during the compaction of nucleosomal arrays into the chromatin fibre. This modified interaction of linker histones with DNA in chromatin differs from that seen with naked DNA in isolation. This implies that protein–protein contacts within chromatin can influence the structure and perhaps the mode of interaction of histone H1 with

DNA (Mirzabekov *et al.*, 1990). Therefore, linker histones might change their position during chromatin isolation procedures that perturb these protein–protein contacts. This important result is indicative of the problems that exist in interrelating work on the nucleosome in isolation with the properties of nucleosomal arrays (Section 2.3.1).

In considering models of higher-order chromatin structure it is important to recognize that there may be considerable redundancy in the protein–DNA and protein–protein interactions necessary to assemble chromatin and the chromosome. We have already discussed the self-assembly of nucleosomal arrays lacking linker histones into folded higher-order structures (Section 2.3.1). Recent experiments have eliminated linker histone genes from the genome of *Tetrahymena* (Shen *et al.*, 1995) or substantially depleted linker histone proteins from *in vitro* systems capable of the assembly of functional chromosomes and nuclei (Ohsumi *et al.*, 1993; Dasso *et al.*, 1994a). *Tetrahymena* strains without linker histones grow normally, but the size of their nuclei increases twofold. This increase in size suggests a less compact packing of DNA in H1-deficient chromatin; however the conclusion remains that linker histones are neither essential for the assembly of a nucleus nor for cell survival. The biochemical support for this conclusion follows from experiments in *Xenopus laevis* egg extracts. *Xenopus* eggs accumulate large stores of material necessary to rapidly assemble a functional embryo following fertilization (Section 3.3). These stores include the four core histones and a linker histone variant called B4 (Dimitrov *et al.*, 1993, 1994). Linker histone B4 is the only linker histone found in the *Xenopus* egg, and is the predominant linker histone present in early embryonic chromatin. *Xenopus laevis* egg extracts retain the capacity to assemble nuclei from sperm chromatin. Egg extracts depleted of histone B4 assemble nuclei, initiate replication and condense their chromosomes in a similar manner to extracts containing linker histone. Thus once again a functional nucleus and presumably any relevant chromatin higher-order structures for nuclear assembly can be assembled in the absence of linker histones (Ohsumi *et al.*, 1993; Dasso *et al.*, 1994a).

Aside from the ribbon and rod-like conformations of chromatin fibres, other electron microscopic images of native chromatin at physiological ionic strengths reveal different structures including those of globular clusters of nucleosomes ('superbeads'). It is also possible to detect bead-like discontinuities in the chromatin fibre in sectioned nuclei. Under certain conditions these superbeads can be remarkably uniform, containing between 8 and 48 nucleosomes. This suggests they represent discrete structural units (Zentgraf and Franke, 1984). However, the failure to find a ubiquitous organization of chromatin into superbeads indicates that all chromatin does not adopt this form of higher-order structure *in vivo*. This irregularity may in fact represent the true state of affairs within the nucleus since, as we will discuss later, it would be surprising if eukaryotic DNA was packaged into structures of a crystalline order and stability.

The simple solenoid proposed by Klug and Koller is not the only helical model for the folding of nucleosomal arrays (Felsenfeld and McGhee, 1986). A major alternative, which has received considerable recent attention, builds on the observation that zig-zag arrays of nucleosomes exist in solutions of low ionic strength and that the appearance of these arrays is dependent on the presence of linker histones. In models developed by Worcel, Woodcock and colleagues, the zig-zag array forms a condensed ribbon containing two parallel rows of nucleosomes, the folding once again mediated by linker histone. Coiling of this ribbon generates the 30-nm fibre (Woodcock *et al.*, 1984). Recent work by Woodcock, van Holde and colleagues has given support to the existence of an asymmetric zig-zag of nucleosomes and linker DNA as a basic principle of chromatin folding (Woodcock *et al.*, 1993; Leuba *et al.*, 1994a; Yang *et al.*, 1994; Bednar *et al.*, 1995). Visualization of the three-dimensional structure of chromatin fibres in sections of nuclei using cryoelectron microscopy reveals chromatin fibres to have very irregular paths with smoothly bending regions interspersed with sharp changes in direction. Within these fibres, a common structural feature is a two-nucleosome wide ribbon that forms a zig-zag with little face to face contact between nucleosomes (Horowitz *et al.*, 1994). A feature of this structural unit is the very straight linker. *In vivo* photofootprinting assays also favour straight linkers (Pehrson, 1989; Pehrson and Cohen, 1992). It has been suggested that models of chromatin fibre structure in which specific nucleosome–nucleosome contacts occur (Thoma *et al.*, 1979) represent *in vitro* artefacts (Woodcock, 1994). Although this represents an extreme view, it serves to emphasize that the organization of the chromatin fibre is very irregular. Other cryoelectron microscopic studies have suggested that extensive interdigitation of adjacent chromatin fibres might occur in the nucleus itself, leading to a fluid organization in which the chromatin fibre does not exist as a distinct entity (McDowall *et al.*, 1986). The existence or stability of the chromatin fibre has important implications for transcriptional regulation in chromatin.

Various spectroscopic techniques have been used to test these different proposals for the compaction of nucleosomal arrays *in vitro*. In view of the *in vivo* results (McDowall *et al.*, 1986; Horowitz *et al.*, 1994), these experiments indicate what structures can form, not of course what structures actually exist in the nucleus itself. Neutron scattering was applied with a similar philosophy to that outlined earlier for the nucleosome (Section 2.2.2), to measure the distribution of DNA and linker histones in the chromatin fibre. Linker histones have been unambiguously placed away from the outside of the chromatin fibre (Graziano *et al.*, 1994). However, as we have discussed the data do not provide information as to where in the nucleosome the linker histones bind, or whether they form a continual structure along the axis of the chromatin fibre. The observed constraints on the dimensions of these chromatin fibres studied by neutron scattering do support solenoid models in which the nucleosomes are positioned radially, like the spokes of a wheel (Suau *et al.*, 1979).

Among the other techniques used to approach higher-order chromatin structure is electric dichroism. This requires chromatin fibres to be orientated in electric fields and the absorbance of polarized ultraviolet light to be measured. Since the path of DNA in the nucleosome and the electric dichroic properties of DNA are well understood, it is possible to calculate the expected properties of any array of orientated nucleosomes. Felsenfeld, Charnay and colleagues showed that a key assumption of the *in vitro* solenoid models was correct, in that the nucleosomes were orientated with their long axes parallel to the fibre (McGhee *et al.*, 1980, 1983a). Fibre diffraction studies also lend further support to the simple solenoid models. Widom and Klug obtained partially orientated chromatin fibres by drawing concentrated solutions of chicken erythrocyte chromatin into capillaries (Widom and Klug, 1985). Analysis of the diffraction data again suggested radial packing of nucleosomes; more importantly, bands that might be predicted by two parallel rows of nucleosomes forming a regular solenoid were absent (Worcel, 1978; Woodcook *et al.*, 1984). Further analysis led Felsenfeld and McGhee to propose a modification of the simple solenoid model, in which linker DNA is not constrained in the centre of the 30-nm fibre but simply follows the path of the chain of nucleosomes itself. In this 'coiled linker' model, the linker DNA is coiled between adjacent nucleosomes (Fig. 2.33; Felsenfeld and McGhee, 1986). If linker histones associate with the nucleosome inside the coils of DNA around the histone octamer (Hayes *et al.*, 1994) they would be ideally placed to provide a surface on which linker DNA could continue to coil (Pruss *et al.*, 1995). *In vitro* reconstitution experiments that show DNase I and hydroxyl radical footprints of linker DNA associated with histone H1 between two nucleosome cores are consistent with a continual supercoiling of the double helix in chromatin (Ura *et al.*, 1995; Fig. 2.34). The existence of an asymmetric nucleosome with the structured domain of the linker

Coiled Linker DNA

Figure 2.33. The coiled linker model of the chromatin fibre, two turns of only one side of the fibre are shown.

histone wrapping DNA around it would also eliminate a central requirement of the simple solenoid model, in which linker histones on the outside of the nucleosome would nucleate fibre assembly through co-operative interactions via the structured domain (see Section 2.3.2; Pruss *et al.*, 1995, 1996). This is because the association of the structured domain of the linker histone within the wraps of DNA to one side of the nucleosome would prevent contacts with the same domain of the linker histone in the next nucleosome.

For any of these models the length of DNA in a nucleosomal repeat is very important. This is because the length of linker DNA constrains the type of arrangement one nucleosome can have relative to the other. In all organisms and tissues the size of DNA that is resistant to micrococcal nuclease digestion in a nucleosome core particle is remarkably constant at about 146 bp. However, the distance between the first cuts by micrococcal nuclease, i.e. the nucleosomal repeat length of DNA in the nucleosomal core particle plus linker DNA, varies greatly. In *Saccharomyces cerevisiae* the nucleosomal repeat length is merely 165 bp pairs, whereas in most mammalian cell lines it is 180–200 bp in length, and grows to almost 260 bp in sea urchin sperm (van Holde, 1988). Variation in the length of linker DNA leads to clear experimental differences in both the efficiency with which different chromatin preparations will compact *in vitro* and in the properties of

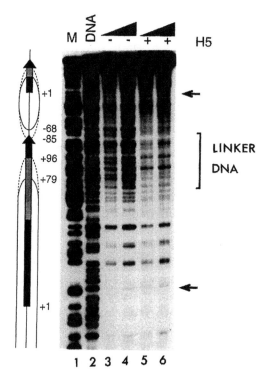

Figure 2.34. DNase I footprinting of dinucleosomes containing histone H5.

A bound or unbound H5 dinucleosome was prepared and digested with DNase I. Individual complexes were isolated by nucleoprotein agarose (0.7%) gel electrophoresis. DNA from these complexes was isolated and analysed by denaturing polyacrylamide (6%) gel electrophoresis. Lanes 1–6, the 5′ end-radiolabelled non-coding strand of the 5S RNA gene is used as a template. Lane 1 shows G-specific cleavage reaction used as markers. Digestion of naked DNA (lane 2), of dinucleosomes (lanes 3 and 4), and of dinucleosomes containing H5 (lanes 5 and 6) is shown, as indicated at the top. Filled triangles indicate increasing DNase I digestion. Small arrows indicate the position of the axes of dyad symmetry of the nucleosomes. The large vertical arrows show the location and orientation of the 5S RNA genes. Grey boxes show internal control regions. Solid and dotted ovals indicate the predominant regions contacted by the nucleosome cores and chromatosome, respectively. The position of linker DNA is indicated.

the resultant chromatin fibre. Langmore and colleagues have shown, using chromatin with long nucleosome repeat lengths, that the diameter of the chromatin fibre increases with DNA linker length (Athey *et al.*, 1990; Williams and Langmore, 1991). It should, however, be noted that Woodcock (1994) detected no variation in chromatin fibre diameter with linker length *in vivo*. At the other extreme, Widom has demonstrated compaction of *Saccharomyces* chromatin with very short repeat lengths.

Nucleosomal arrays containing linkers of less than 20 bp are able to fold into a 30-nm fibre (Lowary and Widom, 1989). More importantly, linker DNA that is very short compared to expectations of DNA persistence length (or 'stiffness') has been shown to bend in chromatin. Histone H1 clearly increases the flexibility of linker DNA by neutralizing negative charge on the phosphodiester backbone (Section 2.3.1). However, it was unclear whether this neutralization would be sufficient for very short linker DNA to bend. Light scattering, which measures the radius of gyration of a particle, and electron microscopy, which gives a direct visualization of particle size, have been used to demonstrate compaction of dinucleosomes (containing histone H1) and dinucleosomal structures (depleted of histone H1). The properties of these defined chromatin substrates were identical to those predicted from the compaction studies of long chromatin with or without histone H1 (Yao *et al.*, 1990, 1991; see also Bartolomé *et al.*, 1994). This flexibility of linker DNA lends strong support to the 'coiled linker' modification of the simple solenoid model proposed by Felsenfeld and McGhee.

Although the available evidence favours the simple solenoid model for the folding of most nucleosomal arrays *in vitro*, it is clear that other ways of folding occur. There are now multiple documented exceptions to the simple 30-nm fibre (Athey *et al.*, 1990; Horowitz *et al.*, 1994). Unlike DNA in the nucleosome, considerable flexibility and hetero-geneity are likely to exist in the higher order structure of chromatin. This complexity in the folding of chromatin is increased by strong influences on the structure due to various post-translational modifica-tions of both core and linker histones that occur during the normal cell cycle and in development (Sections 2.5.2 and 2.5.3).

Summary

Most of the available evidence supports an irregular packing of nucleosome arrays into the chromatin fibre structures (30-nm fibre) *in vivo*. In the solenoidal chromatin fibres assembled *in vitro*, linker DNA appears to be coiled between adjacent nucleosomes. A considerable number of questions remain concerning the relationship between the chromatin fibre structures studied *in vitro* and the compaction of nucleosomal arrays *in vivo*. It is also clear that linker histones are not essential for the assembly of the chromatin higher-order structures necessary for either chromosome or nuclear assembly (Wolffe *et al.*, 1997a).

2.4 CHROMOSOMAL ARCHITECTURE

The chromatin fibre is not necessarily a static, stable structure, as proteins including histones continually equilibrate in and out of these structures (Sections 3.1.1 and 3.1.2). Properties of the fibre can also be changed by modification of its constituent histone proteins or through interaction with structural non-histone proteins that recognize the fibre and package it into a chromosome. The chromatin fibre is organized into large domains potentially separated through interaction with both a nuclear skeleton and a nuclear scaffold or matrix. These chromosomal domains are themselves folded in an ordered manner to form the chromosome.

2.4.1 The radial loop and helical folding models of chromosome structure

Our most thorough understanding of chromosomal organization is for the most condensed and hence most visible of chromosomes, those at metaphase (Rattner and Lin, 1985). Although folding of DNA into nucleosomes leads to a seven-fold compaction in length, and the subsequent folding of arrays of nucleosomes into the chromatin fibre to a further seven-fold compaction, a massive 250-fold compaction of DNA follows the organization of the chromatin fibre into a metaphase chromosome (Earnshaw, 1988, 1991). Two principal models have been proposed to account for this compaction (Fig. 2.35). The first suggests an organization of the fibre into loops that are radially arranged along the axis of the chromosome (Paulson and Laemmli, 1977; Gasser and Laemmli, 1986; Boy de la Tour and Laemmli, 1988). The second suggests a helical folding of the chromatin fibre, followed by a helical folding of the resultant 250-nm fibre (Sedat and Manuelidis, 1978).

These two models have received a great deal of attention in recent years. The evidence for the organization of the chromatin fibre into loops attached to a central axis in normal cells comes from several experimental approaches. Long-standing observations on the morphology of lampbrush chromosomes in amphibian oocytes show a succession of loops emerging from a single chromosomal axis (Section 2.4.3). Worcel and Burgi (1972) developed a model for the *Escherichia coli* chromosome that predicted its organization into independent domains or loops. Worcel extended these studies to interphase chromosomes from *Drosophila* cells. Intact chromosomes were subjected to very mild

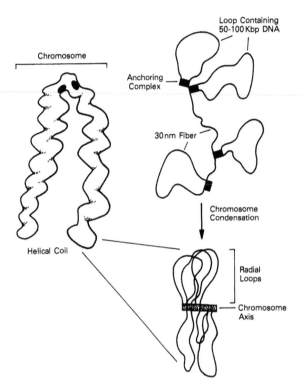

Figure 2.35. The folding of loops of the chromatin fibre into the chromosome.
The chromatin fibre is tethered at anchoring complexes which interact with each other to form the chromosome axis.

digestion with DNase I, so as to produce single-strand nicks. The sizes of the resulting chromosomal fragments were then examined in sedimentation experiments and found to decrease gradually until a plateau value of fragment size was reached in which the complexes contained approximately 85 000 bp of DNA complexed with protein (Benyajati and Worcel, 1976). This suggested that the chromatin fibre was organized into fairly uniform domains containing about 100 kbp of DNA. A second approach involves the analysis of chromosome and chromatin organization by electron microscopy after the fractionation of nuclei that have been extracted of histones by exposure to high salt solution. It is possible to measure directly the length of DNA on the microscope grid from where it exits a residual nuclear structure to where it re-enters this structure. Estimates of loop size between 40 and 90 kbp were obtained consistent with the biochemical measurements (Cook and Brazell, 1975; Paulson and Laemmli, 1977; Jackson *et al.*, 1990).

The development of pulsed field gel electrophoresis (Schwartz and Cantor, 1984) allowed a more systematic analysis of the separation of cleavage sites following mild nuclease digestion of nuclei (Filipski *et al.*, 1990). This technique allows the resolution of DNA molecules of very large size by agarose electrophoresis. Nuclei of eukaryotic cells contain an endonuclease (first responsible for the discovery of the nucleosomal repeat, see Section 2.2.1) that can be activated under controlled conditions (by the addition of exogenous Ca^{2+}/Mg^{2+} to nuclei) and which is believed to have little sequence specificity. It might be expected that cleavage of DNA by this enzyme would be inhibited by folding of DNA into the chromatin fibre, but that structural discontinuities, perhaps where the loop is attached to the chromosome axis, might allow cleavage if the enzyme was activated. Activation of this enzyme followed by resolution of the resultant DNA fragments on pulse field gels would potentially allow loop sizes to be determined. A second approach to this problem makes use of the observation that topoisomerase II is a major component of chromosomes, and potentially the nuclear scaffold (Section 2.4.2). The enzymatic action of topoisomerase II is to introduce a double strand break into DNA which is then resealed (Table 2.1). These double strand breaks can be stabilized by the use of specific drugs (e.g. epipodophyllotoxins) that inhibit the enzyme (Chen *et al.*, 1984). Preferential cleavage sites in nuclei spaced 50–300 kb apart were detected using both the endogenous nuclease or topoisomerase II in the presence of the specific inhibitors of the rejoining of the double helix. Closer analysis revealed a hierarchy of digestion, where the 300 kb cleavage products appeared before those of 50 kb. It has been suggested that the 50 kb intermediate in digestion represents the first level of organization of the chromatin fibre into loops, whereas the 300 kb kinetic intermediate in digestion represents the next level of

Table 2.1 Eukaryotic DNA topoisomerases

	Activity
Topoisomerase I (Topo I)	Removes superhelical turns from DNA in the absence of an energy source by introducing a transient break in one strand and allowing rotation of the broken strand about the intact DNA chain. This results in alterations in linking number by steps of one.
Topoisomerase II (Topo II)	Removes superhelical turns from DNA in an ATP dependent reaction by breaking both strands and allowing rotation of both strands about the intact DNA chain. This results in alterations in linking number by steps of two.

organization. Taken together, these observations establish a strong case for large independent loops (50–100 kb) of the chromatin fibre representing a unit of chromosome structure.

Support for the second model, proposing the folding of the chromatin fibre without specific attachments to an undisrupted chromosome axis, comes from sophisticated high voltage and conventional transmission electron microscopy combined with extensive computer analysis by Sedat, Agard and colleagues. The 'scaffolding' hypothesis would suggest that although the arrangement of loops of the chromatin fibre about the chromosome axis might form distinct patterns, no discrete higher order organization exists above the chromatin fibre. Sedat and Agard demonstrate convincingly that such an order does in fact exist (Belmont et al., 1987, 1989). Early work on large plant chromosomes during the meiotic cycle suggested that coils of chromatin fibres existed in the chromosome (White, 1973). Under certain fixation conditions chromosomes from animal cells could also appear as a spiral or zig-zag fibre (Onnuki, 1968). Sedat and Agard examined native *Drosophila* mitotic chromosomes, observing a hierarchy of higher order chromatin folding patterns for the 30-nm fibre: structures of 50, 100 and 130 nm in diameter were clearly discernible. Although a looping architecture of the 30- and 50-nm fibres could be detected under certain circumstances, the loops were not observed to be consistently orientated radially in three dimensions about any given axis, and no evidence for a central scaffolding was apparent. These observations have now been extended to mammalian cells (Belmont and Bruce, 1994). A new approach to this problem in living cells is to make use of model systems in which repeats (256) of the lac operator are introduced into chromosomes. This segment of DNA is approximately 10.1 kb in length. It can be detected in the chromosome using a fusion of the lac repressor protein with green fluorescent protein (Robinett et al., 1996). Since the green fluorescent protein can be detected *in vivo* without fixation, chromosomal dynamics can be followed directly under physiological conditions. Large-scale chromatin folding is observed that is well above that found in the 30 nm fibre, and that approaches the 100 nm diameter fibre seen in earlier electron microscope images.

These results are not inconsistent with an important role for nonhistone scaffolding proteins in anchoring local loops or domains of chromatin structure, but do appear to rule out a strict radial symmetry or undisrupted central axis for such loops. Most likely a diffuse organization of loops anchored to non-histone proteins actually exists *in vivo* (Fig. 2.35). Consistent with this concept are immunofluorescent experiments in which antibodies against topoisomerase II are distributed in separate 120–200 nm islands (Section 2.4.2). However, Sedat and Agard also found no evidence to support the helical folding of the most condensed form of chromatin, the 130-nm fibre, to form the chromosome. This appears to exclude their earlier models (Sedat and Manuelidis, 1978). Instead they propose a model whereby mitotic chromosomes would progress out of and into interphase through the association and dissociation of distinct 130-nm wide domains of chromatin. Although the exact structure of these large chromatin domains is even more uncertain than that of the 30-nm chromatin fibre, it is clear that non-histone proteins must be important for directing these particular aspects of chromosome organization. Recent biochemical and genetic experiments strongly support this hypothesis (Section 2.4.2).

Summary
Considerable evidence from nuclease digestion, biophysical and morphological studies supports the formation of large loops of the 30-nm chromatin fibre in the eukaryotic nucleus. However, these loops do not appear to originate from a central continuous axis. Instead high resolution light microscopy suggests that the axis can continually assemble and disassemble. Loops of the 30-nm chromatin fibre appear to be organized into even more complex structures to eventually assemble the chromosome.

2.4.2 The nuclear infrastructure, the centromere, telomeres, protein components and their function

Many studies have focused on the non-histone proteins present in the nucleus that generate the infrastructure of chromosomes and of the nucleus itself and the DNA sequences associated with them. The nature of the nuclear skeleton, the nuclear scaffold and the nuclear matrix has been the subject of much debate. The metaphase scaffold of a chromosome initially had a morphological definition as the complex structure at the axis of a mitotic chromosome visualized after swelling and extraction of the histones (Paulson and Laemmli, 1977). Biochemical extraction with high salt (2 M NaCl) or a detergent-like molecule lithium diiodosalicylate (LIS) was used to define the residual nucleoprotein complex at which DNA was attached to the chromosome during interphase (Paulson and Laemmli, 1977; Mirkovitch *et al.*, 1984). This nuclear 'matrix' (after high salt extraction) or 'scaffold' (after LIS extraction) is now known to contain a substantially more complex group of proteins than the metaphase scaffold itself (Gasser *et al.*, 1989). The use of these non physiological extraction procedures was criticized due to their capacity to cause rearrangement of protein–DNA interactions and non-specific aggregation (Cook, 1988). An alternate strategy to study nuclear infrastructure is to encapsulate cells in agarose and to extract most of the chromatin to leave the nucleoprotein complexes essential for nuclear integrity under physiological conditions (Jackson *et al.*, 1988). This last methodology generates a nuclear 'skeleton' that retains the capacity to transcribe and replicate DNA. Recent comparative experiments are consistent with the nuclear matrix or scaffold interacting with gene-poor regions of the genome, whereas the nuclear skeleton interacts with gene rich regions (Craig *et al.*, 1997). It is now clear that the function and composition of the nuclear skeleton is very different from that of the nuclear matrix or scaffold. All define chromosomal domains, yet the reasons for generating attachments to the chromatin fibre will differ.

Whether a nuclear or chromosomal infrastructure existed at all was initially objected to on several grounds. The first objection was that an axial organizing structure within a chromosome had never been observed in normal mitotic chromosomes. This objection can be dealt with by assuming that the structure observed in swollen and extracted chromosomes represents an aggregation of discrete anchoring complexes that may be connected within the chromosome by being assembled on the same DNA molecule (Section 2.4.1). The second problem came from the methodologies used for preparing nuclear scaffold or matrix. These biochemical preparations represent the fraction of cellular proteins that is insoluble under the extraction procedures employed. Why should a biochemist believe that these proteins necessarily had anything to do with chromosomes at all? Perhaps they were just adventitiously precipitated by the extraction procedure used. However, scaffolds and matrices of apparently identical polypeptide compositions are obtained when chromosomes are extracted at high ionic strength (2 M NaCl), with the polyanions dextran sulphate and heparin at low ionic strength (~10 mM NaCl), or with low concentrations (5 mM) of the chaotropic agent lithium diiodosalicylate. It is very unlikely that these different treatments would precipitate the same population of chromosomal non-histone proteins (Earnshaw, 1988).

Several distinct proteins were initially identified in chromosome scaffold preparations. These include the lamins A, B and C which are intermediate filament-like proteins that constitute the nuclear lamina (McKeon *et al.*, 1986; Franke, 1987). This is the fibrillar protein meshwork that lines the nucleoplasmic surface of the nuclear envelope. The nuclear lamina was thought to provide an architectural framework for the attachment of nuclear membranes, nuclear pore complexes and chromatin. However, it is now clear that nuclear pore complexes can assemble into nuclear membrane in the absence of a nuclear lamina (Dabauvalle *et al.*, 1991). The association of lamins with the scaffold fraction then led to the idea that the lamina anchors interphase chromatin to the nuclear envelope and thereby influences higher order chromosome structure (Fig. 2.36). The possible interaction between the nuclear lamina and chromatin is supported by electron microscopic studies which have indicated that chromatin is in intimate contact with the inner surface of the nuclear envelope. Biochemical experiments indicate that solubilization of nuclear membranes with Triton X-100 leaves the lamina in association with chromatin. Gerace and colleagues have made use of an *in vitro* nuclear assembly system (Section 3.1) to examine the reassembly of nuclear membranes and lamina around mitotic chromosomes. They found that lamins A and C specifically bind to mitotic chromosome surfaces and assemble a supramolecular structure (Glass and Gerace, 1990).

Lamin assembly is facilitated by the presence of chromosomes, lending strong support to the hypothesis that the chromosome structure influences lamina structure and nuclear reassembly (Glass and Gerace, 1990). Since chromosomes substantially reduce the concentration of lamins necessary for the assembly of fibrillar networks, it is possible to use this assay to determine what structural features of lamins and chromosomes are necessary for productive interaction (Glass *et al.*, 1993). It was established that the tail domain of the lamin proteins interacted with the core histones (Taniura *et al.*, 1995). Since linker histone incorporation into chromatin is not required for lamina assembly (Dasso *et al.*, 1994a), the site of lamin interaction in chromatin is most probably the core histone tails which protrude at the outside of the nucleosome. Although the assembly of a nuclear

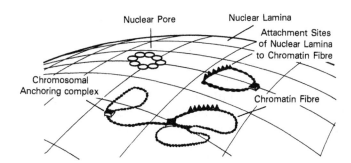

Figure 2.36. Potential association of the chromatin fibre with the nuclear lamina.

lamina is not essential for assembly of the nuclear membrane (Newport *et al.*, 1990; Dabauvalle *et al.*, 1991), integral membrane proteins within the nuclear envelope can interact with both the lamins and chromosomes (Foisner and Gerace, 1993; Collas *et al.*, 1996). This interaction might stabilize the localization of chromatin domains at the nuclear envelope. The association of the nuclear envelope with chromosomes can be regulated through mitotic phosphorylation (Foisner and Gerace, 1993).

A key protein in mediating the association of the nuclear lamina with chromatin is the lamin B receptor. In humans, the lamin B receptor has a domain that lies in the nucleoplasm (208 amino acids) attached to a domain that consists of eight putative transmembrane segments that is an integral component of the inner nuclear membrane (Ye and Worman, 1994, 1996). The lamin B receptor appears to target inner nuclear membrane vesicles to chromatin at the end of mitosis (Collas *et al.*, 1996), and to maintain the attachments of the nuclear lamina and heterochromatin to the nuclear envelope in interphase. The nucleoplasmic domain of the lamin B receptor interacts with a human chromodomain protein, heterochromatin protein 1, HP1 (Ye and Worman, 1996; Yet *et al.*, 1997). HP1 is predominantly found in centromeric heterochromatin in *Drosophila*, but the mammalian homologues are found to be widely dispersed in the chromosome (Horsley *et al.*, 1996; see Section 2.5.6). The association of HP1 with the lamin B receptor explains why transcriptionally inactive heterochromatin might be preferentially localized at the nuclear envelope (Franke, 1974). It also provides a mechanism for compartmentalization of chromatin within the nucleus (Strouboulis and Wolffe, 1996). In this regard, chromatin binding of the lamins could have several interrelated roles. Chromatin might provide a nucleation site for reassembly of the nuclear lamina at the end of mitosis and thus facilitate nuclear envelope reassembly (Glass and Gerace, 1990). Selectivity in the interactions of lamins with particular chromosomal regions might exist such that interactions established during telophase could persist through interphase. This might anchor particular chromosomal domains at the nuclear periphery and directly influence gene activity (Glass *et al.*, 1993).

A second set of proteins found in the scaffold fraction includes Sc (scaffold proteins) I, II and III (Lewis and Laemmli, 1982). The function of Sc III is unknown; however, Sc I (170 kDa) is now known to be topoisomerase II and Sc II is known to be a heterodimeric coiled-coil protein (SMC 1–4 or XCAP-C/E, see later). The enzymatic activity of topoisomerase II is to pass DNA strands through one another. It can cause a double strand break in DNA and rejoin it. (Topoisomerase I can only introduce a single strand break in the double helix, see Table

2.1.) Antibodies to topoisomerase II allowed the demonstration that the protein is an integral component of mitotic chromosomes (Earnshaw *et al.*, 1985; Earnshaw and Heck, 1985; Gasser *et al.*, 1986). Moreover, the efficiency of recovery of total cellular topoisomerase II in the scaffold fraction (>70%) makes it unlikely that the association with the scaffold fraction is accidental. Consistent with the current view of the further folding of the chromatin fibre into the chromosome (Section 2.4.1), immunolocalization data show that topoisomerase II is found in a large number of discrete foci, scattered throughout the axial region of chromosomes. These foci are very uniform in size, suggesting that they represent discrete structural complexes. Each is believed to be an anchoring complex to which chromatin loops are attached. The presence of topoisomerase II in these complexes can be rationalized by the necessity of unravelling DNA knots and tangles that will inevitably be generated during processive enzymatic processes such as replication and transcription (Section 4.3). In fact, if topoisomerase II is inactivated *in vivo* by mutation, the mutant cells die because they cannot separate their chromosomes at the end of mitosis (DiNardo *et al.*, 1984).

The development of *in vitro* nuclear assembly systems (Section 3.3) has allowed a direct demonstration of an essential role for topo II in the chromosome condensation process. The first studies of this type made use of embryonic chicken erythroid cells that provide a source of nuclei from a single lineage that varies in topoisomerase II content (Heck and Earnshaw, 1986). The topoisomerase II content of these nuclei changes dramatically during development with the bulk of the enzyme being lost when mitosis ceases in mature erythrocytes. Chromosome condensation in both mammalian cell and *Xenopus* egg extract systems was found to be closely correlated with the level of topoisomerase II present (Adachi *et al.*, 1991). Hirano and Mitchison (1991, 1993) took a complementary approach towards determining the function of topoisomerase II. They also made use of *Xenopus* egg extracts; however, they initially examined the assembly of higher order chromatin structures on naked DNAs (Section 3.1). Using extracts capable of the assembly of mitotic chromosome-like structures, they found that the addition of a topo II inhibitor (the epipodophyllotoxin, VM26) prevented the final stages of chromatin condensation. Subsequent experiments made use of the remodelling of *Xenopus* sperm chromatin in egg extracts (Section 3.1). Sperm chromatin was incubated in egg extracts arrested at mitosis within the cell cycle (Section 2.5.3). The condensed sperm chromatin was rapidly decondensed and then assembled into entangled chromatin fibres that eventually resolved into highly condensed individual chromosomes.

Both inhibitor and antibody depletion experiments demonstrated that topoisomerase II was required for chromosome assembly and condensation. Within these chromosomes, topo II was found to be distributed throughout the condensed chromosome and not solely restricted to the chromosomal axis. It was shown that all topo II molecules could be extracted from chromosomes by 50 mM NaCl without perturbing their shape (Hirano and Mitchison, 1993). Thus, topo II is essential for chromosome assembly, but does not play a scaffolding role in the structural maintenance of chromosomes assembled *in vitro*. Although the essential role of topoisomerase II in unravelling intertwined DNA molecules within the chromosome is established (Warburton and Earnshaw, 1997), it does not seem to have an essential structural role as a 'building block' of the chromosome. Instead, the enzyme is a very useful marker for the anchoring sites of the chromatin fibre into loops (Section 2.4.1).

It has been suggested that the nuclear matrix or scaffold contains specific DNA sequences known as matrix or scaffold attachment regions (MARs and SARs). Evidence for a function for the sequences in chromosomal dynamics comes from the synthesis of an artificial protein that preferentially binds to these A-T rich sequences (Strick and Laemmli, 1995). The presence of this protein interfered with the chromosomal dynamics that are normally observed during nuclear decondensation or chromosomal condensation in *Xenopus* egg extracts. However, in the absence of the identification of *bona fide* scaffold attachment sequence-binding proteins *in vivo* and examination of their function, an active role for scaffold or matrix attachments in chromosomal function remains speculative. Such attachments might promote transcriptional activity *in vivo* (Poljack *et al.*, 1994; Tompson *et al.*, 1994b). In plants there is compelling evidence that A-T rich segments of the genome that have been biochemically defined as scaffold attachment regions promote the activity of transgenes (Allen *et al.*, 1993, 1996). However it should be noted that in mammalian cells biochemically defined scaffold (or matrix) attachment regions are not preferentially found in gene rich portions of the genome (Craig *et al.*, 1997). Topoisomerase II and histone H1 are believed to bind preferentially to these AT-rich DNA sequences (Adachi *et al.*, 1989; Izaurralde *et al.*, 1989). It has also been suggested that proteins such as HMG I/Y (high mobility group protein, Section 2.5.8) might displace histone H1 selectively from scaffold attachment regions contributing to the local control of transcriptional activity (Zhao *et al.*, 1993). This type of selective association of proteins with scaffold attachment regions remains to be proven to occur *in vivo*. Lysine-rich proteins like histone H1 have long been known to interact with A-T rich DNA (Leng and Felsenfeld, 1966). Hence the significance of *in vitro* binding experiments that enrich lysine-rich proteins bound to A-T rich scaffold attachment regions remains questionable.

The significance of the residual DNA that is retained in scaffold preparations for scaffold function also remains controversial since exchange with free DNA sequences in solution can occur during preparation of the scaffold (Jackson *et al.*, 1990). Substantial evidence exists to suggest that at some time in the cell cycle all DNA in a chromosome will have some attachment to the nuclear infrastructure, as DNA replication occurs within extensive factories with components attached to defined structures (Jackson and Cook, 1986; Amati and Gasser, 1988; Cook, 1991; see Section 3.3).

The Sc II protein identified by Lewis and Laemmli (1982) has a much more active role in directing mitotic chromosome condensation than topo II (Saitoh *et al.*, 1995). ScII is now recognized to be a member of the SMC (stability and maintenance of chromosomes) family of proteins. The function of this family of proteins was first characterized in the yeast, *Saccharomyces cerevisiae* (Strunnikov *et al.*, 1993, 1995; Koshland and Strunnikov, 1996). SMC proteins are conserved in structure from fungi to vertebrates. Each consists of five major regions (Fig. 2.37): a nucleotide-binding region, a region of α-helix with the potential to form a coiled-coil protein–protein interaction domain, a hinge region, a second coiled-coil domain and a C-terminal region. Mutational analysis indicates that all of these domains are required for SMC protein function. The proteins assemble oligomeric structures (Hirano and Mitchison, 1994; Strunnikov *et al.*, 1995). There are at least four different SMC proteins in *S. cerevisiae*, all are essential for viability suggesting that they have essential non-overlapping functions. The phenotypes of SMC mutant cells in yeast show a failure to undergo mitosis with a partial splitting of the undivided nucleus (Samejima *et al.*, 1993). This phenotype is very similar to that of topoisomerase II mutants in yeast (Uemura and Yanagida, 1986; Holm *et al.*, 1985).

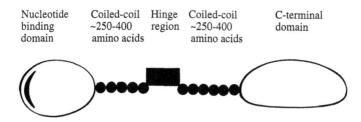

Nucleotide binding domain | Coiled-coil ~250-400 amino acids | Hinge region | Coiled-coil ~250-400 amino acids | C-terminal domain

Figure 2.37. Modular structure of the SMC proteins.

Closer investigation reveals that SMC mutant cells fail to condense and segregate their chromosomes (Strunnikov *et al.*, 1995).

Additional functional information came from an investigation of the properties of vertebrate homologues. Hirano and Mitchison (1994) identified two *Xenopus* chromosome associated proteins, XCAP-C and -E, that appear essential for both chromosome condensation and for maintaining the structural integrity of the chromosome. These proteins are the *Xenopus* homologues of Sc II and share common structural motifs, including amino-terminal nucleoside triphosphate (NTP) binding and central coiled-coil domains (Saitoh *et al.*, 1994). XCAP-C/E is resistant to salt-extraction from chromosomes, but rearranges during salt-treatment from a loose spiral structure to a continual rigid axis running the whole length of the mitotic chromosome (Hirano and Mitchison, 1994). This experiment raises further significant doubts about the nature and significance of such an axis *in vivo* (see earlier). When XCAP-C is depleted from the *Xenopus* egg extracts in which chromosomes are assembled, then chromatin remains as a mass of entangled thin fibres and is not converted into a rod-like chromosomal shape. Hirano and Mitchison speculate that the putative NTP-binding/hydrolysis cycle of XCAP-C/E is important for the dynamic organization of chromosomes. XCAP-C/E is proposed to act as a motor protein that moves along DNA or chromatin (Fig. 2.38). The role of contractile proteins in the structural organization of the interphase nucleus and of metaphase chromosomes still remains largely unknown (Rungger *et al.*, 1979). In an interesting link between higher order chromosome structure and gene activity, Meyer and colleagues found that a gene related to XCAP-C/E is involved in determining the transcriptional activity of chromosomes in the worm *Caenorhabditis elegans* (Chuang *et al.*, 1994). We discuss later in more detail how global changes in chromosomal composition influence gene expression (Section 2.5).

An interesting and separate group of proteins within the chromosome scaffold has been identified primarily through the use of autoantibodies from patients with rheumatic disease. These include the INCENP (inner centromere proteins) and the CENP A, B and C proteins. All of these molecules are now known to associate with the centromere. The centromere is the region of the mitotic chromosome that participates in chromosome movement (Pluta *et al.*, 1995). The mitotic spindle attaches to a specialized structure at the centromere known as the kinetochore. The motor responsible for the movement of the chromosomes towards the spindle poles during mitosis is also located here. The INCENPs (135–150 kDa) bind tightly to mitotic chromosomes between sister chromatids at the centromere and

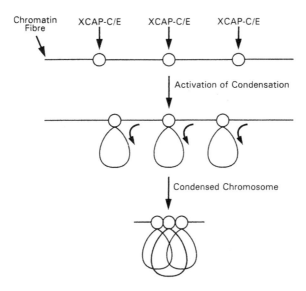

Figure 2.38. A model for the active looping of the chromatin fibre by the XCAP-C/E proteins.
The arrows indicate the movement of the chromatin fibre directed by XCAP-C/E.

wherever else the two chromatids are in close contact (Eckley *et al.*, 1996). Furthermore, the INCENPs are released from a chromosome at the beginning of chromosome separation during mitosis (anaphase), suggesting that they may have a role in regulating sister chromatid pairing (Pluta *et al.*, 1990).

The organization of CENP A, B and C in the centromere is understood in some detail (Fig. 2.39). The centromere can be broken up into three distinct structures: the kinetochore, and the central and pairing domains. Immunological staining reveals that the INCENPs are in the pairing domain. The central domain contains dense chromatin known as constitutive heterochromatin (Section 2.5.6). The kinetochore appears to be anchored to this heterochromatin. The DNA within this heterochromatin is composed primarily of various families of repetitive DNA (satellite DNAs).

The α-satellite family of DNA sequences (comprising 5% of the human genome) is probably present at the centromeres of all human chromosomes. The basic repeat, 171 bp in length, occurs in large arrays of up to 3×10^6 bp long. Nucleosomes have long been known to be rotationally positioned on α-satellite DNA; however, the histone core is believed to adopt several distinct translational positions (Simpson, 1991; Section 2.2.5). An important point is that a single nucleosome is

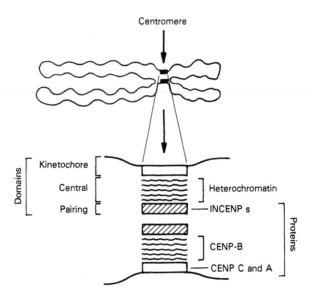

Figure 2.39. Structural domains and protein localization in the centromere of a chromosome.

believed to exist on every 171 bp repeat. This DNA sequence in modified form also provided the foundation for crystallization of the nucleosome core (Luger *et al.*, 1997). Several interesting specialized chromosomal proteins are also known to associate with α-satellite DNA. A 10-kDa protein, HMG-I/Y (high mobility group protein, Section 2.5.8) will bind to α-satellite DNA specifically *in vitro* (Solomon *et al.*, 1986). The HMG-I/Y protein binds in the minor groove of the double helix selectively associating with runs of six or more AT base pairs. In addition HMG-I/Y also probably recognizes certain secondary structural features of the DNA molecule (Section 2.1.1). CENP-B specifically recognizes a 17-bp DNA sequence (5′ CTTCGTTGGAAACGGGA 3′) present in a subset of α-satellite DNA repeats (Masumoto *et al.*, 1989). The DNA binding domain of CENP-B is necessary and sufficient for localization to the centromere *in vivo* (Pluta *et al.*, 1992). CENP-B contains anionic regions rich in aspartic and glutamic acid residues. These are characteristic of many proteins that interact with chromatin. CENP-B is found throughout the centromeric heterochromatin beneath the kinetochore plates (Cooke *et al.*, 1990). Although the exact functions of CENP-B in hetero-chromatin are not known, CENP-B may play a role in the higher order folding of centromeric chromatin through self-assembly mechanisms

(Yoda *et al.*, 1992). It should be noted that although CENP-B is found at all mammalian centromeres examined, the CENP-B recognition element is not found in all α-satellite DNAs (Goldberg *et al.*, 1996; Yoda *et al.*, 1996). It is possible that CENP-B can also be targeted to centromeres through interactions with other protein components.

CENP-A is also specifically associated with centromeric DNA and is especially interesting since it shares homology with core histone H3 (Palmer *et al.*, 1987, 1991) (Section 2.5.1). Recent sequence analysis has revealed that CENP-A is a very specialized H3 variant (Sullivan *et al.*, 1994). It possesses a highly divergent amino-terminal domain and a relatively conserved carboxy-terminal histone-fold domain (Section 2.2.4). The targeting of CENP-A to centromeric DNA is directed by the histone-fold domain; however, the exact mechanism by which this is achieved is, as yet, unknown. Characterization of a homologue in yeast (CSE4, Stoler *et al.*, 1995) establishes that this specialized histone is essential for normal chromosome segregation during mitosis. Genetic evidence suggests that CSE4 heterodimerizes with histone H4 (Smith *et al.*, 1996). This establishes that specialized nucleosomal structures are present at the centromere, in addition to the presence of positioned nucleosomes. Clearly, the centromere represents a highly differentiated chromosomal domain even at the most fundamental level of chromatin structure.

CENP-C is a large 107-kDa protein that is highly hydrophilic and basic (pI = 9.4) (Saitoh *et al.*, 1992). CENP-C is concentrated in a narrow band immediately below the inner kinetochore plate at the interface between the chromosome and the kinetochore. CENP-C has the capacity to bind DNA directly, and is required for normal kinetochore assembly (Tomkiel *et al.*, 1994). CENP-C is found only at the active centromere of a stable dicentric chromosome suggesting a direct role in centromere function (Tomkiel *et al.*, 1994). Other components of the centromere include CENP-E, which resembles microtubule-binding motor proteins (Yen *et al.*, 1991, 1992). CENP-E is a 312 kDa polypeptide with a tripartite structure consisting of globular domains at N- and C-termini separated by a 1500 amino acid α-helical domain that is predicted to form coiled coils. CENP-E colocalizes to the centromere and kinetochore during metaphase, but is released from the centromere at the onset of anaphase when it is degraded (Brown *et al.*, 1994). CENP-E appears to function as a kinetochore motor during the early part of mitosis (Brown *et al.*, 1996a).

The assembly of the centromere and the mechanisms of kinetochore association with chromatin offer perhaps our best opportunity to understand the construction of a specialized chromosomal domain at the biochemical level.

Aside from the centromere, other specialized chromosomal structures associated with non-histone proteins exist. Among the best studied are the telomeres, the ends of eukaryotic chromosomes (Zakian, 1989). These specialized structures protect the chromosomes from exonucleolytic attack, prevent end to end fusion of the chromosomes and promote the complete replication of the ends of the linear DNA molecules present in chromosomes. Telomeric DNA has been studied in many organisms including *Tetrahymena* and yeast. Chromosomes of *S. cerevisiae* terminate in 250–650 bp of the simple repetitive DNA sequence $C_{2-3}A$ $(CA)_{1-6}$. These are binding sites for a non-histone protein known as RAP1 (Conrad *et al.*, 1990). Although some lower eukaryotic telomeres have a nucleosomal organization (Gottschling and Cech, 1984) those of a number of species including *Tetrahymena* and yeast display a protein-dependent protection from nucleases in which the size of the protected structure differs from that anticipated for nucleosomes (< 140 bp). This suggests either that non-histone protein complexes exist in a regular array or that modified nucleosomes exist at the telomere. The RAP1 protein is a participant in the non-histone protein–DNA complexes at the chromosome ends of yeast. RAP1 interacts with yeast telomeres *in vivo* and has been shown to be important for maintenance of telomere length. The protein is abundant (> 4000 copies per cell) and fractionates with the nuclear scaffold. RAP1 has been used as a marker for the localization of yeast telomeres to the nuclear periphery in yeast (Klein *et al.*, 1992; Palladino *et al.*, 1993). This localization depends on proteins SIR3 and SIR4, one of which, SIR3, interacts with the tails of core histone H4 (Johnson *et al.*, 1990, see below). This observation suggested that RAP1, SIR3, SIR4 and the core histones might be involved in the assembly of a specialized nucleoprotein complex at the nuclear periphery in yeast. SIR4 has homology to human lamins, which would be consistent with a localization at the nuclear envelope in yeast (Diffley and Stillman, 1989). The amino-terminal tail domains of histones H3 and H4 are essential for repression of genes placed close to the telomeres in yeast (Hecht *et al.*, 1995). Transcriptional repression at these chromosomal sites also depends on the silent-information regulatory proteins SIR2, SIR3 and SIR4 interacting with each other and with the DNA-binding protein RAP1. Together, they direct both the assembly of hetero-chromatin (Section 2.5.6) and the compartmentalization of yeast chromosomal telomeres to the vicinity of the nuclear envelope.

Mutations in the amino-terminal tail of histone H4 that alleviated silencing can be suppressed by single amino acid substitution in SIR3, suggesting that the two proteins directly interact. Biochemical experiments have confirmed that SIR3 binds directly to the amino-terminal tail of H4, and also to the amino-terminal tail of H3. The data suggest that SIR4 interacts in a similar way with these two histones. The specificity of these interactions was demonstrated by the failure of either SIR3 or SIR4 to interact with the amino-terminal domains of histones H2A and H2B (Hecht *et al.*, 1995). The amino-terminal domains of H3 and H4 are also required for the assembly of SIR3 into telomeric chromatin, and consequently for the association of the telomere with the nuclear envelope.

A model for transcriptional silencing at yeast telomeres predicts that RAP1 interacts with the telomeric repeats and recruits SIR3 and SIR4, which polymerize along nucleosomal arrays through interactions with the amino-terminal tails of H3 and H4 (Fig. 2.40). This model proposes that transcriptional silencing is dependent on the assembly of an extended domain of repressive chromatin structure, where transcription factors and RNA polymerase are excluded both by SIR3 and SIR4 and by the entrapment of this chromatin domain in a perinuclear compartment (see also Section 2.5.6).

Mammalian telomeres are composed of a tandem array of TTAGGG repeats. The length of the telomeric repeat sequences varies between germ cells, tumour cells and somatic cells, indicating that the telomeres are a specialized variable component of the genome. Sperm telomeres may be 10–14 kb in length whereas telomeres in somatic cells are several kilobases shorter and are very heterogeneous in length. Telomere length in tumour cells is even shorter than in normal somatic cells (De Lange *et al.*, 1990). Specialized proteins also recognize mammalian telomeric DNA. One such protein, TTAGGG

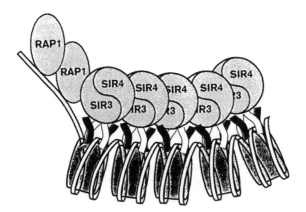

Figure 2.40. Organization of heterochromatin at the yeast telomere. RAP1 targets SIR3 and SIR4 which in turn interact with the amino-terminal tails of histones H3 and H4.

repeat factor (TRF) appears to bind along the entire length of telomeric DNA (Zhong *et al.*, 1992). Short human telomeres (2–7 kb in length) have a very unusual chromatin structure characterized by diffuse micrococcal nuclease digestion patterns (Tommerup *et al.*, 1994). In contrast, longer telomeres (14–150 kb in human, mouse and rat) have a more typical chromatin structure consisting of extensive arrays of close-packed nucleosomes (one every 150–165 bp) (Makarov *et al.*, 1993; Tommerup *et al.*, 1994). Human telomeres also fractionate in the nuclear scaffold fraction (De Lange, 1992) consistent with the concept that telomeres assemble into an extended nucleoprotein complex with a specialized functional role.

To this point we have focused attention on the structure and properties of normal chromosomes such as might be found in any somatic cell. Much of our understanding of chromosomal structure has followed the observation of unusual chromosomes present in particular organisms or tissues during development. These include amphibian lampbrush chromosomes and the polytene chromosomes of insects.

Summary
Biochemical fractionation and immunological localization have defined components required for both the chromosome and nuclear infrastructure. Proteins involved in the large-scale architecture of chromosomes and the nucleus include the lamins, topoisomerase II, SMC proteins and components of the centromere and telomere. These proteins may recognize DNA, other non-histone proteins or histones within the chromatin fibre or some special nucleosome structure.

2.4.3 Lampbrush and polytene chromosomes

Lampbrush chromosomes are found in the oocytes of many animals (Callan, 1986). They contain very transcriptionally active DNA, where loops of DNA emerging from an apparently continuous chromosomal axis are coated with RNA polymerase. Each RNA polymerase is attached to nascent RNA and associated proteins generating a visible 'brush-like' appearance (Fig. 2.41). The axes of lampbrush chromosomes from which the loops project consist visually of linear arrays of compacted beads, known as chromomeres. DNA is concentrated in the chromomeres which represent compacted regions of chromatin. The axis along which the chromomeres exist consists of two distinct strands of chromatin. Gall measured the kinetics of lampbrush

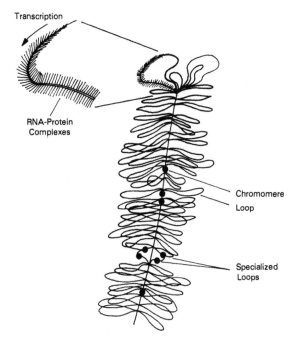

Figure 2.41. A schematic representation of a lampbrush chromosome with an expanded view of a portion of a single loop.
This loop is shown with attacked RNA–protein complexes which grow
larger as transcription proceeds. Occasionally the chromatin in the axis of
the chromosome is condensed to form a chromomere. There are also
specialized loops present with unusual morphologies. These generally
contain special complexes of RNA and protein.

chromosome breakage by DNase I. These experiments led to the
conclusion that there were two DNA duplexes along the chromomere
axis, but only one duplex in each transcriptionally active loop of the
chromosome that emerges from that chromomere axis. Distinctive
loops can be recognized at invariant positions of the chromosomes,
and depend on the DNA sequence contained within each loop (Callan
et al., 1987). Each loop may contain several transcription units and
range in size up to 100 kb. Nucleosomes can clearly be seen within
active transcription units, especially where RNA polymerases are not
so densely packed. An implication of this observation is that
nucleosomal structures must be able to re-form very rapidly following
the passage of RNA polymerase (Section 4.3). In *Xenopus* oocytes,
proteins such as nucleoplasmin associate with lampbrush loops and
might facilitate nucleosome reassembly following transit by RNA
polymerase (Moreau *et al.*, 1986).

Chromomeres occur where there are long regions of inactive chromatin that are compacted into higher order structures, in some respects resembling superbeads (Section 2.3.2). Several groups have prepared antibodies against amphibian oocyte nuclear proteins and have used oocyte sections or isolated lampbrush chromosomes for intranuclear localization. The large size of the chromosomes, their ease of manipulation, and the wealth of morphological detail make them ideal for such studies. Actin, histone H2B, nucleoplasmin (Section 3.2) and RNA-binding proteins have all been localized within loops (Roth and Gall, 1987). Actin filaments may be involved in the extension of the lampbrush chromosome loop away from the chromomere axis (see also Rungger *et al.*, 1979).

The polytene chromosomes of insects have had an equivalent utility for the investigation of chromosomal structure and function. The secretory cells of certain insect larvae have an enormous biological requirement for the accumulation of specific mRNAs. In order to fulfil these demands the cells grow to a large size, carrying out multiple cycles of DNA synthesis without cell division. Eventually, these giant cells contain over a thousand times the quantity of DNA found in a normal nucleus. The chromosomes within these cells are immense. All of the homologous pairs of chromosomes remain side by side forming a single giant chromosome. Like lampbrush chromosomes, it is possible to isolate polytene chromosomes under physiological ionic conditions with higher order structure preserved. Immunological staining has revealed the specific distribution of several non-histone proteins including RNA polymerase II and proteins apparently responsible for the generation of inactive heterochromatin (HP1, see Section 2.5.6). Grossbach and colleagues made use of antibodies against variants of histone H1 (Section 2.5.1) to demonstrate that the localization of particular linker histones can be highly specific for individual chromosomal domains within the polytene chromosomes (Mohr *et al.*, 1989). In contrast, a similar approach with antibodies raised against HMG14 related proteins (Section 2.5.8) leads to the general immunofluorescent staining of transcriptionally active domains (Westerman and Grossbach, 1984). Turner found that individual chromosomal domains within polytene chromosomes can be significantly enriched with forms of histone H4 that have particular states of post-translational modification (Turner *et al.*, 1990, 1992). In related experiments it has been discovered that hyperacetylation of histone H4 on the male X chromosome of *Drosophila* correlates with the increased transcriptional activity necessary for the phenomenon of dosage compensation (Bone *et al.*, 1994). In this phenomenon the transcriptional activity of all genes on the single male X chromosome is increased relative to genes on the other chromosomes, which are present in two copies per cell. These studies strongly suggest that chromosomes can possess a highly selective microheterogeneity in protein composition such that individual chromosomal domains contain particular histone variants or post-translational modification states. Important questions now exist concerning how and why a particular protein or enzymatic activity leading to histone modification is targeted to an individual chromosomal domain.

More detailed ultrastructural analysis especially of the Balbiani ring genes in the polytene larval salivary glands of *Chironomus tentans* has yielded much information. Balbiani ring genes (37 kb in length) constitute well-defined transcription units, with the first engaged RNA polymerase molecule representing the approximate start site of transcription and the last polymerase the site of termination (Bjorkroth *et al.*, 1988; Ericsson *et al.*, 1989). Within each visible puff or expanded region of transcriptionally active chromatin, each Balbiani ring gene forms a loop (rather like in the lampbrush chromosome). The chromatin axis is fully extended during transcription, whereas after the completion of transcription, it coils into a 30-nm chromatin fibre and is finally packaged into a supercoiled loop of the chromatin fibre. Upstream of the start site of transcription is a region free of nucleosome structures, presumably corresponding to the promoter, while further upstream and downstream of the gene are compacted chromatin fibres. More recent experiments using Balbiani ring chromatin indicate that some histone H1 remains on chromatin even while it is actively transcribed (Ericsson *et al.*, 1990). Thus by using immunological staining and keen observation a rather complete ultrastructural picture of the transcription process can be established confirming many biochemical deductions.

Summary
Visualization of large lampbrush and polytene chromosomes provides morphological detail concerning many nuclear processes, especially transcription.

2.5 MODULATION OF CHROMOSOMAL STRUCTURE

Chromosomal structure is not inert. Studies of the molecular mechanisms regulating the condensation and decondensation of chromosomes during the cell cycle demonstrate that gross morphological changes in chromatin structure are driven through reversible modification of chromosomal proteins. Recent progress has defined the molecular machines and events that target chromatin modification as an important component of the transcription regulatory process. This section concerns the structural consequences of chromosomal protein modification, including the significance of histone variants and high mobility group proteins.

2.5.1 Histone variants

Histone genes are invariably present in multiple copies, the level of reiteration varying from two copies of each gene in yeast to several hundredfold in sea urchin. In some organisms, most notably the sea urchin, distinct batteries of different forms of core histone genes are transcribed at precise periods in development (Poccia, 1986). Such variants are particularly prevalent in gametes, where the key function of the histone is to compact the DNA. Metabolic activities involving DNA are generally inhibited in spermatozoa and it is likely that DNA can be compacted in a wide variety of ways. This is because once it is compacted no significant process involving DNA will occur until fertilization (Section 2.5.5). The sea urchin has specialized histone H2A variants (five in all) which have different expression profiles for the cleavage, blastula and gastrula stages of embryogenesis, and four different histone H2B variants that are also developmentally regulated (Table 2.2; see Section 3.4). Sea urchin sperm has an H2B variant which has an amino-terminal tail with a 21 amino acid extension. This tail interacts with linker DNA and may contribute to the unusual stability of sperm nucleosomes (Bavykin *et al.*, 1990; Hill and Thomas, 1990). Differences have also been observed in the stability of nucleosomes containing early and late sea urchin histones (Simpson and Bergman, 1980). Interestingly, a single form of both histone H3 (excluding CENP-A homologues) and histone H4 is present throughout development, reflecting the central role of these histones in nucleosome structure (Section 2.2.4) and chromatin assembly (Sections 3.2). Variations in the primary structure of histones H2A and H2B are likely to alter the compaction of DNA into both the nucleosome and the chromatin fibre. This could be due either to a direct effect on nucleosome structure or an altered binding of histone H1 to the nucleosome core particle (Section 2.3.1).

An evolutionarily conserved variant of histone H2A that is essential for *Drosophila* development is H2A.vD (van Daal and Elgin, 1992). This

Table 2.2 Histone modification in development

Histone	Modification	Variants	Chromatin structure	Functional consequences
H4	Acetylation, prevalent in *Xenopus* and sea urchin eggs and early embryonic chromatin	–	Weakens constraint of linker DNA. Facilitates nucleosome assembly	30 nm fibre destabilized, linker DNA accessible to *trans*-acting factors Facilitates nuclear division
H3	Acetylation (as above)	–	(as above)	(as above)
H2A	–	Seven in sea urchin develop mentally regulated	Early embryonic forms hinder chromatin compaction	(as above)
H2B	–	Four in sea urchin develop mentally regulated	(as above)	(as above)
		Sperm H2B has extended N-terminal tail	Stabilizes nucleosomes and constrains linker DNA	Sperm chromatin rendered inaccessible to *trans*-acting factors and functionally inert
H1	phosphorylation	–	Weakens constraint of linker DNA, creates a paradox, since linker DNA is more accessible, but chromosomes condense	Linker DNA accessible to *trans*-acting factors, transcription facilitated.
		H5 in chicken	Compacts erythrocyte chromatin	DNA made inaccessible to *trans*-acting factors and functionally inert.
		Two in *Xenopus*	Early embryonic chromatin is less compact	Accessible to *trans*-acting factors
			Gastrula chromatin more compacted	Non-essential genes, such as oocyte 5S RNA genes are repressed. Cell cycle becomes longer
		Six in sea urchin, early embryonic forms have SPKK motifs, late embryonic forms do not	Early embryonic chromatin is not compacted	Chromatin becomes more stable as embryogenesis progresses

variant is known as H2A.Z in mammals, H2A.F/Z in chickens, and hv1 in *Tetrahymena* (Harvey *et al.*, 1983; Ernst *et al.*, 1987; van Daal *et al.*, 1988; White *et al.*, 1988; Hatch and Bonner, 1988). This particular H2A variant associates preferentially with actively transcribed chromatin in *Tetrahymena* (Stargell *et al.*, 1993). H2A.Z has an amino-terminal tail that is similar to that of histone H4, and it is post-translationally modified through acetylation to a greater extent than normal H2A. A second conserved variant is H2A.X (Mannironi *et al.*, 1989). This histone has an extended carboxyl-terminal tail beyond the histone-fold domain. Such extensions are not uncommon in variant H2A molecules. Wheat H2A1 has a 19 amino acid carboxyl-terminal extension which has the potential to protect about 20 bp of DNA immediately adjacent to the nucleosome core (Lindsey *et al.*, 1991). This is again consistent with the proposed position of this domain within the nucleosome (Section 2.2.4). Until recently, core histone variants were believed to contain the most extreme changes from a normal histone architecture that might be incorporated into a nucleosome. However, these variants are now recognized as only one example of a family of proteins that share the histone-fold structure and that might be incorporated into a nucleosomal architecture, although these proteins contain wide deviations from normal histone sequences (Baxevanis *et al.*, 1995).

The core histones have evolved from a DNA-binding protein that contained only the three α-helices of the histone-fold domain and lacked any tail domains (Baxevanis *et al.*, 1995). The archaebacterial protein HMf consists of only the histone-fold domain and wraps DNA around itself within nucleosome-like structures (Sandman *et al.*, 1990). The eukaryotic core histones retained this property, but added the capacity of the assembled nucleoprotein complex to interact with other proteins outside the nucleosome through the addition of the tail domains. Other regulatory proteins have made use of their histone-fold domains to confer specialized properties on individual nucleosomes by replacing normal histones within chromatin (Sullivan *et al.*, 1994; Stoler *et al.*, 1995; Shelby *et al.*, 1997). These include the deviant histones: CENP-A, which is found at the centromere, and rat macro H2A (Fig 2.42). The function of macro H2A is not yet known, although an extended C-terminal tail has been added to the histone-fold domain of H2A (Pehrson and Fried, 1992). This tail contains the leucine zipper, which is a dimerization motif often found in transcription factors. Macro H2A might interact with a sequence specific DNA-binding protein to tether a specific nucleosomal structure at a defined site.

More extreme variations from normal histone sequence are found in transcriptional regulatory proteins. These proteins maintain the

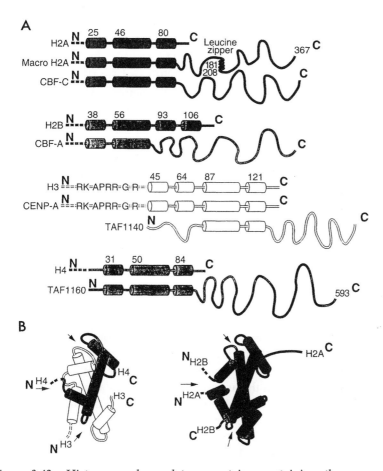

Figure 2.42. Histones and regulatory proteins containing the histone fold.

A. Each core histone is shown in a linear representation with the approximate regions of α-helix shown as cylinders. Numbers indicate the first amino acid relative to the N-terminus of the histone protein for each helix. The related proteins are shown below each histone. Proteins related to histone H2A, histone H2B, histone H3 and histone H4 are shown. Note that the entire N-terminal tail of the core histones is not shown to scale. B. The heterodimerization of histones H3 and H4 and of histones H2A and H2B. The C-terminal structured domain of each core histone is shown in the histone fold. The interaction between the histone-fold domains is described as a 'handshake'. The sites of interaction with DNA within the heterodimer are indicated by the arrows. The position of the N-terminal tails of the core histones are shown as dashed lines. Reproduced, with permission from Wolffe, A.P. and Pruss, D. (1996) *Trends Genet.* **12**, 58–62. Copyright 1996 Elsevier Science Ltd.

histone-fold domain and make use of it both to direct specific protein–protein interactions and to bind DNA (Kokubo *et al.*, 1993; Sinha *et al.*, 1995). Regulatory proteins with these properties include the important components of the basal transcription factor TFIID, known as TATA-binding-protein associated factors $(TAF)_{II}60$ and -40, and the related CCAAT-box-binding proteins, NF-Y (CBF) and HAP2, -3 and -5 (Fig. 2.42). $TAF_{II}40$ resembles H3 and $TAF_{II}60$ resembles H4. Both proteins have extended C-terminal tails that interact with other components of TFIID and transcriptional activators. It has been proposed that $TAF_{II}40$ and $TAF_{II}60$ participate in the assembly of nucleosome-like structures, excluding normal histones from the TATA box, yet maintaining DNA in a semi-compacted state competent for transcription (Kokubo *et al.*, 1993; Nakatani *et al.*, 1996; Xie *et al.*, 1996). Metazoan NF-Y (CBF) and *S. cerevisiae* HAP2, -3' and -5 are highly related trimeric proteins. The evolutionarily conserved peptide sequences of two of the subunits (CBF-A and CBF-C, or HAP3 and HAP5) resemble the histone-fold domains of histones H2B and H2A, respectively (Fig. 2.42). These domains are essential for DNA binding in the presence of the third protein (CBF-B or HAP2) that confers sequence specificity (Sinha *et al.*, 1995). The NF-Y (CBF) and HAP2, -3 and -5 proteins are transcriptional activators.

It appears advantageous for a large number of eukaryotic DNA-binding proteins to retain their histone-like character (Wolffe and Pruss, 1996a). This requirement might be similar to the architectural role of the HMG box in the assembly of large regulatory nucleoprotein complexes (Grosschedl *et al.*, 1994; Section 2.5.8). In certain instances, it is possible to facilitate nucleoprotein-complex assembly through the use of a generic HMG-box protein, such as HMG1, which alters or stabilizes a bend in the double helix. In more specialized cases, a specific transcription factor has evolved that incorporates a structure related to HMG1 that bends DNA, but that also fulfils additional functions by virtue of interaction with other transcription factors. Thus, it might be important to retain DNA in a relatively compact structure within a nucleosome-like architecture, but also useful for the histone-like protein to assume other more specialized functions.

Human centromeric chromatin contains a highly deviant histone called CENP-A. This has significant identity over the histone-fold domain to histone H3, yet has a very different N-terminal tail (Fig 2.42; Section 2.4.2). CENP-A is found within nucleosomes and hetero-dimerizes with H4 (Sullivan *et al.*, 1994; Palmer *et al.*, 1987). Importantly, Sullivan and colleagues discovered that the histone-fold domain of CENP-A targeted the protein to the centromere. This result implies that other histone variants, such as H2A.Z, might be targeted to particular DNA sequences, because the DNA-binding surfaces of the histone-fold domains show a similar number of sequence differences compared to the normal somatic histone H2A as seen between histone H3 and CENP-A (Fig. 2.43).

CENP-A might have arisen through the necessity of having a specialized nucleosomal structure at the centromere where the N-terminal-tail domain makes highly selective contacts with other centromeric components, such as the large hydrophilic DNA-binding proteins CENP-B and CENP-C (Fig. 2.44; Section 2.4.2). Experimental support for this hypothesis comes from genetic experiments in *S. cerevisiae* where a CENP-A homologue, CSE4 is essential for correct sister chromatid segregation (Stoler *et al.*, 1995). This suggests that nucleosomes that include CENP-A will assemble a specialized differentiated chromosomal domain that might serve to facilitate attachment of the kinetochore and the function of the centromere in the segregation of chromosomes. The assembly of a differentiated domain at the centromere establishes an interesting

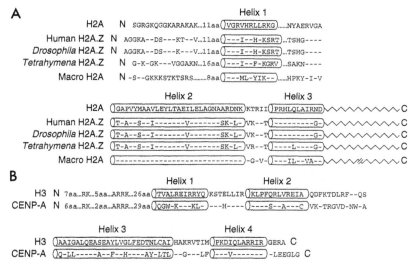

Figure 2.43. Sequence alignments between histones and related proteins.

A. Comparison between histone H2A, human, *Drosophila* and *Tetrahymena* H2A.Z variants, and macro H2A. Identities are indicated by dashes. Gaps to maximize alignment are shown by the dots. The zig-zag lines in the H2A sequences represent the divergent C-terminal tail sequences. Numbers of amino acids are indicated where the sequences are very different (e.g. N-terminal tails). Helical regions are indicated (helix 1, 2 and 3 from the N-terminus). B. Comparison between histone H3 and CENP-A; alignments as in A. Reproduced with permission from Wolffe, A.P. and Pruss, D. (1996) *Trends Genet.* **12**, 58–62. Copyright 1996 Elsevier Science Ltd.

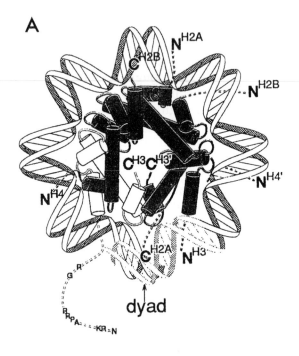

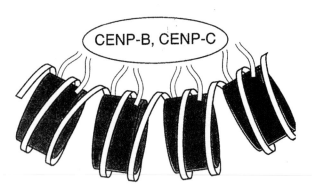

Figure 2.44. The putative arrangement of CENP-A in the nucleosome.
A. The C-terminal histone-fold domains of the core histones. The core
histones are shown in dark shading except for CENP-A, which is shown
unshaded. B. Putative clustering of CENP-A N-terminal tails in a
nucleosomal array. These tails might have protein–protein interactions with
other components of the centromere, such as CENP-B and/or CENP-C.
Reproduced with permission from Wolffe, A.P. and Pruss D. (1996) *Trends
Genet.* **12**, 58–62. Copyright 1996 Elsevier Science Ltd.

precedent for the targeting of core histone variants to particular sites within chromatin.

The form of linker histone varies during sea urchin development and during that of many other organisms, often in a tissue-specific way (Risley and Eckhardt, 1981; Wolffe, 1991a; Khochbin and Wolffe, 1994; Grossbach, 1995; Section 3.4). At least six variants of histone H1 are present during sea urchin embryogenesis. Like histone H2A, there are distinct cleavage stage, blastula and gastrula proteins. Interestingly, these latter proteins do not contain short peptide sequences known as 'SPKK motifs' (see Section 2.5.3). These sequences are found multiple times in the tails of linker histones and are the sites of phosphorylation by the major mitotic kinase in the cell (called the cdc 2 or MPF kinase) (Section 2.5.3). Hence the gastrula form of histone H1 in the sea urchin cannot be phosphorylated by this particular kinase. Once again, an exact role for the different histone H1 variants has not yet been established, but it is clear that the chromatin of the early embryo is less compacted than that of the gastrula (Longo, 1972) and may therefore be more accessible to *trans*-acting factors and more easily replicated. It is also possible that the synthesis of histone H1 variants that cannot be phosphorylated by the MPF protein kinase during embryogenesis reflects a change in the mechanism by which chromosome structure is regulated during the cell cycle (Section 2.5.3).

Linker histones, such as histone H1, have been shown to direct the exact positioning of nucleosomes with respect to DNA sequence (Meersseman *et al.*, 1991). This positioning relies on the sequence and structure selective recognition of DNA by the linker histone, and protein–protein contacts made between the winged helix domain of the linker histone and the histone-fold domains of the core histones (Ura *et al.*, 1995, 1996; Section 2.3.1). The mouse serum albumin enhancer exists in the active state within an array of precisely positioned nucleosome-like particles (McPherson *et al.*, 1993). Specific enhancer-binding factors, including the winged-helix protein HNF3, are part of the nucleosome-like particles and HNF3 can actively direct their positioning with respect to DNA sequence. These observations indicate that HNF3 replaces linker histones within chromatin containing the serum albumin enhancer (McPherson *et al.*, 1993), thereby establishing a precise regulatory nucleoprotein architecture (Fig. 2.45). The replacement of histone H1 by HNF3 would be analogous to the replacement of histone H3 by CENP-A at the centromere. In both instances, a sequence-selective histone-like regulatory protein would direct the assembly of a differentiated chromatin domain. This might encompass a single variant nucleosome, as at the serum albumin enhancer, or long arrays of variant nucleosomes, as at the mammalian centromere.

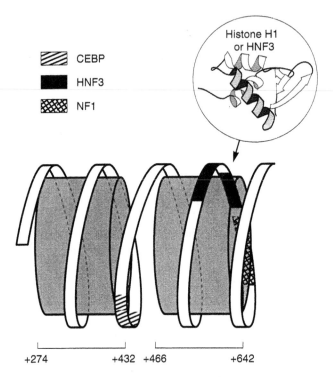

Figure 2.45. A specialized nucleosome on the mouse serum albumin enhancer.

Two nucleosomes are shown positioned on the enhancer of the mouse serum albumin gene (numbers are relative to the 5' end of the albumin enhancer). The boundaries of micrococcal nuclease digestion are indicated by the brackets. The positions of transcription-factor-binding sites are shown as is the potential site of HNF3 or linker histone H1 interaction with the nucleosomal structures. The helix that interacts with DNA is shaded. Reproduced with permission from Wolffe, A.P. and Pruss, D. (1996) *Trends Genet.* **12**, 58–62. Copyright 1996 Elsevier Science Ltd.

What are the advantages for regulatory proteins of maintaining a histone-like structure within a nucleosomal architecture? One immediate advantage is the stability of the nucleosome during the cell cycle. Histones H3 and H4 do not exchange out of chromatin outside of S-phase, moreover, molecular mechanisms exist to reassemble nucleosomes efficiently on newly replicated DNA (Jackson, 1990; Section 4.3). Thus, CENP-A could maintain a very stable association with the centromere even through the replication process. Histones H2A and H2B do exchange out of chromatin, but do so predominantly during transcription (Jackson, 1990). It is very difficult to disrupt core histone interactions within a nucleosome *in vivo*. Regulatory proteins that assemble in association with the core histones will be resistant to displacement from DNA by competing protein–DNA interactions. Linker histones have a much less stable association with nucleosomal DNA. Relatively weak interactions offer the potential for a role in regulatory events where transcription needs to be reversibly activated. For example HNF3 might replace linker histones on the serum albumin enhancer to prevent their repressive influence on transcription (Cirillo *et al.*, 1998; Shim *et al.*, 1998; see Fig. 2.45).

A second advantage of transcription factors resembling the histones lies in the utilization of nucleosomal architecture. As the physical analysis of chromatin clearly demonstrates, nucleosomal arrays do efficiently self-assemble into higher-order chromatin structures. The absence of a nucleosome or the presence of a large non-nucleosomal nucleoprotein complex will probably interfere with this self-assembly process (Hansen and Ausio, 1992). Thus, the requirement for the assembly of higher order chromatin structures might impose constraints on the properties of nucleoprotein complexes in particular chromosomal regions. The easiest way to maintain higher order structure, yet have functional specialization, would be to modify the proteins within a nucleosome, while maintaining the basic function of DNA compaction. A distinct feature of histone interactions with nucleosomal DNA is the exposure of DNA on the surface of the nucleosome. One side of DNA is occluded on the histone surface, but the other is exposed and potentially accessible to other regulatory proteins. Thus, histone-like interactions with DNA could facilitate the assembly of multicomponent complexes (Truss *et al.*, 1995).

Summary

Regulatory proteins exist with strong sequence and structural similarities to the histone proteins. Molecular genetic and cell biological analyses suggest that these proteins are localized at particular sites within the chromosome. Their assembly into nucleosomal structures confers specialized functions to individual chromosomal domains.

2.5.2 Post-translational modification of core histones

Core histones undergo two major post-translational modifications: acetylation and phosphorylation. Both have been the subject of intense interest. Core histone acetylation is now believed to have an integral role in the transcription process because some transcriptional activators have histone acetyltransferase activities (Brownell *et al.,* 1996) and some transcriptional repressors are histone deacetylases (Taunton *et al.,* 1996) (see Section 2.5.4). Acetylation of the four core histones occurs in all animal and plant species examined (Csordas, 1990). The sites of modification are the lysine residues of the positively charged amino terminal tails (Section 2.1.2), where each acetate group added to a histone reduces its net positive charge by 1. The number of acetylated lysine residues per histone molecule is determined by an equilibrium between histone acetylases and deacetylases. Two populations of acetylated histone appear to exist in a particular cell nucleus. For example, in the embryonic chicken erythrocyte, 30% of core histones are stably acetylated while the acetylation status of about 2% changes rapidly. The pattern of specific lysine residues in the histone tails that are acetylated varies between different species. This non-random usage suggests that considerable sequence specificity exists for the relevant acetylases and deacetylases (Turner, 1991). Histone H4 that contains three or more acetylated lysines is described as hyperacetylated, whereas the protein containing one or no acetylated lysines is described as hypoacetylated.

Hyperacetylation of the histone tails leads to relatively subtle changes in nucleosome conformation (Bode *et al.,* 1983; Oliva *et al.,* 1990). However, it appears that the most significant consequences are for protein–protein interactions, either between nucleosomes, with histone H1 or with non-histone proteins. The amino terminal tails of the core histones are accessible to trypsin, suggesting that they are exposed on the outside of the nucleosomal core particle (Section 2.2.4). Chemical cross-linking experiments have shown that weak interactions can occur between the N-terminal domain of H2B and linker DNA, although these experiments used a special sperm H2B with a particularly long N-terminal tail (Bavykin *et al.,* 1990; Hill and Thomas, 1990). The amino termini of histones H3 and H4 also interact with core-particle DNA. High resolution protein nuclear magnetic resonance data indicate association of the amino terminal tails of histones H3 and H4 with DNA in the nucleosome core particle at physiological ionic strength (< 0.3 M NaCl). In contrast, the tails of the normal somatic variants of histones H2A and H2B are relatively mobile at all ionic strengths examined (Cary *et al.,* 1982; but see Lee and Hayes, 1997). The weak interaction of the histone tails with DNA in the nucleosome is reflected in the lack of structural change in the organization of DNA and in the integrity of the nucleosome following their proteolytic removal (Ausio *et al.,* 1989; Hayes *et al.,* 1991b; but see Lilley and Tatchell, 1977). Acetylation or removal of the histone tails from nucleosomes that contain recognition sites for *trans*-acting factors can facilitate factor access to these sites (Lee *et al.*, 1993; Vettese-Dadey *et al.*, 1996). This result suggests that histone acetylation might have a major regulatory role in the transcription process. In addition, the acetylation of the histone tails or removal of the histone tails from arrays of nucleosomes deficient in linker histones, impedes the capacity of these arrays to compact into higher-order structures (Garcia-Ramirez *et al.*, 1992, 1995). These observations suggest that a major influence of the histone tails on chromatin structure may be through their influence on the integrity of the chromatin fibre (Section 2.3.1). Others report that histone hyperacetylation appears to have less effect on the assembly of higher-order structures (McGhee *et al.*, 1983b; Ridsdale *et al.*, 1990). However, these older studies use heterogeneous preparations of chromatin whereas the more recent work of Garcia-Ramirez *et al.*, utilize well-defined preparations of synthetic chromatin reconstituted *in vitro*. Subtle changes in chromatin fibre stability might have large effects on transcription. The formation of compacted structures *in vitro* can severely impede both the initiation of transcription by RNA polymerase and elongation by the polymerase (Hansen and Wolffe, 1992, 1994; see Fig. 2.32).

Induction of histone hyperacetylation with the deacetylase inhibitor sodium butyrate increases the accessibility of DNA in chromatin to DNase I and can improve expression of transfected DNA (Gorman *et al.*, 1983). High levels of histone acetylation improve chromatin solubility, suggesting a reduced tendency to aggregate (Perry and Chalkley, 1981). This also implies that higher order chromatin structures containing acetylated nucleosomes are less stable (Perry and Annunziato, 1989). Maintenance of histone acetylation in nascent chromatin immediately after the replication fork might also contribute to a reduction in the stable sequestration of linker histones (Reeves *et al.*, 1985; Perry and Annunziato, 1989; Ridsdale *et al.*, 1990). However in a purified system histone hyperacetylation does not directly influence the association of histone H1 with nucleosomes (Ura *et al.*, 1994, 1997). Consistent with these *in vitro* observations constitutive hyperacetylation of the core histones does not reduce the association of histone H1 with bulk chromatin (Dimitrov *et al.*, 1993; Almouzni *et al.*, 1994), however it inhibits the complete condensation of interphase chromatin (Annunziato *et al.*, 1988).

Bradbury and colleagues have quantitated the number of super-helical turns introduced into DNA per nucleosome in a purified system, dependent on whether the core histones are acetylated. They found that fewer superhelical turns are introduced by a population of nucleosomes when the core histones are acetylated (Norton *et al.*, 1989,

1990). Detailed analysis reveals that the helical periodicity of DNA in the nucleosome is unchanged by histone acetylation. This suggests, by elimination, that the path of DNA between nucleosomes or writhe of DNA on the histone core is affected by histone acetylation (Bauer *et al.*, 1994). These changes could also have important consequences for *trans*-acting factor access to DNA in the nucleosome. An influence of the histone tails on the path of linker DNA could explain their role in the assembly *in vitro* of stable higher-order chromatin structures (Garcia-Ramirez *et al.*, 1992, 1995).

The generation of antibodies against acetylated histones has allowed a number of general correlations to be made concerning possible functional roles of histone acetylation. In *Tetrahymena*, *Xenopus* and humans there is excellent evidence for the presence of acetylated histones in chromatin immediately following replication (deposition-related) (Lin *et al.*, 1989). It has long been known that histone H4 is diacetylated in the cytoplasm immediately after synthesis (Ruiz-Carrillo *et al.*, 1975). Histone acetylation appears to play an important role in facilitating chromatin assembly (see Section 3.2 for discussion). There is also a strong correlation between histone acetylation and the transcriptional activity of chromatin (Allfrey *et al.*, 1964; Gorovsky *et al.*, 1973; Mathis *et al.*, 1978). In *S. cerevisiae* most of the genome is transcriptionally active and contains hyperacetylated core histones (Clark *et al.*, 1993a). However, there are also transcriptionally inactive domains of chromatin in yeast, such as the silent mating type cassettes and telomeric sequences. These contain histone H4 that is hypoacetylated except at one position lysine 12 (Braunstein *et al.*, 1993, 1996). In higher eukaryotes, acetylation of histone H4 increases during the reactivation of transcription in the initially inactive chicken erythrocyte nucleus following the fusion of the erythrocyte with a transcriptionally active cultured cell to form a heterokaryon (Turner, 1991; Section 3.1). Histone acetylation is particularly prevalent over specific genes that are actively transcribed in erythrocytes (Hebbes *et al.*, 1988). More recent studies have demonstrated convincingly that histone hyperacetylation is actually restricted to the domain of chromatin that contains the potentially active chicken β-globin gene locus (Hebbes *et al.*, 1992, 1994). This result is indicative of a very specific targeting of histone acetyltransferase. Structural data indicate that the hyperacetylated histone tails retain some contact with DNA *in vivo* at the globin gene locus (Ebralidse *et al.*, 1993). Immunolabelling of polytene chromosomes in *Chironomus* and *Drosophila* also reveals a non-random distribution of histone H4 acetylation correlating with transcriptional activity (Turner *et al.*, 1992; Bone *et al.*, 1994). Within female mammals the transcriptionally inactive X chromosome is distinguished by a lack of histone H4 acetylation (Jeppesen and Turner, 1993). Therefore, several independent experimental approaches have shown that actively transcribed and potentially active chromatin domains are selectively enriched in hyperacetylated histones, whereas transcriptionally inactive chromatin contains hypoacetylated histones.

These observations have been dramatically raised in significance following the discovery that transcriptional activators and repressors exist that acetylate or deacetylate the core histones (see Section 2.5.4). Moreover genetic and biochemical experiments have suggested other interesting functions for histone acetylation.

Replacement of all four acetylatable lysines in the H4 tail in *S. cerevisiae*, with arginine such that basic charge is maintained leads to extremely slow growth, whereas substitution with glutamine, which mimics an acetylated lysine leads to a delay in G2/M progression (Meegee *et al.*, 1995). None of these mutations alters the eventual assembly of replicated DNA into nucleosomes, therefore the acetylation and deacetylation of lysines in the H4 tail appear necessary for cell cycle progression itself. The N-terminal tails of histones H3 and H4 have redundant functions in the chromatin assembly process whereas they have quite specific functions in transcription (Wan *et al.*, 1995; Lenfant *et al.*, 1996; Ling *et al.*, 1996). The essential role of specific histone modification might reflect the requirement for structural changes in chromatin necessary for the transcription of genes that regulate or drive the cell cycle. However, the aberrant cell cycle characteristics could also be related to other check-points that monitor chromosome integrity. In particular the mutations in histone H4 increase reliance on DNA-damage-sensitive cell-cycle check-point controls (Meegee *et al.*, 1995), suggesting that increased DNA damage occurs in the H4-mutant cells. How might histone H4 acetylation be involved in cell-cycle check-point control, DNA damage and chromosome repair?

Newly synthesized histones H3 and H4 are acetylated (Ruiz-Carrillo *et al.*, 1975; Chang *et al.*, 1997) and deacetylated shortly after their incorporation into the nascent chromatin assembled immediately after replication (Jackson *et al.*, 1976). The histone acetyltransferase and deacetylase involved in these modifications have been characterized (Taunton *et al.*, 1996; Parthun *et al.*, 1996). These enzymes interact with H3 and H4, and appear to share a common subunit known in mammalian cells as p48/p46. The molecular chaperone involved in the assembly of chromatin on newly replicated DNA is CAF1, which also interacts with H3 and H4 and contains p48/p46 (Verreault *et al.*, 1996; Gaillard *et al.*, 1996). Dynamic alterations in H3 and H4 acetylation might be necessary to drive the exchange of p48/p46 between the acetyltransferase, deacetylase and CAF1. These dynamic transitions will not occur if the lysine residues in histone H4 are mutated (Meegee *et al.*, 1995). A failure to mediate these transitions might in turn impact on the role of CAF1 in the repair of damaged chromatin. For instance CAF1 might be irreversibly sequestered on nascent chromatin and not be available to facilitate chromosomal repair on damaged DNA. Alternatively if DNA damage occurs more readily, because of alterations in chromatin compaction following from the inability to appropriately acetylate or deacetylate histones, then inappropriate sequestration of CAF1 and p48/p46 on damaged DNA might interfere with cell cycle progression due to a decrease in the rate of chromatin assembly (Figure 2.46). It is dangerous for a cell to synthesize naked DNA in the absence of chromatin assembly. This is because of the multiple roles of chromatin both in constraining inappropriate gene activity and in directing the appropriate packaging of DNA into chromosomes. Consequently it might be anticipated that molecular mechanisms will exist to monitor chromosomal integrity.

The association of histone acetylation with transcription (see also Section 2.5.4) and the maintenance of chromosomal integrity point to a

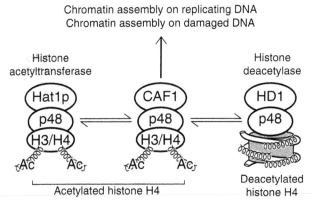

Figure 2.46. A model for the roles of p48 associated proteins. p48 is a component of: (a) a cytoplasmic histone acetyltransferase with hat1p; (b) a chromatin-assembly factor with CAF1; and (c) a histone deacetylase with HD1. Dependent on the subunit composition, this protein will be variously equipped to contribute to all these diverse functions in which the modification state of H4, its cellular localization and deposition in a nucleosome will change as indicated. Hypotheses discussed in the text propose that transitions in histone acetylation might determine the distribution of p48 within these different complexes and their availability for these diverse functions. Reproduced with permission from Wade *et al.* (1997) *Trends Biochem. Sci.* **22**, 128–32. Copyright 1997 Elsevier Science Ltd.

central biological role for this regulatable modification within chromatin. A fusion protein generated by a recurrent chromosomal translocation associated with acute myeloid leukemia incorporates two putative acetyltransferase domains (Borrow *et al.*, 1996; Ogryzko *et al.*, 1996). This suggests that aberrant histone acetylation might contribute to cellular transformation. The two genes fused in this translocation encode the coactivator/histone acetyltransferase CBP (see Section 2.5.4) and MOZ (for *mo*nocytic leukemia *z*inc finger) protein. MOZ contains both the CBP acetyltransferase domain and a region of identity with a yeast protein involved in transcriptional silencing known as SAS2, which shares homology with the Gcn5p protein in the acetyltransferase catalytic domain (Reifsnyder *et al.*, 1996). In addition, the P/CAF acetyltransferase interacts with the p300/CBP acetyltransferase at the same interface as the adenovirus oncoprotein E1A, such that E1A can modulate the association of these proteins (Yang *et al.*, 1996; Ogryzko *et al.*, 1996) and potentially their function. A connection between histone acetylation and cell differentiation has long been known. Histone deacetylase inhibitors such as sodium butyrate and Trichostatin A both promote cell lines to differentiate (Yoshida *et al.*, 1987) and restrict cell transformation (Sugita *et al.*, 1992). These drugs also induce defects during early vertebrate embryogenesis (Almouzni *et al.*, 1994). Clearly inappropriate changes in acetylation patterns might contribute to loss of the differentiated phenotype and cell transformation. How might aberrant acetylation contribute to such events?

Many controls in early vertebrate development depend on the capacity to establish the stable functional differentiation of chromosomal domains: for example the imprinting of chromosomes dependent on parental origin. These epigenetic effects are known to contribute to control of growth and tumorigenesis (Reik and Surani, 1989). Maintenance of histone acetylation states provides an excellent mechanism for the propagation of stable chromosomal imprints determining gene activity (see also Section 4.3). This is because (1) the distributive segregation of nucleosomes during DNA replication will ensure that the parental histone acetylation states are present on both daughter chromatids (Perry *et al.*, 1993), and (2) states of chromosomal acetylation are preserved through mitosis (Lavender *et al.*, 1994). A speculative model for the maintenance of elements of chromatin structure through the cell cycle (Fig. 2.47) would involve a causal role either for histone acetylation states within the nucleosome itself, or for proteins that specifically recognize particular acetylation states and that might segregate in association with the core histones. Strong candidates for such regulatory molecules include the coactiv-

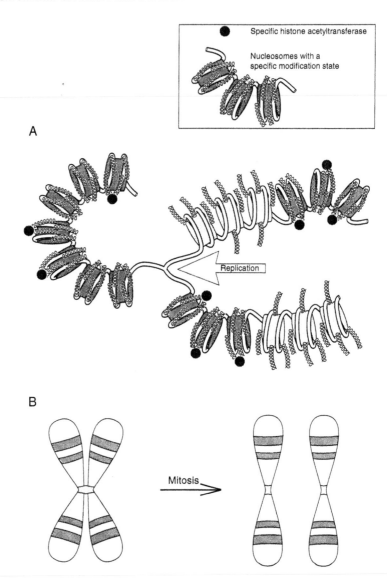

Figure 2.47. A speculative model for the maintenance of acetylation states within chromatin during the cell cycle.
A. Replication leads to the random distribution of parental nucleosomes (dark shading) in small groups to daughter chromatids. Acetylated tail specific histone-binding proteins including coactivators/histone acetyltransferases (circles) might also be distributed to daughter chromatids. New nucleosoms (50% of total) contain diacetylated H4 (light shading); it is possible that histone acetyltransferases segregated with parental nucleosoms will re-establish a predominant acetylation state. B. Domains of chromatin with particular acetylation states are maintained through mitosis. Reproduced with permission from Wade *et al.* (1997) *Trends Biochem. Sci.* **22**, 128–32. Copyright 1997 Elsevier Science LTD.

ators/histone acetyltransferases themselves. Once segregated, the histone acetyltransferases would spread the appropriate state of acetylation over a contiguous imprinted domain of chromatin. Disruption of these imprints by expression or localization of a dysfunctional histone acetyltransferase would therefore be expected to contribute to cellular transformation (Wade *et al.*, 1997).

The second type of core histone modification to receive extensive experimental study is phosphorylation. Histone H3 is rapidly phosphorylated on serine residues within its basic amino terminal domain, when extracellular signals such as growth factors or phorbol esters stimulate quiescent cells to proliferate (Mahadevan *et al.*, 1991). The basic amino terminal domain of histone H3, like that of histone H4, may interact with the ends of DNA in the nucleosomal core particle and therefore perhaps with histone H1 (Glotov *et al.*, 1978). Several studies have suggested a change in either nucleosomal conformation or higher-order structure within the chromatin of the proto-oncogenes *c-fos* and *c-jun* following their rapid induction to high levels of transcriptional activity by phorbol esters (Chen and Allfrey, 1987; Chen *et al.*, 1990). DNase I sensitivity of chromatin rapidly increases and proteins with exposed sulphydryl groups accumulate on the proto-oncogene chromatin. The proteins containing exposed sulphydryl groups include both non-histone proteins, such as RNA polymerase, and molecules of histone H3 with exposed cysteine residues. The histone H3 cysteine residues, the only ones in the nucleosome, are normally buried within the particle. Exposure of the sulphydryl groups might imply a major disruption of nucleosome structure, for example the dissociation of an H2A/H2B dimer might allow access from solution to this region of histone H3. Phosphorylation and acetylation of histone H3 might act in concert to cause these changes. There is likely to be an important yet currently unexplored link between cellular signal transduction pathways and chromatin targets.

In vivo phosphorylation of H4 and H2A occurs in the cytoplasm shortly after histone synthesis (Sung and Dixon, 1970; Ruiz-Carillo *et al.*, 1975; Jackson *et al.*, 1976; Dimitrov *et al.*, 1994). The phosphorylation of these histones, together with the diacetylation of histone H4, may selectively target them to the molecular chaperones involved in nucleosome assembly at the replication fork (Kaufman and Botchan, 1994; Wade *et al.*, 1997). Histone H2A.X is also phosphorylated and has been proposed to have a role in nucleosome spacing during chromatin assembly on naked DNA (Kleinschmidt and Steinbeisser, 1991). H2A.X is a specialized variant synthesized outside of the S phase of the cell cycle (Mannironi *et al.*, 1989), which is stored in *Xenopus* oocytes (Dimitrov *et al.*, 1994). Phosphorylated H2A.X accumulates in decondensing *Xenopus* sperm chromatin such that it eventually represents 50% of the total H2A in the paternal pronucleus. Nevertheless, removal of the modification using phosphatases does not influence the spacing of nucleosomes (Dimitrov *et al.*, 1994). Thus the function of phosphorylation in chromatin assembly remains enigmatic.

Core histones are also methylated on their lysine residues without clearly defined functional consequences. However, it is likely that any inhibition of lysine acetylation would contribute to transcriptional repression. Most methylation in vertebrates occurs on histone H3 at lysines 9 and 27 and histone H4 at lysine 20 (reviewed by Annunziato *et al.*, 1995). Histone H3 is found with up to three methyl groups on each lysine, while lysine 20 in histone H4 maximally contains two methyl groups. Methylation of lysines begins after nucleosome assembly and reaches peak levels at mitosis. In interphase, histone methylation is preferentially targeted to histones H3 and H4 that are already acetylated. This may however only reflect the relative accessibility of the acetylated N-terminal tails. Some evidence suggests that ADP-ribosylation of core histones may lead to localized unfolding of the chromatin fibre. ADP-ribosylation may play a particularly important role in DNA repair (Section 4.4). Here the disruption of chromatin structure cannot always rely on the processive enzyme complexes involved in DNA replication or transcription. The synthesis of long negatively charged chains of ADP-ribose may well facilitate a partial disruption of nucleosomes, presumably by exchange of histones to this competitor polyanion. Histone H2B and especially H2A can also be modified by addition of the small protein ubiquitin (West and Bonner, 1980). Ubiquitin has been found to participate in regulating protein degradation. The protein is covalently attached, via an ATP-dependent reaction, to a protein to be targeted for proteolysis.

Ubiquitin is a 76-amino acid peptide that is attached to the C-terminal tail of histone H2A and H2B. Ubiquitinated H2A is incorporated into nucleosomes, without major changes in the organization of nucleosome cores (Levinger and Varshavsky, 1980; Kleinschmidt and Martison, 1981). Since the C-terminus of histone H2A contacts nucleosomal DNA at the dyad axis of the nucleosome (Guschin *et al.*, 1991; Usachenko *et al.*, 1994), ubiquitination of this tail domain might be anticipated to disrupt the interaction of linker histones with nucleosomal DNA. The bulky ubiquitin adduct might also be anticipated to disrupt higher-order chromatin structures by impeding internucleosomal interactions.

The cell cycle provides an additional useful context with which to consider all these histone modifications and transcription as a whole (Fig. 2.48). RNA polymerase I and II mediated transcription are

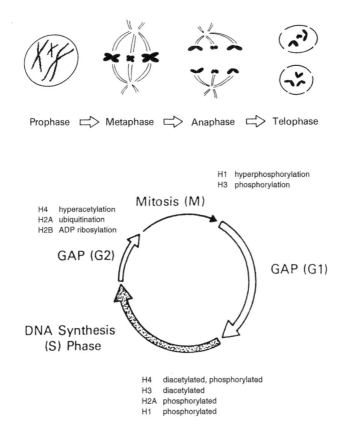

Figure 2.48. A diagram of the cell cycle showing the major changes in histone modification associated with each stage.

severely inhibited during mitosis (Johnson and Holland, 1965; Morcillo *et al.*, 1976). This inhibition involves the post-translational modification of components of the basal transcriptional machinery as well as chromatin structural components (Hartl *et al.*, 1993). During S phase, all the core histones are acetylated but the predominant modifications are mono- and diacetylation of histones H3 and H4 (Waterborg and Matthews, 1982, 1984 see above). In G2, histones H3 and H4 become hyperacetylated and in mitosis all four core histones are deacetylated. With respect to phosphorylation, histone H2A is phosphorylated throughout the cell cycle (Gurley *et al.*, 1978), histone H3 is phosphorylated during mitosis (Paulson and Taylor, 1982) and histone H1 phosphorylation occurs throughout S phase, increases during G2 and becomes maximal at metaphase with 22–24 phosphates per H1 molecule (Mueller *et al.*, 1985; see Section 2.5.3). ADP ribosylation and ubiquitination are present through S phase, becoming maximal in G2 (Kidwell and Mage, 1976; Mueller *et al.*, 1985). These modifications also decline during mitosis. These results indicate the dynamic nature of these modifications, and the necessity of reconfiguring the chromosome continually during the cell cycle.

Summary
Core histone acetylation has important consequences for the organiza-tion of DNA in a nucleosome, potentially loosening interactions at the periphery and probably facilitating the unravelling of higher-order chromatin structure. There is clear evidence that histone acetylation has a role in transcriptional regulation, other roles might include monitoring chromosomal integrity and epigenetic imprinting. Core histone phosphorylation and ubiquitination may also have important structural consequences for nucleosome assembly and integrity.

2.5.3 Linker histone phosphorylation

A widely studied post-translational modification of chromatin is the reversible phosphorylation of histone H1. This modification varies through the mitotic cell cycle (Fig. 2.48). Studies of both the slime mould *Physarum* and mammalian cells in culture show that phosphorylation of H1 is highest in rapidly dividing cells and decreases in non-proliferating cells; levels of histone H1 phosphoryla-tion are lowest in G1 and rise during S phase and mitosis. During mitosis, phosphorylation of histone H1 peaks at metaphase when chromosomes are at their most condensed. These results have led to the suggestion that a causal relationship exists between histone H1 phosphorylation and chromosome compaction (Bradbury *et al.*, 1974).

Histone H1 consists of a globular central domain flanked by lysine-rich highly basic amino terminal and carboxyl terminal tails (Section 2.3.1). The globular domain interacts with DNA in contact with the core histones, whereas the tails bind to linker DNA. Phosphorylation of the histone H1 tails occurs predominantly at conserved (S/T P-X-K/R, serine/threonine, proline, any amino acid, lysine/arginine) motifs of which several exist along the charged tail regions (Churchill and Travers, 1991). The carboxyl-terminal tail has the capacity to adopt an extended α-helical structure (Clark *et al.*, 1988). The central structured domain of histone H1 is not phosphorylated (Langan, 1982). It might be expected that the neutralization of positive charge on the tails would weaken the interaction of histone H1 with the linker DNA. This might be a prerequisite for chromosome condensation, but is also paradoxical since the association of histone H1 with linker DNA has been thought to direct the folding of nucleosomal arrays into the chromatin fibre (Section 2.3.2). The importance of linker histones in the assembly of chromosomes and nuclei has recently been examined through cell biological approaches in which they have been depleted from *Xenopus* egg extracts. It was found that it is possible to assemble mitotic chromosomes and functional nuclei in the complete absence of linker histones (Ohsumi *et al.*, 1993; Dasso *et al.*, 1994b). These experiments unambiguously establish that phosphorylated linker histones are not essential for the chromosome compaction essential for nuclear function.

In order to determine the significance of linker histone phosphorylation for chromosomal function it has been useful to examine systems in which mitosis and chromosome compaction are uncoupled (Roth and Allis, 1992). *Tetrahymena* is a ciliated protozoan in which two distinct nuclei exist differing in structure, function and mitotic behaviour. The somatic macronucleus is responsible for maintaining cell growth, is transcriptionally active and divides amitotically without any apparent condensation of chromatin. Surprisingly, however, macronuclear H1 phosphorylation is controlled through a kinase (the cdc 2 or MPF activity) that is similar to that regulating the cell cycle in normal mammalian cells. Nevertheless, the activity of this kinase and chromosome condensation can be uncoupled. The phosphorylation state of macronuclear histone H1 is highly dependent on cell growth conditions. If the cells are starved, growth ceases and histone H1 is moderately dephosphorylated. More significantly, during conjugation the macronucleus becomes completely inert, chromatin condenses and histone H1 is completely dephosphorylated (Lin *et al.*, 1991). Thus phosphorylation of H1 is *inversely* related to chromosome condensation. In contrast to the macronucleus, the germline micronucleus which is responsible for the sexual cycle is normally transcriptionally silent and undergoes a normal mitotic cycle including the formation of mitotic chromosomes (Gorovsky, 1986).

Several very useful studies correlating nuclear function and histone modification have been carried out comparing these two nuclei within the single *Tetrahymena* cell – a natural heterokaryon (Section 3.1.2). Briefly, special variants of histone H2A are present in the transcriptionally active macronucleus but absent from the micronucleus. Histones are also more extensively acetylated in the macronucleus (Section 2.5.2). The association of the linker histone H1 with chromatin in the macronucleus also decreases with transcriptional activity. This macronuclear histone H1 is highly phosphorylated during vegetative growth. In micronuclei, macronuclear histone H1 is replaced by four specialized linker histone polypeptides (Allis and Gorovsky, 1981; Roth *et al.*, 1988). One of these linker histone polypeptides becomes heavily phosphorylated on transcriptional activation of the micronucleus during the sexual cycle in *Tetrahymena* (Sweet *et al.*, 1996).

A second system in which linker histone phosphorylation can be uncoupled from mitosis concerns the function of the specialized linker histone H5 during development of chicken erythroid cells. During the final stages of chicken erythrocyte development, the nucleus is condensed into inactive heterochromatin due in part to the appearance of histone H5 (Section 2.5.6). Topoisomerase II also becomes much reduced during this differentiative process (Section 2.4.2). Newly synthesized histone H5 is highly phosphorylated, but when the erythrocyte chromatin becomes condensed, histone H5 is quantitatively dephosphorylated. Hence once again dephosphorylation of a linker histone correlates with chromatin compaction (Aubert *et al.*, 1991). That these two events are directly linked receives further support from experiments in which the gene for histone H5 was expressed in fibroblasts. This specialized linker histone would not normally be found in these cells. The accumulation of histone H5 in the fibroblasts inhibited cell growth concomitant with chromatin compaction (see also Sun *et al.*, 1989). Under these circumstances histone H5 was not phosphorylated. Introduction of the protein into transformed cells led to phosphorylation of histone H5. In this case nuclear condensation did not occur and the cells continued to grow and divide. Phosphorylation of the linker histone clearly prevents chromosome folding as might be expected from biophysical analysis (Section 2.3.1).

The final example of a correlation between linker histone phosphorylation and chromatin compaction concerns sea urchin spermatogenesis. Here, dephosphorylation of a sperm-specific histone H1 correlates with chromatin condensation (Green and Poccia, 1985). Following fertilization, sperm histone H1 is phosphorylated in parallel with decondensation of the sperm pronucleus. In all of these examples we see that histone H1 dephosphorylation correlates with chromosome compaction. The co-ordinate phosphorylation of non-histone proteins at the same time as histone H1 during the cell cycle seems more likely to regulate the compaction of chromatin during mitosis (see below).

Phosphorylation of histone H1 has been shown directly to weaken interaction of the basic tails of the protein to DNA. Surprisingly, these changes influence the binding of the protein to chromatin even more than to DNA (Hill *et al.*, 1991). Perhaps phosphorylation of histone H1 is required to weaken the interaction of the linker histone with chromatin and thereby 'loosen' the chromatin fibre to allow other *trans*-acting factors required for gross changes in chromosomal architecture to bind to DNA or the fibre itself. For example, such proteins might include SMC proteins such as XCAP-C/E (see Section 2.4.2).

Characterization of the major mitotic kinase (cdc 2 or MPF) in eukaryotic cells has allowed many of the nuclear events driven by phosphorylation to be defined (Dunphy and Newport, 1988). In the course of these studies it became clear that MPF was similar, if not identical, to the major histone H1 kinase in eukaryotic cells (Langan *et al.*, 1989). During mitosis in higher eukaryotes, MPF induces the ultrastructural changes required for cell division, including nuclear envelope disassembly (nuclear membrane and lamina), chromatin condensation and construction of the mitotic spindle. Although histone H1 becomes hyperphosphorylated during mitosis, it is clearly not the only substrate for MPF during the cell cycle. Newport, Gerace and colleagues have shown that disassembly of the nucleus, nuclear membrane vesicularization, lamin disassembly and chromosome condensation are all independent processes (Ohaviano and Gerace, 1985; Newport and Spann, 1987; Newport *et al.*, 1990). The disassembly of the nuclear lamina appears to be driven by phosphorylation. It is quite possible that phosphorylation of the other proteins found in the nuclear scaffold fraction, including Sc I (topoisomerase II), Sc II (XAP-C/E) and Sc III, may influence chromatin and chromosome folding.

Summary
Linker histone phosphorylation in systems uncoupled from mitosis leads to decondensation of chromatin. Consequently the increase in phosphorylation during mitosis is paradoxical and of unresolved functional significance.

2.5.4 Activators and repressors that make use of chromatin modifications to regulate transcription

The acute regulation of transcription in response to the addition or withdrawal of inductive agents, such as hormones and nutrients, requires the rapid turn-on and turn-off of gene activity. This type of gene regulation is particularly prevalent in the yeast, *Saccharomyces cerevisiae* where the entire organism is continually responding to changes in the environment. The elegant genetic dissection of transcriptional control circuits in *S. cerevisiae* has recently been complemented by substantial progress in our understanding of the biochemistry of transcriptional regulation in metazoans. Remarkable similarities have been found to exist between *S. cerevisiae* and metazoans in the molecular mechanisms used to reversibly regulate gene activity in response to diverse signalling pathways. These mechanisms have been found to employ the modification of chromatin as a central element in gene control.

Central to this grand unification of mechanism is the observation that diverse sequence-specific transcription factors such as steroid and nuclear hormone receptors, Mad/Max, c-Jun/v-Jun, C-Myb/v-Myb, c-Fos, MyoD and CREB utilize a limited number of transcriptional coactivators and/or corepressors to effect their regulatory functions. Thus, the coactivators and corepressors integrate diverse regulatory signals to determine gene activity. Transcriptional coactivators and corepressors have multiple activities that together contribute to their regulatory function. They have the capacity to interact both with the regulatory domains of sequence-specific transcription factors and with the basal transcriptional machinery (Section 4.1). In addition coactivators and corepressors directly modify the chromatin environment within which the transcriptional machinery functions. In fact it appears that the transcriptional machinery requires a chromatin environment in which to function most effectively. Coactivators and corepressors have been found to utilize chromatin to amplify the range of transcriptional regulation far beyond what might be achieved on naked DNA (Wolffe *et al.*, 1997b).

Early experiments established the existence of large molecular machines with the dedicated function of disrupting chromatin structure and facilitating transcription (reviewed by Peterson and Tamkun, 1995). Most recently the covalent modification of the core histone proteins that assemble DNA into nucleosomes has been recognized as having a major role in transcriptional regulation. Multiple coactivators have been shown to possess histone acetyltransferase activity (Brownell *et al.*, 1996; Mizzen *et al.*, 1996; Ogryzko *et al.*, 1996; Yang *et al.*, 1996), and corepressors have been identified that recruit histone deacetylases (Alland *et al.*, 1997; Heinzel *et al.*, 1997; Kadosh and Struhl, 1997; Laherty *et al.*, 1997) and that selectively bind to deacetylated histones (Edmondson *et al.*, 1996).

Genetic selections and screens for yeast mutants that influence the transcription of genes limiting for growth have been successful in defining many structural and regulatory components of the transcriptional machinery. These include proteins with well defined functions such as RNA polymerase II subunits and the Tata binding protein TBP (Arndt *et al.*, 1989; Eisenmann *et al.*, 1989). Other genes were identified whose functional roles in transcription were less clearly defined. Among this latter group were positive regulators of both the HO endonuclease gene required for mating type switching and of the SUC2 (invertase gene) (Neigeborn and Carlson, 1984; Stern *et al.*, 1984).

SWI (switch) and *SNF* (sucrose non-fermenting) genes have been found to encode proteins that together assemble a large multisubunit complex required for the regulation of a specific group of inducible genes (Cairns *et al.*, 1994; Peterson *et al.*, 1994). A major clue to the molecular mechanisms by which the SWI/SNF activator complex functions came from a genetic screen for mutations of genes that would allow transcription from the HO gene in the absence of specific SWI genes (Herskowitz *et al.*, 1992). These studies identified the SIN genes (*SWI IN*dependent). SIN 1–4 have been found or inferred to have a direct impact on chromatin structure and function. A simple model would predict that the SWI/SNF activator complex functions by overcoming the repressive effects of the SIN gene products on transcription (Fig. 2.49). Consistent with this hypothesis: *in vivo* experiments in *S. cerevisiae* suggest that the SWI/SNF activator complex activates transcription by altering chromatin structure (Hirschhorn *et al.*, 1992); and *in vitro* experiments using purified SWI/SNF complex indicate that stoichiometric amounts of SWI/SNF complex can alter histone–DNA interactions in the nucleosome (Côté *et al.*, 1994). The subsequent analysis of the SIN gene products and other repressive components of the yeast transcriptional machinery and their effects on chromatin have been particularly informative.

SIN1p is a highly charged nuclear protein, somewhat similar in sequence to the vertebrate HMG1/2 proteins and containing the HMG box domain (Kruger and Herskowitz, 1991, Section 2.5.8). The carboxy-terminal domain of RNA polymerase II has been proposed to antagonize the repressive effects of SIN1 on transcription (Peterson *et al.*, 1991). However, mutations in SIN1 decrease expression of some genes, suggesting a more complex structural role for this protein in chromatin. In vertebrates, HMG1/2 have been shown to replace linker histones within the nucleosome and to directly influence the transcription process in a chromatin context (Nightingale *et al.*, 1996; Ura *et al.*, 1996).

SIN2p is either histone H3 or H4 (Kruger *et al.*, 1995). These histones assemble a tetramer $(H3/H4)_2$ that forms the foundation of

REPRESSORS

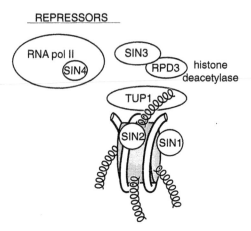

Figure 2.49. Repressive components of chromatin that have been defined at a genetic level.
These include potential structural components of the nucleosome such as SIN1p (resembles HMG1), and known structural proteins such as SIN2p (histones H3 and H4). Other repressive components target the modification of chromatin through known mechanisms such as SIN3p (recruits the RPD3p histone deacetylase) or through unknown mechanisms, such as SIN4p. Additional repressive components such as Tup1p make contact with the core histones and direct chromatin organization (see text for details). Reproduced with permission from Wolffe *et al.* (1997) *Genes to Cells* **2**, 291–302. Copyright 1997 Blackwell Science Limited.

nucleosomal architecture (Section 2.2.4). The SIN mutations cluster in β-bridge motifs within the heterodimer of histones H3 and H4 (Fig. 2.50). Because of the juxtaposition of two (H3, H4) heterodimers at the dyad axis of the nucleosome, the SIN mutations have the potential to disrupt histone–DNA interactions involving the central turn of DNA at the dyad axis. This has a major impact on the integrity of the nucleosome (Kurumizaka and Wolffe, 1997; Wechser *et al.*, 1997).

SIN3p is a large 175 kDa polypeptide containing four paired amphipathic helices, and is proposed to interact with *bona fide* yeast DNA binding proteins (Wang *et al.*, 1990; Wang and Stillman, 1990). Targeting of SIN3p by fusion to a DNA binding domain will direct transcriptional repression (Wang and Stillman, 1993). Genetic experiments indicate a close functional relation between SIN3p and another transcriptional regulatory protein RPD3p (Vidal and Gaber, 1991; Vidal *et al.*, 1991). RPD3p and SIN3p participate in the same transcriptional regulatory functions and appear to be components of one pathway. Both genes are required for the regulation of inducible genes responding to external signals (*PHO5*), cell differentiation (*SPO11* and *SPO13*) and cell type (*HO*, *TY2* and *STE6*). It has recently

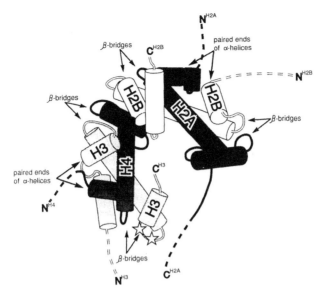

Figure 2.50. Heterodimers of H2A, H2B and H3, H4 with the location of the SIN2 mutations indicated (stars).

been shown that transcriptional repression by the sequence specific DNA binding protein Ume6p involves recruitment of a complex containing the SIN3p corepressor and the RPD3p histone deacetylase to target promoters (Kadosh and Struhl, 1997). In RPD3p-deficient strains both gene activation and repression can be impaired by as much as five fold, thereby leading to a 25-fold reduction in the range of transcriptional regulation for particular genes (Vidal and Gaber, 1991).

RPD3p is very similar in sequence to the mammalian histone deacetylase (Taunton *et al.*, 1996). Subsequent work by Grunstein and colleagues have confirmed this identity, characterizing a 350 kDa histone deacetylase complex (HDA) from yeast nuclei containing three polypeptides, HDA1, HDA2 and HDA3 (Carmen *et al.*, 1996). The HDA1 polypeptide is very similar in sequence to RPD3p, and antibodies against both proteins immunoprecipitate histone deacetylase activity (Rundlett *et al.*, 1996). Mutations in either HDA1p or RPD3p, that lead to a general increase in the acetylation of histones H3 and H4 increase transcriptional repression at the telomeres of yeast chromosomes (Rundlett *et al.*, 1996; De Rubertis *et al.*, 1996). These results could either reflect a direct role for high histone acetylation levels in gene repression at the telomeres, or more likely a consequence of a general increase in the access of transcription factors to non-productive sites in chromosomes following an increase in acetylation levels. A general titration of the transcriptional machinery might account for the difficulty in transcribing telomeric genes. These complications in interpretation are unfortunately inherent to genetic approaches to transcriptional control in chromatin. To understand molecular mechanism, biochemical data are necessary to complement genetics.

SIN4p is a 111 kDa protein that shares no significant homology with other proteins in the database (Jiang and Stillman, 1992). Mutations in SIN4p alleviate the repression of some genes such as *HO*, and activate others such as *HIS4* and *MATα2* (Jiang and Stillman, 1995; Wahi and Johnson, 1995). These mutations have phenotypes similar to those observed in strains with histone mutations suggesting a modulatory role for SIN4p in organizing chromatin. SIN4p is a component of the mediator complex within the RNA polymerase II holoenzyme (Li *et al.*, 1995; Section 4.1), however this does not exclude functions relevant to chromatin organization. The loss of SIN4p has no effect on nucleosome positioning, but does lead to a striking increase in the sensitivity of chromatin to digestion by micrococcal nuclease (Macatee *et al.*, 1997). The molecular basis for this transition in chromatin structure is unknown.

TUP1p is a yeast global transcriptional repressor that is also required for the repression of the SUC2 (invertase) gene (Carlson *et al.*, 1984; Williams and Trumbly, 1990). Tup1p is a 713 amino acid protein: the C-terminal domain contains eight repeats of a 43 amino acid sequence rich in aspartate and tryptophan (WD-40) repeats (Fong *et al.*, 1986). These WD-40 repeats facilitate the targeting of TUP1p to particular promoters through interaction with the DNA sequence-specific α2 repressor (MATα2p) (Komachi *et al.*, 1994; Section 3.3.4). The N-terminal domain of TUP1p also interacts with proteins. Here there are two regions of defined function: the first 72 amino acids interact with SSN6p – a large 107 kDa phosphoprotein that contains 10 tandem copies of a tetratricopeptide repeat (Schultz and Carlson, 1987), this region also facilitates multimerization of Tup1p (Tzamarias and Struhl, 1994). Both TUP1p and SSN6p contribute to the establishment of a repressive chromatin structure targetted by the DNA bound MATα2p/MCM1p complex (Cooper *et al.*, 1994). The MATα2p/MCM1p complex had been shown by nuclease mapping studies to direct the specific positioning of nucleosomes (Roth *et al.*, 1990; Shimizu *et al.*, 1991). This positioning is dependent on interactions with the N-terminus of TUP1p and histones H3 and H4 (Edmondson *et al.*, 1996). Amino-terminal mutations in histones H3 and H4 that interfere with this interaction with TUP1p relieve the repression of genes by MATα2p/MCM1p. Changing specific lysines (12 and 16) to glutamines, which resemble the consequences of acetylation in the N-terminal tail of histone H4, interfere with Tup1 binding (Edmondson *et al.*, 1996) suggesting that Tup1 prefers to bind to deacetylated histone H4.

These results provide a firm genetic and biochemical basis for considering a specific role for the active modification and organization of chromatin in transcriptional control. This conclusion is further substantiated by the definition of a distinct activator complexes in *S. cerevisiae*.

The *GCN5p/ADA2p/ADA3p activator complex* was identified using a distinct type of genetic screen carried out by Guarente and colleagues to identify mutations in genes that confer resistance to the toxic chimeric transcriptional activator GAL4-VP16 (Berger *et al.*, 1992). Genes identified by this screen might be involved in facilitating gene activation by the VP16 acidic activation domain. In this way two 'adaptor' proteins, ADA2p and ADA3p were identified that were proposed to bridge interactions between activation domains and the basal transcriptional machinery (Guarente, 1995). A comparable mutation in the gene *GCN5* impaired the activation of transcription by the transcription factor GCN4p (Georgakopoulos and Thireos, 1992). Subsequent genetic and biochemical experiments established that GCN5p/ADA2p/ADA3p exist as a coactivator complex in yeast (Georgakopoulos *et al.*, 1995; Marcus *et al.*, 1994; Horiuchi *et al.*, 1995) and that the ADA2p interacts with both acidic activation domains and TBP (Barlev *et al.*, 1995).

The GCN5p/ADA2p/ADA3p coactivator is a histone acetyltransferase (Brownell *et al.*, 1996) that selectively modifies lysine 16 in the N-terminal tail domain of histone H4 (Kuo *et al.*, 1996). This property suggested for the first time that coactivators have the capacity to directly modify the chromatin template in order to facilitate transcription. *GCN5* is not an essential gene in yeast, however the capacity to induce gene expression by GCN4p is reduced by 60% if *GCN5* is non-functional. This suggests that like the histone deacetylases (HDA1p and RPD3p), the individual histone acetyltransferases may not be essential in yeast. This might reflect the presence of numerous genes with overlapping functions, and/or merely that the modification of chromatin structure is only one contributor to transcriptional regulation. The existence of multiple potentially redundant histone acetyltransferases is substantiated by recent observations in metazoans.

There is excellent precedent for pioneering experimental work in *S. cerevisiae* leading to the recognition of comparable regulatory mechanisms in metazoans. The identification of the *SWI/SNF activator complex* (Peterson and Herskowitz, 1992) offered insight into potential regulatory roles for related proteins in *Drosophila* (Tamkun *et al.*, 1992). It was also shown that metazoan regulatory proteins including the glucocorticoid receptor introduced into yeast could make use of the SWI/SNF activator complex to activate synthetic promoters containing their recognition elements (Laurent and Carlson, 1992; Yoshinaga *et al.*, 1992; Laurent *et al.*, 1993). Mammalian homologs of components of the SWI/SNF complex were character-ized (Khavari *et al.*, 1993; Muchardt and Yaniv, 1993; Muchardt *et al.*, 1995; Chiba *et al.*, 1994). These proteins hSNF2α and hSNF2β, possess amino terminal proline- and glutamine-rich regions which resemble transcriptional activation domains. Their capacity to interact with other components of the transcriptional machinery including the glucocorticoid and oestrogen receptor is shown by their activation of transcription in transient co-transfection assays which may be largely independent of chromatin-mediated effects (Muchardt and Yaniv, 1993; Chiba *et al.*, 1994).

Evidence for the targeted disruption of chromatin by the mammalian SWI/SNF complex has remained elusive. A 100-fold molar excess of the 2×10^6 Da SWI/SNF complex can disrupt a synthetic nucleosome core (containing 0.1×10^6 Da of histone) *in vitro* (Imbalzano *et al.*, 1994). It has also been suggested that the RNA polymerase II holoenzyme contains SWI/SNF and might remodel chromatin (Wilson *et al.*, 1996). If this was true, the targetting problem would be solved. However SWI/SNF is present at only 200 copies per yeast or mammalian cell (Côté *et al.*, 1994; Gerald Crabtree, Stanford University, personal communication), therefore only a small subset of total holoenzyme would contain SWI/SNF and recent purification schemes have not detected SWI/SNF in holoenzyme preparations (Cairns *et al.*, 1996). Moreover recent experiments suggest that the yeast RNA polymerase II holoenzyme might under certain circum-stances disrupt chromatin independent of the presence of SWI/SNF (Gaudreau *et al.*, 1997). Crabtree and colleagues have identified multiple complexes containing mammalian SWI/SNF homologs (Wang, W. *et al.*, 1996a,b) and suggested the existence of developmen-tally distinct functions. Certain cell lines lack hSNF2α and β (also known on hbrahma and brahma related gene 1, respectively) entirely, indicating that they are not essential for cell viability (Wang, W. *et al.*, 1996b). The association of hSNF2α and β with chromosomes is modulated during the cell cycle and following cell transformation (Muchardt *et al.*, 1996), leading to the suggestion that these proteins might be involved in the control of cell growth. The exact functions of the metazoan SWI/SNF complex remain to be determined, however a connection to transcriptional regulation for a subset of this diverse family of complexes seems to be a reasonable speculation (Wang, W. *et al.*, 1996a,b).

Histone acetyltransferases: PCAF, p300, TAF$_{II}$250. The discovery that *S. cerevisiae* GCN5p has histone acetyltransferase activity (Brownell *et al.*, 1996) led to the recognition that comparable regulatory mechanisms might exist in metazoans (Yang *et al.*, 1996b). A human homolog of GCN5p known as *p300/CBP associated factor* (PCAF) acetylates histones (Yang *et al.*, 1996), as does p300/CBP itself (Ogryzko *et al.*, 1996). p300/CBP serves as an integrator to mediate regulation by a wide variety of sequence-specific transcription factors (Kamei *et al.*, 1996) including steroid and nuclear hormone receptors, c-Jun/vJun, cMyb/v-Myb, c-Fos and MyoD (Janknecht and Hunter, 1996). To strengthen the analogy with the GCN5p/ADA2p/ADA3p complex, p300/CBP has a domain highly similar to part of ADA2p and associates with PCAF, the homolog of GCN5p (Yang *et al.*, 1996). Most recently a component of the DNA-binding basal transcription factor TFIID has also been shown to have histone acetyltransferase activity (Mizzen *et al.*, 1996). TAF$_{II}$250 is the architectural core of TFIID interacting with the other TAFs (TBP associated factors) as well as with TBP (Section 4.1). TAF$_{II}$250 is required for the activation of particular genes indicative of coactivator function, and associates with components of the basal transcriptional machinery such as TFIIA, TFIIE and TFIIF (Dikstein *et al.*, 1996). In addition, TAF$_{II}$250 functions as both a kinase and a histone acetyltransferase (Mizzen *et al.*, 1996; Dikstein *et al.*, 1996). Although GCN5p and PCAF are related proteins, there is no significant sequence identity or known structural similarity with p300/CBP or TAF$_{II}$250. Thus diverse proteins in metazoans (and potentially in *S. cerevisiae*) possess histone acetyltransferase activity. In an interesting link between the mammalian SWI/SNF activator complex, monoclonal antibodies against p300 immunoprecipitate a complex of at least seven cellular proteins (Dallas *et al.*, 1997). Within this complex is TBP, TAF$_{II}$250 and hSNF2β (BRG1) suggesting that functions of histone acetyltransferases might be linked to those of other activators that contend with chromatin.

Histone deacetylase and mammalian SIN3. The purification of the mammalian histone deacetylase and the recognition of the similarities to *S. cerevisiae* RPD3p (Taunton *et al.*, 1996) has provided considerable insight into transcriptional repression in metazoans. The first direct evidence for mammalian homologs of RPD3p being involved in transcriptional repression came from two hybrid screens indicating that the transcriptional regulatory factor YY1 interacted with mouse and human RPD3p (Yang *et al.*, 1996a). The fusion of mammalian RPD3p to a targeted DNA binding domain directed transcriptional repression by more than 10-fold. Mutations in a glycine rich domain of YY1 that directs binding to RPD3p could abolish transcriptional repression by YY1 suggesting that YY1 negatively regulates transcription by tethering RPD3. YY1 is a mammalian zinc-finger transcription factor (Shi *et al.*, 1991) that is proposed to regulate cell growth and differentiation (Shrivastava and Calame, 1994).

A second well-defined protein complex that influences cell growth and differentiation in mammalian cells is the Mad-Max heterodimer (Lahoz *et al.*, 1994; Chen *et al.*, 1995; Hurlin *et al.*, 1995). Max is a widely expressed sequence-specific transcriptional regulator of the basic region-helix-loop-helix-leucine zipper family (bHLH-ZIP). Max heterodimerizes with the Myc family of bHLH-ZIP proteins including Myc, Mad and Mxi-1 (Ayer *et al.*, 1993; Zervos *et al.*, 1993). While the Myc-Max complex activates transcription and transformation, the Mad-Max complex represses these events. Eisenman and colleagues identified two mammalian proteins mSin3A and mSin3B that interact with Mad and that have striking homology to *S. cerevisiae* Sin3p including the four paired amphipathic helix (PAH) domains. The association between Mad-Max and mSin3A and B requires the second PAH domain. Mutations in this domain eliminate the interaction with mSin3A and prevent the Mad-Max complex from repressing transcription (Ayer *et al.*, 1995). The next step was to establish that the mSIN3 proteins interact with the mammalian histone deacetylases. Mad, mSIN3 and the mammalian histone deacetylases coimmunoprecipitate (Laherty *et al.*, 1997; Alland *et al.*, 1997). The third PAH domain of mSIN3 interacts with the mammalian RPD3p homologs and can confer transcriptional repression when attached to a DNA binding domain. More subtle mutational analysis suggests that the cell transformation and transcriptional repression suppressed by the Mad-Max complex depend on distinct domains of the mSIN3 proteins (Alland *et al.*, 1997). However, an active role for histone deacetylation in transcriptional control is demonstrated by the use of deacetylase inhibitors such as Trichostatin A (Yoshida *et al.*, 1990), that abolish Mad's ability to repress transcription. The existence of a conserved transcriptional repression mechanism that utilizes SIN3p and histone deacetylase emphasizes the significance of the chromatin environment for transcriptional control. Histone deacetylation directs the assembly of a stable repressive chromatin structure (Fig. 2.51).

A role for chromatin had already been established in the control of transcription by the thyroid hormone receptor (Wong *et al.*, 1995, 1997a). These studies provide a useful example of how the histones

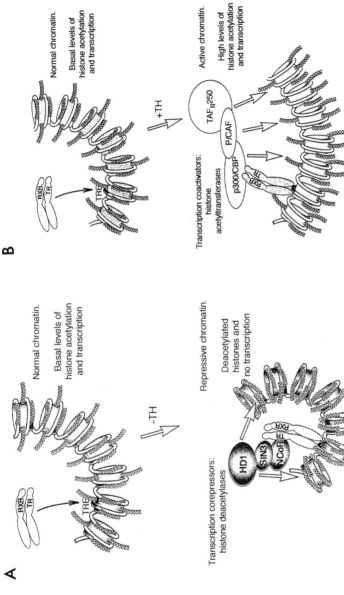

Figure 2.51. How the thyroid hormone receptor makes use of chromatin to regulate transcription. A. Normal chromatin has a basal level of histone acetylation and transcriptional activity. The binding of the thyroid hormone receptor (TR/RXR) to a thyroid response element (TRE) on a positioned nucleosome in the absence of thyroid hormone (TH) leads to the recruitment of a corepressor complex (NCoR, Sin3, HD1) to direct histone deacetylation. B. The binding of the thyroid hormone receptor to a TRE in the presence of ligand leads to the recruitment of the co-activator complex (p300/CBP, p/CAF, TAF$_{II}$250) that directs histone acetylation and facilitates transcription. Reproduced with permission from Wolffe *et al.* (1997) *Genes to Cells* **2**, 291–302. Copyright 1997 Blackwell Science Limited.

can contribute to gene regulation. The assembly of minichromosomes has been utilized within the *Xenopus* oocyte nucleus to examine the role of chromatin in both transcriptional silencing and activation of the *Xenopus* TRβA promoter. Transcription from this promoter is under the control of thyroid hormone and the thyroid hormone receptor (Ranjan *et al.*, 1994), which exists as a heterodimer of TR and RXR. Microinjection of either single-stranded or double-stranded DNA templates into the *Xenopus* oocyte nucleus offers the opportunity for examination of the influence on gene regulation of chromatin assembly pathways that are either coupled or uncoupled to DNA synthesis (Almouzni and Wolffe, 1993a). The staged injection of mRNA encoding transcriptional regulatory proteins and of template DNA offers the potential for examining the mechanisms of transcription factor-mediated transcriptional activation of promoters within a chromatin environment. In particular, it is possible to discriminate between pre-emptive mechanisms in which transcription factors bind during chromatin assembly to activate transcription, and post-replicative mechanisms in which transcription factors gain access to their recognition elements after they have been assembled into mature chromatin structures. TR/RXR heterodimers bind constitutively within the minichromosome, independently of whether the receptor is synthesized before or after chromatin assembly. Rotational positioning of the TRE on the surface of the histone octamer allows the specific association of the TR/RXR heterodimer *in vitro* (Wong *et al.*, 1997b). The coupling of chromatin assembly to the replication process augments transcriptional repression by unliganded TR/RXR without influencing the final level of transcriptional activity in the presence of thyroid hormone.

The molecular mechanisms by which the unliganded thyroid hormone receptor makes use of chromatin in order to augment transcriptional repression also involve mSin3 and histone deacetylase (Alland *et al.*, 1997; Heinzel *et al.*, 1997). The unliganded thyroid hormone receptor and retinoic acid receptor bind a corepressor NCoR (Horlein *et al.*, 1995). NCoR interacts with Sin3 and recruits the histone deacetylase (Alland *et al.*, 1997; Heinzel *et al.*, 1997, see Fig. 2.51). All the transcriptional repression conferred by the unliganded thyroid hormone receptor in *Xenopus* oocytes (Wong *et al.*, 1995, 1997a) can be alleviated by the inhibition of histone deacetylase using Trichostatin A (Fig. 2.52) indicative of an essential role for deacetylation in establishing transcriptional repression in a chromatin environment.

The addition of thyroid hormone to the chromatin bound receptor leads to the disruption of chromatin structure (Wong *et al.*, 1995, 1997a). Chromatin disruption is not restricted to the receptor binding

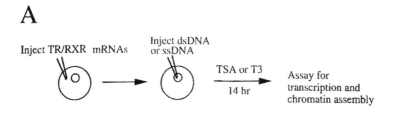

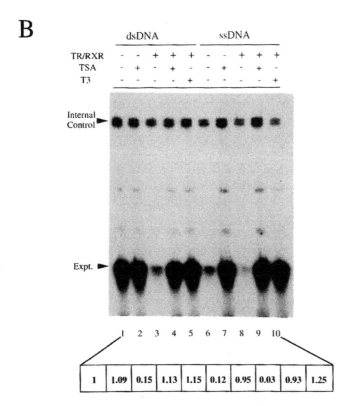

Figure 2.52. The histone deacetylase inhibitor TSA blocks the transcriptional regulation by TR/RXR.

A. Group of oocytes were first injected with (+) or without (-) TR/RXR mRNAs and then injected with dsDNA or ssDNA of pTRβA as indicated. The oocytes were treated with (+) or without (–) TSA (5 ng/ml) or T3 (50 nM) overnight. B. RNA was then prepared from the injected oocytes and the transcription from TRβA promoter was analysed by primer extension (Expt). The internal control represents the primer extension product derived from endogenous storage pool of histone H4 mRNAs which serves as an isolation and loading control. Levels of transcription from pTRβA were quantitated by phosphorimaging and normalized against the internal control. The level of transcription from control dsDNA of pTRβA was designated as 1 (lane 1) and the other lanes were compared to it.

site and involves the reorganization of chromatin structure in which targeted histone acetylation by the PCAF and p300/CBP activators may have a contributory role (Yang *et al.*, 1996b; Ogryzko *et al.*, 1996, see Fig. 2.51). Recently, yet another targeted coactivator with histone acetyltransferase activity has been discovered (Chen *et al.*, 1997). The exact coactivators that function in *Xenopus* are being defined. It is possible to separate chromatin disruption from productive recruitment of the basal transcription machinery *in vivo* by deletion of regulatory elements essential for transcription initiation at the start site and by the use of transcriptional inhibitors (Wong *et al.*, 1995, 1997a). Therefore chromatin disruption is an independent hormone-regulated function targeted by DNA-bound thyroid hormone receptor. It is remarkable just how effectively the various functions of the thyroid hormone receptor are mediated through the recruitment of enzyme complexes that modify chromatin. These results provide compelling evidence for the productive utilization of structural transitions in chromatin as a regulatory principle in gene control (Wolffe, 1997).

The genetic, biochemical and cell biological evidence that we have outlined provides a substantial rationale for considering chromatin structural proteins as integral components of the transcriptional machinery. It is important to recognize that chromatin structure is not necessarily static and obstructive to transcription but provides a means of display for the DNA template that determines function. Variation in the quality of histone–DNA interactions and in the three-dimensional path of the double helix can directly influence transcription (Ura *et al.*, 1997; Schild *et al.*, 1993; see Sections 4.1 and 4.2). Conformation is a well-known determinant of enzymatic activity, alterations in chromatin conformation may well determine transcriptional activity.

The facts that: (1) core histone acetylation greatly facilitates the access of transcription factors to DNA in a nucleosome; (2) transcriptional repressors recruit histone deacetylases; and (3) transcriptional coactivators are histone acetyltransferases leads to a model for transcriptional regulation in which the recruitment of repressors could direct the local stabilization of repressive histone–DNA interactions and where the recruitment of activators could destabilize these interactions (Fig. 2.53). Repressive nucleosomes might prevent either the association or function of the basal transcriptional machinery on a particular promoter. However, it is important to note that certain transcriptional regulators such as the thyroid hormone receptor can bind to their recognition elements in a nucleosome even when the histones are deacetylated (Wong *et al.*, 1995, 1997a) and nucleosome assembly is not always repressive (Schild *et al.*, 1993).

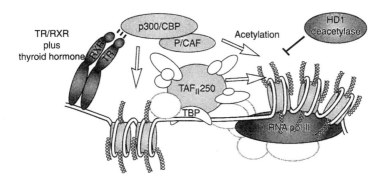

Figure 2.53. Transcriptional regulation in chromatin.
Hormone bound thyroid hormone receptor recruits a co-activator complex (p300/CBP, P/CAF) that retains chromatin in an 'open' configuration and a functional transcriptional machinery associated with the promoter. This complex counteracts the continued activity of the histone deacetylase (HD1).

The recruitment of histone deacetylase by chromatin bound repressors will potentially eliminate basal levels of histone acetylation and impede the recruitment or function of the basal transcriptional machinery. Targeted acetylation provides a means of allowing the basal machinery to displace nucleosomes, assemble a functional transcription complex and never have to deal with chromatin again. For example we can propose three steps in the regulation of transcription by thyroid hormone receptor: (1) thyroid hormone receptor binds to chromatin on the surface of a positioned nucleosome and facilitates the assembly of a repressive chromatin structure; (2) in response to hormone, the receptor recruits molecular machines or enzymes that disrupt local chromatin structure; (3) the hormone-bound receptor and associated activators facilitate the recruitment and activity of the basal transcriptional machinery to further activate transcription. Additional interesting possibilities include the regulated association and activity of histone acetyltransferases and deacetylases within a common complex. In this way transcriptional activity could be continually modulated through variation in chromatin conformation (Fig. 2.53).

Summary
Metazoans and yeast use enzymes that modulate histone acetylation and nucleosomal integrity in order to regulate transcription. Repressor complexes deacetylate histones and stabilize nucleosomes. Activator complexes acetylate histones and disrupt nucleosomes. Variation in

chromatin structure makes a major contribution to gene regulation. Other enzymatic complexes and molecular machines including SWI/ SNF also make use of chromatin to control transcription.

2.5.5 Remodelling of chromatin during spermatogenesis

Histones represent only one way of packaging DNA such that it can fit into the volume of a nucleus. There are many possible ways of rendering DNA compact in a reversible fashion. It is a measure of the major role of histone structure in many other nuclear processes (Sections 2.5.4 and 4.1) that histones have been so conserved through evolution. Perhaps the best example of the reversible compaction of DNA by multiple pathways concerns the condensation of DNA into sperm nuclei during spermatogenesis. This provides an excellent example of roles for histone variants, post-translational modification of histones and non-histone proteins in regulating chromosome structure and function.

The easy availability of pure populations of spermatozoa led to the early realization that the types of proteins in the sperm nucleus could vary greatly in different organisms. These proteins have been divided into certain classes. One class, the protamines, comes in two types: one type is rich in polyarginine tracts (4–6 residues), punctuated with proline, and potentially phosphorylatable serine and threonine residues; the other type is rich in cysteine. Both types of protamine are small (3000–5000 Da) and highly basic. Protamine–DNA complexes often represent the final state of chromatin in fish and mammalian sperm nuclei. However, during the process of spermatogenesis other proteins can transiently replace the histones (Poccia, 1986).

The transition proteins replace histones during the initial stages of condensation of chromatin in spermiogenesis and are later replaced by protamines, which are the only basic nuclear structural proteins in the sperm of most mammals. The transition proteins presumably facilitate the replacement of histones by protamines, although little is known about their specific function. The amino acid sequence of a mouse transition protein (TP 2) suggests two domains of protein structure: an amino-terminus which, like the protamines, is rich in cysteine, proline and phosphorylatable serine and threonine; and a highly basic carboxyl-terminus rich in arginine (Kleene and Flynn, 1987; Luerssen et al., 1989). The transition proteins are of special interest because they are the only proteins that are proven to dissociate histones from DNA in a physiological context by competitive mechanisms without the use of molecular machines such as coactivators and SWI/SNF or the progression of RNA or DNA polymerase. Several mechanisms have been proposed. First, the electrostatic binding of the transition proteins to DNA should be intermediate between histones and protamines because the concentration of basic amino acids near the carboxyl-terminus is greater than that of histones and less than in protamines. Second, the putative DNA-binding domain contains aromatic amino acids: phenylalanine and tyrosine. The tyrosine residues in the transition proteins have been shown to intercalate between the bases of DNA lowering its thermal stability (Singh and Rao, 1987). Intercalation of aromatic amino acids can induce bends or kinks in native DNA, and these bends might alter the path of DNA around the histone core and potentially destabilize the nucleosome.

Special histone variants exist that are specific for the testis. In sea urchin sperm, specific variants exist of histones H2B and histone H1. Both molecules are considerably larger and richer in arginine than their relatives in somatic cells. Most of the extra size is due to amino-terminal extensions. These additional amino acids increase the interaction of both histones with linker DNA (Bavykin *et al.*, 1990; Hill and Thomas, 1990). It is possible that they might also further facilitate the stabilization and condensation of chromatin by creating *inter* strand linkages between chromatin fibres. Phosphorylation of the histone H1 tails regulates their interaction with DNA (Hill *et al.*, 1991). As chromatin is compacted, the tails are dephosphorylated (Section 2.5.3). Testis-specific histone variants also exist in mammals. Variants of histone H1, histones H2A and H2B accumulate during meiotic prophase. As all of these transitions in chromatin structure occur after replication, the movement of a processive enzyme complex along the DNA duplex is not a prerequisite for the remodelling of chromosomal structure (Sections 3.1.1 and 3.1.2). In the rat, the core histone variants are heavily acetylated prior to their dissociation from DNA, which is driven by the accumulation of the transition proteins. Finally, accumulation of the protamines, whose binding is also regulated by their phosphorylation state, leads to the progressive displacement of the transition proteins and the nucleus is completely condensed. The problem of decondensing the sperm nucleus following its arrival in the cytoplasm of the fertilized egg is considered later (Section 3.1).

Why is sperm chromatin condensed through such a special mechanism? Suggestions include the protection and stability of the genetic material DNA in such structures. DNA in the nucleus of a spermatozoon is much less accessible to nucleases than in a somatic cell. It is also much more resistant to physical and chemical perturbants. A concomitant effect of chromatin compaction, as normal histones are replaced in chromosomes, is the suppression of gene activity. It is possible that the compaction of DNA into the sperm nucleus not only renders the DNA inaccessible to RNA polymerases but also erases the developmental history of a chromosome. Specifically the *trans*-acting factors responsible for directing specific events in the nucleus could be displaced. Evidence to support this concept comes from an analysis of DNase I hypersensitive sites in chromatin (Section 4.2) which are lost in sperm for all genes. However, those genes that are constitutively expressed are marked during spermatogenesis by hypomethylation at sites of future hyper-sensitivity (Groudine and Conkin, 1985).

Summary
Spermatogenesis requires the packaging of DNA into an inert chromatin structure such that DNA can be unfolded rapidly following fertilization. A variety of proteins have been found that accomplish this packaging, probably because little metabolic activity involving DNA occurs in the sperm nucleus. Histones are removed through modifications such as acetylation and competed away from DNA by very basic proteins such as arginine-rich transition proteins or protamines.

2.5.6 Heterochromatin, position effect, locus control regions and insulators

Early studies by cytologists led to the realization that some chromosomal regions have properties distinct from the rest of the chromosome (Henikoff, 1990, 1994; Pardue and Hennig, 1990). Large segments of chromatin were found to be highly condensed and to replicate late in S-phase. Geneticists determined that these chromo-somal regions, which they called heterochromatin, did not participate in meiotic recombination. However, heterochromatin does have significant genetic effects. The most common observed consequence of heterochromatin formation is the repression of transcription either in heterochromatin itself or in regions of chromatin that lie adjacent to the heterochromatin domain. The variability in gene expression at the border of the heterochromatin is described as 'position effect variegation'. Three explanations have been offered for this phenom-enon. The first is that special proteins, such as HP1 (see below), exist that cause heterochromatin to adopt its distinct structure, and these proteins can 'spill over' into regions of normal chromatin. The second is that heterochromatin represents the sequestration of chromosomal domains in specialized nuclear compartments from which the transcriptional machinery is excluded. The third applies only to *Drosophila* and other insects with polytene chromosomes in that, following placement adjacent to heterochromatin, a gene will undergo fewer rounds of replication than would normally occur (Spradling and Karpen, 1990). Fewer copies of the gene would cause a concomitant reduction in transcription. Most investigators accept that heterochromatin-specific proteins can diffuse on to normal chromatin and thereby influence gene expression of juxtaposed genes. Although initially defined from work on insects, there is an increasing body of evidence that heterochromatin domains exist in all eukaryotic chromosomes including those of yeast. Moreover, it is clear that metazoans have made use of this type of chromosomal organization for the developmental regulation of gene expression (Singh, 1994).

An approach to the molecular basis of heterochromatin formation has been to look for mutations in *Drosophila* or more recently in *S. cerevisiae*, that enhance or suppress position effects on gene expression (modifiers of position effect variegation). Modifiers of position effect variegation would include mutations in the genes encoding the chromosomal proteins involved in forming heterochromatin (Renter and Spierer, 1992). Support for this approach comes from the observation that chromosomal deletions that reduce the number of copies of histone genes in *Drosophila* reduce the influence of position effects on gene expression (Moore *et al.*, 1983). A role for histones is consistent with the observation that chemicals that maintain histones in a hyperacetylated state suppress position effect (Mottus *et al.*, 1980). Using this approach a gene encoding a non-histone protein (HP1, heterochromatin protein 1) has been identified (James and Elgin, 1986). Mutation of the HP1 gene reduces position effects on gene expression (Elgin, 1990). HP1 is preferentially associated with the heterochromatin regions of polytene chromosomes. Other proteins homologous to HP1 include Polycomb, which is also chromatin associated and is known from genetic experiments to influence the expression of many genes in normal chromatin (Zink and Paro, 1989). Neither HP1 nor Polycomb interact with DNA directly but presumably recognize some aspect of nucleosome or chromatin fibre structure. Any such direct interaction is however yet to be defined. Polycomb and HP1 share a common amino acid sequence known as the chromodomain (*chromatin modification*). This domain is highly conserved through evolution and can be found in the *S. pombe* SWI6 gene. The SWI6 protein is involved in the assembly of the chromosomal domain containing the transcriptionally silent mating type cassettes (Klar and Bonaduce, 1991; Singh, 1994; Lorentz *et al.*, 1992, 1994). The chromodomain does seem to have a role in the subnuclear compartmentalization of chromodomain proteins such as HP1 and Polycomb. Mammalian homologs of HP1 are found in subnuclear structures or foci termed 'polymorphic interphase karyosomal associations' (Saunders *et al.*, 1993). Polycomb has a similar distribution in *Drosophila* nuclei. Recent studies have established that the chromodomain family of proteins comprising more than 40 members (Aasland and Stewart, 1995; Koonin *et al.*, 1995) can be subdivided into two major groups. Proteins such as HP1 contain both an N-terminal chromodomain and a C-terminal 'shadow' chromodomain (Epstein *et al.*, 1992; Aasland and Stewart, 1995). The N-terminal chromodomain appears to direct binding to heterochromatin whereas the C-terminal 'shadow' chromodomain determines nuclear localization and assists in binding to chromatin (Messmer *et al.*, 1992; Platero *et al.*, 1995; Powers and Eissenberg, 1993). The second group of proteins including Polycomb and SWI6 rely on interactions with other proteins to target association with particular chromatin domains (Lorentz *et al.*, 1994; Ekwall *et al.*, 1995).

The structure of the chromodomain was recently solved using nuclear magnetic resonance (NMR) (Ball *et al.*, 1997). The chromodomain has strong homology to two archaebacterial DNA binding proteins (Baumann *et al.*, 1994; Edmondson *et al.*, 1995). However, the eukaryotic chromodomain does not interact with DNA (Ball *et al.*, 1997) and appears to be involved in protein–protein interactions (Le Douarin *et al.*, 1996). Each chromodomain consists of an N-terminal three-stranded anti-parallel β sheet which folds against a C-terminal α-helix. The presence of both a chromodomain and a shadow chromodomain are thought to allow proteins such as HP1 to function as adaptors in the assembly of large multicomponent proteins.

The Polycomb-group (PcG) of proteins are found to be non-randomly associated with over 100 chromosomal domains within polytene chromosomes (Zink and Paro, 1989; De Camillis *et al.*, 1992; Rastelli *et al.*, 1993). Of particular interest are the homeotic genes that are regulated by PcG proteins. How PcG proteins are directed to particular regions such as the homeotic loci of the chromosome remains substantially unknown. Mutations in the genes encoding the PcG proteins generally lead to the activation of target genes. Changes in the expression of homeotic genes that control the segmental identity of the insect body following mutation of the PcG proteins have been studied in detail (Moehrle and Paro, 1994; Pirotta and Rastella, 1994).

The general properties of 'modifier' genes that enhance or suppress position effect variegation provide some insight into how PcG proteins might work to change chromosomal structure. The phenotypes of modifier mutations show a strong dependence on the number of gene copies within the cell (Locke *et al.*, 1988; Reuter *et al.*, 1990). This has led to the proposal that the modifier proteins that assemble heterochromatin act through simple mass action. The more modifier protein there is in the nucleus the more repressive heterochromatin will be assembled. It is probable that certain modifier proteins co-operate to assemble multimeric complexes that alter the structure of entire chromosomal domains (Fig. 2.54) (Moehrle and Paro, 1994). The distribution of Polycomb along the chromosomal domain including the bithorax complex which encodes three homoeotic genes supports this hypothesis. Orlando and Paro (1993) found that the Polycomb protein is associated with transcriptionally inactive chromatin over more than 200 kb of DNA, while it is absent from regions of gene activity. The boundaries of Polycomb-associated chromatin and Polycomb-free chromatin are very distinct, which implies that regulatory elements exist that initiate and terminate Polycomb binding (Fig. 2.54).

Recent genetic approaches have delineated regulatory elements within genes such as *engrailed* that confer silencing in response to Polycomb (Polycomb response elements, PREs). Each PRE is several hundred base pairs in length and can independently direct position effect variegation in the presence of Polycomb (Chan *et al.*, 1994; Rastelli *et al.*, 1993). The sequences of individual PREs show little

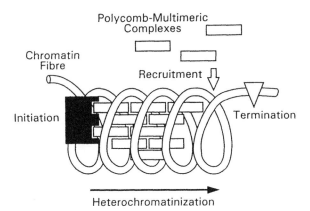

Figure 2.54. A model for the initiation and termination of Polycomb-group protein mediated changes in chromosomal organization.

homology to each other suggesting that features of DNA structure or diverse DNA binding proteins could target Polycomb mediated silencing. For the PRE in the *Ubx* gene a binding site for the GAGA factor has been defined suggesting that this regulatory factor has a role in facilitating Polycomb function (Pirotta, 1997). The GAGA factor has been implicated in the remodeling of chromatin (Tsukiyama *et al.*, 1994; Section 4.2).

An important aspect of the functional role of the PcG proteins is that they are not involved in establishing the expression state of a particular gene but in the maintenance of the repressed state through replication and chromosomal duplication. Since they act through large multicomponent complexes it is possible that they might subdivide through a replicative event, thereby maintaining a repressive chromatin structure (Fig. 2.55). Specialized mechanisms appear to exist within the chromosome to enable the removal of Polycomb from chromatin and the resetting of chromatin to a transcriptionally competent state (Tamkun *et al.*, 1992). The molecular machines involved in this process include the Brahma protein which is structurally related to a component of the yeast general activator complex SNF2/SWI2 (Section 2.5.4). Evidence for competition between transcriptional activators and repressors such as Polycomb comes from experiments in which PRE/Polycomb-mediated transcriptional silencing can be overcome by the expression of high levels of Gal4p (Zink and Paro, 1995). Immunostaining of polytene chromosomes revealed that the Polycomb complex was removed from the vicinity of the Gal4p binding site without the need for DNA replication and cell division.

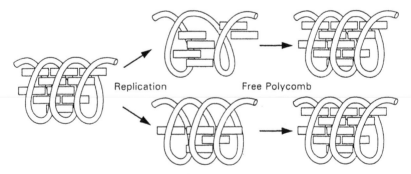

Figure 2.55. A model for the maintenance of repressive chromosomal structures by the Polycomb-group proteins.
Replication disperses Polycomb-group proteins to both daughter chromatids which then sequester free Polycomb-group proteins from the nucleoplasm to reassemble repressive chromosomal structures.

Gottschling and colleagues have successfully established that position effect variegation also occurs in the chromosomes of *S. cerevisiae*. When a gene is located near a telomere, transcriptional activity is reduced (Gottschling *et al.*, 1990). Transcriptional repression at yeast telomeres appears to be due to the assembly of a distinct chromatin structure that initiates at the telomere. The sequence of the DNA is important since internal tracts of telomeric DNA have the capacity to also act as silencers of transcription in *S. cerevisiae* (Stavenhagen and Zakian, 1994). Mutations in the amino-terminal tails of histones H3 and H4 relieve transcriptional silencing (Aparicio *et al.*, 1991; Thompson *et al.*, 1994a). Histone H4 at the telomeres is hypoacetylated and mutations in the amino terminal acetyltransferases also relieve silencing (Aparicio *et al.*, 1991; Braunstein *et al.*, 1993). DNA methyltransferases expressed in yeast have restricted access to telomeric chromatin compared to most of the chromosome (Gottschling, 1992). This implies that telomeric chromatin is either more compacted or is sequestered away from freely diffusible *trans*-acting factors in the nucleoplasm. There are also several similarities to properties of yeast and *Drosophila* heterochromatin. Genes within *Drosophila* heterochromatin are normally maintained in a stably repressed state, but will occasionally escape repression (Henikoff, 1990), and similar phenomena occur in yeast (Gottschling *et al.*, 1990).

The efficiency of transcriptional repression decreases when a gene is placed further from the *S. cerevisiae* telomere (over a 10–20 kb range; Renauld *et al.*, 1993). This result supports the idea that a repressive chromatin structure originates at the telomere and spreads along the chromosome. The expression of a protein SIR3 influences the extent of silencing at the telomeres, suggesting that it is essential for the assembly of repressive chromatin (Renauld *et al.*, 1993). SIR3 and the histone tails are also involved in the sequestration of yeast chromosomal telomeres at the periphery of the nucleus (Palladino *et al.*, 1993; Thompson *et al.*, 1994a). Thus telomeric silencing could be related to the sequestration of this portion of the chromosome within a transcriptionally incompetent compartment of the nucleus adjacent to the nuclear envelope (Franke, 1974).

Many of the modifiers of position effect in yeast chromosomes are shared between the telomeres of the chromosomes and the silent mating type cassettes in yeast. Mutations in four silent information regulator genes, SIR1, 2, 3 and 4, cause derepression of the silent mating type cassettes (Laurensen and Rine, 1992). Mutations in three of these, SIR2, 3 and 4, influence telomeric repression (Aparicio *et al.*, 1991). The assembly of specialized chromosomal domains at the silent mating type cassettes is reflected in the limited access of restriction enzymes compared to bulk chromatin (Loo and Rine, 1994a). The silencers at the mating type cassettes are required after each round of replication to re-establish the transcriptionally repressed state (Holmes and Broach, 1996). This indicates that repressive chromatin structures do not self-template the reassembly of a repressive structure. It is possible that both the telomeres and the mating type cassettes are sequestered in specialized nuclear structures or compartments in which the transcriptional machinery does not function efficiently.

Recent results also implicate heterochromatin-mediated silencing mechanisms in the nucleolus, the site of ribosomal RNA transcription. SIR2, SIR4, H2A and H2B have been found to influence the transcription of genes requiring RNA polymerase II that are integrated into ribosomal DNA (Bryk *et al.*, 1997; Smith and Boeke, 1997). Remarkably, silencing also appears to control aging in yeast. Interference with the targeting of silencing complexes by expression of mutant forms of SIR4 allows yeast to live longer (Kennedy *et al.*, 1997). Nucleolar organization is implicated in this phenomenon because mutant SIR4 proteins localize to the nucleolus. In the fission yeast *Schizosaccharomyces pombe*, the determinants of gene silencing and heterochromatin assembly at the silent mating type cassettes are shared with the centromeres (Thon and Klar, 1992; Klar, 1992; Allshire *et al.*, 1994, 1995).

The existence of these diverse sites of heterochromatin-mediated silencing has led to the idea that there is a competition for limiting components and that the telomere might act as a molecular sink to form a reservoir of silencing factors (Maillet *et al.*, 1996; Gotta *et al.*, 1996). Alternatively individual targeting factors such as SIR1 at the mating type loci might modulate the stability of heterochromatin assembly.

The heterochromatin structures that control transcription are likely to be dynamic, and ubiquitination may have a role in regulating silencing. A de-ubiquitinating enzyme Ubp3 binds to SIR4p and mutations in the Ubp3 gene increase silencing at telomeres and mating type loci (Moazed and Johnson, 1996). Mutations in the yeast SAS acetyltransferases also influence silencing at the telomeres and mating type loci (Reifsnyder *et al.*, 1996). Histone modification is likely to have a significant role in the silencing process.

In an apparently related phenomenon, components of the yeast origin recognition complex (ORC) that normally regulate the site and frequency of chromosomal initiation of replication, direct transcriptional silencing within the same chromosomal domain (Bell *et al.*, 1993a,b). The molecular mechanism exerting this effect is unknown. It is possible that a gene adjacent to the ORC is directed to reside in a replication competent but transcriptionally incompetent environment. Alternatively, perhaps replication and ORC function are necessary to remodel chromatin from an active to a repressed state (Almouzni and Wolffe, 1993a; Almouzni, 1994), as replication is required to repress the mating type cassettes (Miller and Nasmyth, 1984).

Position effects in mammalian chromosomes have been a recurrent problem for transgenic research since highly variable levels of transcriptional activity follow from the random introduction of reporter genes into the genome (Forrester *et al.*, 1987; Grosveld *et al.*, 1987; Stief *et al.*, 1989). These effects can be relieved by the introduction of a *locus control region* (LCR) that exerts a dominant transcriptional activation function over a chromatin domain (10–100 kb). The mechanism of this activation function remains to be determined; however, communication between LCRs, enhancers and promoters either directly or through modifications of chromatin structural components are favoured hypotheses (Felsenfeld, 1992; Wijgerde *et al.*, 1995; Section 4.2). Recent evidence suggests that both the locus control regions and enhancers act in *cis* to actively suppress position-effect variegation (Walters *et al.*, 1996; Festenstein *et al.*, 1996). In this regard locus control regions basically function as operationally defined 'powerful' enhancers. The coexistence of heterochromatin domains that can transmit repressive effects and the definition of the extensive long-range activation function of LCRs emphasize the necessary compartmentalization of the chromosome into discrete functional units. These discrete functional units are prevented from influencing each other in a natural chromosomal context (Fiering *et al.*, 1995). This is due in part to the existence of special chromosomal regions that prevent the transmission of chromatin structural features associated with the boundaries of repressive or active domains. These specialized chromosomal regions are known as insulators (Chung *et al.*, 1993; Wolffe, 1994a).

The original evidence for an insulator function within the chromosome came from genetic experiments in *Drosophila*. Each boundary of the 87A7 heat-shock locus is defined by a pair of nuclease hypersensitive sites bordering a 250–300 bp segment of DNA. These specialized chromatin structures (scs) are located at the junctions between the decondensed chromatin of the transcriptionally active 87A7 heat-shock locus and adjacent condensed chromatin. The scs were found to have three functional properties: (1) they establish a domain of independent gene activity at many distinct chromosomal positions; (2) scs elements are necessary at each edge of the domain; and (3) the elements are independently neither inhibitory nor stimulatory to transcriptional activity within the domain (Kellum and Schedl, 1991, 1992). Subsequent work found that the introduction of an scs element between an enhancer and a promoter blocked communication between the two elements. Thus the scs elements prevent both the transmission of repressive effects on transcription from proximity to heterochromatin and the transmission of stimulatory effects on transcription from an enhancer. How this insulation is achieved is unknown; moreover, the nature of the nucleoprotein complex assembled on the scs elements has yet to be defined. Fortunately, similar phenomena have been described associated with a well-defined nucleoprotein complex between the *Drosophila* suppressor of hairy-wing protein (su(Hw)) and the gypsy retrotransposon (Corces and Geyer, 1991).

Insertion of a gypsy element as far as 10–30 kb from a promoter can cause a mutant phenotype (Jack *et al.*, 1991). The mutant phenotype requires the su(Hw) protein to interact with the inserted gypsy element at a 350 bp region containing 12 copies of a 10-bp sequence separated by AT-rich sequences. The su(Hw) protein has a molecular weight of 100 kDa, and its sequence includes several motifs characteristic of eukaryotic transcription factors, including 12 zinc fingers, a leucine zipper and two acidic domains (Parkhurst *et al.*, 1988; Fig. 2.56). The complex of the su(Hw) protein with a gypsy element has many of the properties of an insulator element: the complex blocks enhancer activity when placed between an enhancer and a promoter (Holdridge and Dorsett, 1991; Geyer and Corces, 1992), and when the complex is placed at the boundaries of a gene-containing fragment, the gene is protected from the repressive effects of heterochromatin on transcription (Roseman *et al.*, 1993).

Although in certain circumstances the su(Hw) protein does not independently stimulate or repress transcription of a reporter gene, the su(Hw) protein can occasionally function as a transcriptional activator (Corces and Geyer, 1991). This suggests that the function of an insulator may be conferred on sequences by DNA-binding proteins that might under other circumstances have more conventional roles in the transcription process. Although it is clear that the su(Hw) protein does not bind to scs elements (Roseman *et al.*, 1993) it seems likely that these elements will form large nucleoprotein complexes with a similar composition.

How might insulators function? Any models must explain how characteristics of both repressive or active chromatin are restricted to particular chromosomal domains. Several models have been suggested to explain the activities of LCRs and enhancers (Fig. 2.57) (Section 4.1). These elements might function as entry points for transcription factors, RNA polymerase or other components of the transcription machinery, which might then track along the DNA until reaching the promoter. Alternatively the LCR or enhancer complex might associate with the promoter complex by stable looping of the

DNA: *gypsy* Fragment

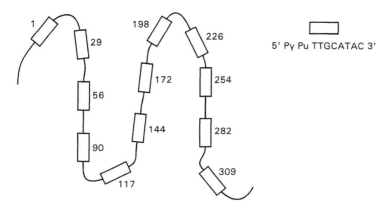

5' Py Pu TTGCATAC 3'

Protein: su(Hw)

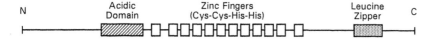

Nucleoprotein Complex Model

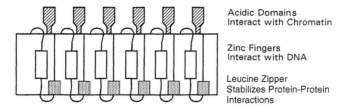

Acidic Domains
Interact with Chromatin

Zinc Fingers
Interact with DNA

Leucine Zipper
Stabilizes Protein-Protein
Interactions

Figure 2.56. Anatomy of an insulator.
The complex of the *Drosophila* su(Hw) protein with a region of the gypsy transposable element has the properties of an insulator (see text). The model for the interaction between the su(Hw) protein and gypsy 10 bp repeats is based on experimental data (see text for details).

intervening DNA or chromatin, forming a complex that increases the efficiency with which RNA polymerase is recruited and used. Another possibility is that the LCR or enhancer complex might cause the gene to assemble into a chromatin structure capable of being transcribed through its association with nuclear compartments (or organelles) that act as transcription factories (Section 2.5.9), or that associate with

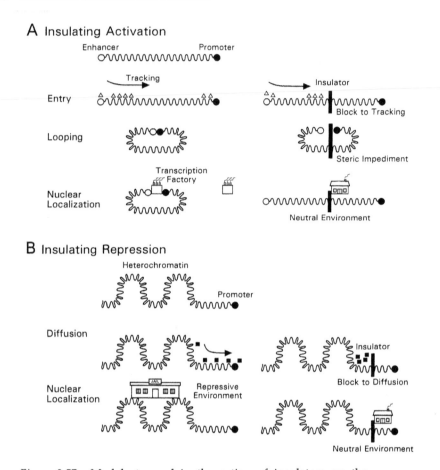

Figure 2.57. Models to explain the action of insulators on the activities of locus control regions/enhancers and heterochromatin (see text for details).

proteins that modify repressive chromatin structure by disrupting nucleosomes (Section 2.5.4). Similar models can explain the repressive effects of heterochromatin. Repressive chromatin proteins, such as the *Drosophila* HP1 protein or histone deacetylases, may undergo local diffusion enlarging the heterochromatin domain. Alternatively, heterochromatin may be sequestered in a transcriptionally incompetent region of the nucleus (see above).

In considering these models, it is important to recognize that the eukaryotic nucleus is a highly organized structure in which DNA is compacted by its association with histone proteins into nucleosomes and the chromatin fibre (Section 2.5.9). Although it is possible that insulators prevent protein tracking or diffusion between active and repressive domains, it is difficult to envisage how this might occur in the nucleus – where DNA segments separated linearly by many kilobases can be juxtaposed by folding of the DNA helical axis in three dimensions – without invoking some specific attachment of inactive chromatin domains, insulators and active chromatin domains to a nuclear framework. Similar attachments might be required to prevent the juxtaposition, as a result of DNA looping, of LCRs, enhancers and promoters. Perhaps the most economical suggestion is that insulators are nucleoprotein complexes that associate neither with regions or structures in the nucleus where 'transcription factories' load on to DNA (Jackson *et al.*, 1993), nor with regions or structures from which the transcriptional machinery is excluded (Palladino *et al.*, 1993). Instead, the insulators might associate with distinct 'neutral' nuclear structures. The 'neutral' nuclear structures would tether promoter elements where the transmissible activating effects of enhancers or silencing effects of heterochromatin could not occur. This absence of transmissible effects could be accounted for by the exclusion from the 'neutral' nuclear structures of particular transcriptional coactivators normally associated with communication between promoters and enhancers, or of chromatin modification proteins or enzymes normally associated with heterochromatin.

Examples of mammalian chromosomal regions that contain heterochromatin are the centromere and the telomere (Section 2.4.2). The telomeres of mammalian chromosomes have an unusual chromatin structure in which nucleosomes are closely packed with a repeat length of 157 bp (Makarov *et al.*, 1993). Mammalian telomeres consist of the sequence $(TTAGGG)_n$ repeated for 10–100 kbp (Section 2.4.2). Heterochromatin at the centromere contains tandemly repeated simple sequence 'satellite' DNA, for example the α-satellite DNA at the human centromere. This α-satellite heterochromatin appears to play a structural role by mediating attachment of the kinetochore (Section 2.4.2). The inactive X chromosome of female mammals is also heterochromatic. The properties of the inactive X chromosome have been the focus of much interesting research.

Female mammalian embryos begin development with two active X chromosomes; however, very early in embryogenesis almost all of the genes on one of the two X chromosomes become inactivated. This transcriptional inactivation is concomitant with the chromosome both taking on the appearance of heterochromatin and also becoming late replicating during S-phase. Although in eutherian (placental) mammals the initial choice between inactivation of the maternal or paternal X chromosome is random, once established in a repressed state the same

X chromosome will be inactivated after every cell division. This is an excellent example of the establishment and maintenance of a chromosomal state of determination (Section 4.3). Abnormalities in X-chromosome inactivation have allowed the definition of the X inactivation center that is required to present for inactivation to occur (Willard *et al.*, 1993). Within the X inactivation center is the *Xist* gene which is the key regulator of inactivation (Brown *et al.*, 1991; Brockdorff *et al.*, 1991). Remarkably the gene does not encode mRNA, but instead a long, untranslated RNA that remains associated with the inactive X chromosome (Brockdorff *et al.*, 1992; Brown *et al.*, 1992). This result means that the inactive X chromosome actually has a transcriptionally active *Xist* gene at the X-inactivation center. The *Xist* RNA through some unknown mechanism has a causal role in directing the heterochromatinization of the inactive X-chromosome. The process of heterochromatinization leads to several differences between active and inactive X-chromosomes including DNA methylation and histone acetylation status. DNA methylation is a characteristic of inactive promoters in eukaryotic chromosomes (Section 2.5.7). Many genes in the inactive X chromosome are heavily methylated in contrast to the active X chromosome (Grant and Chapman, 1988). However, the kinetics with which particular sites within the X-linked genes become methylated during the differentiation of embryonic female somatic cells do not always correlate with the timing of transcriptional inactivation (Lock *et al.*, 1987). Although it remains to be determined if a subset of key sites around regulatory elements is always methylated before transcription is repressed it seems probable that other mechanisms must supplement any influence of DNA methylation on transcription.

Heterochromatin normally replicates late during S-phase. Replication timing has been proposed as a determinant of transcriptional activity (Section 4.3). Genes that replicate late during S-phase might do so under conditions of limiting transcription factors, or might be assembled into a repressive chromatin structure using components translated only late in S-phase. The active X chromosome normally replicates early in S-phase whereas the inactive X chromosome replicates late (Takagi, 1974). However, female lymphoma cell lines have been isolated in which the opposite occurs (Yoshida *et al.*, 1993). Thus the inactive X chromosome does not have to replicate late in S-phase in order to be transcriptionally quiescent.

The formation of local repressive chromatin structures in which key genetic regulatory elements are rendered inaccessible to transcription factors by inclusion within positioned nucleosomes is an important mechanism for transcriptional repression (Section 4.2). Promoters in the inactive X chromosome appear to be incorporated into positioned nucleosomes whereas promoters in the active X chromosome are free of such structures and have transcription factors bound to them (Riggs and Pfeifer, 1992). Thus specific local chromatin structures clearly appear to have a role in regulating differential gene activity between the two X chromosomes. Nevertheless, a causal relationship between chromatin structure and transcription is yet to be established. Nucleosomes on the active X chromosome contain predominantly acetylated histones whereas those on the inactive X chromosome are not acetylated (Jeppesen and Turner, 1993) (Section 2.4.3). Thus the establishment and maintenance of specific chromatin structures containing modified histones is an excellent candidate mechanism for establishing and maintaining differential expression of genes between the two X chromosomes. Differential methylation and replication timing may serve to stabilize these different states of gene activity.

Occasionally, an entire nucleus will become heterochromatinized, one example being the inactivation of the erythrocyte nucleus in chicken. Here the special linker histone variant H5 accumulates, which represses transcription and compacts nucleosomal arrays very effectively (Sections 2.5.3 and 3.1.2). Histone H5 is more arginine rich than the normal linker histone H1 found in somatic cells. This increase in arginine content probably strengthens the interaction of H5 with DNA and stabilizes chromatin structure.

Summary

Several different proteins have been found that mediate the assembly of an inert chromatin state resistant to the transcriptional and recombinational machinery, known as heterochromatin. Heterochromatin is important because its formation influences the transcription of genes both within and adjacent to it – a phenomenon known as 'position effect'. This repressive influence is believed to occur either through nuclear compartmentalization or through lateral diffusion of the proteins or RNAs responsible for an additional stabilization of chromatin structure. Insulator elements exist that prevent heterochromatin exerting a repressive effect on the expression of a gene.

2.5.7 DNA methylation and chromatin

The covalent modification of DNA provides a direct and powerful mechanism to regulate gene expression (Kass *et al.*, 1997a).

Considerable experimental evidence supports the existence of such a mechanism in the majority of plants and animals (Bird, 1986, 1995; Szyf, 1996; Yoder *et al.*, 1997). The genome of an adult vertebrate cell has 60-90% of the cytosines in CpG dinucleotides methylated by DNA methyltransferase (Riggs and Porter, 1996). This modification can alter the recognition of the double helix by the transcriptional machinery and the structural proteins that assemble chromatin (Nan *et al.*, 1997; Kass *et al.*, 1997b). How these events might together contribute to gene control is the major theme of this section.

DNA methylation could control gene activity either at a local level through effects at a single promoter and enhancer, or through global mechanisms that influence many genes within an entire chromosome or genome (Tatte and Bird, 1993). An attractive suggestion is that DNA methylation evolved as a host-defence mechanism in metazoans to protect the genome against genomic parasites such as transposable elements (Yoder *et al.*, 1997). An increase in methyl-CpG correlates with transcriptional silencing for whole chromosomes, transgenes, particular developmentally regulated genes and human disease genes (Li *et al.*, 1993; Szyf, 1996). All of these systems exhibit epigenetic effects on transcriptional regulation in which identical DNA sequences are differentially utilized within the same cell nucleus. These patterns of differential gene activity are clonally inherited through cell division. Because specific methyl-CpG dinucleotides are maintained through DNA replication, DNA methylation states also provide an attractive mechanism (epigenetic mark) to maintain a particular state of gene activity through cell division and, thus, to contribute to the maintenance of the differentiated state (Holliday, 1987). We discuss how the molecular mechanisms that accomplish this important goal might also involve the assembly of specialized chromatin structures on methylated DNA.

Saccharomyces cerevisiae and *Drosophila melanogaster* live without any detectable methyl-CpG in their genomes. DNA-methylation-dependent gene regulation is not necessarily essential for cell division or metazoan development, because other gene regulatory mechanisms can compensate for the lack of DNA methylation in these organisms. In the mouse, primordial germ cells, embryonal stem cells and the cells of the blastocyst also progress through the cell cycle and divide without detectable DNA methylation (Jaenisch, 1997). Nevertheless, once embryonic stem cells begin to differentiate, normal DNA methylation levels are essential for individual cell viability (Panning and Jaenisch, 1996). The *de novo* methylation of CpG dinucleotides is a regulated process (Szyf, 1996). In the embryo, normal DNA methylation levels are essential for post-gastrulation development (Li *et al.*,

1992). Jaenisch has proposed that DNA methylation has no role in cell viability in mammalian embryonic lineages including the germ line, but that it has an important role in the differentiation of somatic cells. In complex organisms, such as vertebrates, that contain a large number of tissue-specific genes, DNA methylation provides a mechanism to turn-off permanently the transcription of those genes whose activity is not required in a particular cell type (Bird, 1995). This stable silencing of a large fraction of the genome would allow the transcriptional machinery to focus on those genes that are essential for the expression and maintenance of the differentiated phenotype. Consistent with this hypothesis, the inhibition of DNA methyltransferase activity with 5-azacytidine leads to the activation of several repressed endogenous genes (Jones, 1985). How might CpG methylation contribute to this global control of gene activity?

The most direct mechanism by which DNA methylation could interfere with transcription would be to prevent the binding of the basal transcriptional machinery and of ubiquitous transcription factors to promoters. This is not a generally applicable mechanism because some promoters are transcribed effectively as naked DNA templates independent of DNA methylation (Busslinger *et al.*, 1983; Iguchi-Ariga *et al.*, 1989; Kass *et al.*, 1997b). Certain transcription factors (e.g. the cyclic AMP dependent activator CREB) bind less well to methylated recognition elements, however the reduction in affinity is often insufficient to account for the inactivity of promoters *in vivo* (Hoeller *et al.*, 1988; Weih *et al.*, 1991). It seems unlikely that DNA methylation would function to repress transcription globally by modifying the majority of CpGs in a chromosome, if the only sites of action are to be a limited set of recognition elements for individual transcription factors.

The second possibility is that specific transcriptional repressors exist that recognize methyl-CpG and, either independently or together with other components of chromatin, turn off transcription. This mechanism would have the advantage of being substantially independent of DNA sequence itself, thereby offering a simple means of global transcriptional control. It would be especially attractive if the methylation-dependent repressors work in a chromatin context because then DNA could maintain the nucleosomal and chromatin fibre architecture necessary to compact DNA (Jost and Hofsteenge, 1992; McArthur and Thomas, 1996). Moreover, because chromatin assembly also represses transcription, methylation-dependent repression mechanisms would add to those already in place.

Bird and colleagues have identified two repressors MeCP1 and MeCP2 that bind to methyl-CpG without apparent sequence specificity (Meehan *et al.*, 1989, 1992; Lewis *et al.*, 1992). Like DNA methylation itself, MeCP2 is dispensable for the viability of embryonic stem cells, however it is essential for normal embryonic development. Consistent with the capacity of methylation-dependent repressors to operate in chromatin, recent studies indicate that MeCP2 is a chromosomal protein with the capacity to displace histone H1 from the nucleosome (Nan *et al.*, 1996). Moreover, MeCP2 contains a methyl-CpG DNA-binding domain, which might alter chromatin structure directly, and a repressor domain, which might function indirectly to confer long-range repression *in vivo* (Nan *et al.*, 1993, 1997). The capacity for MeCP2 to function in chromatin explains several phenomena connected with unique aspects of chromatin assembled on methylated DNA.

A role for specialized chromatin structures in mediating transcriptional silencing by methylated DNA has been suggested by several investigators. High levels of methyl-CpG correlate with transcriptional inactivity and nuclease resistance in endogenous chromosomes (Antequera *et al.*, 1989, 1990). Methylated DNA transfected into mammalian cells is also assembled into a nuclease-resistant structure containing unusual nucleosomal particles (Keshet *et al.*, 1986). These unusual nucleosomes migrate as large nucleoprotein complexes on agarose gels. These complexes are held together by higher-order protein–DNA interactions despite the presence of abundant micrococcal nuclease cleavage points within the DNA. Individual nucleosomes assembled on methylated DNA appear to interact together more stably than on unmethylated templates (Keshet *et al.*, 1986). Each nucleosome normally contains an octamer of core histones (H2A, H2B, H3 and H4) around which is wrapped approximately 160 bp of DNA, and a single molecule of histone H1, which constrains the linker DNA between adjacent nucleosomes (Section 2.1). The replacement of histone H1 with MeCP2 is a possible explanation for the assembly of a distinct chromatin structure on methylated DNA (Nan *et al.*, 1997).

The accessibility of chromatin to nucleases could also be affected directly by the stability with which the histones interact with DNA within the nucleosome. DNA methylation does not influence the association of core histones with the vast majority of DNA sequences in the genome (Felsenfeld *et al.*, 1983; Englander *et al.*, 1993). However, for certain specific sequences, such as those found in the Fragile X mental retardation gene 1 promoter, methylation of CpG dinucleotides can alter the positioning of histone–DNA contacts and the affinity with which these histones bind to DNA (Godde *et al.*, 1996; Section 2.2.5). The exact chromatin structure found *in vivo* can also be a consequence of gene activity. Linker histones, such as H1, are relatively deficient on the transcribed region of genes (Kamakaka and Thomas, 1990). So it is not surprising that transcriptionally inactive chromatin containing methyl-CpG should show an increase in the abundance of histone H1, whereas DNA sequences lacking methyl-CpG are deficient in H1 (Ball *et al.*, 1983; Tazi and Bird, 1990). *In vitro* studies indicate that histone H1 can interact preferentially with methylated DNA under certain conditions, although there is no measurable preference for the assembly of H1 into a nucleosomal architecture containing methylated DNA (Levine *et al.*, 1993; Campoy *et al.*, 1995; Nightingale and Wolffe, 1995). Recent *in vivo* studies indicate that rather than functioning as a general transcriptional repressor, histone H1 is highly specific with respect to the genes whose activity it regulates (see Section 2.5.1). It seems probable that the major differences between chromatin assembled on methylated versus unmethylated DNA will be determined by the inclusion of methylation-specific DNA-binding proteins, such as MeCP2.

There are features of transcriptional repression dependent on methylated DNA that can be explained by methylation-specific repressors operating more effectively within a chromatin environment. Transcriptional repression is strongly related to the density of DNA methylation (Boyes and Bird, 1992; Hsieh, 1994). There is a non-linear relationship between the lack of repression observed at low densities of methyl CpG and repression at higher densities. These results led to the demonstration that local domains of high methyl-CpG density could confer transcriptional repression on unmethylated promoters in *cis* (Kass *et al.*, 1993). This observation is consistent with MeCP2 containing DNA binding and transcriptional repression domains (Nan *et al.*, 1997). Thus MeCP2 does not necessarily have to function by occluding regulatory elements from the transcriptional machinery (by binding to promoter sequences itself), but might bind to a methyl-CpG sequence at one place on a DNA molecule and then use the repression domain to silence transcription at a distance. Chromatin assembly itself might promote this 'action at a distance' by juxtaposing MeCP2 and the regulatory elements under control through the compaction of intervening DNA (Fig. 2.58).

Early experiments using the microinjection of templates into the nuclei of mammalian cells suggested that the prior assembly of methylated, but not unmethylated, DNA into chromatin represses transcription (Buschhausen *et al.*, 1987). The importance of a nucleosomal infrastructure for transcriptional repression dependent on DNA methylation was reinforced by the observation that immediately after injection into *Xenopus* oocyte nuclei, methylated and unmethylated templates both have equivalent activity (Kass *et al.*,

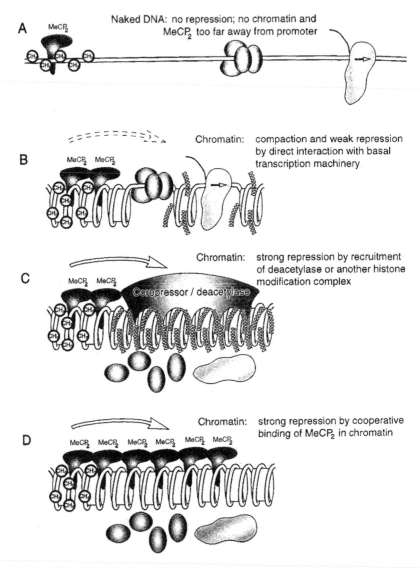

Figure 2.58. Models for molecular mechanisms in which DNA methylation and chromatin co-operate together to direct transcriptional repression (see text for details).

1997b). However, as chromatin is assembled, the methylated DNA is repressed with the loss of DNase I hypersensitivity and the loss of engaged RNA polymerase. The requirement for nucleosomes to exert efficient repression can be explained in several ways. The repression domain of MeCP2 might recruit a co-repressor complex that directs the modification of the chromatin template into a more stable and transcriptionally inert state (Fig. 2.58). One potential candidate corepressor for MeCP2 is the SIN3-histone deacetylase complex, because inhibition of histone deacetylation can reverse some of the transcriptional repression conferred by DNA methylation (Jones *et al.*, 1998). MeCP2 also copurifies with SIN3 and histone deacetylase in *Xenopus* oocytes (Jones *et al.*, 1998) and MeCP2 and SIN3 interact in *in vitro* binding studies (Nan *et al.*, 1998). Alternatively, like histone H1, MeCP2 might bind more efficiently to nucleosomal rather than to naked DNA. Any co-operative interactions between molecules could propagate the association of MeCP2 along the nucleosomal array even into unmethylated DNA segments (Fig. 2.58). This latter mechanism is analogous to the nucleation of heterochromatin assembly at the yeast telomeres by the DNA-binding protein RAP1, which then recruits the repressors SIR3p and SIR4p that organize chromatin into a repressive structure (Grunstein *et al.*, 1995; Hecht *et al.*, 1996). All of these potential mechanisms could individually or together contribute to the assembly of a repressive chromatin domain. Although these molecular mechanisms are speculative, they illustrate the advantages of a nucleosomal infrastructure. It should be noted that MeCP2 can repress transcription in an *in vitro* extract, although this might be by direct occlusion of transcription factor binding over the methylated promoter.

If methylated DNA directs the assembly of a specialized repressive chromatin structure, it might be anticipated that the transcriptional machinery will have less access to such a structure than the orthodox chromatin assembled on unmethylated promoters and genes. Activators such as Gal4-VP16 can normally penetrate a preassembled chromatin template to activate transcription, even in the presence of histone H1 (Laybourn and Kadonaga, 1992). However once chromatin has been assembled on methylated DNA, Gal4-VP16 can no longer gain access to its binding sites and activate transcription. This suggests that the specialized features of chromatin assembly on methylated DNA provide a molecular lock to silence the transcription process permanently (Siegfried and Cedar, 1997). This capacity of DNA methylation to strengthen transcriptional silencing in a chromatin context could be an important contributor to the separation of the genome into active and inactive compartments in a differentiated cell.

DNA methyltransferase maintains the methyl CpG content of both daughter DNA duplexes following replication (Holliday, 1987). Methyltransferase localizes to the chromosomal replication complex and maintenance methylation takes place less than one minute after replication (Leonhardt *et al.*, 1992; Gruenbaum *et al.*, 1983). By contrast, chromatin assembly takes 10–20 minutes in a mammalian tissue culture cell (Cusick *et al.*, 1983). Histone deposition occurs in stages, and it is not until a complete histone octamer is assembled with DNA that histone H1 is stably sequestered (Worcel *et al.*, 1978). Comparable limitations might restrict the stable association of methylation-specific repressors. This would account for the lag time before methylated DNA is repressed following injection as a naked template into the nuclei of mammalian tissue culture cells or *Xenopus* oocytes (Kass *et al.*, 1997b; Buschhausen *et al.*, 1987).

A significant feature of transcriptional repression on methylated DNA is that it is not only time dependent but also potentially dominant (Kass *et al.*, 1997b). Thus, at early times when chromatin assembly is incomplete, the transcriptional machinery has the potential to associate with methylated regulatory DNA. As chromatin structure matures, the basal transcriptional machinery is potentially erased from the template. This provides a general mechanism for the global silencing of transcription dependent only on DNA methylation state (Fig. 2.59), although Gal4-VP16 cannot function if chromatin is assembled on methylated DNA *before* exposure to the activator. If a very strong activator, such as Gal4-VP16, is present *during* chromatin assembly then transcriptional activity can resist methylation-dependent transcriptional silencing (Fig. 2.60). Therefore, under certain circumstances, regulatory nucleoprotein complexes might be assembled that resist this powerful silencing mechanism. Such a mechanism has been suggested to be dependent on SP1 sites in the promoter of a housekeeping gene in the mouse (adenine phosphoribosyl-transferase) that is maintained in a methylation-free state (Macleod *et al.*, 1994). For example, if components of regulatory complexes could bind to DNA immediately after replication with reasonable efficiency and before DNA methyltransferase can begin to modify the template, then they might prevent DNA methylation around their binding sites. These sequences might then become progressively demethylated and eventually resist transcriptional repression (Fig. 2.60). This would provide a mechanism for the demethylation of regulatory DNA in particular differentiated cell lines.

Other mechanisms might contribute to the maintenance of transcriptional repression through DNA synthesis. The assembly of a specialized chromatin structure on methylated DNA might result in the presence of additional proteins (e.g. MeCP2) and histone modifications (e.g. histone deacetylation) that could be maintained in daughter chromatids. Nucleosomes segregate dispersively in small groups to daughter DNA molecules at the replication fork (Sogo *et al.*, 1986). Particular modified histones and repressors such as MeCP2

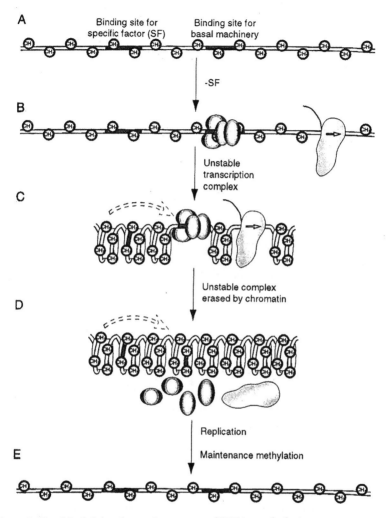

Figure 2.59. Model for the maintenance of DNA methylation state and transcriptional silencing through the replication process.

would be anticipated to segregate within the nucleosomal context (Perry *et al.*, 1993). These proteins could therefore provide at least 50% of the chromatin proteins necessary to restrict transcription. Their continued presence on DNA could help to re-establish transcriptional repression on both daughter chromatids through any of the mechanisms illustrated in Fig. 2.58. Therefore, demethylation alone might be insufficient to relieve transcriptional repression until successive cell divisions eventually unravel the repressive chromatin structure.

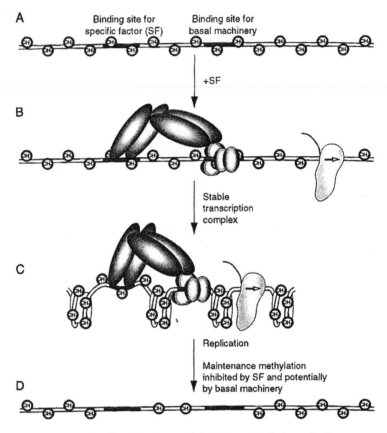

Figure 2.60. Model for the loss of DNA methylation during replication.

Although I focus on molecular mechanisms that might influence DNA methylation and gene expression in dividing cells, DNA demethylation is also important in non-dividing terminally differentiated cells. Under these circumstances demethylation at particular promoters must occur in the absence of replication (Sullivan and Grainger, 1986; Saluz *et al.*, 1986). Presumably mechanisms must also exist to destabilize any repressive chromatin structure associated with methylated DNA in order to allow the demethylation machinery access to the template.

The importance of DNA methylation and methylation-specific DNA-binding proteins for the viability of a differentiated mammalian somatic cell is well documented. An attractive explanation for the importance of DNA methylation is that it helps to turn off transcription from the large number of genes not required in a particular differentiated cell through global mechanisms. The major problem is in determining how this global repression might first be achieved and then maintained through successive cell generations? We have described evidence for methylation-specific and chromatin-dependent transcriptional repression mechanisms operating *in vivo*. Recent studies discussed here suggest that these two mechanisms operate together to regulate gene expression more tightly. There is now excellent precedent for transcriptional activators and repressors operating most effectively in a nucleosomal environment. Clearly, future experiments should explore how MeCP2 is incorporated into a nucleosomal array, together with the physical and functional consequences of this inclusion. The potential interaction of the MeCP2 repression domain with co-repressor complexes that might modify chromatin is also an area of active interest. Finally, a mutual dependence on DNA methylation and chromatin assembly for transcriptional silencing provides a potential mechanism not only for the stable propagation of the repressed state through cell division, but also for the targeted demethylation of promoter DNA. The availability of replication systems capable of propagating methylation states as well as directing chromatin assembly will allow this model to be tested directly (Harland, 1982).

Summary
DNA methylation has an essential regulatory function in mammalian development, serving to repress non-transcribed genes stably in differentiated adult somatic cells. Recent data implicate transcriptional repressors specific for methylated DNA and chromatin assembly in this global control of gene activity. The assembly of specialized nucleosomal structures on methylated DNA helps to explain the capacity of methylated DNA segments to silence transcription more effectively than conventional chromatin. Specialized nucleosomes also provide a potential molecular mechanism for the stable propagation of DNA methylation-dependent transcriptional silencing through cell division.

2.5.8 The HMGs and related proteins

Chromatin structure can be modified by the selective association of abundant non-histone proteins that interact with DNA histone complexes. Primary among these are the high mobility group proteins. Early methodologies for the fractionation of the linker histone H1 employed perchloric acid extraction of chromatin. Several other proteins were also found to be solubilized during this process. Later it was noticed that extraction of chromatin at moderate ionic strengths (0.35 M NaCl) released similar proteins. The addition of trichloroacetic acid (2%) to these salt-extracted proteins separated them into an insoluble fraction of large proteins (low-mobility group, LMG, when molecular size was assayed by gel electrophoresis) and a soluble fraction of small proteins (a group of high-mobility proteins during electrophoresis, HMG) (Johns, 1982). Four major proteins are found in the HMG group. These fall into two classes: HMG1 and 2 are one pair of homologous proteins (~29 000 Da in size); HMG14 and 17 are the other (10 000–12 000 Da in size). The content of HMG14 and 17 in chromatin may range up to 10% of DNA weight, similar to that of histone H1. In addition there are also several minor HMG proteins, for example HMG-I/Y which binds to the α-satellite sequences in the centromere (Bustin and Reeves, 1996; Section 2.4.2).

The genes encoding all four of the major HMGs have been cloned. The HMG14 and 17 proteins are highly conserved from human to chicken, certain basic stretches of amino acids being completely identical. These amino terminal basic regions are believed to interact with nucleosomal DNA. HMG14 and 17 also have an acidic carboxyl-terminal tail (Srikantha et al., 1988). Both proteins bind selectively to nucleosomal DNA in preference to naked DNA of a comparable length. It appears that two HMG molecules can bind per core particle (Mardian et al., 1980; Sandeen et al., 1980; Crippa et al., 1992). Surprisingly both HMG-14 and HMG-17 bind as homodimers to nucleosomes but do not interact together directly suggesting that they induce specific allosteric transitions in the nucleosome core to promote this selective association (Postnikov et al., 1995). One model based on chemical cross-linking suggests that the HMG14 and 17 proteins can interact with DNA where it exits and enters the nucleosome (Shick et al., 1985; Alfonso et al., 1994). Incorporation of HMG14 and 17 alters the stability of protein–DNA interactions at the nucleosome boundaries. This leads to a change in micrococcal nuclease digestion and potentially in the spacing of nucleosome cores (Crippa et al., 1993; Tremethick and Drew, 1993). In this respect HMG14 and 17 function in the nucleosome very much like some models for linker histone function (Pruss et al., 1995). Like acetylation or phosphorylation of the core histones, interaction of HMG14 and 17 at the nucleosomal boundaries is likely to modify histone H1 interaction and hence higher-order chromatin structure. Although definitive proof is lacking, considerable circumstantial evidence suggests that HMG14 and 17 are involved in potentiating the transcription of genes *in vivo* (Einck and Bustin, 1985; Section 4.3).

Recent experiments have examined the consequences for basal transcription of incorporating HMG14 and 17 into chromatin during the assembly process (Crippa *et al.*, 1993; Tremethick, 1994; Trieschmann *et al.*, 1995a; Paranjape *et al.*, 1995). Deletion mutagenesis of HMG-14 and HMG-17 suggests that the negatively charged C-terminal region of the proteins is required for transcriptional enhancement (Trieschmann *et al.*, 1996). In general, there is a modest 10-fold increase in transcription that appears related to a more accessible chromatin structure. The observation that RNA polymerase II elongation is stimulated by inclusion of HMG14 into chromatin is also consistent with a less compact, more accessible chromatin environment in which transcription can occur more efficiently (Ding *et al.*, 1994). A role for HMG14 and 17 in transcription would explain their recruitment to chromatin domains associated with elongating RNA polymerase within polytene chromosomes (Section 2.4.3).

Rather more functional information is available concerning the other pair of HMG proteins 1 and 2. These proteins have attracted a great deal of attention since conserved amino acid sequence motifs within these proteins are also found in transcription factors (Grosschedl *et al.*, 1994). HMG1 and 2 have a basic amino terminus and an acidic carboxyl terminus. The carboxyl terminus influences DNA binding selectivity (Wisniewski and Schulze, 1994). The basic region contains two HMG boxes. Each HMG box is a protein domain consisting of an L-shaped arrangement of three α-helices containing two independent DNA-binding surfaces (Read *et al.*, 1993; Weir *et al.*, 1993). A single HMG domain may cover 20 bp at a specific binding site and potentially distort the DNA molecule through as much as 130°. HMG1 and 2 also recognize unusual DNA structures such as cruciform DNA, which may simply reflect the presence of multiple DNA-binding sites on the same protein (Bianchi *et al.*, 1989; Lilley, 1992). Linker histones have similar properties (Varga-Weisz *et al.*, 1994). It has been suggested that this reflects the affinity of linker histones and potentially HMG1 and 2 for DNA where it enters and exits wrapping around the histone octamer. At this site DNA might cross-over itself (Section 2.3.1). Recent experiments demonstrate that HMG1 can replace linker histones in chromatin (Nightingale *et al.*, 1996; Ura *et al.*, 1996). Specific HMG domain proteins have now been defined that control lymphoid transcription, mating-type switching (SIN1, see Section 2.5.4) and mammalian sex determination (Pentiggia *et al.*, 1994; Werner *et al.*, 1995). Proteins with a single HMG domain associate with DNA sites relatively weakly, probably because of the energy required to direct the distortion of inflexible DNA. However, other proteins often contain several HMG domains, which form more stable complexes with DNA.

Most notable among the sequence-specific HMG domain transcription factors is the protein *upstream binding factor* or UBF, involved in the transcriptional regulation of mammalian ribosomal RNA genes (Jantzen *et al.*, 1990). The UBF protein contains five HMG domains flanked by an amino-terminal dimerization motif and an acidic carboxyl-terminal tail. Any adjacent pair of HMG domains will bind to DNA; however, the selectivity of binding is conferred by the three domains closest to the amino terminus (Le Blanc *et al.*, 1993; Hu *et al.*, 1994). Each UBF dimer contains 10 HMG domains, a binding site that potentially includes up to 200 bp of DNA. This extended region contains a site of DNA distortion every two turns of the double helix as a consequence of the binding of an HMG domain. Because these sites occur on the same face of the helix, DNA is bent into a superhelical turn around the contiguous HMG domains (Bazett-Jones *et al.*, 1994). Deoxyribonuclease I digestion of UBF–DNA complexes reveals a 10- to 11-bp periodicity of cleavage that is reminiscent of the access that this enzyme has to DNA wrapped around the histones within the nucleosome (Dunaway, 1989; Section 2.2.3). Looping appears to be facilitated not only by DNA deformation directed by the HMG domains but also by protein–protein interactions between the acidic carboxyl-terminal tail and the HMG domains toward the amino terminus of the UBF molecule.

How might the wrapping of DNA by UBF facilitate the transcription process? Like other eukaryotic genes, transcription by RNA polymerase I requires TBP, which in this system is a component of a sequence-specific transcription factor SL1. UBF and SL1 appear to bind co-operatively to the ribosomal promoter to form a stable complex that recruits RNA polymerase (Bell *et al.*, 1988). Two binding sites for SL1 are separated by 120 bp within the DNA wound around UBF. These sites function co-operatively, are separated by an integral number of helical turns of DNA, and remain exposed to the solution within the UBF–ribosomal promoter complex. UBF provides the correct scaffolding for productive interaction between individual SL1 molecules bound at the two recognition sites within each complex (Fig. 2.61). In this way UBF increases the probability and stability of transcription complex formation (McStay *et al.*, 1997).

The architectural role proposed for the UBF transcription factor has been seen with other HMG domain proteins (Giese *et al.*, 1992; Ferrari *et al.*, 1992). An HMG domain architectural transcription factor LEF-1

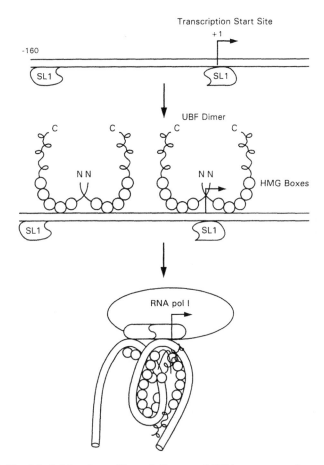

Figure 2.61. Model for the coiling of ribosomal RNA promoters by UBF to facilitate interaction between separated SL1 binding sites.

(lymphoid enhancer-binding factor, Grosschedl *et al.*, 1994) interacts with a cytoplasmic protein, β-catenin which links the cadherin cell adhesion molecule to the cytoskeleton (Behrens *et al.*, 1996). LEF-1 and β-catenin bind together to DNA and induce a specific bend in the double-helix. If LEF-1 and β-catenin are coexpressed in *Xenopus* embryos the axis of the embryo is duplicated reflecting aberrant cell signalling. Thus HMG domain proteins can be components of signal transduction pathways from cell adhesion components to the cell nucleus. In several similar cases, the association of the HMG domain with DNA directs the assembly of clusters of transcription factors bound to DNA into precise higher-order nucleoprotein complexes (Sections 4.1).

The role of the relatively abundant HMG1 and 2 proteins themselves in chromatin remains enigmatic. It has been proposed that these proteins might promote nucleosome assembly (Bonne-Andrea *et al.*, 1984) or prevent nucleosome assembly (Waga *et al.*, 1989). They have been reported to facilitate transcription (Tremethick and Molloy, 1986) and to inhibit transcription (Ge and Roeder, 1994a; Stelzer *et al.*, 1994). HMG1 will repress transcription selectively from nucleosomal templates by positioning nucleosomes and restricting octamer mobility (Ura *et al.*, 1996). The proteins may function as stable components of nucleoprotein structures (Paull *et al.*, 1993) or transiently as assembly factors required to bend DNA and then dissociate (Travers *et al.*, 1994). HMG1 appears to function in an analogous manner to the sequence-specific HMG domain proteins in facilitating the binding of the progesterone receptor transcription factor to its recognition element (Onate *et al.*, 1994). Normally HMG1 and 2 are associated with a relatively minor fraction of chromatin (< 5%; Goodwin *et al.*, 1977; Isackson *et al.*, 1980). It has been proposed that HMG1 might be capable of functionally replacing linker histones within chromatin (Jackson *et al.*, 1979; Nightingale *et al.*, 1996). This idea receives support from the observation that early embryonic chromatin in *Drosophila* and *Xenopus* is highly enriched in HMG1- and 2-like proteins (Dimitrov *et al.*, 1994; Ner and Travers, 1994). This strongly suggests that HMG1 and 2 primarily fulfil a structural role in chromatin.

A negative regulator of inducible transcription in yeast known as SIN1 (Kruger and Herskowitz, 1991) (Section 2.5.4) has a very similar structure to HMG1. It is possible that SIN1 might help to fulfil the regulatory role of linker histones (Wolffe, 1994b).

Sequences rich in glutamic and aspartic acid residues (Section 2.4.2), are found in HMG1 and 2, the centromeric protein CENP-B, chromatin assembly proteins N1/N2, nucleoplasmin (Section 3.2), and topoisomerase I. These anionic regions have been postulated to have the capacity to interact directly with histones, since N1/N2 interacts specifically with histones H3/H4 and nucleoplasmin with H2A/H2B. This property may account for the role of HMG1 and 2 in nucleosome assembly. The physiological significance of this assembly activity is unknown (Section 3.2). It has also been postulated that the acidic regions found within transcription factors might cause a local destabilization of nucleosome structure, perhaps by competing with DNA for interaction with histones, especially histones H2A/H2B.

Summary
The HMG14 and 17 proteins have a higher affinity for nucleosomal DNA than for naked DNA. They may influence the folding of chromatin and indirectly increase the accessibility of regulatory complexes to RNA polymerase. The incorporation of HMG14/17 into chromatin may also facilitate progression of RNA polymerase through nucleosomal arrays.

The HMG1 and 2 proteins are representative of a large family of DNA-binding proteins some of which interact with DNA specifically. They have a highly conserved DNA-binding domain and a domain of acidic amino acids. HMG1 and 2 appear to have a structural role within chromatin and may under certain circumstances substitute for linker histones in the nucleosome.

2.5.9 Functional compartmentalization of the nucleus

Functional compartmentalization in the eukaryotic cell is readily accepted from observation of membrane-bounded organelles that can be fractionated and their properties determined in isolation. The existence of discrete compartments within a given organelle is less immediately apparent, but none the less real. Within the eukaryotic nucleus several independent approaches point to the compartment-alization of particular activities such as transcription, RNA processing and replication. Chromosomes are revealed to occupy defined territories and to represent highly differentiated structures. The numerous activities that use DNA and RNA as a template occur with a defined spatial and temporal relationship. A remarkable fusion of methodologies including cell biology, molecular genetics and bio-chemistry has contributed to the recognition of nuclear architecture as defined by function (Strouboulis and Wolffe, 1996).

The most dramatic example of the compartmentalization of nuclear function is seen with replication. Analysis of DNA synthetic sites with bromodeoxyuridine or biotinylated dUTP reveals only 150 foci of incorporation within each nucleus during S phase (Nakamura *et al.*, 1986). The foci are clearly defined with a clear and relatively uniform separation from each other. When replication initiates, these foci are small and appear as 'dots', as time progresses they become more diffuse (Manders *et al.*, 1992; O'Keefe *et al.*, 1992). These foci contain accumulations of the proteins necessary for replication: DNA polymerase α, PCNA, and RP-A as well as regulatory molecules such as cyclin A, cdk2, and RPA70 (Adachi and Laemmli, 1992; Hozak *et al.*,

1993; Cardoso *et al.*, 1993; Sobczak-Thepot *et al.*, 1993). Immunolabel-ling synthetic sites with gold particles suggests that nascent DNA is extruded from the replication foci (Hozak *et al.*, 1993). This implies that DNA moves through a fixed architecture containing the molecular machines directing replication. The advantages of the compartmentalization of DNA replication include a concentration of the necessary regulatory, structural and enzymatic components required to duplicate both DNA and chromosomal structure. The staged assembly of a functional replication elongation complex occurs within a defined macromolecular complex, this allows many check points and controls to be built into the initiation of replication (Almouzni and Wolffe, 1993b).

The essential role of nuclear architecture in determining the functional properties of DNA is perhaps most apparent in connection with chromosomal replication in *Xenopus laevis* eggs. Injection of prokaryotic DNA into an egg or incubation of the DNA in an egg extract leads to the assembly of a pseudonucleus competent to replicate DNA (Forbes *et al.*, 1983; Blow and Laskey, 1986; Section 3.1). Importantly, replication is regulated spatially in that it occurs at discrete sites containing clusters of replication forks (Cox and Laskey, 1991). There is a remarkable similarity between the number and distribution of replication 'foci' in the pseudonuclei and those observed in replicating eukaryotic nuclei in tissue culture cells (Nakamura *et al.*, 1986; Mills *et al.*, 1989). The assembly of functional replication origins is not necessarily dependent on defined DNA sequences in the chromosomes, but on features of nuclear architecture that can be assembled even on prokaryotic DNA. The implication is that general features of nuclear architecture can impose a particular function, in this case that of replication. It should be noted that the early *Xenopus* embryo is a special case in which normal somatic controls might have been relaxed. Chromatin loop attachments to the chromosomal axis and the number of chromosomal origins of replication are much more frequent in the chromosomes of early embryonic nuclei in *Xenopus* compared to somatic cell nuclei (Laskey *et al.*, 1983; Micheli *et al.*, 1993). Only when the cell cycle lengthens at the mid blastula transition are normal controls established (see Hyrien *et al.*, 1995).

In normal somatic nuclei the replication foci do not all engage in replication simultaneously, some are utilized early in S-phase and others late in S-phase (Nakamura *et al.*, 1986; Ariel *et al.*, 1993). This reflects differential replication timing, which is an important regula-tory step in maintaining local chromatin organization and gene activity (Wolffe, 1991c). The molecular mechanisms controlling the differential utilization of origins are presently unknown (Gilbert, 1986; Guinta and Korn, 1986; Wolffe, 1993). However, comparable phenomena occur in yeast (Newlon *et al.*, 1993), where origin utilization is found to be dependent on chromosomal position. This is an important area for future study.

These observations on DNA replication, the dependence on nuclear architecture and the movement of the DNA strand through a fixed site have led to speculation that comparable phenomena also govern transcription (Cook, 1994; Hughes *et al.*, 1995). The concentration of RNA polymerase within defined nuclear compartments together with other components of the transcriptional machinery lends some support to these ideas (see later). RNA polymerase may be much less mobile and chromatin more mobile than generally considered. An important additional point is that the assembly of a particular nucleoprotein architecture that favors one biological process, e.g. replication, might exert an exclusionary or repressive influence on another, e.g. transcription (see Wansink *et al.*, 1994). In fact in the *S. cerevisiae* chromosome, components of the Origin Recognition Complex required for replication exert a silencing effect on transcription (Fox *et al.*, 1993; Section 2.5.6).

Ribosomal gene transcription, rRNA processing and preribosomal particle assembly occur in the nucleolus (Scheer and Benavente, 1990). All of these events involve the assembly of macromolecular machines that localize within this specialized nuclear compartment. The molecular mechanisms that direct particular proteins and enzyme complexes to this compartment and retain them there are largely unknown (see Hatanaka, 1990). One simple hypothesis is that the majority of the nucleolar architecture is generated from the activities of the transcriptional machinery itself which assembles reiterated regulatory nucleoprotein complexes on rDNA. Ribosomal RNA genes are tandemly arrayed with approximately 250 copies of a 44 Kb repeat in humans (Scheer and Benavente, 1990). Thus, more than 10×10^6 bp of rDNA and associated proteins could provide the framework for the nucleolus. Once transcription itself is in progress, additional features of nucleolar architecture would potentially follow from the accumulation and activities of the molecular machines that process pre-rRNA and that assemble ribosomes.

Morphologically the nucleolus has three major organizational areas: (1) the nucleolar fibrillar centers, which are surrounded by (2) a dense fibrillar region, and (3) the granular region. Numerous localization studies using specific antibodies and hybridization probes indicate that the nuclear fibrillar centers are the sites where the ribosomal RNA genes, RNA polymerase I, the class I gene transcription factor UBF and topoisomerase I are localized (Scheer and Rose, 1984; Raska *et al.*, 1989; Rendon *et al.*, 1992; Thiry, 1992a,b). Accumulation of these particular macromolecules leads to the inference that the nucleolar fibrillar centres are the assembly sites for the regulatory nucleoprotein complexes that direct transcription (Fig. 2.62). The dense fibrillar component that surrounds the nucleolar fibrillar center consists of nascent ribosomal RNA and associated proteins. It is in the dense fibrillar component that RNA precursors such as [^{3}H]uridine or biotinylated ribonucleotides are initially found on pulse labelling (Thiry and Goessens, 1992). Specfic hybridization probes localize unprocessed nascent transcripts and associated processing machinery to the dense fibrillar component (Ochs *et al.*, 1985; Kass *et al.*, 1990; Puvion-Dutilleul *et al.*, 1991). Mature 28S and 18S rRNA, partially processed transcripts and intermediates in ribosome assembly are found in the granular region. These assembly intermediates are visualized as particles 15–20 nm in diameter. Movement of preribosomal subunits from the nucleolar granular region to the cytoplasm might be facilitated by proteins that move within specific pathways or tracks from the nucleolus to the nuclear envelope (Meier and Blobel, 1992).

This hierarchical organization of the nucleolus with particular morphologically distinct compartments reflecting accumulations of specialized molecular machines and their substrates provides an extremely useful model with which to consider the functional compartmentalization of mRNA synthesis, processing and export.

The synthesis of mRNA within the nucleus and the subsequent delivery of the mature transcript to the translational machinery within the cytoplasm also involves the concerted and co-ordinated activities of multiple molecular machines. Transcription requires the assembly of a regulatory nucleoprotein complex, containing the promoter region, associated coactivators and the RNA polymerase holoenzyme. The polymerase must initiate RNA synthesis and traverse the gene. The pre-mRNA must be processed through the addition of a $m^7G(5')pp$ cap, removal of introns (splicing) and polyadenylation. These various biochemical events are interdependent since transcription by RNA polymerase II is a prerequisite for both efficient splicing and polyadenylation (Sisodia *et al.*, 1987), and the 5' cap and associated proteins also facilitate splicing and mRNA export (Izaurralde *et al.*, 1994; Lewis *et al.*, 1995).

The transcriptional machinery that synthesizes pre-mRNA localizes with the perichromatin fibrils found at the boundaries of condensed chromatin domains. Perichromatin fibrils are nuclear ribonucleoprotein complexes with a diameter varying from 3 nm to 20 nm. They are

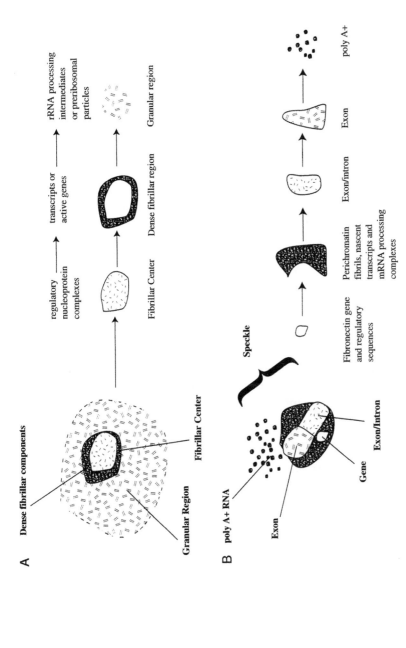

Figure 2.62. Transcription compartmentalization.
A. A transcription domain in the nucleolus. A fibrillar centre is shown surrounded by dense fibrillar components and the granular domain. A model of activities in these domains is presented. B. A transcription domain for RNA polymerase II. A diagram of the structure and a model of activities in various compartments are presented. Reproduced with permission from Strouboulis, J. and Wolffe, A.P. (1996) *J. Cell Sci.* **109,** 1991–2000. Copyright 1996 The Company of Biologists Limited.

enriched in nascent pre-mRNA radiolabelled with [^{3}H]uridine or bromouridine, and fibril density correlates with transcriptional activity (Bachellerie *et al.*, 1975; Fakan, 1994; van Driel *et al.*, 1995). Components of the splicing machinery are found with the perichromatin fibrils (Fakan *et al.*, 1984) consistent with the assembly of the splicing machinery initiating at the site of transcription. This is visualized through immunofluorescent probe detection of the splicing machinery as diffuse nucleoplasmic staining (Spector, 1993). Considerable morphological and molecular biological evidence indicates that splicing occurs concomitant with transcription (Beyer and Osheim, 1988; Le Maire and Thummel, 1990; Wuarin and Schibler, 1994).

The localization of specific transcripts such as fibronectin pre-mRNA using hybridization probes reveals elongated 'tracks' or more compact 'dots' at one or two discrete sites corresponding to the chromosomal copies of the gene (Xing *et al.*, 1993, 1995; Huang and Spector, 1991). Simultaneous RNA and DNA hybridization demonstrates that transcribing genes directly localize with the RNA tracks or dots, with the gene at one end of the track (Xing *et al.*, 1993, 1995; Xing and Lawrence, 1993). Moreover probes for introns only detect the track near the gene, suggesting that splicing occurs along the track. These results suggest a model of cotranscriptional assembly of the splicing machinery onto pre-mRNA at the perichromatin fibrils, with splicing continuing as the pre-mRNA is released from the gene.

RNA tracks can have a very close association with discrete structures known as interchromatin granules or 'speckles' (Xing *et al.*, 1993). These speckles are sites at which the splicing machinery accumulates together with intron containing pre-mRNA and polyadenylated mRNA (Spector, 1990; Fu and Maniatis, 1990; Carter *et al.*, 1991, 1993; Visa *et al.*, 1993a; Wang *et al.*, 1991). Thus speckles potentially represent sites of pre-mRNA processing and of mature mRNA accumulation in the nucleus.

Lawrence and colleagues have suggested that actively transcribing genes have a non-random association with the speckles (Xing and Lawrence, 1993; Lawrence *et al.*, 1993; Xing *et al.*, 1995), the implication of this association being that in certain instances speckle structures represent the sites of transcription itself. An immediate limitation to this latter hypothesis is that there are only 20–50 speckles scattered in a punctate distribution throughout the nucleus. Clearly not every active gene can be associated with these structures. However, out of ten transcribing genes investigated in the Lawrence laboratory, seven are associated with speckles (Xing *et al.*, 1995). It has also been hypothesized that these transcription domains might contain several actively transcribing genes at any one time (Xing *et al.*, 1993; Xing and Lawrence, 1993). In contrast to this view, the bulk of nascent pre-mRNA labelled with bromouridine accumulates in a distinct pattern which does not correspond to that of speckles containing snRNPs (Jackson *et al.*, 1993; Wansink *et al.*, 1993). Adenoviral and actin transcripts can be visualized as discrete dots in the nucleus, with no apparent association with snRNP speckles (Zhang *et al.*, 1994). However, a more detailed study indicates a clear association between actin transcripts and speckles (Xing *et al.*, 1995). An additional complication is that not all RNA detected in these assays is pre-mRNA, but an ill defined proportion might correspond to a large pool of stable nuclear polyadenylated RNA involved in structural functions within the nucleus (Huang *et al.*, 1994; Mattaj, 1994).

Much of the evidence presented in support of the association between sites of transcription and speckles is based on the transcription of very active genes (e.g. collagen, which comprises 4% of total mRNA in fibroblasts; and the induced expression of the *fos* gene after serum starvation). It could be envisaged that the speckles, rich in splicing components, act as processing factories closely associated only with the most actively transcribing genes.

Once synthesized and assembled with the splicing machinery, the pre-mRNA has to reach the nuclear envelope and enter the cytoplasm. The pre-mRNA is packaged not only with the splicing apparatus but also with heterogeneous nuclear ribonucleoproteins (hnRNPs). These proteins provide the 'workbench' on which mRNA is processed. Like the packaging of DNA with histones, the resulting architectures are important for the maturation of mRNA (Dreyfuss *et al.*, 1993). The preparation of nuclear matrix, a process that removes the vast majority of chromatin from the nucleus, retains pre-mRNA, hnRNPs and some elements of the splicing machinery (Huang and Spector, 1991; Mattern *et al.*, 1996). This is indicative of both the abundance and major structural role for hnRNPs in the nucleus. In general, hnRNPs are diffusely distributed throughout the nucleoplasm, however a subset overlap snRNP speckles, and some even shuttle with mRNA into the cytoplasm before returning to the nucleus (Visa *et al.*, 1996; Pinol-Roma and Dreyfuss, 1992). The movement of a specific pre-mRNA from gene to cytoplasm has been reconstructed based on the export pathway of the Balbiani ring (BR) pre-mRNP particles in the dipteran *Chironomus tentans* (Mehlin and Daneholt, 1993). BR genes are easily visualized as two giant puffs in the polytene chromosomes of the salivary glands. The large nascent transcripts from these puffs are assembled with hnRNPs and the splicing machinery during transcription to form a thin fibre, which, with elongation, becomes thicker and bends into a ring-like structure. The mature pre-mRNP granule, now thought to contain spliced RNA is released into the nucleoplasm making its way to the nuclear envelope, where it positions itself against a nuclear pore and becomes elongated into a rod-shaped structure which goes through the pore in a 5'-head-first manner. As the pre-mRNP granule emerges on the cytoplasmic side it immediately becomes associated with ribosomes (Mehlin and Daneholt, 1993). The important point for this discussion is that the entire process occurs within precise nucleoprotein architectures.

How does the pre-mRNA reach the nuclear membrane from the sites where transcription takes place? Estimates of the rate of movement of pre-mRNA have been made using a highly expressed hybrid gene in *Drosophila* salivary glands, which gives a strong signal at the site of transcription and a more diffuse channel-like network pattern throughout the nucleoplasm (Zachar *et al.*, 1993). These results have been interpreted to suggest that simple diffusion alone could account for the dispersal of mRNA. However, there is also evidence to suggest that pre-mRNA movement through the nucleoplasm occurs in a directed fashion. For example, the need for particular structural features in the pre-mRNAs for their movement into the cytoplasm (Elliot *et al.*, 1994) is inconsistent with a simple diffusion model. Furthermore, the tight association of transcripts, hnRNPs, and functional processing components (e.g. the splicing machinery) with the nuclear skeleton and nuclear matrix argues against the pre-mRNA being freely diffusible in the nucleoplasm. Huang and Spector (1991) were able to visualize 'tracks' corresponding to *fos* gene transcripts frequently extending to the nuclear envelope and exiting over a limited area. This would appear to give credence to one aspect of the gene gating model proposed by Blobel (1985). This postulates that due to overall three-dimensional architectural constraints in the nucleus, genes will associate with a specific region of the nuclear envelope, hence their transcripts are 'gated' to exit at a defined set of nuclear pores. Though the organization of genes relative to the nuclear envelope remains largely unproven, the findings of Huang and Spector (1991) are consistent with the 'gating' hypothesis. In contrast, Lawrence and colleagues saw no significant evidence for their transcript 'tracks' making contact with the nuclear envelope (Xing and Lawrence, 1993; Xing *et al.*, 1995). Collagen mRNA was visualized as 'studding' or 'encircling' the nuclear envelope (Xing *et al.*, 1995), indicative of an exit at many nuclear pores. It has been suggested that the lack of tracks visibly extending to the nuclear envelope may be due to the fact that somewhere along the transport pathway, pre-mRNA rapidly disperses in many directions (Xing and Lawrence, 1993; Xing *et al.*, 1995). Once again evidence of specificity in a nuclear process is indicative of a high degree of structural organization. Not all mRNAs might require such specificity in their export pathway.

The process of mRNA synthesis has many parallels with that of rRNA: sites of synthesis can be visualized, the pre-mRNA is packaged with processing machinery cotranscriptionally, and then for certain mRNAs processing within a defined structure takes place before release for export from the nucleus. The various structures visualized reflect the molecular machines active at those sites (Fig. 2.62). Moreover the structures are dynamic with a constant vectorial flow from the sites of synthesis to the next step on the way to the cytoplasm. Components can also recycle between the different functional compartments (e.g. Pinol-Roma and Dreyfuss, 1992). Everything happens as a nucleoprotein complex that is visually identified as a functional and morphologically discrete compartment. Does it matter that transcription and splicing/processing occur in particular domains? The advantages of compartmentalization are similar to those discussed earlier for replication. There is a concentration of the necessary regulatory, structural and enzymatic components required to transcribe or splice mRNA. The organization of the components within an architectural framework provides many more opportunities for regulation compared to a freely diffusible state. Clearly transcription and splicing can occur in dilute solutions (1 μg/ml) within an *in vitro* reaction tube, however the efficiency with which these events occur is much less than that achieved *in vivo*. Organization and channelling of macromolecules from one site of enzymatic activity to another within a specific architecture is clearly advantageous within a nucleus containing nucleoprotein at >50mg/ml.

The functional organization of the chromosome into discrete domains has been increasingly recognized through experiments in yeast and *Drosophila* that have made use of the phenomenon of position effect variegation (Schaffer *et al.*, 1993; Section 2.5.6). Early cytological experiments demonstrated the positioning of telomeres at the nuclear envelope in salivary gland cells of salamanders (Rabl, 1885). The telomeres of *Drosophila* polytene chromosomes and those of *Schizosaccharomyces pombe* chromosomes in G2 phase of the cell cycle also show comparable localization of the telomeres at the nuclear periphery (Hochstrasser *et al.*, 1986; Funabiki *et al.*, 1993). Advances in confocal immunofluorescence microscopy and molecular genetics have allowed the demonstration that two proteins, silent information regulators (SIR)3 and 4 are required for the perinuclear localization of *Saccharomyces cerevisiae* telomeres (Palladino *et al.*, 1993). These proteins are also required for the heritable inactivation of genes within specific chromosomal domains located at the silent mating type loci and telomeres of *S. cerevisiae*. Thus a connection is made between the location of a particular chromosomal territory in the nucleus and transcriptional repression *per se* (Section 2.5.6).

Experiments designed to examine the localization of active genes in the nucleus clearly demonstrate that these are predominantly found within the nuclear interior (Spector, 1993). Early suggestions that active genes were preferentially located at the nuclear periphery are probably based on experimental artefact (Hutchison and Weintraub, 1985). With respect to specific active genes, Lawrence and colleagues have suggested that some genes occupy non random positions (Lawrence *et al.*, 1993). For example: three active genes with very different localizations are the whole EBV genome and the *neu* oncogene (transcriptionally active) which are positioned within the inner 50% of the nuclear volume, whereas the dystrophin gene is at the extreme nuclear periphery. However, three inactive genes (albumin, cardiac myosin heavy chain and neurotensin) all localize in constitutive heterochromatin at the nuclear periphery or near the nucleolus (Xing *et al.*, 1995). UV microirradiation and *in situ* hybridization experiments extend the experiments examining telomeres or specific genes to suggest that individual chromosomes occupy broad, but discrete territories within the nucleus (Cremer *et al.*, 1993; Heslop-Harrison and Bennett, 1990; van Driel *et al.*, 1995).

The spatial relationship between chromosome territories and other subnuclear compartments has been investigated by Cremer and colleagues (Zirbel *et al.*, 1993). It was shown that the splicing machinery subcompartments were associated with the periphery of chromosome territories and were excluded from their interior (Zirbel *et al.*, 1993). Similarly, a specific gene transcript visualized as an RNA track was shown to be preferentially localized on the surface of the chromosome territory and a very limited survey of the localization of individual genes again placed them to the exterior of chromosomal territories (Zirbel *et al.*, 1993; Cremer *et al.*, 1993). On the basis of the above evidence, Cremer and colleagues have postulated that the interchromosome space excluded by the chromosomal territories defines an interconnected functional compartment for transcription, splicing, maturation and transport (Fig. 2.63). This compartment is intimately associated with actively transcribing genes localized on the surface of the territory, presumably on extended loops (Zirbel *et al.*, 1993; Cremer *et al.*, 1993). This is an intuitively appealing model which potentially encompasses the observations of Lawrence and colleagues regarding the non-random distribution of genes. However, the nature of the functional interface between an active gene in a chromosome

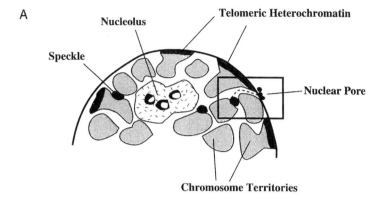

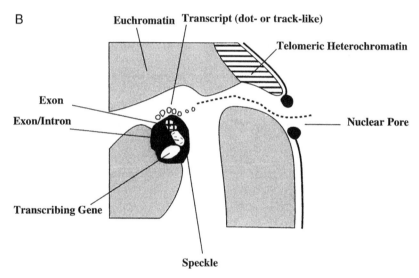

Figure 2.63. Chromosomal territories in the nucleus.
A. A cross section of the nucleus indicating the organization of chromosomes and transcriptionally active and inactive chromatin. B. An expanded view of a section of the nucleus showing a speckle and the transcript path (dotted line) to the nuclear pore. Reproduced with permission from Strouboulis, J. and Wolffe, A.P. (1996) *J. Cell Sci.* **109**, 1991–2000. Copyright 1996 The Company of Biologists Limited.

territory and the interchromatin compartment remains unclear, primarily due to our poor understanding of the higher order organization and compaction of DNA into chromosomes.

The compartmentalization of transcription, RNA processing and replication components within the nucleus lends credence to the existence of specialized functional roles for any morphologically distinct structure in which a protein of interest accumulates. Nuclear bodies, originally described at an ultrastructural level, represent such structures in search of a function.

Coiled bodies Morphologically defined as a tangle of coiled threads (Monneron and Bernhard, 1969), coiled bodies are often associated with the periphery of nucleoli. A number of nucleolar proteins and RNAs important for rRNA modification and processing are found in the coiled body including fibrillarin and the U3 small nucleolar RNA (Jimenez-Garcia *et al.*, 1994). snRNPs including U7 snRNA also accumulate in this structure, as do specialized proteins such as p80-coilin (Frey and Matera, 1995; Bohmann *et al.*, 1995). Accumulation of these proteins suggests a role in RNA processing, however rRNA and mRNA have not been detected in these structures (Huang *et al.*, 1994; Jimenez-Garcia *et al.*, 1994). Coiled bodies are dynamic, they disassemble at mitosis and reassemble in G1 during the cell cycle, they also increase in abundance in response to growth stimuli (Carmo-Fonseca *et al.*, 1993; Lamond and Carmo-Fonseca, 1993).

The nuclei of amphibian oocytes contain structures known as C-snurposomes with many similarities to mammalian coiled bodies (Bauer *et al.*, 1994; Gall *et al.*, 1995). These nuclear organelles are found attached to the histone gene loci of lampbrush chromosomes, and since U7 snRNA is involved in the 3'-end modification of histone pre-mRNAs, it has been suggested that this is one function of the C-snurposome and of coiled bodies (Gall *et al.*, 1995; Frey and Matera, 1995). Alternatively the C-snurposome or coiled bodies might represent assembly sites for the molecular machines that process various RNAs (Bohmann *et al.*, 1995).

PML (promyelocytic leukemia) nuclear bodies Acute PML is a haemopoietic malignancy that is most often associated with a t(15; 17) chromosome translocation which results in an inframe fusion of the *PML* gene to that of the retinoic acid receptor α (RARα) (Warrell *et al.*, 1993). The PML protein itself contains a zinc binding RING finger, two cysteine rich domains and a C-terminal coiled-coil domain (Lovering *et al.*, 1993; Reddy *et al.*, 1992; Perez *et al.*, 1993). In cells from normal individuals, PML predominantly accumulates in a novel nuclear body consisting of a dense fibrillar ring surrounding a central core (Koken *et al.*, 1994; Weis *et al.*, 1994; Dyck *et al.*, 1994). PML is also found with U1 snRNA and p80-coilin in a distinct compartment or zone surrounding interchromatin granules or speckles (Visa *et al.*, 1993b; Puvion-Dutilleul *et al.*, 1995). PML nuclear bodies are dynamic with respect to the cell cycle and there appears to be a correlation between their prominence and proliferative states (Koken *et al.*, 1995; Terris *et al.*, 1995). Viral infections can disrupt PML bodies (for example Puvion-Dutilleul *et al.*, 1995; Kelly *et al.*, 1995). For adenovirus, this disruption may be a crucial step for replication of the viral genome (Doucas *et al.*, 1996). Treatment of cells with interferon induces PML expression and represses viral replication, thus offering an additional contribution to antiviral activity (Guldner *et al.*, 1992; Stadler *et al.*, 1995).

When PML is fused to RARα in patients with acute promyelocytic leukemia the normal distribution of PML in defined nuclear bodies is disrupted and a 'micropunctate' pattern is observed (Weis *et al.*, 1994; Koken *et al.*, 1994; Dyck *et al.*, 1994). Any wild-type PML protein is also sequestered into this micropunctate pattern. Treatment of cells with retinoic acid facilitates the restoration of the normal nuclear body distribution of PML through a mechanism that is not understood (Dyck *et al.*, 1994; Weis *et al.*, 1994; Koken *et al.*, 1994). This correlates with the fact that patients with acute promyelocytic leukemia go into remission following treatment with retinoic acid. It has been suggested that wild-type PML functions to suppress growth, slowing down the growth rate of transformed cell lines and suppressing their tumorigenicity (Koken *et al.*, 1995). The molecular mechanisms by which this would be achieved remains unknown.

WT1 nuclear domains   Wilm's tumor is a childhood kidney malignancy frequently associated with congenital urogenital abnormalities, thus indicating underlying developmental deficiencies. It is a complex genetic disease, with at least three genetic loci contributing to it. So far only one gene has been isolated, the WT1 tumor suppressor gene (Hastie, 1994). Mice homozygous for the WT1 knock-out die before day 15 of gestation with a clear failure to develop kidneys, gonads and a normal mesothelium (Kreidberg *et al.*, 1993).

The product of the WT1 gene was originally thought to be a transcription factor since it contains an N-terminal proline/glutamine-rich domain, frequently associated with transcriptional activators, and four C-terminal zinc-fingers, very closely related to those found in transcription factors such as Sp1, EGR1 and EGR2. WT1 has been shown to bind to a GC-rich motif *in vitro* and to repress transcription in transient transfection assays in promoters that contain this motif. As yet, no physiological gene target for WT1 activity has been identified.

The subnuclear localization of WT1 in a mouse mesonephric cell line, as well as in fetal kidney and testis, showed a distinct punctate pattern as well as a diffuse nucleoplasmic staining (Larsson *et al.*, 1995; Englert *et al.*, 1995). Double staining clearly showed that WT1 did not occupy the nuclear transcription factor domains that were stained by Sp1. Instead, by using an anti-Sm antibody that detects snRNPs, WT1 was shown to colocalize with the snRNP 'speckles' (Larsson *et al.*, 1995). WT1 was also co-immunoprecipitated with an anti-p80-coilin antibody, suggesting that WT1 is also present in coiled bodies. One recent study in transfected osteosarcoma cells showed that WT1 colocalized with only a subset of the snRNP speckles which did not stain with a monoclonal antibody against the essential non-snRNP splicing protein SC-35 (Englert *et al.*, 1995). This raises the possibility that the WT1-rich speckles constitute a novel nuclear subcompartment that also contains snRNPs. The distribution of WT1 in the nucleus has been shown to be a dynamic one which paralleled, to a large extent, that of snRNPs. For example, microinjection in the nucleus of anti-snRNA oligonucleotides (Larsson *et al.*, 1995) or heat shock (Charlieu *et al.*, 1995) led to an identical rearrangement for both WT1 and snRNPs in both cases. In contrast, treatment with actinomycin D, a transcriptional inhibitor, led to a re-distribution of WT1 that over-lapped that observed for p80-coilin, a coiled body hallmark. In both cases, the proteins were seen to be re-distributed around the nucleolus remnants, as opposed to the majority of snRNPs which concentrated in large foci (Larsson *et al.*, 1995).

Alternative splicing gives rise to four WT1 isoforms dependent on the inclusion or omission of two motifs: 17 amino acids encoded by exon 5 included N-terminally to the zinc-fingers, and/or the KTS motif (for lysine-threonine-serine) included between zinc-fingers 3 and 4. The presence or absence of the KTS motif significantly affects the DNA binding properties of WT1, with the +KTS WT1 isoforms having a significantly lower affinity for binding to DNA. In addition, the +KTS isoforms were seen to distribute mostly in a speckled pattern, whereas the –KTS isoforms were distributed more diffusely (Charlieu *et al.*, 1995). DNA binding-defective –KTS isoforms accumulate in the speckles, thus suggesting a correlation between DNA-binding affinity and subnuclear localization (Larsson *et al.*, 1995; Charlieu *et al.*, 1995). Moreover, DNA-binding –KTS isoforms are sequestered into a speckled distribution by a WT1 mutant lacking the zinc-finger domain (Englert *et al.*, 1995). These findings are significant, since naturally occurring dominant negative mutations that give rise to developmental abnormalities, have been mapped within the zinc-finger domain (Hastie, 1994). These are likely to affect not only DNA binding by the mutant WT1, but also the subnuclear distribution of any wild-type –KTS protein.

This work therefore indicates the existence of a clear subcompart-mentalization of WT1 isoforms relating to WT1 function. The dynamic association of WT1 with the splicing machinery, suggests a previously unappreciated role for WT1 in post-transcriptional gene regulation, as well as in transcription *per se* (Charlieu *et al.*, 1995).

The experimental data discussed here illustrate the diversity of nuclear events in their structural context. There are two major landmarks in the nucleus: (1) the nuclear envelope, associated lamina and nuclear pores (Dingwall and Laskey, 1986; Gerace and Burke, 1988) which marks the outer boundary (Section 2.4.2), and (2) the chromosomes within the interior which represent the reason for the nucleus to exist through transcription and replication of DNA. These sites of chromosomal activity are non-randomly distributed with respect to the nuclear envelope. Unknown features of nuclear architecture direct the spatial arrangement of replication foci (Cox and Laskey, 1991). A major unresolved issue at this time is whether there is an underlying structure within the nucleus that directs the spatial distribution of functional compartments (Section 2.4.2). Chromatin is potentially mobile, moving through the replication foci during S-phase (Hughes *et al.*, 1995). The chromosome provides the template for replication and transcription, but is clearly not a static structure that continually maintains one particular architecture. DNA is packaged with chromatin structural proteins in a way that allows transcription and replication to occur within the functionally differentiated structures.

The molecular machines that transcribe and replicate DNA, as well as those that regulate these events are so extensive that it appears probable that concentrations of these machines together with the associated RNA and DNA, account for many of the structures that can be morphologically distinguished in the nucleus. This is particularly apparent for the nucleolus, where distinct domains are visualized representing: (1) regulatory nucleoprotein complexes controlling transcription, (2) the active transcriptional machinery itself and associated transcripts and (3) transcripts in the process of being assembled into functional ribonucleoprotein complexes. On a more local scale the same domains are visualized for RNA polymerase II transcripts (Fig. 2.62). Ribonucleoprotein complexes clearly account for many morphologically distinct structures in the nucleus, especially in the interchromosomal domain (Cremer *et al.*, 1993). This is true both for a normal nucleus within a somatic cell and for the enormous nucleus of the amphibian oocyte. It is, therefore, not surprising that many attempts to characterize a nuclear matrix at a biochemical level reveal ribonucleoprotein as a major structural component (Mattern *et al.*, 1996). Nuclear structures such as the coiled body may in fact be sites of assembly of the ribonucleoprotein needed to process other ribonucleoprotein complexes.

One role for the chromosome in the overall organization of the nucleus that emerges from these studies is the segregation and dispersal of the DNA template within a particular territory or nuclear domain. There is a directionality in this organization, since telomeres are orientated towards the nuclear periphery and most active genes appear to have an interior location. There are suggestions that this organization might have consequences for the release of mRNA to the cytoplasm and subsequent utilization of the transcripts. This is an important topic for future investigation.

Molecular genetics defines disease genes involved in acute pro-myelocytic leukemia and Wilm's tumor. Expression of these genes leads to the accumulation of proteins in nuclear bodies or domains of unknown function: the PML nuclear body and the WT1 domain. Comparable compartmentalization is visualized for other regulatory molecules such as the transcription factors: the glucocorticoid and mineralocorticoid receptor (van Steensel *et al.*, 1996). These concentrations of specific proteins are dynamic and might represent sites to which particular signal transduction pathways are channelled within the nucleus. Alternatively this compartmentalization might reflect roles for these proteins that are yet to be defined. For example, a potential role for WT1 in post-transcriptional gene regulation emerges from the colocalization of this protein with speckles. The clear significance of these gene products for human disease should further stimulate research on novel functions for nuclear organelles.

Summary

The nucleus has a structure far removed from an amorphous bag of chromosomes. Recent applications of cell biology and molecular genetics have built an image of nuclear organization in which the molecular machines involved in transcription, RNA processing and replication assemble morphologically distinct nuclear organelles with defined functional properties. These observations indicate a very high level of structural organization for the various metabolic activities occurring within the nucleus. Novel regulatory functions may exist that are inherent to nuclear architecture itself. The nuclear components and structures are assembled and are utilized with a precise temporal and spatial order. Effective nuclear metabolism appears to require a high degree of organization. It is likely that much insight into both transcription and replication will follow from definition of what this organization is, and how it is assembled and regulated.

CHAPTER THREE

Chromatin and Nuclear Assembly

Our knowledge of chromatin structure is largely dependent on the analysis of relatively homogeneous populations of both large and small chromosomal fragments. The methodological approach has been to take the chromosome apart progressively and to examine its constituents. However, in order to understand a complex multicomponent structure completely, we must also be able to reassemble it from its constituents. Considerable progress has been made towards understanding the organization of chromosomes and nuclei through attempts to reconstruct them both *in vivo* and *in vitro*.

3.1 INTERACTIONS BETWEEN NUCLEAR STRUCTURE AND CYTOPLASM

Much of our understanding of nuclear assembly comes from experiments pioneered on the large eggs and oocytes of the frog *Xenopus laevis*. The large size and easy availability of these eggs and oocytes make them a particularly suitable target for the microinjection of macromolecules (Gurdon, 1974). However, there are significant physiological differences between eggs and oocytes. An oocyte is developing into an egg cell, and it is located in the ovary, closely surrounded by several thousand follicle cells and cannot be fertilized. During the 6 or more months of oogenesis the chromosomes of an oocyte are very actively transcribed, sometimes as lampbrush chromosomes (Section 2.4.3). Fully grown oocytes respond to the hormone gonadotropin by undergoing maturation. This involves the breakdown of the membrane and lamina surrounding the oocyte nucleus (the germinal vesicle), release of the oocyte from the ovary (ovulation), as well as the completion of the first meiotic division with arrest at the second meiotic metaphase. When released from the frog, the egg can be fertilized, and if so will develop very rapidly (3 days) into a free-swimming tadpole. This rapid development is due in part to the large stores of nuclear components sequestered in the egg during oogenesis. The great majority of microinjection experiments are carried out on fully grown oocytes, removed from the ovary of a female, or on recently fertilized eggs.

3.1.1 Nuclear transplantation, remodelling and assembly

The first experiments to broadly address directed alterations in nuclear architecture *in vivo* under controlled conditions were those that examined the consequences of introducing the nuclei of somatic cells into *Xenopus* eggs and oocytes. When the nucleus of a single somatic cell is injected into an enucleated egg, this nucleus will promote development of the egg into an embryo and sometimes into a tadpole or adult frog (Gurdon, 1974). However, the more differentiated the cell from which a donor nucleus is taken the more unlikely it is that correct development will proceed. For example, it has still not proven possible to obtain a normal adult animal by the transplantation of the nucleus of an adult somatic frog cell into a frog egg. These nuclei have, however, yielded swimming tadpoles with functional differentiated cells of most kinds. This approach has also successfully yielded transgenic *Xenopus laevis* embryos from eggs transplanted with nuclei from cultured somatic cells previously transformed with exogenous DNA (Kroll and Gerhart, 1994). In the past year, the research of Wilmut and colleagues have broken through this conceptual and technical barrier to successfully clone a mammal using nuclei of adult somatic cells from mammary epithelium transplanted into a sheep egg (Wilmut *et al.*, 1997). The significance of these results for our purposes is that they suggest that nuclei from differentiated somatic cells are totipotent (Di Bernardino, 1987). Totipotency requires that whatever genetic mechanisms operate to direct differentiation, they are completely reversible. This has important implications for the role of chromatin structure and *trans*-acting factor-DNA interactions in establishing and maintaining stable states of differentiated gene activity. Although in certain instances cell specialization is coupled to the loss of DNA from the eukaryotic cell or the irreversible rearrangement of DNA sequences, these phenomena are not generally observed (Klobutcher *et al.*, 1984; Hood *et al.*, 1985). Experiments in other systems including heterokaryons, cell hybrids and tumor cells, also point towards the reversibility of the differentiated state under certain conditions (Section 3.1.2).

The pioneering studies on the behaviour of somatic nuclei transplanted into eggs revealed that these nuclei swelled over 60-fold after injection. Furthermore, this change in nuclear volume correlated with the capacity of these nuclei to initiate DNA synthesis (Graham *et al.*, 1966). Subsequent experiments have suggested that the major failure of differentiated cell nuclei to fulfil a complete developmental programme after transplantation is a consequence of chromosomal damage. This appears to be due to the premature initiation of cell cycle directed changes in chromosomes not appropriately organized for these highly regulated events to proceed correctly (Di Bernardino, 1987). The orchestrated exchange of somatic nuclear proteins for egg cytoplasmic components takes time and it is the failure to effect the restructuring of chromatin, chromosomes and nuclei before cell division that most probably leads to chromosomal damage and the developmental abnormalities apparent in many nuclear transplant embryos. In this regard the mammalian cloning experiments provide additional insight (Wilmut *et al.*, 1997). These investigators made use of adult somatic cells that they synchronized in G0 – a quiescent state within the cell cycle. This state of quiescence is normally achieved by starving cells for serum, causing cells in G1 to leave the cell cycle. This exit can be reversed by adding back serum in culture, or evidently by transplanting a G0 cell nucleus into the egg. This might facilitate the remodelling of chromatin by attuning the nuclear and cytoplasmic cell cycles just before entry into S phase. DNA replication itself will further facilitate the disruption of regulatory nucleoprotein complexes. Therefore, by using the strategy of Wilmut and colleagues, an embryonic chromosomal structure might be established before cell division occurs, thereby preventing chromosomal damage. If this hypothesis is true, then simple manipulations to somatic cell nuclei that would facilitate nuclear remodelling would greatly facilitate the efficiency of animal cloning.

Early experiments examining specific gene activity within somatic nuclei transplanted into frog eggs revealed that nucleoli disappeared and previously active ribosomal RNA genes were inactivated (Gurdon and Brown, 1965). Nucleoli are an example of the compartmentalization of a particular biosynthetic event, the synthesis of rRNA, to a specific nuclear structure (Section 2.5.9). The inhibition of ribosomal

RNA transcription in eggs clearly demonstrated the capacity of egg cytoplasm to influence nuclear function. As development of the embryo containing the transplanted nucleus proceeds, the ribosomal genes were reactivated and nucleoli reappeared. This influence of cytoplasm on nuclear function could be better understood once it was shown that a considerable movement of proteins from the egg cytoplasm to the somatic nucleus occurred following transplantation (Merriam, 1969; Barry and Merriam, 1972). This movement was concomitant with nuclear swelling and with a significant reduction in the amount of heterochromatin within the nucleus. The capacity of nuclei to restructure in this way was much more effective for egg as opposed to oocyte cytoplasm (Gurdon, 1968, 1976).

Both the enlargement of somatic nuclei following transplantation into oocytes and their concomitant increase in transcriptional activity are found to be enhanced by rupture of the large oocyte nucleus (the germinal vesicle). This suggested that large stores of nuclear components (including the chaperone nucleoplasmin) were present in this enormous nucleus (50 μm diameter). HeLa cell nuclei can enlarge over 500 times in oocytes. Although substantial quantities of both histone and non-histone protein are taken up by the enlarging nuclei, over 75% of pre-existing nuclear protein was lost (Gurdon *et al.*, 1976). Together, these observations represent the evidence for a comprehensive remodelling of nuclear structures by *Xenopus* egg and oocyte cytoplasm.

These results were placed into the context of events at fertilization by the observation that *Xenopus* sperm nuclei increase in volume over 50 times in egg cytoplasm within 30 min of fertilization. The relatively slow increase in nuclear volume following transplantation of somatic nuclei into oocytes was shown to be coincident with an increase in transcriptional activity of the nuclei. Unlike eggs, nuclei transplanted into oocytes do not replicate their DNA. Transcriptional activity was reflected in an enlargement of the oocyte nucleoli and enhancement of ribosomal RNA synthesis (Gurdon, 1976).

Substantial progress in this research area followed from the development of *in vitro* systems competent to remodel nuclei. Lohka and Masui observed that gentle centrifugation of eggs led to the stratification of the egg contents (Lohka and Masui, 1983; Fig. 3.1). Following this it was relatively easy to isolate a fraction greatly enriched in egg cytoplasm. An important advantage of this protocol was that it enabled egg cytoplasmic preparations to be made that were essentially free of yolk. Yolk contains highly phosphorylated poly-peptides that exert a strong inhibitory effect on any process involving nucleic acid (Wolffe and Schild, 1991).

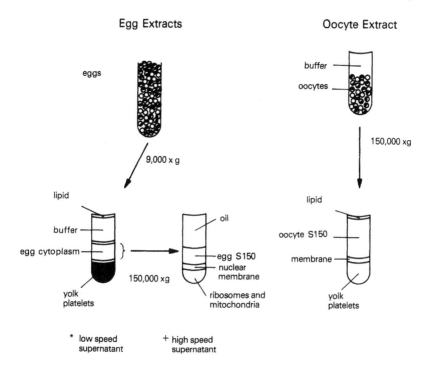

Figure 3.1. Preparation of chromatin assembly extracts from *Xenopus* eggs and oocytes. A single centrifugation step (oocyte) or two centrifugation steps (egg) are shown.

Lohka, Masui and colleagues found that *Xenopus* sperm nuclei incubated in *Xenopus* egg extracts decondensed and initiated DNA replication (Lohka and Masui, 1983, 1984; Fig. 3.2). Comparable systems are being developed using *Drosophila* (Berrios and Avilion, 1990; Ulitzer and Gruenbaum, 1989). Subsequent work has defined the decondensation process and subsequent assembly of 'normal nuclear architecture' as a multistage process. *X. laevis* sperm nuclei contain normal histones H3 and H4, but have most of their H2A and H2B and all of their H1 replaced by sperm-specific proteins (Risley, 1983; Dimitrov *et al.*, 1994). These sperm-specific proteins (z and y, Fig. 3.3) dissociate from the sperm chromatin in the *Xenopus* egg extract (Wolffe, 1989a; Philpott *et al.*, 1991). This dissociation may be facilitated by the phosphorylation of these small basic proteins, as occurs in mammalian sperm (Banerjee *et al.*, 1995). It also appears that nucleoplasmin facilitates both the removal of the sperm-specific proteins and the deposition of histones H2A/H2B (Philpott and Leno, 1992). Both histone H2A and the variant H2A.X are deposited

5 min	30 min	120 min

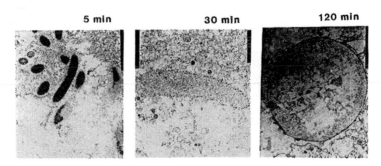

Figure 3.2.   Remodelling of *Xenopus* sperm nuclei during incubation in the *Xenopus* egg extract.
Electron micrographs of sperm nuclei incubated in the extract for the times indicated are shown.

by nucleoplasmin. H2A.X is derived from protein synthesized and stored in the developing oocyte. This synthesis is distinct from the normal cell-cycle regulated synthesis of histones. Two other proteins aside from the core histones are incorporated into sperm chromatin during assembly of the pronucleus, the linker histone B4 and HMG2 (Fig. 3.3). Both of these proteins are present in amounts such that the majority of nucleosomes would be expected to contain a molecule of B4 and/or HMG2. The expression of the histone H1-like protein, B4, is restricted to oogenesis and early embryogenesis (Smith *et al.*, 1988; Dimitrov *et al.*, 1993; Hock *et al.*, 1993; Cho and Wolffe, 1994). If B4 is depleted from the *Xenopus* egg extract, chromatosome length DNA (166 bp) no longer accumulates during micrococcal nuclease digestion of pronuclear sperm chromatin. This result indicates that histone B4 is functioning like a true linker histone in the paternal pronucleus.

HMG1/2-like molecules accumulate in *Xenopus* oocytes (Weisbrod *et al.*, 1982; Kleinschmidt *et al.*, 1983). This serves as a storage pool for the protein incorporated into the paternal pronucleus. It is possible that HMG1 and 2 might be capable of replacing histone H1 in chromatin (Jackson *et al.*, 1979). In normal somatic nuclei, HMG1 and 2 are associated with a relatively minor fraction of chromatin (<5%) (Goodwin *et al.*, 1977; Isackson *et al.*, 1980). The discovery that sperm chromatin contains significantly more abundant HMG1/2-like molecules makes a structural role for HMG1 and 2 much more probable. The deficiency of normal somatic histone H1, and the abundance of histone B4 and HMG2 in the pronucleus, might reflect a functional requirement for a decondensed chromatin structure to facilitate rapid replication and chromatin assembly.

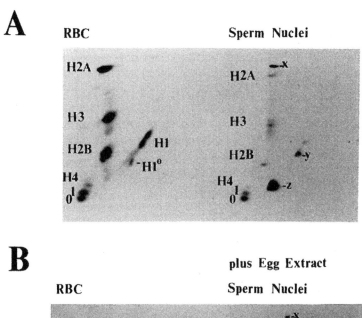

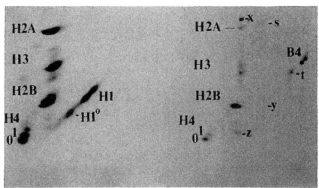

Figure 3.3. Remodelling of *Xenopus* sperm chromatin into the paternal pronucleus.

Xenopus red blood cell (RBC) histones are used as markers. The composition of sperm chromatin before (A) and after (B) remodelling in *Xenopus* egg extract is shown: t, HMG2; s, histone H2A.X; z and y, sperm-specific basic proteins; x, a dimer of H3 generated artefactually during sample preparation (see Dimitrov *et al.*, 1994 for a similar experiment and details).

It has proven possible to make use of the remodelling of sperm chromatin in the *Xenopus* egg extract to examine whether particular proteins are necessary to assemble higher order aspects of chromosome structure. Depletion of histone B4 from the extract does not influence either pronuclear assembly or the condensation of chromatin into mitotic chromosomes (Ohsumi *et al.*, 1993; Dasso *et al.*, 1994a). This result dramatically illustrates the fact that linker histones are not necessary for chromosome condensation (Section 2.5.3). It also implies that either the 30-nm chromatin fibre does not require linker histones for assembly or that aspects of chromatin fibre structure dependent on linker histones are not necessary for the assembly of chromosomes. This represents the first stage of decondensation of sperm chromatin. The second stage requires the assembly of a nuclear membrane on the surface of the remodelled chromosomes. Once a nuclear membrane is assembled, the nucleus swells to the normal volume expected for a nucleus within the *Xenopus* early embryo (see Fig. 3.1). What determines this volume is unknown.

A major change associated with the remodelling of somatic nuclei in *Xenopus* egg cytoplasm is the selective release from chromatin of somatic variants of linker histones H1 and H1°. This selective release is particularly surprising because the proteins histone B4 and HMG1 are rapidly and efficiently incorporated into chromatin (Fig. 3.4). Investigating the molecular basis of this remodelling event provides some insight into the special features of the egg that make it possible (Dimitrov and Wolffe, 1996). Histone B4 and HMG1 form much less stable complexes with chromatin than normal somatic histone variants such as H1, H1° and H5 (Nightingale *et al.*, 1996; Ura *et al.*, 1996). This relative instability most probably reflects the fact that histone B4 is much less basic than H1, H1° and H5 (Doenecke and Tonjes, 1986; Smith *et al.*, 1988; Wells and Brown, 1991; Dimitrov *et al.*, 1993; Khochbin and Wolffe, 1994). Thus the replacement of H1 and H1° with B4 and HMG1 is at variance with their relative affinities for naked DNA or mononucleosomal DNA (Nightingale *et al.*, 1996).

Histone B4 is stored in the *Xenopus* oocyte, and is incorporated into chromatin through the mid-blastula transition (Dimitrov *et al.*, 1993, 1994). However, depletion of histone B4 from the egg extract does not influence the efficiency of release of histone H1 from the chromatin of somatic nuclei demonstrating that simple competition between histone B4 and histone H1 for association with nucleosomal DNA cannot explain the replacement phenomenon. The decrease in basic character and the destabilization of DNA binding concomitant with linker histone phosphorylation (Hill *et al.*, 1991) might have been a major driving force for their release from chromatin. However, phosphorylation of the somatic linker histones is not required for release. These results lead to the hypothesis that protein–protein interactions, perhaps with molecular chaperones stored in the egg might have a dominant role in determining the stability of protein incorporation into chromatin or dissociation from chromatin.

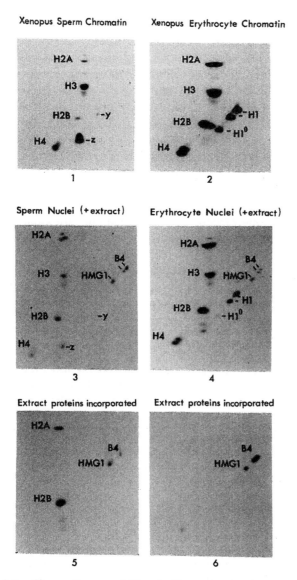

Figure 3.4. Chromatin remodelling in *Xenopus* egg extract.
The protein composition of *X. laevis* sperm chromatin (Panel 1) and *X. laevis* erythrocyte chromatin (Panel 2) before remodelling is shown following two-dimensional gel electrophoresis and staining with Coomassie Blue. The core histones H2A, H2B, H3 and H4 are indicated as are linker histones H1 and H1°. Within sperm chromatin two sperm specific basic proteins are indicated by z and y. Panels 3 and 4 show remodelled sperm and erythrocyte chromatin after addition of the egg extract. The location of histone B4 and HMG1 is shown. Panels 5 and 6 show the newly synthesized radiolabelled proteins present in the egg extract that are taken up by sperm chromatin and erythrocyte chromatin respectively during remodelling in the extract. The positions of H2A, H2B, B4 and HMG1 are indicated.

Nucleoplasmin again serves as the archetypical molecular chaperone in remodelling somatic nuclei. Nucleoplasmin is phosphorylated on maturation of an oocyte to an egg (Sealy *et al.*, 1986). The phosphorylated form of nucleoplasmin is more efficient in removing sperm specific basic proteins concomitant with the decondensation of sperm chromatin (Leno *et al.*, 1996). Nucleoplasmin has a major role in the removal of somatic linker histones from chromatin in the egg. Egg cytoplasm depleted in nucleoplasmin no longer releases H1 from chromatin, readdition of nucleoplasmin to the depleted extract restores H1 release. The presence of nucleoplasmin in *Xenopus* eggs and the specialized function of this protein in associating with arginine rich sperm-specific basic proteins (Hiyoshi *et al.*, 1991) or somatic linker histone variants potentially explains the capacity of the *Xenopus* egg to render sperm and somatic nuclei capable of new transcriptional programs (Gurdon, 1974, 1976).

Chromatin assembly has the dual function of packaging DNA into a compact structure and of contributing to transcriptional regulation. Recent attention has focused on molecular machines that remodel chromatin structure and render promoters accessible to the transcriptional machinery (Varga-Weisz *et al.*, 1995; Wall *et al.*, 1995; reviewed by Kingston *et al.*, 1996; Section 2.5.4). A simple way to remodel chromatin is to reverse the process of nucleosome assembly. Histone H1 is the last protein to be sequestered during chromatin assembly *in vivo* (Worcel *et al.*, 1978). Thus the dissociation of histone H1 would represent the first step in chromatin remodelling. Removal of histone H1 has been observed during the activation of chromatin during glucocorticoid receptor-induced transcription of the mouse mammary tumor virus long terminal repeat (Bresnick *et al.*, 1992; see Section 4.2). Such a transition in chromatin composition would directly increase the mobility of histone octamers relative to DNA sequence and facilitate transcription (Ura *et al.*, 1995). The removal of somatic linker histone variants from chromatin mediated by nucleoplasmin provides a mechanism by which global chromatin remodelling might be achieved, leading somatic nuclei to acquire transcriptional pluripotence following transplantation into the egg.

The remodelling of sperm and somatic nuclei in *Xenopus* egg extracts strongly resembles events that occur in the living egg. However, unlike the *in vivo* case, *in vitro* it is possible to dissect the process in some detail. Methodologies have been pioneered by examining the reorganization of specialized nuclei in *Xenopus* egg extracts. An important result is that it is also possible to assemble nuclei *de novo* on naked DNA. Newport and colleagues have followed this process in some detail (Newport, 1987). Building upon the observation that the injection of bacteriophage λ DNA into *Xenopus* eggs resulted in the efficient assembly of nuclei, the analogous process was followed *in vitro*. Immediately after addition to the extract, the λ DNA appeared as long decondensed strands. After 20–30 min, the DNA was condensed into structures resembling the chromatin (30-nm) fibre and DNA became even more condensed over the following 40 min. At this time a nuclear envelope appeared around the chromatin and nuclear volume increased 20–30-fold concomitant with decondensation of the chromatin (Fig. 3.5).

Topoisomerase II inhibitors were found to prevent nuclear formation by preventing the formation of highly condensed chromatin. Laemmli and colleagues specifically depleted topoisomerase II from *Xenopus* extracts using antibodies (Section 2.4.2). In these depleted extracts the capacity of the chromatin fibre to condense was severely inhibited. Readdition of purified topoisomerase II (from yeast) to the depleted extract restored its capacity to convert decondensed chromosomes into their mitotic state. Titration of the amount of topoisomerase II revealed that the high topoisomerase II concentration of mitotic extracts needs to be matched with the yeast enzyme in order to rescue the assembly potential of extracts depleted for this enzyme. One topoisomerase II protein per 2 kb of DNA leads to partial condensation, whereas three times as much leads to complete condensation of chromosomes (Luke and Bogenhagen, 1989; Adachi *et al.*, 1991). These experiments dramatically confirm the central role of topoisomerase II in mediating the DNA rearrangements necessary for chromosome folding (see also Hirano and Mitchison, 1991).

The next experiments examined the assembly of the nuclear lamina and membrane. *Xenopus* embryonic nuclei appear to only contain a single lamin species, L_{III} (Stick and Hansen, 1985). Presumably this simplifies the structure and properties of the lamina. DNA at the highly condensed stage of nuclear reconstitution does not initially have detectable amounts of lamin associated with it. However, eventually a lamina and nuclear envelope form around the condensed chromatin and appear to promote the subsequent swelling of the nuclear structure. Nuclear membranes are stored in the *Xenopus* egg, free of the lamina and chromatin as 'annulate lamellae' (Fig. 3.6). Assembly of the nuclear membrane appears to be an essential prerequisite for subsequent nuclear events such as the initiation of DNA replication (Blow and Laskey, 1986; Leno and Laskey, 1991). The nuclear lamina is presumed to have a key role in determining chromosomal architecture as a component of the nuclear scaffold (Section 2.4.2).

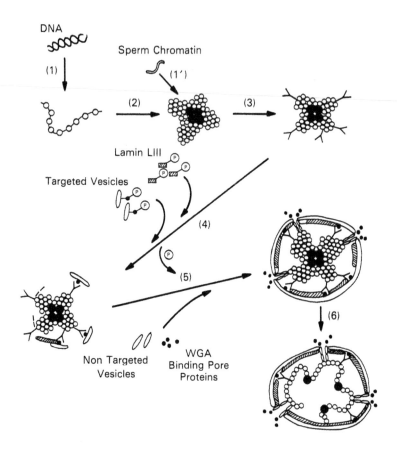

Figure 3.5. A sequential model for DNA organization and nuclear membrane formation in a *Xenopus* egg extract.

In step (1) protein free DNA is assembled into nucleosomes. Step (2) leads to rearrangement and condensation of chromatin in a process involving topoisomerase II. At this step it is assumed that scaffold proteins bind and organize the chromatin into loop domains. In step (1′) sperm chromatin is rearranged and the protamines are exchanged for histones H2A/H2B. Step (3) corresponds to modification of chromatin such that it will bind targeted vesicles in step (4). The surface of chromatin becomes coated with newly assembled immature nuclear pore complexes or 'pre-pores'. In step (4), the chromatin receptor on the vesicles becomes dephosphorylated and can then interact with its binding site on chromatin. These binding sites occur on average every 100 kb on DNA. Lamins upon dephosphorylation polymerize to form the nuclear lamina. The process of lamina formation and membrane formation are independent in this system. In step (5) the membrane vesicles fuse with non-targeted vesicles to form a double membrane envelope. Fusion of vesicles requires GTP. Nuclear pre-pore complexes mature and include the N-acetylglucosamine modified pore proteins. In step (6) the nucleus grows further and chromatin decondensation occurs in a process probably mediated by topoisomerase II activity.

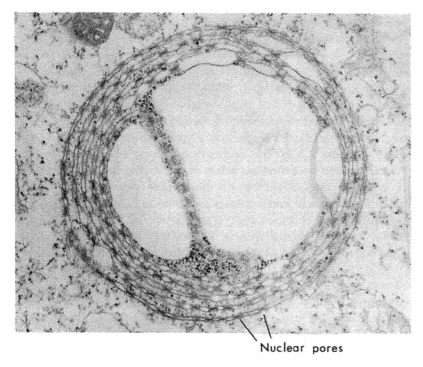

Nuclear pores

Figure 3.6. Electron micrograph of annulate lamellae in *Xenopus* eggs.
These are storage structures containing predominantly nuclear membrane and nuclear pores.

Aside from bacteriophage λ, other prokaryotic DNA templates are assembled into nuclear structures by the *Xenopus* egg extract. Importantly, it is possible to assemble on these apparently non-specific DNAs the appropriate nuclear architecture to facilitate the initiation and completion of semiconservative DNA replication. Laskey and colleagues have demonstrated using confocal scanning laser microscopy that replication takes place at discrete chromosomal sites in *Xenopus* sperm nuclei incubated in the egg extract. The number and nuclear distribution of these sites is similar to those observed for somatic cells in tissue culture. Each site (100–300 per nucleus) contains up to 300–1000 tightly clustered replication forks. Laskey demonstrated that comparable replication factories are assembled on bacteriophage λ DNA in the egg extract. These results establish the important fact that this highly regulated functional chromosome structure can be established *de novo* independent of any precise eukaryotic DNA sequence or pre-existing chromosomal or nuclear structure (Cox and Laskey, 1991, Section 2.5.9).

The nuclei assembled from *Xenopus* sperm chromatin have proven remarkably useful in exploring the signal transduction pathways that regulate nuclear events during the cell cycle. It is possible to prepare egg extracts that recapitulate the cell cycle *in vitro* (Hutchison *et al.*, 1988; Murray and Kirschner, 1989). These egg extracts continually oscillate between S-phase and M-phase (Section 2.5.3) periodically converting the key regulatory protein kinase (p34^{cdc2} – cyclin B) into an active form so that it functions as a *mitosis promoting factor* or MPF kinase (Section 2.5.3). The addition of sperm chromatin to these extracts leads to the following series of events: the sperm chromatin is decondensed, remodelled into the pronucleus and replicates. After 90 min, the extract will enter mitosis as visualized by nuclear envelope breakdown and chromosome condensation. This recapitulation of normal nuclear events during the cell cycle allows the signal transduction events controlling the entire process to be dissected. There are a number of checkpoint controls that regulate the onset of mitosis. One of these involves the necessary completion of DNA replication. Failure to complete replication prior to chromosome condensation will lead to severe chromosomal damage. The onset of mitosis is inhibited by the presence of unreplicated DNA (Dasso and Newport, 1990). This coupling is sensitive to drugs (e.g. caffeine and okadaic acid) that influence kinase activity. Unreplicated DNA acts to maintain the MPF kinase in a phosphorylated, inactive state. The signalling system that leads to MPF phosphorylation depends on a protein known as *regulator of chromosome condensation 1* (RCC1) (Dasso, 1993). In *Xenopus*, RCC1 is an abundant chromatin-binding protein present at approximately one molecule per nucleosome (Dasso *et al.*, 1992). If nuclei are assembled in extracts depleted of RCC1 they do not enlarge to the normal extent and they are defective for replication. RCC1 interacts not only with chromatin, but also with a protein TC4/Ran (Bischoff and Ponstingl, 1991). TC4/Ran is a *ras*-like GTPase which is mainly nuclear. RCC1 acts as a guanine nucleotide exchange protein for TC4/Ran, suggesting that this activity might function in the checkpoint control signalling pathway. This hypothesis is confirmed by incubation of egg extracts with mutant TC4/Ran proteins that block the guanine nucleotide exchange activity of RCC1. These mutant TC4/Ran proteins prevent nuclear assembly and DNA replication (Dasso *et al.*, 1994b). These results demonstrate that TC4/Ran and RCC1 play a critical role in several nuclear functions independent of the regulation of MPF activity. It will be of great interest to determine how the activity of RCC1 is controlled and what other proteins interact with Ran in the control of nuclear assembly and disassembly during the cell cycle.

Summary
The first evidence for a remodelling of chromatin structure with important functional consequences came from the microinjection of somatic nuclei into the oocytes and eggs of amphibians. Cell-free preparations from these cells have proven extraordinarily useful to dissect the determinants of nuclear assembly. The assembly of nuclei *in vitro* does not require any specific eukaryotic DNA sequence, but does require DNA. A nucleus is built from the initial assembly of simple structures to the final assembly of more complex ones. The assembly of chromatin precedes the assembly of a nuclear lamina and nuclear membrane. Not until the nuclear membrane is complete will the nucleus swell to its final volume. Once the complete nucleus is assembled, regulated events such as the initiation of DNA replication can take place, even on prokaryotic DNA templates. Proteins such as RCC1 and Ran play an important role in signal transduction pathways co-ordinating nuclear events with the cell cycle.

3.1.2 Heterokaryons

An approach conceptually related to the reprogramming of somatic nuclei following their introduction into the cytoplasm of a *Xenopus* egg or oocyte, is the fusion of two distinct somatic cells to form a single cell with two different nuclei bathed in a common cytoplasm (a heterokaryon). Early experiments had shown that gene expression in the donor cells could be dramatically changed following formation of a heterokaryon. This work was among the first to suggest the existence of specialized *trans*-acting factors in eukaryotes capable of repressing or inducing the expression of differentiated functions, i.e. regulating differential gene expression (Ephrussi, 1972; Ringertz and Savage, 1976). A common observation was that a gene normally only active in a differentiated cell was inactivated upon fusion with a different differentiated or an undifferentiated cell. Somatic cell hybrids in which the two nuclei of the heterokaryon fuse often lose chromosomes in culture. This type of phenomenon led to the attribution of individual repressive effects to particular chromosomes. Very occasionally gene activation was observed in cell fusion experiments. For example, extensive experiments in heterokaryons have clearly shown that fusion of one differentiated cell (a muscle cell) with a different cell in which muscle genes are not normally expressed (a human amniocyte) leads to the activation of muscle genes in the amniocyte (Blau *et al.*, 1983). This result suggests that factors capable of activating genes can either exchange freely between nuclei or also exist in excess within the cytoplasm. The activation of differentiated genes in a non-differentiated cell is rapid (within 2 days) and does not require cell division or DNA replication. This implies that genes can be activated (at some level) without requiring replication events. This result is identical in principle to the activation of genes following the introduction of somatic nuclei into *Xenopus* oocyte cytoplasm (Section 3.1.1).

The maintenance of specialized cellular phenotypes through a dynamic interplay between positive and negative regulatory molecules could involve either direct interactions by complementing a particular deficiency in one of the cell types in a heterokaryon (Baron, 1993; Blau, 1992), or it could involve indirect effects. Such indirect effects might occur when a positive regulator induces other cell specific transcription factors that in turn might activate a diverse group of downstream genes (Hardeman *et al.*, 1986). This latter mechanism appears to operate when erythroid cells are fused with non-erythroid cells (Baron and Maniatis, 1986, 1991; Baron and Farrington, 1994). Certain experiments fuse erythroid cells with embryonic stem cells deficient in a key transcriptional regulator of the globin genes (GATA1) (Evans and Felsenfeld, 1989, 1991), yet the nuclei of the embryonic stem cells can still be reprogrammed to express their globin genes in the heterokaryons. This indicates that erythroid cells contain the complement of factors necessary not only to activate the globin genes but also upstream regulators such as GATA1 (Baron and Farrington, 1994).

Experimental results with heterokaryons and *Xenopus* eggs have been interpreted as providing evidence for a continuous regulation of a plastic differentiated state. Implicit in this model is the idea that all genes are continually regulated by *trans*-acting factors that can either activate or repress genes (Chiu and Blau, 1984; Blau *et al.*, 1985; Blau and Baltimore, 1991). For certain genes this is clearly true; however, it has also been shown that a considerable remodelling of chromosomal structure occurs in *Xenopus* egg and oocyte cytoplasm. A similar albeit less impressive remodelling of chromosomes occurs in heterokaryons. For example, the nuclei of chicken erythrocytes consist predominantly of heterochromatin containing the specialized linker histone H5 (Section 2.5.6). In heterokaryons formed by fusion of chicken erythrocytes with proliferating mammalian cells, the chicken erythrocyte nuclei once again become transcriptionally active. This process is accompanied by decondensation of chromatin, enlargement of the nucleus and the appearance of nucleoli. Transcription and replication of these nuclei are activated. The enlargement of the chicken erythrocyte nucleus is due to a massive, but selective, uptake of mammalian nuclear proteins including RNA polymerases. Histone H5 is partially lost from the chicken erythrocyte nucleus and partially taken up by the mammalian nucleus in the heterokaryon (Ringertz *et al.*, 1985). Histones H2A and H2B also exchange under these circumstances, but not histones H3 and H4. These results might be expected considering the relative affinity of the histones for DNA and their organization in the nucleosome (Sections 2.2.2 and 2.2.4). This reorganization is independent of replication. It is therefore clear that chromosome structure is quite dynamic, with some histones (H1, H2A, H2B) continually exchanging with a free pool of proteins in the cytoplasm.

Several experiments suggest that at physiological ionic strength histone H1 rapidly exchanges into and out of the chromatin fibre (Caron and Thomas, 1981). Presumably this dynamic property of the chromatin fibre and the nucleosome would eventually allow many *trans*-acting factors to gain access to their cognate DNA sequences (Section 4.2.3). An important and unresolved question is whether this access is unlimited or whether access is restricted by chromosomal organization. A quantitative determination of whether the level of transcriptional activity following *de novo* activation of a gene in a heterokaryon is identical to the transcription of the same gene in a differentiated cell has not yet been made. Of course in *Xenopus* egg cytoplasm, such equivalent activation does occur in order for correct development to proceed through to the tadpole stage; however, here nuclear reprogramming is more rapid and is likely to be facilitated by DNA replication (Section 4.3.1).

Summary
Heterokaryons are formed following the fusion of two different cells such that two distinct nuclei exist in a common cytoplasm. They have been useful in demonstrating a plasticity in gene expression. Previously active or repressed genes can have their expression states changed following cell fusion. Chromatin structure is revealed to be able to change in a reversible way.

3.2 CHROMATIN ASSEMBLY

The molecular mechanisms that underly gene regulatory phenomena depend on the capacity not only to replicate DNA but also to duplicate chromatin and chromosomal structure. Most chromatin and chromosome assembly occurs during S-phase. Insights into chromatin assembly might also point to the molecular mechanisms necessary to reverse the process for example during transcriptional activation.

3.2.1 Chromatin assembly on replicating endogenous chromosomal DNA *in vivo*

A large fraction of histone synthesis in somatic cells (unlike *Xenopus* oocytes and egg) is coupled to DNA replication during the S-phase of the cell cycle. As the DNA content of the chromosome is doubled, so must the protein component be duplicated to reconstruct two daughter chromosomes. A normal stoichiometry of the core histones within chromatin is essential for cell division to proceed correctly (Meeks-Wagner and Hartwell, 1986). Several experiments in which newly synthesized chromatin (nascent chromatin) was fractionated from pre-existing chromatin (old chromatin) have shown that chromatin assembly *in vivo* is a staged process. Worcel exploited the increased susceptibility of nascent chromatin to nucleases to fractionate nascent chromatin from *Drosophila* embryos. Newly synthesized DNA was enriched for newly synthesized histones H3 and H4, while newly synthesized histones H2A/H2B and H1 associate with chromatin that had properties similar to those of bulk non-replicating chromatin. Worcel concluded that newly synthesized histones associated with newly synthesized DNA in a sequential order: histones H3 and H4 are deposited first, then histones H2A and H2B, and finally histone H1 (Senshu *et al.*, 1978; Worcel *et al.*, 1978; see also Cremisi and Yaniv, 1980; Fig. 3.7).

Experiments with mammalian cells have reached remarkably similar conclusions with respect to the staged assembly of chromatin *in vivo*. The process of chromatin maturation on newly synthesized DNA takes several minutes even in a rapidly proliferating mammalian cell. Nascent DNA is assembled into a structure that is more sensitive to nucleases than mature chromatin. The loss of this nuclease-sensitive conformation takes over 10–20 min (Kempnauer *et al.*, 1980; Cusick *et al.*, 1983). Chalkley and Jackson suggested that the initial rapid deposition of histones H3 and H4 on newly synthesized DNA reflects the nuclease-sensitive stage, whereas the subsequent deposition of histones H2A and H2B correlates with the appearance of regular nucleosomal arrays and nuclease resistance (Smith *et al.*, 1984). The sequential sequestration of histones is clearly related to the structure

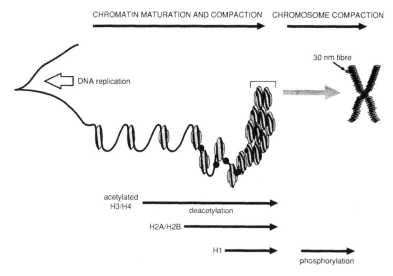

CHROMATIN MATURATION AND COMPACTION   CHROMOSOME COMPACTION

30 nm fibre

DNA replication

acetylated
H3/H4
                    deacetylation

H2A/H2B

H1
                    phosphorylation

Figure 3.7. Chromatin assembly on replicating DNA.
Maturation of chromatin is shown from left to right. Binding of $(H3/H4)_2$
tetramers to nascent DNA is followed by H2A/H2B dimers and finally
histone H1. Phosphorylation of histone H1 correlates with chromosome
compaction at mitosis.

of the nucleosome, since histones H3 and H4 form the core of the
structure, whereas histones H2A and H2B bind at the periphery of the
nucleosome and histone H1 can only associate in its proper place once
two turns of DNA are wrapped around the core histones (Sections
2.2.2, 2.2.3 and 2.2.4). We see later that this two-stage maturation
process has important implications for the accessibility of *trans*-acting
factors to newly synthesized DNA (Section 4.3.1).

Newly synthesized histones that are destined for deposition on
nascent DNA possess transient post-translational modifications that
distinguish them from the 'old histones' of bulk chromatin. One
modification studied in some detail is the diacetylation of histone H4
at specific lysine residues. This is true not only for newly synthesized
H4 in mammalian cells, but is also the form in which histone H4 is
stored in *Xenopus* oocytes, the form present in cleavage stage
embryonic chromatin in the sea urchin and in replicating nuclei in
Tetrahymena (Csordas, 1990). Approximately 30–60 min after deposition during chromatin assembly, the diacetylated histone H4 is
deacetylated to its mature form. Histone H3 is also acetylated, but is
deacetylated more rapidly (1–2 min) than histone H4 in nascent
chromatin. Acetylation may actually facilitate deposition of histones
H3 and H4 at the replication fork. Moreover, acetylation of the histone

tails may partially account for the lack of compaction of nascent chromatin, the reduction in histone H1 content, the accessibility of this chromatin to nucleases and perhaps to *trans*-acting factors (Section 2.5.2). Experiments in yeast indicate that the histone H3 and H4 N-terminal tails are important for the physiological assembly of chromatin (Ling *et al.*, 1996). Although deletion of both N-termini is lethal, the presence of either tail domain allows nucleosome assembly and cell viability. Grunstein and colleagues propose that acetylation of the H4 tail may provide a signal to allow the appropriate targeting and assembly of newly synthesized histones into nucleosomes.

Annunziato and colleagues have examined the significance of acetylation for nascent chromatin using the deacetylase inhibitor sodium butyrate. This inhibitor has no effect on the DNA replication process; however, newly synthesized histone H4 retains its nascent, diacetylated form. Normally, nascent chromatin would mature to a nuclease resistant form after 10–20 min. In the presence of butyrate, nascent chromatin never achieves the nuclease resistance of bulk chromatin. The nascent chromatin does, however, assemble correctly spaced nucleosomes, demonstrating that acetylated histones can be assembled into chromatin resembling that found in a normal chromosome. Subsequent experiments have shown the observed difference in nuclease sensitivity between nascent and bulk chromatin to be related to the binding of histone H1 (Perry and Annunziato, 1989). Histone H4 deacetylation appears to be required for this linker histone to bind *in vivo* within newly assembled chromatin (but see Ura *et al.*, 1994, 1997). Consistent with other observations (Section 3.4.2) these results demonstrate that histone H1 is not required for the assembly of spaced nucleosomal arrays.

Histone H1 itself is either unmodified or moderately phosphorylated in nascent chromatin. As chromatin matures, histone H1 becomes more highly phosphorylated (Jackson *et al.*, 1976). As we have discussed (Section 2.5.3), the effect of phosphorylating histone H1 is most likely to weaken its binding to linker DNA by altering the level of charge neutralization of linker DNA. Consequently, we might expect the dephosphorylated histone H1 to exhibit a tighter interaction with DNA even though the highly acetylated state of histone H4 might prevent correct interaction with the core histones.

An important consideration for the assembly of chromatin within the endogenous chromosome is the fate of pre-existing chromatin structures (Section 4.3.2). A major question initially concerned whether histones present in the nucleosome remained together on nascent DNA and whether nucleosomes were randomly or conservatively segregated to daughter DNA strands. DNA replication requires the transient unwinding of duplex parental DNA into two single-stranded regions. Although histones will associate with single-stranded DNA (Almouzni *et al.*, 1990a) they do not assemble nucleosomes. This property coupled to the competing protein–DNA interactions involved in DNA synthesis at the replication fork probably accounts for nucleosome disruption.

The histone octamer is not stable at physiological ionic strengths in the absence of DNA (Eickbusch and Moudrianakis, 1978). Jackson (1987, 1990) determined that a substantial fraction of histone octamers fell apart during replication into dimers (H2A, H2B) and tetramers ((H3, H4)$_2$). Histones released from the parental chromatin during replication *in vitro* can easily be sequestered on to competitor DNA (Gruss *et al.*, 1993). *In vivo* these histones are sequestered on to daughter DNA molecules close to the replication fork (Sogo *et al.*, 1986; Perry *et al.*, 1993).

Newly synthesized dimers (H2A, H2B) can be sequestered together with old tetramers from pre-existing nucleosomes, mixing old and new histones into a single structure. With a certain frequency old tetramers will mix with old dimers (Leffak, 1984). However, the disruption of pre-existing nucleosomal structure at the replication fork, coupled to dissociation of the histones from DNA, is entirely consistent with the generally dispersive segregation of these histones to both daughter DNA duplexes (Cusick *et al.*, 1984; Sogo *et al.*, 1986; Burhans *et al.*, 1991; Fig. 3.8). Importantly, the incorporation of pre-existing histone tetramers ((H3, H4)$_2$) into nascent chromatin provides a means of maintaining and propagating a stable state of gene activity (Section 4.3.2). The old H3 and H4 present in the nascent chromatin retain their pre-existing post-translational modification state (Perry *et al.*, 1993). This will differ from that of newly synthesized H3 and H4 and can potentially influence subsequent transcription of the associated DNA (Section 4.2.3). The dispersive segregation of 'old' histones coupled to maintenance of their pre-existing states of modification provides a molecular mechanism whereby the potential for transcriptional activity or repression might be propagated through replication. This type of effect is known as an epigenetic imprint.

Not all chromatin assembly in the cell has to occur during replication. Normally, the vast majority of histone synthesis is coupled to DNA synthesis. However, Bonner, Zweidler and colleagues identified specific variants of histones H2A and H2B encoded by distinct genes (e.g. H2A.X, Section 2.5.1) that are synthesized in G1 and G2 as well as in S-phase. This constitutes a basal level of histone synthesis which varies from organism to organism. In mammalian

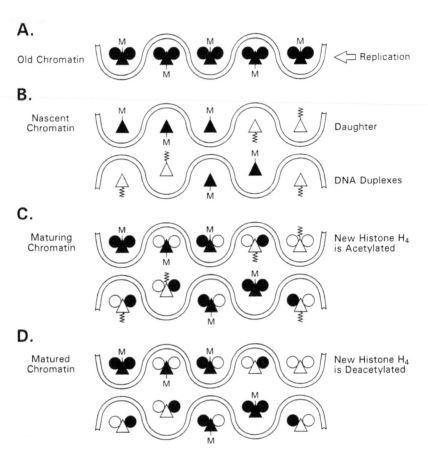

A.

Old Chromatin

Replication

B.

Nascent Chromatin

Daughter

DNA Duplexes

C.

Maturing Chromatin

New Histone H₄ is Acetylated

D.

Matured Chromatin

New Histone H₄ is Deacetylated

Figure 3.8. Dispersal and reassembly of old histones (black) during replication.
New histones are the open symbols. Triangles are (H3, H4)₂ tetramers, circles are (H2A, H2B) dimers, acetylation is indicated by the zig-zag lines. Potential modifications of the old histones are indicated (M).

cells it can constitute approximately 4% of total histone production. The histones synthesized in G1 enter chromatin in the absence of replication, demonstrating an active and efficient exchange process with pre-existing histones (Zweidler, 1980; Bonner *et al.*, 1988; Section 3.1.2). Chalkley and colleagues have shown that histones H1, H2A and H2B synthesized in the presence of inhibitors of DNA synthesis will exchange with pre-existing histones in nucleosomes whereas histones H3 and H4 will not (Louters and Chalkley, 1985). It has been suggested that this exchange is facilitated by transcription (Jackson, 1990).

Summary
Chromatin assembly *in vivo* is coupled to replication. It is possible to divide this process into two stages: the initial deposition of histones H3/H4 followed by the sequestration of histones H2A/H2B and H1. 'Old' histones present in pre-existing nucleosomes mix with newly synthesized histones in nascent chromatin. Newly synthesized histone H4 is acetylated when it is initially incorporated into chromatin, and the progressive deacetylation of H4 correlates with the sequestration of histone H1 and chromatin maturation.

3.2.2 Chromatin assembly *in vitro*

Major advances in understanding chromatin assembly processes *in vitro* followed from the discovery that the *Xenopus* egg contained large stores of histones which could provide a source of material for chromatin assembly. It was possible to prepare extracts of *Xenopus* eggs, mix in purified homogeneous DNA and assay for chromatin assembly. The definition of the chromatin structure of the SV40 minichromosome provided a reference product for assaying nucleosome assembly on a small closed circular DNA molecule. Laskey and colleagues prepared homogenates of *Xenopus* eggs in which SV40 DNA would be assembled into a minichromosome under apparently physiological conditions. The assembly of chromatin was found not to be co-operative, nor did it require synthetic processes such as DNA replication or protein synthesis. Chromatin assembly on double-stranded DNA was, however, disturbingly slow requiring over an hour to generate a nucleosomal array. *In vivo* chromatin assembly would have to be much more efficient as in the early embryo the whole genome is assembled into chromatin within a cell cycle time of less than 30 min. Biochemical fractionation of the extract led to the recognition that at least three protein components were required for chromatin assembly on non-replicating DNA: histones, topoisomerases and nucleoplasmin (Laskey *et al.*, 1978; Earnshaw *et al.*, 1980; Laskey and Earnshaw, 1980).

Nucleoplasmin is an acidic thermostable protein which promotes nucleosome assembly from purified histones and DNA, but only when present in excess over histones. As we have discussed (Section 3.1), nucleoplasmin has a much more specialized storage role for histones H2A and H2B. It releases these histones *in vivo* when it is phosphorylated following maturation of the oocyte to the egg. More importantly, it exchanges histones (H2A, H2B) for sperm-specific basic proteins during the remodelling of *X. laevis* sperm nuclei (Philpott and Leno, 1992) (Section 3.1.1). The interaction of nucleoplasmin with histones in the absence of DNA led to the use of the term chaperone to describe its properties. In effect, nucleoplasmin prevented the uncontrolled association of histones with the double helix. Nucleoplasmin is the most abundant protein in the nuclei of *Xenopus* oocytes representing up to 10% of the total nuclear protein (Krohne and Franke, 1980; Mills *et al.*, 1980). Nucleoplasmin is rich in acidic amino acids (~20% aspartic and glutamic acid), and the phosphorylation that occurs on oocyte maturation makes the protein even more acidic. The active form of nucleoplasmin for nucleosome assembly is a pentamer of 22 kDa subunits. Interestingly, nucleoplasmin does not appear to interact directly with DNA or chromatin, but will bind histones *in vitro*.

Nucleoplasmin was cloned and sequenced revealing that the acidic amino acids form several clusters (Dingwall *et al.*, 1987). One region contains 17 acidic amino acids (15 glutamic acid and two aspartic acid) out of 20 in the middle of the monomer. The acidic character of nucleoplasmin led Stein and colleagues to demonstrate that polyglutamic and polyaspartic acid can greatly facilitate nucleosome assembly at physiological ionic strength. Moreover, Stein showed that polyglutamic acid could actually cause histones to associate as stable octamers at physiological ionic strength in the absence of DNA. These observations make the polyacidic tracts obvious candidates for the histone binding site in nucleoplasmin (Stein *et al.*, 1979). Anionic regions are a common feature of chromatin-associated proteins (Earnshaw, 1987).

Although it was originally thought that nucleoplasmin was the only molecular chaperone required for nucleosome assembly, this soon proved not to be the case. Kleinschmidt and colleagues identified a second protein within the nuclei of *Xenopus* oocytes called N1/N2 (Kleinschmidt *et al.*, 1986) that forms a specific complex with only histones H3 and H4. N1/N2 was cloned and found to be a 64 774 Da protein, also containing polyacidic tracts. One tract contains 18 glutamic acid and three aspartic acid residues out of a run of 31 amino acids. This particular site has been shown by mutagenesis to be important for the binding of histones H3/H4 to N1/N2 (Kleinschmidt *et al.*, 1985; Kleinschmidt and Seiter, 1988).

The preparation of specific antibodies to both nucleoplasmin and N1/N2 allowed the resolution of their respective roles in nucleosome assembly in *Xenopus* egg extracts. Histones H3/H4 are found to co-immunoprecipitate with N1/N2, whereas histones H2A/H2B are found to co-immunoprecipitate with nucleoplasmin. Subsequent work showed that each complex transfers histones to DNA separately. Histones H3/H4 have to associate with DNA before nucleoplasmin can deposit histones H2A/H2B (Dilworth *et al.*, 1987; Kleinschmidt *et al.*, 1990). This ordered sequestration of histones is similar to that occurring at the replication fork *in vivo*, and *in vitro*, although different proteins appear to be involved in directing nucleosome assembly at the fork. In fact any nucleosome assembly process will require the ordered addition of histones (H3, H4)$_2$ and (H2A, H2B). An important point from these observations is that although many polyanions such as polyglutamic acid, proteins containing polyacidic tracts (like HMG1 or CENP-B) or even ribonucleic acid can assemble nucleosomes, this assembly process may not be physiologically relevant (Nelson *et al.*, 1981; Bonne-Andrea *et al.*, 1984). Nucleoplasmin and other acidic proteins can also associate with histones to help disassemble nucleosomes *in vitro* with unclear biological significance for transcription in an *in vivo* context (Erard *et al.*, 1988; Chen *et al.*, 1994a). Indeed, although nucleoplasmin and N1/N2 can function as nucleosome assembly proteins in *Xenopus* eggs, it is unclear whether they actually fulfil these roles in a physiological context. Nucleoplasmin-like proteins have been purified from *Xenopus* somatic cells, but have not been characterized at the molecular level (Cotten and Chalkley, 1987). Questions concerning the role of nucleoplasmin and N1/N2 follow from the observation that the chromatin they assemble from their complexes with histones is not physiologically spaced (one nucleosome is present every ~160 bp, not every 180–190 bp). In addition the efficiency of chromatin assembly is low and the rate of chromatin assembly is slow. *In vivo* chromatin assembly processes generate nucleosomes with a physiological spacing (one every 180–190 bp). Moreover, *in vivo* chromatin assembly is efficient and fast. The true physiological chromatin assembly pathway is coupled to replication, and it seems probable that nucleoplasmin and N1/N2 primarily serve a storage role for the histones.

Worcel and colleagues adapted the procedure of Lohka and Masui for extract preparation in the absence of yolk for *Xenopus* oocytes (Glikin *et al.*, 1984). Their high speed supernatant of oocytes (after centrifugation at 150 000 g, known as the oocyte S150) has been used in a number of highly significant experiments.

An important aspect of these *Xenopus* chromatin assembly systems is that the resultant nucleosomal arrays are physiologically spaced (i.e. one nucleosome every 180–190 bp) identically to their spacing in *Xenopus* chromosomes. As we will discuss, the failure to assemble spaced nucleosomal arrays presents an important limitation to chromatin assembly using purified components (Section 3.3).

Although the exact spacing of nucleosomes is influenced by the excess of histones over DNA in the extract, close packed arrays of nucleosomes are difficult to form.

The assembly of spaced nucleosomes initially presented an enigma to investigators as histone H1 is not present in the minichromosomes assembled *in vitro*. Previous work using synthetic DNA templates (oligo(dA.dT)) and purified histones suggested that histone H1 was involved in determining nucleosome spacing (Section 3.3). Consistent with the absence of histone H1 from the oocyte S150 assembly system, but also in keeping with the results using purified components, Worcel and colleagues were able to show that exogenous histone H1 could influence the spacing of nucleosomes on DNA. Nucleosomal repeats as long as 220 bp could be obtained (Rodriguez-Campos *et al.*, 1989). This probably reflects a non-physiological inclusion of more than one histone H1 molecule per nucleosome, since it is only when more than one linker histone is incorporated that this spacing is achieved *in vivo* (Weintraub, 1978; Bates and Thomas, 1981). Subsequent work determined that a variant linker histone (B4) exists in these extracts; however, removal of B4 does not influence nucleosome spacing (Ohsumi *et al.*, 1993; Dimitrov *et al.*, 1994). This suggests that linker histones are not necessary to space nucleosomes in a physiological context.

Worcel also studied the possible role of histone acetylation in the spacing of nucleosomes. Although histone H4 was deacetylated during the maturation of chromatin in the *Xenopus* oocyte S150, this deacetylation process could be completely inhibited without influencing nucleosome spacing (Shimamura and Worcel, 1989). Thus, these experiments with oocyte extracts eliminate two major candidates for spacing nucleosomes: linker histones and the acetylation of histone H4.

Worcel and colleagues discovered that ATP was required for the assembly of a physiologically spaced nucleosomal array (Glikin *et al.*, 1984). This result led to the suggestion that topoisomerase II, which requires ATP as an energy source for activity, might have a major role in the assembly process. However, subsequent work demonstrated that topoisomerase I has the predominant role in chromatin assembly (Almouzni and Méchali, 1988a; Wolffe *et al.*, 1987; Annunziato, 1989). Importantly, hydrolysis of ATP or other nucleotides was found not to be necessary to provide an energy source for chromatin assembly. Other explanations for the ATP requirement in chromatin assembly include the inhibition of phosphatases, for example this might lead to the maintenance of phosphorylation of histone H2A.X (Kleinschmidt and Steinbeisser, 1991), of nucleoplasmin (Sealy *et al.*, 1986; Almouzni *et al.*, 1991) or of HMG14/17-like proteins (Tremethick and Frommer, 1992; Tremethick and Drew, 1993). Any of these phosphorylated proteins might have a role in nucleosome spacing. Importantly none of these phosphorylation events is required for the assembly of spaced nucleosomes in the remodelling of the *X. laevis* sperm pronucleus (Dimitrov *et al.*, 1994). This suggests that any phosphorylation requirement for nucleosome spacing on naked DNA may occur during the assembly of the histone tetramer $(H3/H4)_2$ on to DNA. This is because a major difference between chromatin assembly on naked DNA versus pronuclei is that histones H3/H4 pre-exist within sperm chromatin. Moreover, there is too little H2A.X in pronuclei to account for all nucleosome spacing and HMG14/17 appear naturally deficient in *Xenopus* egg extract (Crippa *et al.*, 1993).

Similar experiments were performed using preparations derived from *Xenopus* eggs. Eggs have several advantages over oocytes for the preparation of chromatin assembly extracts. The egg is a well-defined biochemical entity, whereas an oocyte may be within one of several developmental stages each containing a distinct store of macromolecules (Section 3.1.1). Aside from the physiological integrity of the egg compared to an oocyte, the molecules involved in chromatin assembly may be more active when isolated from eggs. This is simply because the egg has molecules in a physiologically primed state for rapid cell division and nuclear assembly.

Egg extracts have the capacity to carry out second strand synthesis of a single-stranded DNA molecule (Méchali and Harland, 1982). This replication process resembles the enzymatic activities associated with lagging strand synthesis at the replication fork in the *Xenopus* embryo. Most importantly, the synthesis of duplex DNA is very efficient and rapid, occurring at comparable rates to those observed *in vivo*. Using this system, Almouzni and colleagues were able to show that nucleosome assembly was much more rapid on the replicating template than on non-replicating duplex DNA mixed with the extract. This observation resolved the paradox concerning the slow assembly of chromatin on duplex DNA using other egg and oocyte systems. These systems did not use the appropriate molecular chaperones for assembly. *Xenopus* homologs of the mammalian chromatin assembly factor 1, CAF1, which is the chaperone that links chromatin assembly to replication, most probably function during second strand synthesis (Krude and Knippers, 1993) (see later). Dissection of the molecular basis of replication-coupled chromatin assembly in *Xenopus* revealed that the process was staged on the replicating template such that histones H3 and H4 were rapidly deposited on DNA to form an intermediate complex. Following assembly of histone H3/H4 tetramers, histones H2A/H2B were also preferentially sequestered on to the replicated DNA (Almouzni and Méchali, 1988b; Almouzni *et al.*, 1990a,b). This preferential association of histones with the replicating template is not due to a selective affinity for single-stranded DNA, but is instead coupled to the replication machinery itself most probably through *Xenopus* CAF1. This observation is important because *in vivo* chromatin assembly is coupled to replication and needs to occur very rapidly within the embryonic cell cycle.

Nucleosome assembly systems derived from mammalian cells were developed some years after the *Xenopus in vitro* systems. These experiments are based on an *in vitro* replication system developed by Kelly and colleagues that makes use of the simian virus 40 origin of replication and a viral protein, T antigen (Li and Kelly, 1984). Whole cell extracts have been studied that are dependent on DNA replication for efficient nucleosome assembly. Rather less work has been carried out on the replication independent system. Nucleosome assembly in mammalian whole cell extracts is also promoted by the replication process. Like the *Xenopus* egg and oocyte extracts, the assembly of regular arrays of nucleosomes does not require histone H1, but does require exogenous Mg^{2+}/ATP.

As discussed earlier, SV40 replicates in the nucleus of the host cell as a circular chromosome whose nucleosome structure and histone composition are identical with those of the host. With the exception of T antigen, the only viral protein required for replication, all of the enzymology of replication and the assembly of the minichromosome are identical to normal cellular processes (Stillman, 1989). The original extracts used for replication contained predominantly cytoplasmic components. Stillman and colleagues observed that the addition of an extract of mammalian cell nuclei promoted the assembly of nucleo-somes on the replicating SV40 DNA (Stillman, 1986). This nucleosome assembly was dependent upon and occurs concomitantly with DNA replication. Subsequent experiments fractionated the nuclear extract and showed that it contained a single component required for the replication-dependent nucleosome assembly (Smith and Stillman, 1989, 1991a,b). This nuclear protein, called chromatin assembly factor 1 (CAF1), was purified to homogeneity and found to consist of five polypeptides of 150, 62, 60, 58 and 50 kDa. CAF1 is not related to either nucleoplasmin or N1/N2, although some of the polypeptides are phosphorylated (Kaufman *et al.*, 1995). Like chromatin assembly in *Xenopus* extracts, nucleosome assembly on the replicating SV40 template is a stepwise process (Fotedar and Roberts, 1989). During the first step, CAF1 targets the deposition of newly synthesized and acetylated histones H3 and H4 to the replicating DNA. This reaction is dependent on and coupled to DNA replication. In the second step, the histone H3/H4 complex is converted into a mature nucleosome by the sequestration of histones H2A/H2B. This latter process can occur after replication is complete (Gruss *et al.*, 1990; Verreault *et al.*, 1996).

Knippers and colleagues extended these observations on CAF1 to show that the chromatin assembly factor is a stable constituent of native minichromosomes (Krude *et al.*, 1993). CAF1 functions to facilitate chromatin assembly not only on duplex DNA but also during complementary DNA synthesis using single-stranded templates (Krude and Knippers, 1993). These results provide an important link between the evidence for replication-coupled chromatin assembly in *Xenopus* extracts (Almouzni and Méchali, 1988b) and replication coupled events in the mammalian system (Krude and Knippers, 1993).

Mammalian cell-free preparations will also assemble chromatin independent of the replication process (Banerjee and Cantor, 1990; Banerjee *et al.*, 1991). Chromatin assembly on double-stranded DNA requires ATP hydrolysis for the assembly of a physiologically spaced nucleosomal array. Banerjee and colleagues found that histone phosphorylation was essential for this nucleosome spacing during assembly on naked DNA. This result is in contrast to the lack of a phosphorylation requirement for the assembly of spaced nucleosomal arrays during the remodelling of *Xenopus* sperm chromatin (Dimitrov *et al.*, 1994). However, the specific requirement for H2A phosphorylation to increase linker length from 0 to 45 bp is similar to the effect of H2A.X phosphorylation during chromatin assembly on naked DNA in *Xenopus* extracts (Kleinschmidt and Steinbeisser, 1991). Phosphorylation of H3 and H1 was without effect on nucleosome spacing in the mammalian system, but did correlate with the condensation of chromatin (Section 2.5.3). Since many individual components are simultaneously modified in these chromatin assembly assays it is as yet difficult to attribute a particular structural consequence to a single modification event. More detailed analysis with purified systems will be necessary before rigorous conclusions can be made.

Very similar results to those obtained with *Xenopus* oocyte and egg extracts and mammalian extracts were achieved with cell-free preparations from *Drosophila* eggs and embryos (Becker and Wu, 1992; Kamakaka *et al.*, 1993). These extracts have been widely used to examine the consequences of chromatin assembly for the transcription process (Section 4.2). An important concern in interpreting the results of these experiments is the composition of chromatin assembled by these extracts. The chromatin will resemble that typically present in the oocyte or early embryo, not that present in normal nuclei within somatic cells. At least for *Xenopus* where it has been defined: the assembled minichromosomes will be enriched in phosphorylated and acetylated core histones, including variants such as H2A.X. In addition, linker histone variants such as B4 and HMG2 will be incorporated. These differences from normal somatic chromatin should be considered in considering the accessibility of chromatin assembled in the extract systems to transcription factors.

Kadonaga and colleagues have carefully dissected the molecular mechanism of chromatin assembly in the *Drosophila* extracts. They have defined numerous histone chaperones including *Drosophila* variants of nucleoplasmin (Ito *et al.*, 1996b), nucleosome assembly protein 1 (NAP-1) (Ito *et al.*, 1996a), chromatin assembly factor-1 (CAF-1) (Kamakaka *et al.*, 1996; Tyler *et al.*, 1996) and a novel ATP-utilizing chromatin assembly and remodelling factor, ACF (Ito *et al.*, 1997). CAF-1 and ACF together with histone chaperones can direct the complete assembly of the core histones into nucleosomes (Ito *et al.*, 1997). Importantly Kadonaga and colleagues are able to obtain selective chromatin assembly on replicating templates *in vitro* under conditions of limiting CAF-1 (Kamakaka *et al.*, 1996). This reflects the physiological coupling of chromatin assembly to the replication process. How CAF-1 is targeted to replicating templates remains unclear, however the fact that CAF-1 recognizes sites of DNA damage suggests that single-stranded DNA, DNA breaks or other aspects of nascent DNA structure might facilitate preferential association (Gaillard *et al.*, 1996). ACF contains the *Drosophila* ISWI protein as a component (Ito *et al.*, 1997). This protein is a component of several nucleosome remodelling complexes (Section 2.5.4) such as NURF (Tsukiyama and Wu, 1995; Tsukiyama *et al.*, 1995) and CHRAC (Varga-Weisz *et al.*, 1995). In fact ACF will act together with *Drosophila* NAP-1 to remodel chromatin and activate transcription (Ito *et al.*, 1997). However, in the absence of evidence for targeted chromatin disruption by sequence-specific *trans*-activators the biological significance of this transcriptional activation remains unclear. What is clear is that the principal actors in the assembly of chromatin are now well defined in the *Drosophila in vitro* system.

Summary

Biochemical fractionation of *Xenopus* egg extracts led to the discovery of molecular chaperones: nucleoplasmin and N1/N2. Both are acidic proteins that interact with histones to form specific complexes. Nucleoplasmin binds histones H2A/H2B and N1/N2 binds histones H3/H4. It appears that these proteins most probably fulfil storage functions for histones in the oocyte and egg.

Whole egg and oocyte extracts assemble correctly spaced nucleo-somal arrays. Linker histones and histone H4 acetylation are not involved in the spacing phenomenon. The core histones alone appear to have the capacity to assemble spaced nucleosomes under physiological conditions. How this is achieved is unknown. DNA replication in whole egg extracts has been shown to facilitate nucleosome assembly. Chromatin assembly is a staged process where deposition of histone tetramers ((H3, H4)$_2$) precedes that of (H2A, H2B) dimers.

Mammalian whole cell extracts are capable of replicating duplex DNA containing a viral origin of replication in the presence of a single viral replication protein. In this system nucleosome assembly is coupled to replication through the molecular chaperone CAF1. Sequestration of both histones H3/H4 and histones H2A/H2B is strongly favoured on replicating DNA. *Drosophila in vitro* chromatin assembly systems are currently the best defined at a biochemical level. Replication-dependent chromatin assembly has also been recon-stituted in the *Drosophila* system dependent on CAF-1.

3.3 EXPERIMENTAL APPROACHES TOWARDS THE RECONSTITUTION OF TRANSCRIPTIONALLY ACTIVE AND SILENT STATES

A major reason for wishing to understand chromatin assembly and chromatin architecture is to appreciate how the many structures within the chromosome might exert significant regulatory roles in establishing and maintaining particular states of gene activity. Approaches to this problem are diverse, however all are limited by the extent to which the structure of chromatin can be defined and manipulated to determine the impact on transcription.

3.3.1 Purified systems

Once it was realized that high salt concentrations (> 1.2 M NaCl) would dissociate nucleosome core particles into their components of DNA and histones, it was soon established that the process was reversible. This reversibility has been extremely useful in establishing the organization of both DNA and histones in the nucleosome (Section 2.2.5). These reconstitution strategies were exploited extensively to examine the sequence determinants of nucleosome positioning (FitzGerald and Simpson, 1985; Shrader and Crothers, 1989, 1990; Wolffe and Drew, 1989). Separation of H3/H4 tetramers from H2A/H2B dimers allowed the demonstration that histones H3/H4 recognize DNA sequence-directed nucleosome positioning elements even when they first associate with DNA at high ionic strength ($\sim$1 M NaCl) (Hansen *et al.*, 1991; Bashkin *et al.*, 1993). Reconstitution of nucleosomes on short DNA fragments less than 200 bp in length led to the following conclusions. The histone tetramer (H3, H4)$_2$ wraps 120 bp of DNA around it at least transiently before two dimers of histones (H2A, H2B) can bind to either side of the tetramer–DNA complex and extend core histone–DNA contacts over 160 bp (Hayes *et al.*, 1990, 1991b). Histones (H2A, H2B) need to be sequestered before linker histones can associate, further extending and stabilizing histone–DNA interactions (Hayes and Wolffe, 1993; Hayes *et al.*, 1994). Interactions with the core histone tails are not necessary for sequestration of the linker histones (Ura *et al.*, 1994), nor are the tails of the linker histone necessary for nucleosome assembly (Hayes *et al.*, 1994).

Although useful for examining the organization of DNA with a single histone octamer or tetramer, reconstitution of chromatin from high salt has several disadvantages. The most notable failing of this methodology is the inability of nucleosomes assembled in this way to space themselves correctly. Instead of the physiological spacing found in the chromosome, histone octamers reconstituted by dialysis from high salt concentrations pack together on DNA as closely as possible. This does not mean that all nucleosomes form adjacent to each other, only that when they are adjacent there is no linker DNA. In general, one nucleosome is found every 150–160 bp, whereas *in vivo* nucleosomes are spaced every 180–190 bp (depending on tissue and organism). This close packing appears to be a consequence of maximizing not only strong protein–DNA interactions, but also protein–protein interaction along the DNA backbone. *In vivo* nucleosomes are assembled in stages, the histones are extensively acetylated and thus have relatively weak protein–DNA contacts. In contrast, following dialysis from high salt, an entire unmodified histone octamer is deposited on to DNA. Since there is very little linker DNA between these close-packed nucleosome structures, histone H1 cannot bind correctly. Consequently histone H1 interacts with long exposed stretches of free DNA forming aggregates or causing DNA to precipitate. Nevertheless, under highly controlled conditions, Stein and colleagues were able to show that linker histones influence the spacing of histone octamers, at least on highly flexible oligo(dA.dT) DNA molecules (Stein and Bina, 1984; Section 2.1.1).

Simpson and colleagues took a novel approach to this question by constructing a template that avoids all the problems we have just discussed and assembles correctly spaced chromatin following dialysis of histone and DNA from high to low salt concentrations. This difficult task was accomplished by spacing the nucleosome positioning elements in the sea urchin 5S RNA gene in tandem arrays, with repeat lengths spanning the range of most cellular chromatins. The long tandemly repeated (> 50) DNA fragments were then excised and reconstituted with core histones. The DNA sequence directed nucleosome positioning signals were of sufficient strength to overcome the tendency of histone octamers to close pack (Simpson *et al.*, 1985). We have already discussed the utility of these spaced nucleosomal arrays in examining their salt-dependent compaction into the chromatin fibre (Section 2.3.1). More recent studies have demonstrated that histone H1 binds correctly to nucleosome arrays (Meersseman *et al.*, 1991; Ura *et al.*, 1995, 1997). Bradbury and colleagues made use of an analysis of minor translational positions of histone octamers on the sea urchin 5S DNA to show that histone H1 could influence translational position. Histone H1 was mixed with the chromatin at high salt concentrations (0.5 M NaCl) and dialysed to low ionic strength leading to its association with chromatin without causing precipitation. This system offers considerable hope for determining the structural consequences of reconstituting histone H1 into chromatin.

Obviously, the mechanism of chromatin assembly *in vivo* differs considerably from the *in vitro* dialysis of mixtures of histone and DNA from high to low salt. However, there are several examples of macromolecules that perhaps resemble salt in their effects and interaction with the core histones. These acidic macromolecules such as polyglutamic acid have been used to assemble chromatin. In many ways they resemble physiological nucleosome assembly factors. At physiological ionic strengths (0.2 M NaCl) nucleosome assembly normally occurs very inefficiently; however, it can be accomplished by renaturing core histones into H3/H4 tetramers and H2A/H2B into dimers by prolonged dialysis. These proteins are then titrated over another protracted period with a considerable excess of naked DNA. The addition of limiting histones prevents the formation of large aggregates and the precipitation of the histone–DNA complex (Ruiz-Carrillo *et al.*, 1979). Obviously, this process has to be considerably improved upon *in vivo*. In order to understand chromatin assembly *in vivo* we have to introduce molecular chaperones to facilitate the interaction between histones and DNA. We discuss in some detail the utility of these nucleosomal structures reconstituted using purified components in characterizing the many ways that nucleosome assembly can either restrict or facilitate transcription factor access (Sections 4.1 and 4.2).

Summary
It is possible to assemble nucleosomal structures by the progressive dialysis of mixtures of DNA and histones from high to low salt. Unless special nucleosome positioning sequences are used the nucleosomes will pack together very closely, unlike the correct physiological spacing seen in the chromosome. Constructs that do position nucleosomes allow histone H1 to associate, thereby assembling completely synthetic chromatin. These templates have been useful in determining the roles of purified histones in transcriptional control.

3.3.2 Chromatin assembly and transcription in *Xenopus* eggs and oocytes

In a series of important experiments Woodland and colleagues demonstrated that *Xenopus* eggs acquire a huge store of core histones during the growth of the oocyte within the ovary. This store of 140 ng of core histone is enough to make over 20 000 nuclei (Woodland and Adamson, 1977). These core histones are stored in specialized complexes in the oocyte nucleus. In contrast, somatic histone H1 is severely deficient in eggs (Wolffe, 1989a; Hock *et al.*, 1993; Dimitrov *et al.*, 1993, 1994) and is replaced with an embryonic variant (Smith *et al.*, 1988; Dimitrov *et al.*, 1994) with specialized functions (Section 2.5.1).

When purified closed circular viral DNA molecules became available in the 1970s, the *Xenopus* oocyte was used as a living test-tube to investigate what functions were encoded by the viral genome. The first experiments showed that viral DNA (SV40) injected in the oocyte cytoplasm was degraded, whereas viral DNA injected into the oocyte nucleus was maintained in a stable supercoiled state (Wyllie *et al.*, 1977, 1978). Each nucleosome is known to stabilize a single supercoil in a closed circular DNA molecule (Section 2.2.3). Furthermore, the injected DNA was transcribed, and assembled into a nucleoprotein complex whose buoyant density and regularly repeated particulate structure was that expected for chromatin. More extensive analysis revealed that each oocyte nucleus could convert a mass of DNA equivalent to 1000 diploid *Xenopus* nuclei into chromatin. This is a surprisingly low efficiency of chromatin assembly considering the large stores of histones in the nucleus. However, oocytes are adapted to the storage of nuclear components, whereas eggs are adapted to the assembly of nuclear structures. For example, chromatin assembly on non-replicating DNA in the oocyte is a non-physiological process, whereas chromatin assembly on replicating DNA in the fertilized egg is a physiologically necessary event. As we will discuss, if DNA capable of some replication is injected into an oocyte nucleus, chromatin assembly is much more efficient.

Transcription of the viral DNA injected into oocytes occurred only under circumstances in which chromatin was formed. In this respect it is of interest that linear DNA does not form chromatin containing regularly spaced nucleosomes as efficiently as negatively supercoiled DNA in *Xenopus* oocyte nuclei, and is transcribed much less efficiently (Mertz, 1982; Harland *et al.*, 1983). Negatively supercoiled DNA is preferentially assembled with histone octamers, because the association of each octamer with closed circular DNA will introduce a positive superhelical turn into the linker DNA (Section 2.2.3). It is more energetically favourable for octamers to bind to negatively superhelical DNA since the resultant positive supercoils in free DNA will be cancelled out by the existing negative supercoils (Clark and Felsenfeld, 1991). These correlations between chromatin assembly and transcription, which probably reflect the constraints of assembling any multicomponent protein complex on DNA, were next extended by examining the chromatin organization of specific DNA sequences containing well-studied cellular genes.

Worcel and colleagues examined the organization into chromatin of the histone genes of *Drosophila* and *Xenopus* and the 5S RNA genes of *Xenopus*, after injection into oocyte nuclei as closed circular plasmid DNAs (Gargiulo and Worcel, 1983; Gargiulo *et al.*, 1985). Indiret end-labelling was used to map nuclease cuts on the assembled chromatin with great precision (Fig. 3.9). It was found that changes in the amount of DNA injected into the oocyte nucleus influenced both the structure and expression of the assembled chromatin. Minichromosomes with a spaced array of nucleosomes, in which the distance between nucleosomes was identical to that found in the native chromosome, could be assembled under appropriate conditions. These parameters corresponded exactly to those for maximal transcription of the 5S RNA genes. Clearly transcription of eukaryotic genes is not incompatible with the assembly of chromatin.

The correct separation between nucleosomes on small circular DNA molecules (one nucleosome per 180–190 bp) suggested that higher-order chromatin structures such as the chromatin fibre were not required for the formation of correctly spaced arrays. This is because

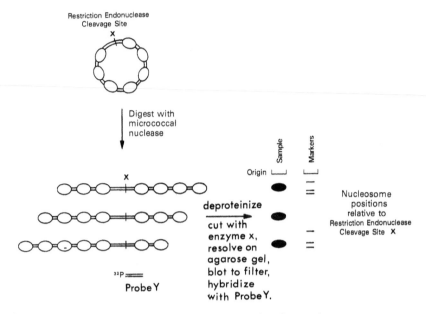

Figure 3.9. Methodology for detecting specific chromatin structures, on closed circular plasmid DNA molecules.

the plasmids used in these studies (3.5 kb in length) are too short to assemble more than 3–4 turns of a solenoidal chromatin fibre. Topological constraints make the formation of such a fibre unlikely. Surprisingly, nucleosome positioning was not detected on the *Xenopus* 5S RNA or *Drosophila* histone gene in these experiments (Section 2.2.5). However, DNase I hypersensitive sites did form 5' and 3' to the *Xenopus* histone gene, suggesting that under appropriate conditions specific chromatin structures could actually form on these artificial templates (Section 4.2.4). These hypersensitive sites depend on the association of specific non-histone proteins with the promoter.

The transcriptional activity of chromatin containing spaced nucleosomes should be contrasted with the inactivity of the same genes (5S RNA) assembled with nucleosome structures in which no linker DNA is present (Weisbrod *et al.*, 1982). Under these circumstances nucleosomes are described as close packed (one nucleosome per 160 bp). Thus incorrectly assembled chromatin structures are sometimes incompatible with transcription (Section 4.2.2).

Experiments examining the time course of chromatin assembly under optimal conditions demonstrated that over 30 min were required to partially assemble double-stranded DNA into a spaced array of nucleosomes in the oocyte nucleus (Ryoji and Worcel, 1984). This suggests that the organization of a nucleosomal array is a slow process. This inefficient assembly of nucleosomal arrays on double-stranded DNA in the oocyte nucleus reflects the non-physiological nature of the chromatin assembly process (see section 3.2.2). *In vivo* the vast majority of chromatin assembly normally occurs on replicating DNA during S-phase. This replication-coupled chromatin assembly can be reproduced in *Xenopus* oocytes by the injection of single-stranded DNA into the nucleus. Complementary second-strand synthesis occurs on the single-stranded template using similar enzymes to those involved in lagging strand synthesis at the chromosomal replication fork. Chromatin assembly occurs much more rapidly on the replicating template than on double-stranded DNA (Almouzni and Wolffe, 1993a). This rapid chromatin assembly is seen in the almost complete supercoiling of a replicating template compared to a double-stranded DNA molecule injected into the nucleus (Fig. 3.10). Many of these observations concerning chromatin assembly in the nucleus of the living oocyte can be reproduced *in vitro* (Section 3.2). The use of cloned DNA as

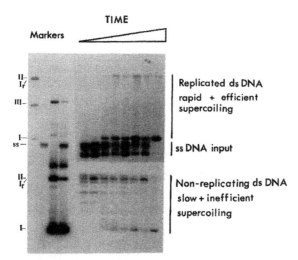

Figure 3.10. Chromatin assembly is very rapid and efficient on 'replicating' single-stranded DNA compared to non-replicating double-stranded DNA.

Each negative supercoil indicates assembly of a nucleosome, the more DNA that is fully supercoiled (Form I), the more nucleosomes are assembled on that DNA. Here mixtures of long replicating M13 DNA and short duplex pUC9 DNA are used (see Almouzni and Wolffe, 1993b for details of this type of experiment). M13 DNA is added in single-stranded form (ss input) and converted into double-stranded forms in the oocyte nucleus after micro-injection.

templates for chromatin assembly has a great potential for correlating structure and function.

Replication coupled chromatin assembly following the microinjection of single-stranded templates into oocytes nuclei invariably leads to the repression of basal transcription (Almouzni and Wolffe, 1993a; Wong *et al.*, 1995; Landsberger *et al.*, 1995). Some transcription factors can associate with their recognition elements in chromatin in spite of replication – coupled chromatin assembly. These include NF-Y on the *Xenopus* hsp70 promoter which contributes to the assembly of a DNase I hypersensitive site (Fig. 3.11) even though the hsp70 promoter is inactive in the absence of heat shock. The experiment shown in Fig. 3.11 also indicates that blocking histone deacetylation with the inhibitor Trichostatin A (TSA) is without any consequences for the assembly of the DNase I hypersensitive site. The thyroid hormone receptor, TR/RXR provides another example since it binds to DNA on the surface of a positioned nucleosome and again assembles a DNase I

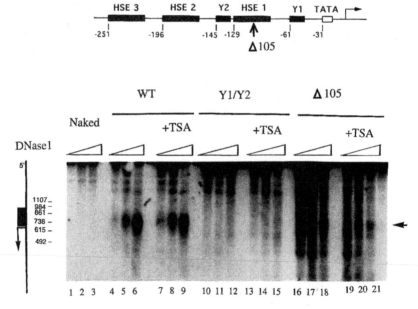

Figure 3.11. The Y-box element is essential for the assembly of a DNase I hypersensitive site on the *Xenopus* hsp70 promoter.
DNase I hypersensitivity is not influenced by the inhibition of histone deacetylase using Trichostatin A (TSA). The horizontal arrow indicates the DNase I hypersensitive site. A scheme of the promoter shows the positions of the heat shock response elements (HSEs), the Y-boxes (Y1 and Y2), the TATA box (TATA), the start site of transcription (hooked arrow) and the position of the 5' deletion Δ 105.

hypersensitive site in the chromatin of the *Xenopus* TRβA gene (Section 4.2). These DNase I hypersensitive sites act as targets for other regulatory factors whose capacity to associate with chromatin depends on signals such as heat shock for the heat shock transcription factor (HSF). The association of HSF with the hsp70 promoter depends on NF-Y and modifies the structure of the DNase I hypersensitive site (Fig. 3.12). The role of HSF is presumably to recruit coactivators that remodel chromatin and facilitate transcription. The addition of progesterone to mature the oocyte into an egg shows that without HSF the DNase I hypersensitive site is lost, whereas in the presence of HSF it is retained (see Section 4.1.4). This is a consequence of the increase in efficiency of

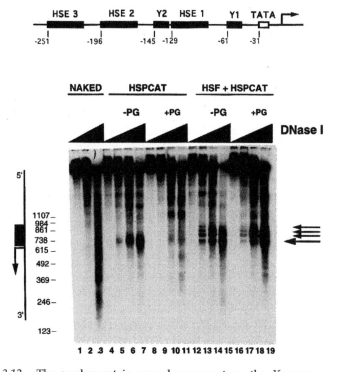

Figure 3.12. The nucleoprotein complex present on the *Xenopus* hsp70 promoter that assembles the DNase I hypersensitive site is remodelled during oocyte maturation into an egg.
Compare lanes 4–7 without progesterone (–PG) with lanes 8–11 plus progesterone (+PG). Progesterone is a hormone that controls the maturation process. The addition of *Xenopus* heat shock transcription factor (HSF) alters the organization of the DNase I hypersensitive site. This new structure resists remodelling during maturation. A scheme of the promoter shows the positions of the heat shock response elements (HSEs), the Y-boxes (Y1 and Y2), the TATA box (TATA) and the start site of transcription (hooked arrow).

chromatin assembly that occurs when an oocyte becomes an egg (Landsberge and Wolffe, 1997, Section 3.2.2). A large component of transcriptional regulation on the hsp70 and TRβA promoters following replicative chromatin assembly is to relieve the repression due to the presence of the histones (Landsberger and Wolffe, 1995a; Wong *et al.*, 1995). Some other transcription factors such as Gal4-VP16 working on artificial templates can both relieve chromatin-mediated repression and activate transcription (Almouzni and Wolffe, 1993a). However, this has not yet been observed with natural transcription factors and natural chromatin templates in the oocyte system.

Insight into the molecular mechanisms necessary for transcriptional regulation in chromatin comes from the use of histone deacetylase inhibitors such as Trichostatin A and manipulation of the abundance of *Xenopus* histone deacetylase itself. As we have discussed histone H4 is stored in a diacetylated form in oocytes and deacetylated following chromatin assembly. Addition of TSA prevents this deacetylation from occurring, which in turn prevents the repression of transcription associated with replicative chromatin assembly (Wong *et al.*, 1998). This result re-emphasizes the key role that chromatin assembly has in repressing basal transcription. Manipulating the abundance of histone deacetylase in oocytes reinforces this conclusion. Increasing the expression of *Xenopus* histone deacetylase represses transcription from double-stranded DNA templates microinjected into oocyte nuclei, but only after a physiological density of nucleosomes has been assembled. This is a relatively slow process requiring incubation periods of several hours for microinjected double-stranded DNA. Histone deacetylase can facilitate transcriptional repression in a dominant and non-targeted process. Moreover transcriptional repression by histone deacetylase requires a chromatin substrate.

Mutational analysis of the hsp70 and TRβA promoters indicates that the addition of TSA relieves the requirement for the sequence specific transcriptional activators such as HSF or hormone-bound TR/RXR for transcription, however TSA does not remove the requirement for any component of the basal transcriptional machinery. It is possible that HSF and hormone-bound TR/RXR recruit transcriptional coactivators with the capacity to acetylate histones and overcome the activity of the histone deacetylase (Section 2.5.4).

In contrast to the relative efficiency with which a wide variety of promoters are transcribed in the oocyte nucleus, the microinjection of duplex DNA into embryos leads to transcriptional quiescence prior to the mid-blastula transition (Prioleau *et al.*, 1994, 1995; Bendig, 1981; Rusconi and Schaffner, 1981; Krieg and Melton, 1985; Almouzni and Wolffe, 1995; Landsberger and Wolffe, 1997). However, due to the toxicity of exogenous DNA for subsequent embryogenesis, generally much less DNA (~300 pg) is typically injected into fertilized eggs compared to oocytes (~3 ng). This facilitates the assembly of templates into both chromatin and 'pseudo-nuclei'.

It has been suggested that a simple titration of core histones regulates transcription during early embryogenesis and accounts for transcriptional quiescence (Prioleau *et al.*, 1994, 1995; see also Edgar *et al.*, 1986). Titration and readdition of core histones within the developing embryo shows that these proteins have a necessary, but far from sufficient role in generating a transcriptionally inactive state prior to the MBT (Almouzni and Wolffe, 1995; Landsberger and Wolffe, 1997). An artifical increase in DNA content within embryos prior to the MBT will lead to the detection of significant transcription from certain genes (Priouleau *et al.*, 1994; Newport and Kirschner, 1982a,b). Nevertheless significant transcriptional activation of the microinjected *c-myc* promoter prior to the MBT occurs with core histones in considerable mass excess (> 24-fold) over DNA. Thus it is possible that titration of another more limiting transcriptional repressor or some other independent process determines basal transcriptional activity. Several independent observations suggest that this is in fact the case.

An important consideration concerning the simple histone titration model for transcriptional activation during embryogenesis is whether the transcription obtained in the presence of exogenous DNA prior to the MBT reflects basal or regulated transcription. Under normal developmental conditions only certain class II genes are activated at the MBT, others remain repressed (Krieg and Melton, 1985). In a detailed study it was found that a simple titration of the core histones by the addition of a mass of DNA equivalent to that found at the MBT (25 ng) would activate exogenous class III gene transcription, but not exogenous class II promoters, including the CMV immediate early promoter or the adenovirus E4 promoter (Almouzni and Wolffe, 1995). This result is consistent with that of Lund and Dahlberg who found that the transcriptional activation of exogenous *Xenopus* U1 genes prior to the MBT would not be achieved by a simple increase in the DNA content of the embryo (Lund and Dahlberg, 1992). Taken together these results demonstrate that the activation of *c-myc* transcription prior to the MBT is a promoter-specific phenomenon.

The key step for transcription of several promoters following injection into developing *Xenopus* embryos prior to the MBT is the recruitment of the basal transcriptional machinery to the TATA homology. This can be accomplished either by titration of chromatin and addition of exogenous TBP, or by expression of a transcriptional activator that can penetrate chromatin, with binding sites next to the promoter (Almouzni and Wolffe, 1995). Presumably such activators are either not required for transcription of genes like *c-myc* specific activators for genes of this type are present in excess within the pre-MBT embryo (Priouleau *et al.*, 1994). The capacity of transcriptional activators to overcome repressive influences on transcription prior to the MBT demonstrates that the class II gene basal transcriptional machinery including endogenous TBP is fully competent in the cleavage embryo. Therefore, the basal transcriptional machinery must be prevented from stably associating with promoter elements by inhibitory proteins. Since exogenous TBP alone does not overcome this inhibition, the inhibitory proteins must be dominant over the basal class II transcriptional machinery (Fig. 3.13). The reversible nature of protein–DNA interactions in the egg environment is illustrated by the remodelling of regulatory nucleoprotein complexes on the *Xenopus* hsp70 promoter during meiotic maturation of the oocyte into an egg (Landsberger and Wolffe, 1997; see Section 4.1).

Summary

Microinjection of exogenous DNA into *Xenopus* oocyte nuclei leads to its slow assembly into nucleosomal arrays, some of which can be quite specific in their organization over promoter elements. The coupling of DNA replication to chromatin assembly facilitates the assembly process in oocyte nuclei. Microinjection of DNA into eggs leads to a more rapid assembly of templates into chromatin with distinct transcriptional consequences.

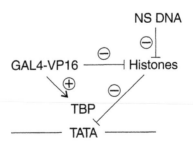

Figure 3.13. Model for transcriptional control in the cleavage *Xenopus* embryo (see Almouzni and Wolffe, 1995).
The default state in the cleavage *Xenopus* embryo is for the core histones and other chromatin components (Histones) to repress transcription, either non-specific DNA (NS DNA) in the presence of exogenous TATA-binding protein (TBP) or transcriptional activators (GAL4-VP16) can overcome this repression.

3.3.3 Chromatin assembly on DNA introduced into somatic cells

Although work with *Xenopus* eggs and oocytes provided our initial insights into how chromatin is assembled using exogenous DNA, obviously the general relevance and applicability of these results had to be established in somatic cells. There are several ways of introducing exogenous DNA into cells in culture or more interestingly into *Drosophila* or mammalian embryos. We have already discussed the first approach of fusing cells together to form somatic cell hybrids or heterokaryons. In recent times microinjection of *Drosophila* and mammalian eggs has yielded much important information concerning chromosomal structure and nuclear processes. However, these studies were pioneered by the transient or stable transformation of mammalian cells with plasmid DNAs or by infection with viral DNA.

The small size of the SV40 genome (5243 bp in length) and the early availability of information concerning DNA sequence and gene organization made it a convenient model for studying the structure and function of chromatin. Late in infection the SV40 genome is organized into nucleosomal arrays as a minichromosome, which can be isolated from infected cells under appropriate conditions. One region of the minichromosome (ORI) contains several important recognition sites for *trans*-acting factors: the origin of replication, the binding sites for the viral regulatory protein (T-antigen) and the promoters driving transcription of early and late SV40 mRNAs (Fig. 3.14). The chromosomal organization of this region was recognized as

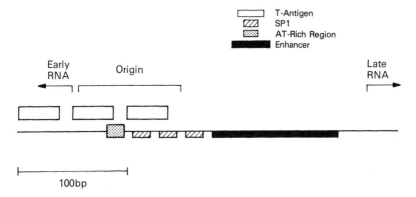

Figure 3.14. DNA sequence organization of the SV40 ORI region. Binding sites for many *trans*-acting factors are compressed into this small segment of DNA. The origin and sites of transcription initiation are indicated as are the major binding sites for *trans*-acting factors.

being important for the processes of DNA replication and transcription, and was shown to differ from the rest of the minichromosome.

The first experiments used nucleases to digest the minichromosome. The DNA sequence in the ORI region is preferentially cut in the nuclei of infected cells with DNase I (a hypersensitive site, Section 4.2.4). This suggested that DNA in the ORI region is more accessible to DNA-binding proteins than in the rest of the minichromosome. Chemical carcinogens such as psoralen that interact with the free DNA duplex by intercalation, bind preferentially to the ORI region in infected cells. Also consistent with a preferential accessibility of the ORI region is its rapid digestion with a variety of endonucleases in isolated minichromosomes. Finally, electron microscopy reveals that 20–25% of the isolated minichromosomes contain a nucleosome-free region (or gap) of approximately 350 bp covering the ORI region. This gap therefore represents the best early documentation of nucleosome positioning on regulatory DNA (Varshavsky *et al.*, 1978; Jacobovits *et al.*, 1980; Robinson and Hallick, 1982; Cereghini and Yaniv, 1984; Section 4.2.3).

An important question that is still to be clearly resolved is whether the nucleosome-free region at the ORI was imposed upon the chromosome by the interaction of *trans*-acting factors with their cognate DNA sequences, or whether histone–DNA interactions determine the placement of nucleosomes (sequence-directed positioning, Section 2.2.5). Recent studies have shown that although minor preferences exist in nucleosome positioning in the presence of only histones and DNA, other components are required to create a nucleosome-free gap (Weiss *et al.*, 1985; Ambrose *et al.*, 1989). We will see that the influence of non-histone proteins on nucleosome positioning is a common feature of chromatin structure that has important functional consequences (Section 4.2.5). In contrast to the failure of the ORI DNA sequence to exclude nucleosomes, similar reconstitutions of purified histones and DNA revealed a very favourable sequence for forming a nucleosome at the opposite end of the minichromosome. This is the region of the SV40 minichromosome where both replication and transcription terminate (Poljak and Gralla, 1987; Hsieh and Griffith, 1988). It is also possible that a nucleosome at this position might assist the termination process by causing processive enzyme complexes to pause (Sections 4.3.1 and 4.3.4). These early experiments with SV40 minichromosomes provide strong support to the hypothesis that *trans*-acting factors function in an organized chromosomal environment.

Several genes or promoter elements have been introduced into SV40 minichromosomes and the influence of chromatin structure on gene expression examined. Although in general repressive effects on transcription are observed, the structural basis of this repression has not yet been determined (Lassar *et al.*, 1985). The advantage of viral genomes for the analysis of the interrelationship between chromatin structure and function is that they contain replication origins. This means that the viral minichromosome can be studied as an episome, i.e. without being integrated into the cellular chromosomes. Moreover, as chromatin assembly is coupled to DNA replication (Section 3.2), the influence and access of *trans*-acting factors on nucleosome organization will be more like that within the true chromosomal context (Section 4.3.1). The only reservation about studies with small viral genomes is that possible regulatory effects dependent on higher-order chromatin structure are unlikely to be observed. Again this is due to topological constraints in folding small DNA circles in the chromatin fibre. Viral genomes other than SV40 have been very useful for the detailed analysis of chromosomal influences on transcription. These include the bovine papilloma virus (BPV)-based episomes that have been used to investigate the molecular mechanisms by which glucocorticoid receptor activates gene expression (Ostrowski *et al.*, 1983; Archer *et al.*, 1992) (Section 4.2.4). Here, the interplay between specific chromatin structures and transcription factors has been rigorously documented.

Viral episomes have been used to establish the organization of *trans*-acting factors and nucleosomes on specific DNA sequences. As we will see, there is often interaction between these two components (Section 4.2.3). However, most scientists have investigated gene regulation by transiently transfecting cloned DNA without a viral origin of replication into eukaryotic cells. Under these conditions these investigators have rarely investigated repressive or stimulatory effects that might be attributed to chromatin. Importantly, the observed regulation due to enhancers, promoters or other elements often does not reflect the range of response observed in the natural chromosomal context (Section 4.2.4). Early studies suggested that transfected plasmids that did not integrate into the natural chromosomal context would be assembled into nucleosomal arrays, even if they did not replicate in the cell (Camerini-Otero and Zasloff, 1980). However, more extensive studies by Howard and colleagues demonstrated that the efficiency of chromatin assembly depended on the transfection conditions. The amount of DNA transfected into cells, the method of compacting DNA prior to transfection and the efficiency with which particular DNA sequences partitioned to the nucleus greatly affected chromatin assembly. Over 80–90% of nuclear plasmid material might not be assembled into chromatin, however some templates and protocols generate nucleosomal arrays (Reeves *et al.*, 1985). Consequently, it is not surprising that the regulation of transiently transfected DNA does not always completely reflect that found in the natural chromosomal context.

An important example of the differences between transfected and episomal templates concerns the regulation of the mouse mammary tumor virus (MMTV) long terminal repeat (LTR) promoter (Archer·*et al.*, 1992). When this regulatory DNA is present within a stably replicating episomal template it is assembled into a positioned array of nucleosomes. Transcription factors such as nuclear factor 1 (NF1) cannot gain access to their binding sites due to the specific association of these sites with positioned nucleosomes and the promoter is transcriptionally silent (Fig. 3.15). A complex series of events initiated by the binding of the specialized transcription factor known as the glucocorticoid receptor is necessary to remodel chromatin structure and allow NF1 to gain access to its binding site (Section 4.2.3). This remodelling constitutes much of the transcriptional regulatory process at the MMTV LTR. When this regulatory DNA is transiently transfected into the same cells the template no longer replicates. Now when protein–DNA interactions and transcriptional activity are assayed *in vivo* a very different result is obtained. Nucleosomal arrays are not assembled on the transfected template, NF1 binds to its recognition site and transcription has a low level of basal activity. Thus, the episomal template recapitulates normal transcriptional regulation similar to that seen with the endogenous chromosomal MMTV LTR, whereas significant differences are seen with the

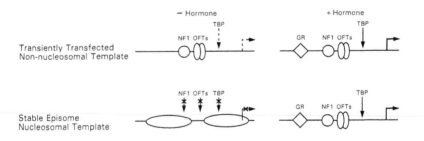

Figure 3.15. A comparison between transcription factor interactions on transiently transfected, non-nucleosomal DNA containing the MMTV promoter with those on a stably replicating episomal template fully assembled into nucleosomes (ellipsoids).

Transcription factor association, nuclear factor 1 (NF1), octamer transcription factor (OTF), glucocorticoid receptor (GR) and TATA binding protein (TBP) are shown in the presence or absence of hormone. Hormone binding activates the GR protein to interact with DNA. Transcription initiation (hooked arrow) frequency is indicated by varying the strength of line.

transfected template. Clearly, the method of introducing exogenous DNA into mammalian cells is important for subsequent studies of transcriptional regulation.

It is possible to select for the stable chromosomal integration of transfected DNA in somatic cells using genetic markers conferring resistance against certain drugs. These studies have revealed that gene expression varies depending on the site of integration within the chromosome. This is another manifestation of a position effect (Section 2.5.6). For example, an intact single copy of the immunoglobulin K gene containing 1.5 kb of upstream and 8.5 kb of downstream flanking sequences exhibited a 100-fold variation in transcriptional activity dependent on chromosomal position. This probably reflects the proximity of the integrated DNA sequence to heterochromatin (Section 2.5.6) or other repressive influences such as silencers (Section 4.1.2). Expression of the stably integrated DNA was found to be only a quarter of that expected for the endogenous gene when a large number of transformed cell lines were examined. Clearly a significant factor required for efficient gene expression lies in having the correct chromosomal organization (Blasquez *et al.*, 1989).

The correct chromatin organization for gene expression has only been determined for a few genes (Section 4.2.3). Although the position of individual nucleosomes has generally not been examined in this type of experiment, investigators have searched for sequence-specific influences on the organization of nucleosomal arrays and the chromatin fibre. Specific DNA elements distant from the gene itself appear to confer locus control functions or insulation from chromosomal position effects (Sections 2.5.6 and 4.2.4). At a more fundamental level, removal of all non-coding sequences other than the promoter from the immunoglobulin K gene in stable integrants leads to a reduction of transcriptional activity to only 2% that of wild type. These experiments introduce the role of elements other than the promoter that act at a distance to influence transcription initiation (Sections 2.5.6 and 4.1.2) and that may also influence chromosomal organization (Sections 2.5.6 and 4.2.4).

The method of choice for introducing DNA into mammalian chromosomes, such that it will be both expressed and regulated in the correct way, is to microinject it into mammalian eggs. Similar experiments have been carried out in *Drosophila*. This approach allows both the stable integration of the microinjected DNA into the chromosome and facilitates the analysis of the function of various *cis*-acting elements in regulating genes *in vivo* in the normal developmental environment (Grosveld *et al.*, 1987; Kellum and Schedl, 1991). An additional advantage is the capacity to assess the tissue-specific expression of genes integrated at identical chromosomal positions in each of the different cell types present in the animal. These studies have facilitated the definition of regulatory elements (insulators and locus control regions, Section 2.5.6). These DNA sequences apparently shield the gene from the influence of chromosomal position, i.e. position effects do not occur.

Summary

The assembly of exogenous DNA into chromatin in somatic cells depends both on the organization of the DNA and how it is introduced. Use of constructs containing origins of replication (e.g. viral DNAs) leads to chromatin structures that resemble those found in the chromosome itself. Microinjection of DNA into fertilized eggs (transgenic experiments) has led to the definition of distinct DNA elements known as locus control regions and insulators that may organize whole domains of chromatin.

3.3.4 Yeast minichromosomes

A particularly attractive system for examining the influence of chromatin structure on the function of DNA is yeast (Grunstein *et al.*, 1992). Experimental work with yeast has many advantages especially for molecular biologists and geneticists, among these being the existence of small (< 1500 bp) extrachromosomal plasmids that replicate autonomously. Particular experimental attention has been given to a plasmid present at about 100 copies/cell known as TRP1ARS1 (1453 bp in length). This plasmid consists of one gene coding for N-(5′-phosphoribosyl)-anthranilate isomerase (TRP1), a sequence containing a replication origin (ARS1), and part of the Ga13 promoter (UNF) (Fig. 3.16). Simpson, Thoma and colleagues have determined the chromatin structure of TRP1ARS1 in some detail (Thoma *et al.*, 1984; Thoma and Simpson, 1985; Thoma, 1986; Simpson, 1991). Nucleosome position has been determined by the nuclease accessibility of isolated minichromosomes. Four different regions of chromatin structure have been defined. The UNF region contains three nucleosomes that have very strong DNA sequence-directed positions (Section 2.2.5), which are not changed by insertion of DNA fragments of various lengths into the plasmid. In contrast, four loosely positioned nucleosomes are found on the TRP1 gene. These nucleosomes can easily rearrange following insertion of additional DNA.

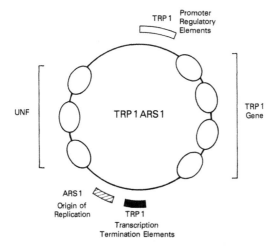

Figure 3.16. The structure of the TRP1ARS1 minichromosome. Nucleosomes (open ellipsoids) and key DNA sequences are shown. The region originally called UNF, for unknown function, is now known to be part of the GAL3 promoter.

Two nucleosome-free regions are also present (like the gap in the SV40 minichromosome) which are hypersensitive to nuclease digestion. These sequences include the TRP1 promoter and the ARS origin of replication. Although lacking centromeres and telomeres, the TRP1ARS1 minichromosome therefore contains many elements associated with normal cellular chromosomes.

Among the DNA sequences inserted into the TRP1ARS1 plasmid was a sea-urchin 5S RNA gene, on which base pair resolution of the rotational positioning of DNA on the histone core had originally been defined (Section 2.2.5). A nucleosome formed including this sequence, with exactly the same position in the yeast minichromosome as observed *in vitro*. This important observation shows that yeast histones *in vivo* are able to recognize the same DNA sequence-directed nucleosome positioning elements as chicken histones *in vitro* (Thoma and Simpson, 1985). Similar experiments with yeast genomic sequences inserted into the TRP1ARS1 plasmid consistently reveal the same nucleosome positioning to occur on the episome as seen in the chromosome (Simpson, 1991). In certain instances, unstable nucleosomes such as those on the TRP1 gene may have their positions influenced by the organization of other regions of the episome into chromatin (Thoma and Zatchej, 1988). More recent studies have taken these observations further to dissect the contributions of specific *trans*-acting factors to nucleosome positioning.

A segment of DNA containing the binding site for a *trans*-acting factor (the α2 repressor) was inserted into the TRP1ARS1 plasmid. The α2 repressor is a protein produced by α-mating type cells that binds to DNA as a complex with a second protein, MCM1, to repress transcription of genes normally expressed only in α-mating type cells (Kelcher *et al.*, 1988). When α2 was not present the chromatin structure of the inserted DNA segment appeared to be random. No nucleosome positioning was apparent on micrococcal nuclease digestion. However, the presence of α2 led to a dramatic change in chromatin structure (Roth *et al.*, 1990). The entire minichromosome, except for the nucleosome-free region around the origin of replication, was organized into an array of precisely positioned nucleosomes. An interaction between the α2 repressor and the nucleosome was suggested by the capacity of the protein to move a nucleosome adjacent to it apparently without any influence from the underlying DNA sequence. Protein–protein contacts between the α2 repressor, the Tup1 repressor and histones are now biochemically and genetically defined (Section 4.2.5). These studies have been extended to show that the association of the α2 repressor with its binding site will cause a nucleosome to be positioned over the TATA box of several yeast chromosomal α-mating type specific genes *in vivo* (Shimizu *et al.*, 1991). In three genes repressed by α2, the binding site for α2/MCM1 lies about 100 bp from the TATA box. This spacing has been suggested to occur in order for the TATA box to be incorporated around the dyad axis of the nucleosome positioned by α2/MCM1. Since this is the region of the nucleosome with the most stable histone–DNA interactions it might be expected to be the most repressive towards the transcription process. The demonstration of specific interactions between non-histone proteins and histones that contribute to nucleosome positioning introduces a new level of complexity and specificity to the assembly of chromatin and chromosomes (Section 4.2.5).

The functional consequences of the specific organization of chromatin structure by α2/MCM1 remain controversial. Simpson and colleagues found that two general transcriptional repressors of yeast known as SSN6 and Tup1 (Keleher *et al.*, 1992) are required together with α2 and MCM1 to direct the local positioning of nucleosomes (Cooper *et al.*, 1994; Edmondson *et al.*, 1996; Fig. 3.17). This result suggests that the general repression of transcription in yeast might involve the assembly of repressive chromatin structures. Nevertheless, moving the TATA box of a gene repressed by α2/MCM1 away from contact with histones does not prevent transcriptional repression from occurring (Patterton and Simpson, 1994). This result demonstrates that local nucleosome positioning over the TATA box by α2/MCM1 is not necessary to establish the repressed state. It remains

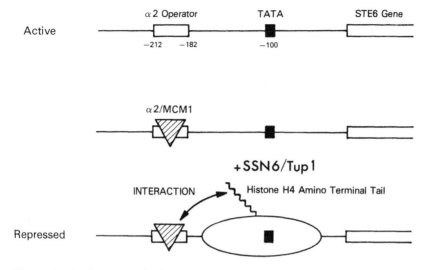

Figure 3.17. Repression of transcription of the STE6 gene is correlated with chromatin organization directed by the α2-MCM1 complex (triangle).
This interaction is mediated through the amino-terminal tail of histone H4. The general transcription repressors SSN6/Tup1 are important for chromatin organization. Base pair positions are indicated relative to the gene.

possible that α2/MCM1, SSN6, Tup1 and the histone H4 amino terminal tails are involved in the assembly of an extended repressive chromosomal domain. The nucleosome positioned next to the α2/MCM1 binding site might be only one component of this domain (Simpson *et al.*, 1993). Alternatively the SSN6-Tup1 complex targeted by α2/MCM1 to yeast promoters may function to repress transcription by mechanisms independent of chromatin structure that involve direct interactions with components of the basal transcriptional machinery (Keleher *et al.*, 1992; Herschbach *et al.*, 1994). A limitation of yeast chromatin research until very recently (Schultz *et al.*, 1997) has been the lack of biochemical systems capable of efficient minichromosome assembly *in vitro*, such that the role of individual proteins in transcriptional activation or repression could be tested. Yeast minichromosomes assembled *in vivo* are difficult to isolate for biochemical experiments *in vitro* (Dean *et al.*, 1989; Pederson *et al.*, 1986).

Summary
Yeast minichromosomes have provided major insights into nucleosome positioning. DNA sequence and *trans*-acting factor directed

nucleosome positioning have been defined. It has been established that specific interactions exist between the histone proteins and *trans*-acting factors responsible for directing nucleosome position.

3.4 MODULATION OF THE CHROMOSOMAL ENVIRONMENT DURING DEVELOPMENT

Developmental events provide many useful examples of how different forms of chromatin are assembled and how these transitions in chromatin can influence nuclear events (Patterton and Wolffe, 1996). Unicellular eukaryotes such as the ciliate *Tetrahymena* exhibit several distinct patterns of differential patterns of gene activity associated with developmental events. Metazoans such as the common experimental animals: *D. melanogaster*, the sea urchin, *Xenopus laevis* and the mouse all exhibit global changes in gene activity during early development. These organisms rapidly establish cell lineage and tissue specific patterns of gene expression early in embryogenesis. These committed states are stably maintained throughout adult life. In general it appears that the global changes in gene activity seen during early embryogenesis are dependent on developmentally regulated changes in the type or modification of histone or other basic chromatin protein. In contrast the selective repression of genes in particular differentiated cell types appears dependent on proteins influencing higher-order chromatin structure, such as those of the chromodomain class. This latter type of mechanism might also account for the phenomenon of imprinting or epigenetic determination in which genes of identical sequence are differentially expressed in the same cell dependent on the history of the chromosomes in which they are found (Section 2.5.6).

3.4.1 *Tetrahymena*

Changes in nucleosomal structure are a ubiquitous feature of developmental programs associated with the global control of nuclear activity. Excellent examples are found in the work of Gorovsky, Allis and colleagues who have systematically characterized transitions in nucleosomal proteins associated with development in *Tetrahymena thermophila*, a ciliated protozoan. *T. thermophila* has two types of nuclei:

a diploid micronucleus that represents the germ line of the cell, and a polyploid somatic macronucleus that controls the phenotype of the cell (Gorovsky, 1973). During the sexual cycle known as conjugation, cells of opposite mating types fuse. This leads the germ-line micronucleus to undergo meiosis, exchange and fertilization resulting in the appearance of a diploid zygotic nucleus in each of the original mating cells (Nenney, 1953; Martindale *et al.*, 1982). The prophase of meiosis for the micronuclei is characterized by meiotic DNA recombination and repair as well as a brief period of transcription (Allis *et al.*, 1987). At the end of the mating process, the diploid zygotic nucleus divides to generate a new macronucleus and micronucleus. The decision to differentiate into a macro- or micronucleus depends on their exact positions in the cell. The pre-existing 'old' or senescent macronucleus is transcriptionally inactivated and eliminated, leading the transcriptional needs of the cell to be assumed by the developing 'new' macronuclei (Wenkert and Allis, 1984). Changes in nuclear function during this ordered developmental pathway are dramatically correlated with transitions in nucleosomal composition (Table 3.1).

Differences exist between the chromosomal composition of the macro- and micronucleus throughout the life cycle of *Tetrahymena*. Macronuclei contain histone H1, whereas micronuclei contain three distinct basic polypeptides, including an HMG1-like protein (Schulman *et al.*, 1991; Wu *et al.*, 1994). None of these proteins are essential for the assembly of functional nuclei (Shen *et al.*, 1995; see also Scarlato *et al.*, 1995). Macronuclei are enriched in several components characteristic of transcriptionally active chromatin including acetylated core histones (Chicoine and Allis, 1986), a histone H2A variant called hv1 containing an N-terminal tail domain similar to that of histone H4, and phosphorylated histone H1 (Stargell *et al.*, 1993; Roth

Table 3.1. Developmental regulation of chromatin composition in *Tetrahymena*

Macronucleus	Macronucleus	Micronucleus
Transcriptionally active	Senescent, transcriptionally inactive	Transcriptionally inactive
Phosphorylated H1 Acetylated H4 hv1 (H2A.Z) Acetylated hv1 HMG B	Dephosphorylated H1 Tailless core histones	Specialized linker binding protein. Unmodified core histones No hv1

et al., 1988). These proteins and modifications are normally lacking from the micronucleus. In *Drosophila* the hv1 homolog known as H2A.vD is absolutely essential for early development, embryos in which this gene is 'knocked out' die (van Daal and Elgin, 1992; van Daal *et al.*, 1988) (Section 2.5.1). The initial cell–cell interactions of *Tetrahymena* conjugation also trigger the rapid induction of a distinct HMG 1-like protein, HMG B, which accumulates both in the 'old' macronucleus and in the micronucleus (Schulman *et al.*, 1991; Wang and Allis, 1992). The appearance of HMG B and some hv1 in micronuclei correlate with their brief period of transcriptional activity during meiotic prophase. All of these modifications are predicted to destabilize nucleosomes and chromatin. This would lead DNA within the chromosome to be more accessible to transcription factors and more easily traversed by processive enzyme complexes such as DNA and RNA polymerases. Finally, during the senescence of 'old' macronuclei at the end of conjugation, the core histone N-terminal tails are proteolytically removed and histone H1 is dephosphorylated. Removal of the core-histone tails will prevent any interaction with regulatory proteins. Dephosphorylation of the linker histone tail domains will stabilize their binding to nucleosomal DNA and facilitate the assembly of higher order chromatin structures. These events correlate with the inhibition of transcription and the condensation of macronuclear chromatin (Lin *et al.*, 1991).

These observations indicate the plasticity of chromatin composition, although some nucleosomal proteins are selectively incorporated into chromatin during nuclear replication, others exchange into chromatin independent of the replication process (Wu *et al.*, 1988; Wang and Allis, 1992), others are modified independent of replication (Lin *et al.*, 1991). It is clear that changes in the chromosomal environment, especially in the micronucleus early in conjugation and in the senescent macronucleus, are tightly coupled to states of nuclear activity. Comparable changes occur during metazoan development.

3.4.2 *Xenopus* and sea urchin

In the sea urchin and *Xenopus laevis*, stores of core histones are sequestered in the egg to facilitate the efficient assembly of chromatin during the rapid cell division events following fertilization (Salik *et al.*, 1981; Adamson and Woodland, 1974). The stored histones also serve to remodel sperm chromatin once it enters the egg cytoplasm (Poccia *et al.*, 1981; Poccia and Green, 1992; Dimitrov *et al.*, 1994). Sperm chromatin is highly specialized and the paternal genome is packaged by a wide variety of proteins including specialized histones (Strickland *et al.*, 1980; Jutglar *et al.*, 1991), small core-histone like proteins (Abé and Hiyoshi, 1991) and protamines (Marushige and Dixon, 1969). All that these proteins have in common is their capacity to reversibly compact DNA into a transcriptionally inert state. They are all removed and replaced by core histones within the egg cytoplasm. In *Xenopus* the replacement of sperm specific basic proteins by histones H2A and H2B is facilitated by molecular chaperones such as nucleoplasmin (Philpott *et al.*, 1991; Philpott and Leno, 1992; Section 3.1).

Histone acetylation appears to have a functional role in coupling chromatin assembly to the replication process (Kaufmann and Botchan, 1994). Histone H4 is stored in the diacetylated form in both the *Xenopus* and sea urchin egg. During early embryogenesis these stores of histone H4 are utilized for chromosome assembly and the diacetylated form is progressively converted into the deacetylated form (Dimitrov *et al.*, 1993; Chambers and Shaw, 1984). The level of deacetylation of histone H4 appears to be inversely related to the rate of nuclear assembly and cell division during development. Acetylation of histone H4 in early embryogenesis may reduce chromatin condensation within the chromosomes of embryonic nuclei, accounting in part for the large nuclear volume (Gurdon, 1976).

Histone hyperacetylation has been shown to facilitate the access of *trans*-acting factors to DNA (Lee *et al.*, 1993; Section 2.5.2). Since the level of acetylation of the core histones is controlled by an equilibrium between histone acetyltransferases and deacetylases (Turner, 1993), it is normally possible to induce histone hyperacetylation using inhibitors of deacetylation such as sodium butyrate (Candido *et al.*, 1978) or Trichostatin A (Yoshida *et al.*, 1995). However in *Xenopus*, hyperacetylated histones only accumulate in the presence of these inhibitors for the first time during early gastrulation (Dimitrov *et al.*, 1993). This implies that histone acetyltransferase activity is developmentally regulated. Hyperacetylation of the core histones correlates with the capacity to selectively induce transcriptional activity (Khochbin and Wolffe, 1993) and leads to a delay in the completion of gastrulation (Almouzni *et al.*, 1994; Fig. 3.18). *Xenopus* embryos maintained in Trichostatin A have defects in mesoderm formation. In the starfish *Asterina pectinifera*, the inhibition of histone deacetylation leads to the arrest of development at the early gastrula stage before mesenchyme formation (Ikegami *et al.*, 1993). Thus the capacity to modulate histone acetylation levels appears important in establishing stable states of differential gene activity during gastrulation (Almouzni *et al.*, 1994).

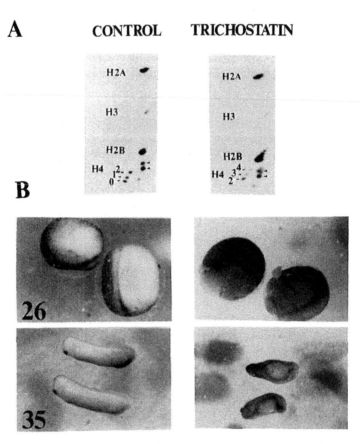

Figure 3.18. Histone hyperacetylation leads to developmental abnormalities during *Xenopus* embryogenesis.
A. Histone isolated from control neurula embryos labelled *in vivo* with [³H]Lys and [³H]Arg or grown in the presence of Trichostatin (30 nM) are shown resolved on two-dimensional gels. The number of acetylated lysine residues per histone H4 molecule is indicated. The arrows indicate two proteins whose mobilities do not change in the presence of inhibitors of histone deacetylase. B. Embryos were grown under the conditions described in A and photographed at the stages indicated. The appearance of the embryos was highly homogeneous; pictures representative of each incubation condition are presented (see Almouzni *et al.*, 1994 for details).

In the growing *Xenopus* oocyte, transcription is very active (Section 2.4.3). A large fraction of RNA synthesis occurs within specialized lampbrush chromosomes (Callan *et al.*, 1987; Anderson and Smith, 1978). Following oocyte maturation into an egg, transcription from the endogenous chromosomes is inhibited (La Marca *et al.*, 1975). Fertilization initiates eleven cycles of cell division, which occur in the absence of zygotic transcription (Newport and Kirschner, 1982a). At the mid-blastula transition (4000 cell stage), the cell cycle lengthens and a complex program of transcription is initiated (Lund and Dahlberg, 1992; Rupp and Weintraub, 1991; Krieg and Melton, 1987). Various models have been proposed to explain this global control of gene expression. The simplest model is that the DNA synthesized during embryogenesis titrates an inhibitory component of chromatin, perhaps the large excess of core histones stored in the embryo, and that once the histones are all sequestered onto DNA transcriptional programs will initiate (Newport and Kirschner, 1982b; Prioleau *et al.*, 1994). Substantial evidence suggests that more complex mechanisms also contribute to this major developmental transition. Injection of exogenous DNA, or arrest of DNA cleavage does not influence the temporal control of snRNA, tRNA or rRNA synthesis (Lund and Dahberg, 1992; Takeichi *et al.*, 1985; Shiokawa *et al.*, 1989). The transcriptional activation of exogenous DNA by titration of histones is a gene specific phenomenon, indicative of a key role for particular transcription factors in the process (Almouzni and Wolffe, 1995; Section 3.3). A limitation in transcription factor activity in the early embryo is also suggested from experiments that indicate chromatin assembly to have a key role in repressing basal but not activated transcription (Almouzni and Wolffe, 1993a, 1995). Early experiments actually failed to observe a heat-shock response in *Xenopus* oocytes (Bienz, 1986) simply because basal transcription was not repressed through chromatin assembly (Landsberger *et al.*, 1995).

The chromosomal environment in the early *Xenopus* embryo is also unusual in many other respects. The entire genome is being replicated every 30 min and chromatin assembly is coupled to replication, which tends to repress basal transcription (Almouzni *et al.*, 1990a,b, 1991; Almouzni and Wolffe, 1993a, 1995), DNA replication will also disrupt pre-existing transcription complexes (Wolffe and Brown, 1986). The reassembly of transcription complexes has a considerable lag time (Bieker *et al.*, 1985). Mitotic events that influence transcription will also occur every 30 min. Mitosis leads to the modification of the transcriptional machinery (Hartl *et al.*, 1993), condensation of the chromosome into a transcriptionally incompetent environment (Hirano and Mitchison, 1994; Chuang *et al.*, 1994), and the displacement of both transcription factors and RNA polymerase (Hershkovitz and Riggs, 1995; Prescott and Bender, 1962). Thus many events in the nucleus have the potential to influence transcription. These variables may have distinct consequences for transcription from genes within the endogenous chromosomes compared to those within exogenous plasmid DNAs microinjected into the egg cytoplasm (Landsberger and Wolffe, 1995b).

In the sea urchin, stores of nuclear components do not apparently inhibit zygotic transcription. Both the maternal and paternal pronuclei are transcriptionally active following fertilization (Brandhorst, 1980; Poccia *et al.*, 1985). The cleavage nuclei are also actively transcribed (Wilt, 1963, 1964; Kedes and Gross, 1969). However, the pattern of gene expression changes dramatically between the early blastula and gastrula. The majority of mRNAs expressed during sea urchin embryogenesis begin to accumulate at very early blastula stages and are then spatially restricted during gastrulation (Kingsley *et al.*, 1993). A comparable restriction occurs in *Xenopus* between the MBT and gastrulation. At the mid-blastula transition (MBT) many cell lineage specific genes are pleiotropically transcribed, only during gastrulation are cell lineage specific patterns established (Wormington and Brown, 1983; Rupp and Weintraub, 1991). These transitions in gene expression are associated with major changes in the type of histone present within the chromosome (Newrock *et al.*, 1977; Dimitrov *et al.*, 1993, 1994).

Distinct core histones and linker histones accumulate in chromatin as embryogenesis proceeds in the sea urchin. Maternal pools of cleavage stage (CS) histones CSH1, CSH2A and CSH2B replace the specialized sperm histones following fertilization (Poccia *et al.*, 1981; Section 2.5.1). After the 16-cell stage, the CS variants are replaced with newly synthesized 'α' or 'early' histone H1, H2A and H2B variants (Newrock *et al.*, 1977). In the sea urchin, the rate of histone biosynthesis is equivalent to that necessary to package newly synthesized DNA into chromatin, thus pre-existing histones are rapidly diluted during the early cell cycles (Goustin and Wilt, 1981; Moav and Nemer, 1971). The late β, γ and δ sea urchin histone genes begin to be transcribed in the blastula and continue to be active in adult cells, whereas transcription of the α histone genes is repressed (Maxson *et al.*, 1983; Knowles and Childs, 1984). Thus δ and γ variants of histone H2B, β, δ and γ variants of H2A and β and γ variants of H1 accumulate in chromatin as development proceeds. The expression of CS and early histone gene sets during oogenesis and early embryogenesis is uncoupled from DNA replication, whereas the expression of the late histone genes is temporally linked to DNA replication.

A common theme in sea urchin development is the variation in histones H1, H2A and H2B compared to histones H3 and H4. This reflects the essential structural and signal transduction role of histones H3 and H4 in the nucleosome. Histones H1, H2B and H2B readily exchange out of the nucleosome *in vivo* (Louters and Chalkley, 1985). These proteins do not influence the rotational positioning of DNA on the surface of the histone-octamer, but they do influence the accessibility of nucleosomal DNA to *trans*-acting factors (Hayes *et al.*, 1991b; Dong and van Holde, 1991; Hayes and Wolffe, 1992; Lee *et al.*, 1993; Wolffe, 1989a; Ura *et al.*, 1995). Thus it can be expected that modulation in the structural properties of the nucleosome through variation in the structure of the H2A/H2B dimer and of H1 may effect nuclear functions such as transcription and replication in early sea urchin development. Conclusive proof for such an essential role for a linker histone variant (H1 in *Xenopus*, see below) and for a histone H2A (H2AvD in *Drosophila*, see earlier) has been obtained.

In *Xenopus*, changes in histone gene expression are much more modest than those in the sea urchin. It was found that the bulk of *Xenopus* core histone genes are constitutively active in the oocyte and in the developing embryo following the mid-blastula transition (Perry *et al.*, 1986). Oocytes do accumulate large quantities of the specialized variant histone H2A.X (Kleinschmidt and Steinbeisser, 1991; Dimitrov *et al.*, 1994). This protein accumulates in early embryonic chromatin but is rapidly diluted as cell division proceeds. Linker histones show major alterations in abundance during *Xenopus* embryogenesis (Khochbin and Wolffe, 1994). The *Xenopus* egg stores large quantities of HMG1 and a linker histone B4 (Kleinschmidt *et al.*, 1985; Smith *et al.*, 1988; Dimitrov *et al.*, 1994). HMG1 can functionally replace linker histones in binding to nucleosomal DNA. HMG1-like proteins are also abundant in early embryonic chromatin from *Drosophila* (Ner and Travers, 1994). Aside from a structural role in chromatin HMG1 can also influence the functionality of the basal transcriptional machinery (Ge and Roeder, 1994a; Shykind *et al.*, 1995). Thus the transcriptional quiescence in the early *Xenopus* and *Drosophila* embryo (Brown and Littna, 1966; Edgar and Shubiger, 1986) might be a consequence of the enrichment of HMG1 within chromatin. Linker histone B4 is oocyte specific in expression (Cho and Wolffe, 1994) and is much less basic than somatic histone H1 (Dimitrov *et al.*, 1993). Nucleosomal arrays containing histone B4 would be expected to be less stable than those containing histone H1. Histone B4 accumulates in early embryonic chromatin, but is progressively diluted by somatic histone H1 as development proceeds (Bouvet *et al.*, 1994). This somatic histone H1 is synthesized from large maternal stores of histone H1 mRNA that are only released for translation after fertilization (Woodland *et al.*, 1979; Flynn and Woodland, 1980; Tafuri and Wolffe, 1993; Bouvet and Wolffe, 1994). Somatic histone H1 begins to substantially replace histone B4 at the mid-blastula transition, replacement is complete at the end of gastrulation (Dimitrov *et al.*, 1993; Hock *et al.*, 1993) (Fig. 3.19). This transition from the maternal histone B4 to a somatic histone

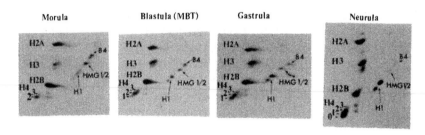

Figure 3.19. Changes in linker histone variants during the early development of *Xenopus laevis*.
Autoradiograms of two-dimensional gels of chromatin associated proteins synthesized during early development are shown. The positions of the core and linker histones are indicated.

H1 is remarkably conserved from the sea urchin *Psammechinus miliaris* to *Xenopus laevis* (Mandl *et al.*, 1997).

The progressive accumulation of histone H1 in *Xenopus* embryonic chromatin correlates with the transcriptional repression of many genes between the mid-blastula transition and neurulation (Wolffe, 1989b; Andrews *et al.*, 1991; Bouvet *et al.*, 1994). The precocious expression of histone H1 will cause a large family of oocyte specific 5S rRNA genes to become transcriptionally repressed (Bouvet *et al.*, 1994). Inhibition of H1 synthesis sustains 5S rRNA gene transcription (Bouvet *et al.*, 1994; Kandolf, 1994). This result establishes that changes in chromosomal composition can directly influence the transcriptional activity of specific genes during metazoan development (Sections 2.5.1 and 4.1.5). The accumulation of histone H1 in *Xenopus* chromatin also correlates both with the restriction in cell lineage specific gene expression through gastrulation (Rupp and Weintraub, 1991; Steinbach *et al.*, 1997) and in the capacity of embryonic nuclei to revert to totipotency (Gurdon, 1962, 1963). An active role of histone H1 in restricting the developmental window in which *Xenopus* ectodermal cells can be induced to form mesodermal tissue, i.e. their state of commitment has now been established (Steinbach *et al.*, 1997). It is, therefore, possible to speculate that somatic linker histones have an essential role as repressors of transcriptional programs in maintaining stable states of gene activity.

3.4.3 The mouse

Considerable circumstantial evidence supports an important role for chromatin structure in the global control of transcriptional regulatory mechanisms during early embryogenesis in the mouse. Extensive changes in nuclear activity occur during early mammalian embryogenesis (Telford *et al.*, 1990). Immediately after fertilization, development in the mouse is directed by maternally inherited mRNAs and proteins. However, the zygotic genome is transcriptionally active as early as G2 of the first cell cycle (Latham *et al.*, 1992; Ram and Schultz, 1993; Matsumoto *et al.*, 1994; Christians *et al.*, 1995). Nuclear transplantation experiments indicate that changes in chromatin structure occur immediately after the 2-cell stage. This is because nuclei of early 2-cell mouse embryos will readily support normal embryonic development when transplanted into enucleated 1-cell embryos, whereas nuclei from more advanced embryos do not (McGrath and Solter, 1984; Robl *et al.*, 1986; Howlett *et al.*, 1987). Thus beyond the 2-cell stage chromatin begins to differentiate. In an interesting parallel to comparable events in *Xenopus* (see earlier) somatic histone H1 is first detectable in the mouse embryo at the 4-cell stage (Clarke *et al.*, 1992).

The major transcriptional activation of the endogenous mouse chromosomes begins in 2-cell embryos at the end of G1, concurrent with DNA replication. A further significant increase in transcription occurs during G2 when maternal mRNAs are degraded. This pattern of gene activity is controlled by a zygotic clock (Schultz, 1993). This clock is largely independent of cell division or the ratio of DNA to cytoplasm. It regulates the transcriptional activity of exogenous DNA microinjected into mouse embryos as well as that of the endogenous chromosomes (Martinez-Salas *et al.*, 1989; Wiekowski *et al.*, 1991). One function of the zygotic clock may be the regulation of nuclear architecture. Core histone acetylation changes dramatically within chromatin during the first few cell divisions of mouse embryogenesis. Chromatin-containing acetylated histone H4 becomes enriched at the nuclear periphery when the zygotic genome is strongly activated at the 2-cell stage (Worrad *et al.*, 1995). Inhibition of histone deacetylase using Trichostatin A increases the efficiency of gene expression. Acetylated chromatin localizes with RNA polymerase II suggesting that it represents the site of active transcription. This localization of acetylated chromatin to the nuclear periphery is lost in the 4-cell embryo and during subsequent development. An important conclusion from these experiments is that the functional compartmentalization of the nucleus occurs very early in mouse embryogenesis (Thompson *et al.*, 1995).

De Pamphilis and colleagues have examined the transcriptional activity of exogenous DNA microinjected into developing mouse embryos. They have found that the promoters of microinjected plasmid DNAs are strongly repressed in 2-cell embryos, but that this repression can be relieved by embryo-specific enhancers (Martinez-Salas *et al.*, 1989; Mélin *et al.*, 1993). Promoter activity can also be relieved by the inhibition of histone deacetylases using sodium butyrate. If histones are hyperacetylated then the need for an enhancer to achieve full promoter activity is substantially reduced (Majumder *et al.*, 1993; Wiekowski *et al.*, 1991). This result implicates histone acetylation as a regulatory factor in the 2-cell mouse embryo.

Further evidence for an influence of nuclear environment on gene function follows from the microinjection of templates into paternal and maternal pronuclei. Promoters are repressed when injected into the paternal pronucleus. This transcriptional repression is not relieved by the inhibition of histone deacetylases. In contrast, although the same templates are repressed following injection into maternal pronuclei, transcription is stimulated by histone hyperacetylation (Wiekowski *et al.*, 1993). Thus rather like the macro- and micronuclei of *Tetrahymena* (see earlier), the paternal and maternal pronuclei provide very different environments for transcription. A challenge for the future is to determine the potential role of these different nuclear environments in the imprinting of paternally and maternally derived chromosomes (Pfeifer and Tilghman, 1994).

3.4.4 Developmental regulation of long-range chromatin structure

The interrelationship between changes in chromosomal composition and function is less clear for long-range chromatin organization than for nucleosomal DNA. This is simply because the nature of long-range chromosomal structure is currently much less well defined than nucleosomal organization (Section 2.5.6). It is also important to recognize that the changes in nucleosomal organization outlined in the preceding section will also have direct consequences for overall nuclear structure and function. The most dramatic example of this phenomenon occurs during erythropoiesis in birds and amphibians when the whole erythrocyte nucleus becomes heterochromatinized due to the accumulation of linker histone variants (H5 in chicken, H1° in *Xenopus*, Appels and Wells, 1972; Rutledge *et al.*, 1988; Khochbin and Wolffe, 1993). The accumulation of these linker histone variants leads to the arrest of transcription (Hentschel and Tata, 1978), and of cell proliferation (Aubert *et al.*, 1991; Sun *et al.*, 1989).

During early *Xenopus* embryogenesis there are major changes in the long-range organization of chromosomes. As we have discussed, the oocyte genome is organized into loop domains (Callan *et al.*, 1987). A loop domain organization is maintained through early development, however the size of the loops changes dramatically. There is a progressive thickening and shortening of metaphase chromosomes as development proceeds from the blastula embryo to the swimming tadpole (Micheli *et al.*, 1993). Concomitant with this change in chromosomal morphology the average length of DNA loops increases by more than 60%. Moreover, an analysis of chromosome length values at different developmental stages suggests that chromosomal shortening is simultaneous for all chromosomes in an individual cell, but occurs asynchronously in different embryonic cell types. These cell-specific phenomena might reflect the differentiation of chromosomal structures concomitant with determinative events. The size of DNA loops correlates with the size of replicons, i.e. the distance between adjacent origins of replication (Buongiorno-Nardelli *et al.*, 1982). During *Xenopus* embryogenesis, prior to the mid-blastula transition replication origins are closely spaced. Once the cell cycle lengthens, origins become more dispersed (Graham and Morgan, 1966). It has been proposed that the attachment of DNA loops to the chromosomal scaffold corresponds to their association to replication factories (Cook, 1991). Changes in chromosomal organization during early embryogenesis would offer a useful system to explore the developmental control of replicon size and of replication. These observations illustrate the dramatic changes in chromosome loop domain organization that can occur during early development.

The majority of the evidence for higher order chromatin structures having a role in the determinative events of development is based on genetic analysis. We have discussed the role of the histones, HP1 and proteins of the Polycomb group in position effect, and the evidence for the existence of insulators that separate functional domains of chromatin (Section 2.5.6). Here we discuss how chromodomain proteins might function in a developmental context. Proteins of the Polycomb group have important roles in homeotic gene regulation. The homeotic genes of the *Antennapedia* complex and the *bithorax* complex need to be differentially expressed in the developing *Drosophila* embryo in order to establish correct segmental identity (Lewis, 1978; Kaufman *et al.*, 1980). The segmental identity of the nuclei within a *Drosophila* embryo begins to be established even before the transition from syncytial to cellular blastoderm (Illmensee, 1978; Kauffman, 1980). It is at this time that heterochromatin and associated chromodomain proteins can first be cytologically detected. Homeotic genes are initially activated in the correct spatial domains by the interplay of transcriptional regulators, encoded by maternally expressed genes and early acting segmentation genes (Harding and Levine, 1988; Irish *et al.*, 1989). However, having once established a pattern of differential *Antennapedia* and *bithorax* gene complex expression within the *Drosophila* embryo, the transacting factors decay away. Nevertheless, the homeotic genes need to remain differentially active throughout development in order to ensure proper segmental identity (Morata *et al.*, 1983). The Polycomb group of proteins is an essential part of the maintenance mechanism that ensures segmental identity by maintaining silent homeotic genes in the repressed state through the rest of development. The Polycomb group proteins are not required to establish normal patterning of homeotic gene expression early in embryogenesis (Kuziora and McGinnis, 1988; Franke, 1991). However, at later embryonic stages when homeotic genes should be maintained in a silent state, mutation of the *Polycomb* gene leads to relief of homeotic gene repression (Beachy *et al.*, 1985; Carroll *et al.*, 1986; Celnikov *et al.*, 1990). Regulation of the *Drosophila* homeotic genes by the Polycomb group proteins is similar to the regulation of *Xenopus* 5S rRNA genes by histone H1 in that differential transcription factor association establishes a pattern of differential gene activity that is then maintained and enhanced through the assembly of a specific chromatin structure.

Although the exact structural consequences of the association of the Polycomb group proteins with chromatin are not known, modifiers of position effect influence the homeotic genes regulated by the Polycomb group (Grigliatti, 1991). A model for heterochromatin formation proposes that the component structural proteins co-operate in large complexes which can self-assemble to package large chromosomal domains (Locke *et al.*, 1988). Consistent with this hypothesis, the Polycomb protein assembles into multimeric complexes (Franke *et al.*, 1992) and is associated with inactive chromatin over large chromosomal domains (Orlando and Paro, 1993). In contrast the Polycomb protein is absent from transcribed regions (Section 2.5.6).

Immunofluorescent techniques have been used to examine the distribution of proteins within chromosomal domains in the interphase nucleus. These studies complement the more structural and genetic analysis of domain organization (Laemmli *et al.*, 1978; Saitoh and Laemmli, 1994 see Section 2.4.1). Clear evidence emerges for the existence of specific compartments within the nucleus (see also Cook, 1991) at which particular proteins are clustered into foci. The Polycomb protein is found to be present within such foci within the nuclei of *Drosophila* tissue culture cells (Messmer *et al.*, 1992). This subnuclear localization depends on the integrity of the chromodomain, and if localization is lost, so is the repressed state of the homeotic genes. Thus, the Polycomb protein might function by assembling a repressive chromatin domain that is tethered to the nuclear matrix in a transcriptionally incompetent nuclear compartment (Section 2.5.9).

Opposing the repressive effects of Polycomb group proteins on homeotic gene expression are the trithorax group of activators (Kennison, 1993). Experiments on the gene *brahma*, a member of the trithorax group lend additional support to the hypothesis that alterations in chromatin structure have an important maintenance role in sustaining patterns of homeotic gene transcription. Mutations in *brahma* strongly suppress mutations in *Polycomb*. It appears that *brahma* is necessary to relieve the repressive influence of *Polycomb* on homeotic gene expression after embryogenesis (Kennison and Tamkun, 1988; Tamkun *et al.*, 1992). The Brahma protein has similarities to the yeast SWI/SNF general activator complex (Section 2.5.4). This complex of more than 10 proteins (total of 2×10^6 Da) assists a wide variety of *trans*-activators in facilitating the transcriptional activation process (Cairns *et al.*, 1994). However, mutations within the histone fold domain of the core histone proteins H3 and H4 relieve the requirement for the SWI/SNF complex (Winston and Carlson, 1992; Herskowitz *et al.*, 1992; Wolffe, 1994c). This suggests that one role of SWI/SNF and potentially the Brahma protein is to disrupt locally repressive chromatin structures.

Although the long-range chromatin organization of the mammalian embryo is even less well understood than that of either *Drosophila* or *Xenopus*, considerable evidence suggests that aspects of chromosomal structure will influence gene expression during early mammalian development (Fundele and Surani, 1994). We have already discussed how nucleosomal composition changes during early mouse development, the differences in nuclear environment between paternal and maternal pronuclei and the correlations with transcriptional activity. Related epigenetic phenomena are the silencing of one of the two wild-type copies of a gene within a diploid nucleus during early development: these include X chromosomal inactivation in females and the parental imprinting of selected autosomal genes (Pfeifer and Tilghman, 1994). Much attention has focused on differential DNA methylation as a marker for imprinting, however it still remains unclear whether DNA methylation acts as a primary determinant of differential gene activity in these cases (Groudine and Conkin, 1985 see Section 2.5.7) or whether it simply reflects changes in chromatin structure that determine differential activity and that restrict the access of DNA methyltransferases to genes (Gottschling, 1992). DNA binding proteins exist that can selectively recognize methylated DNA and direct transcriptional repression (Boyes and Bird, 1991, 1992;

Section 2.5.7). Alternatively the association of histones and the assembly of repressive nucleosomal structures might differ between methylated and unmethylated DNA (Kass *et al.*, 1993).

Many genes in the inactive X chromosome are heavily methylated, in contrast to those of the active X chromosome (Grant and Chapman, 1988). However, the kinetics with which X-linked genes become methylated during the differentiation of embryonic female somatic cells do not always correlate with the timing of transcriptional inactivation (Lock *et al.*, 1987). Thus other mechanisms must supplement any direct influence of DNA methylation on transcription within the inactive X chromosome, i.e. methylation might serve to maintain a state of repression, but not to establish repression. In contrast, if methylation is impaired in the developing mouse embryo by restricting the activity of DNA methyltransferase, major changes occur in the expression pattern of imprinted autosomal genes (Li *et al.*, 1993). A reduction in DNA methylation leads to the activation of the normally silent paternal allele of *H19*, the repression of the normally active paternal allele of the *Igf2* gene, and the repression of the normally active maternal allele of the *Igf2r* gene. These results dramatically demonstrate a key role for DNA methylation in imprinting, however they also show that there is no simple relationship between the methylation state of a gene and transcriptional activity. Determination of the nucleoprotein organization of imprinted autosomal genes having different levels of methylation will help to establish the molecular basis for the imprinting phenomenon.

Summary

Chromosomes are highly differentiated structures. Their precise structure varies depending upon the functional requirements of the nucleus and cell. A unifying feature of early metazoan development is the remodelling of chromatin structure from fertilization to the terminal differentiation of particular cell types. Chromatin assembled immediately after fertilization contains specialized proteins (HMG1, and variant core and linker histones) or post-translational modifications (acetylated core histones) that are known to assemble less stable nucleosomal structures or nucleosomal arrays. This may facilitate both *trans*-acting factor access to regulatory sequences and the rapid movement of replication forks during the early cleavage cycles. However, there are no simple rules concerning the contribution of these specialized chromatin structures to the transcription process. In *Xenopus* and *Drosophila* transcription in the early embryo is very low. In the sea urchin and the mouse, transcription is much more active. These variations will depend not only on the quantity and modification of chromatin structural proteins, but also on comparable variations in general and specific transcription factors. Nuclear organization and cell division frequency will also have an important influence on the transcription process especially the timing between replication and mitotic events.

As embryogenesis proceeds patterns of gene activity become progressively more restricted. This restriction correlates with the appearance of the histone variants found in normal somatic cells. Particular histone variants can potentially be causal for the repression of particular genes. These developmental transitions that correlate with cell determinative events may be established in large part by variation in *trans*-acting factor abundance or activity between different cell types, but they appear to be 'locked in' or maintained by transitions in chromatin structure. This stabilizing role might be dependent on histone H1, Polycomb or DNA methylation dependent on the repressed gene and organism. It is important to emphasize that the nucleoprotein complexes regulating gene expression will be specific, whether or not these complexes involve histones, HMGs or elements of higher order chromatin structure will depend on the particular gene examined (Section 4.2).

There is much to be discovered concerning how chromatin and chromosomal environments influence gene expression during development. The developmental literature contains many experiments involving the introduction of exogenous template DNA into embryos. In general this exogenous DNA only partially recapitulates the pattern of gene expression obtained from the same gene within the endogenous chromosomes (Davidson, 1986). These discrepancies are not due to differences in regulatory sequences, they are most probably due to differences in the packaging of those regulatory sequences into nucleoprotein complexes (see Archer *et al.*, 1992). This should be seen as an opportunity to determine the influence of chromosomal packaging on gene expression by attempting to recreate aspects of the chromosomal environment on exogenous DNA. This might require the physiological packaging of DNA into chromatin during replication, it might require chromodomain proteins such as Polycomb or specialized linker histone variants to be made available. Having established a repressive chromosomal architecture, other molecular machines like those containing the brahma protein or the replication fork might be necessary to relieve this repression and activate transcription.

CHAPTER FOUR

How do Nuclear Processes Occur in Chromatin?

At first sight the folding of DNA into a chromosome presents many impediments to any potential metabolic process requiring access to the double helix. Even though DNA is severely compacted, complex events such as replication, transcription, recombination and repair must occur efficiently in a chromatin environment. Evolution has been remarkably successful in shaping chromatin such that it does not prevent *trans*-acting factors from gaining access to specific DNA sequences or hinder polymerases from progressing along the chromatin fibre. We see that eukaryotic *trans*-acting factors have evolved to operate in a chromatin environment and that histones have evolved to let them function.

4.1 OVERVIEW OF NUCLEAR PROCESSES

Most regulated events involving DNA offer what appears to be a bewildering complexity of specific DNA sequences (*cis*-acting elements) and proteins (*trans*-acting factors) controlling a particular process. Although the individual proteins and DNA sequences regulating events differ for DNA replication, recombination, repair or transcription certain general principles apply to each. For eukaryotes, the regulation of transcription has by far the best understood molecular mechanisms; however, much of our insight into the control of metabolic events involving DNA was first established in a variety of prokaryotic systems.

4.1.1 The problem of specificity

The conventional approach to dissecting a complex process follows from the biochemical fractionation of crude extracts *in vitro*. The molecular dissection of the chromatin assembly process can be categorized in this way (Section 3.2). Kornberg and colleagues have explored in some considerable detail the molecular mechanisms controlling the highly regulated initiation of chromosomal replication in *Escherichia coli* (Kornberg, 1988; Kornberg and Baker, 1991). This event normally occurs once per cell generation at a single site selected from the entire *E. coli* genome (4×10^6 bp). This unique chromosomal origin (oriC, 245 bp in length) is recognized by a sequence-specific DNA-binding protein, DnaA, that associates with four non-contiguous 9 bp repeats. As the DnaA protein functions to determine specifically the site at which replication will initiate, it can be described as a specificity factor. The protein can interact with individual 9 bp repeats, but only when four are placed in the correct positions relative to each other can a large complex of DnaA protein and the oriC DNA sequence be formed. It is this large nucleoprotein complex that is recognized by the other proteins required for replication.

The DnaA protein not only interacts with DNA but also associates through direct protein–protein contacts with other DnaA molecules. This results in a co-operative association of 20–30 DnaA molecules with oriC. DNA is wrapped around the complex of DnaA protein rather like it is around the core histones (Section 2.2.2). The DnaA protein–DNA complex formed at oriC facilitates a specific duplex opening reaction in an adjacent AT-rich DNA sequence. The DnaC protein mediates the association of the DnaB helicase with this single-stranded AT-rich sequence unwinding a substantial segment of the double helix, at which replication enzymes such as DNA primase and polymerase begin to act. Interestingly, the protein most analogous to a histone in *E. coli*, the HU protein, facilitates the formation of a functional nucleoprotein complex at oriC, as do topoisomerases. The role of these proteins is probably to facilitate the correct topological arrangement of DNA for the subsequent replication events.

A similar pattern of events occurs during the initiation of replication in bacteriophage λ (Echols, 1986). The O-protein interacts as a dimer with four repeated sequences (18 bp). Electron microscopy reveals that a specific nucleoprotein structure is assembled at the λ origin, with DNA wrapped on the outside (the O-some). When the O-some is assembled on a negatively supercoiled DNA template, structural changes are induced in a 40 bp AT-rich sequence immediately adjacent to the recognition sites for the O-protein (Schnos *et al.*, 1988). Once again other proteins recognize the O-some and direct the initiation of replication (Fig. 4.1). The bacteriophage λ P protein interacts with the *E. coli* DnaB protein and with the O-some. The DnaB protein is the primary replicative DNA helicase of *E. coli* (LeBowitz and McMacken, 1986); however, it requires the *E. coli* DnaJ and DnaK proteins to also bind to the O-some before it can begin to unwind the DNA template (Learn *et al.*, 1993). Reconstruction of the normal physiological control of replication *in vitro* at the λ origin also requires inclusion of the histone-like HU protein (Mensa-Wilmot *et al.*, 1989). Presumably the wrapping of DNA by the HU protein into nucleosome-like structures (Rouviere-Yaniv *et al.*, 1979; Broyles and Pettijohn, 1986) establishes a level of supercoiling in the closed circular DNA molecules used for replication more typical of that found *in vivo*. Site-specific recombination by bacteriophage λ also employs proteins (Int) that bind at multiple sites (the att P site) to arrange DNA into a specific nucleoprotein complex (the intasome). A specialized HU protein, IHF (integration host factor), facilitates the recombination process both by assisting Int binding and by bending DNA (Yang and Nash, 1989).

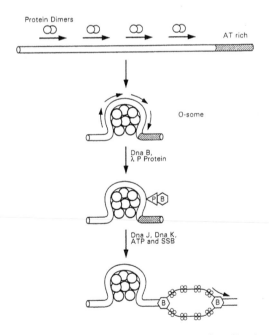

Figure 4.1. Formation of the O-some at the origin of replication of bacteriophage λ.

Several general rules follow from these analyses. All of these processes require exceptional precision, as do similar events in eukaryotic cells. Although DNA-binding proteins can interact with these specific sites with high affinity ($K_D = 10^{-9} - 10^{-13}$ M) they also bind DNA non-specifically ($K_D = 10^{-6} - 10^{-3}$ M). Thus the basis of the exceptional precision of replication and recombination events in *E. coli* is unlikely to follow from the binding of a *single* protein to a *single* DNA site. Instead, as illustrated by the examples, the co-operative interaction of a particular protein (the DnaA or O-proteins) with multiple sites over a 200–300 bp region of DNA is required for initiation of these processes. The precise organization of DNA into these complexes is necessary for other proteins to recognize the origin or integration site. This precise organization also requires the mediation of proteins that alter DNA conformation (HU or IHF) and remove topological constraints to the folding of DNA (topoisomerases).

As discussed by Echols (1986), not all metabolic events involving DNA require the level of precision or regulation inherent to chromosomal replication in *E. coli*. Transcription events in *E. coli* do not need a comparable level of accuracy since the occasional erroneous initiation event is unlikely to kill the cell (Reznikoff *et al.*, 1985). However, the human genome of 3×10^9 bp is roughly a thousand times greater in size and consequently more complex than that of *E. coli*. It is probable that the human genome contains over a thousand times more low affinity or non-specific protein binding sites than seen in *E. coli*. This number of sites will in principle have to be scanned through non-specific protein–DNA interactions before the correct binding region is found and specific binding occurs. The promoters of eukaryotic genes are not three orders of magnitude less accurately regulated than their prokaryotic counterparts. Therefore, it is safe to assume that the transcription process in eukaryotes has a level of precision comparable to prokaryotic replication or recombination.

The precision of regulated events in eukaryotic cells is determined both by following the prokaryotic paradigm and utilizing multiple high-affinity sequence-specific DNA-binding proteins that recognize multiple related sequence elements, and by masking many of the non-specific sites by folding them into chromatin (Lin and Riggs, 1975; Section 4.2.1). The first examples of multiple binding sites within DNA for particular sequence recognition proteins regulating replication or transcription in eukaryotic cells were determined in viral systems. The simian virus 40 genome has three binding sites for T-antigen at the origin of replication (ORI), a nucleosome free region in the minichromosome. Scanning transmission electron microscopy reveals that trimers and tetramers of T-antigen bind at each of these sites (Mastrangelo *et al.*, 1985). Like DnaA, T-antigen also causes significant changes in local DNA structure at the origin (Challberg and Kelly, 1989). Other viral genomes have a similar requirement for multiple binding sites to be occupied by a particular virally encoded protein in order to initiate replication. The nuclear antigen (EBNA-1) protein of Epstein–Barr virus has six specific binding sites in the viral replication origin region. Thus eukaryotic viruses are likely to regulate replication in a comparable way to *E. coli.*

In *Saccharomyces cerevisiae* chromosomes a multiprotein complex has been identified that interacts with over 150 bp of DNA at several functionally defined origins of chromosomal replication (or autonomously replicating sequences, ARS) (Bell and Stillman, 1992). This origin recognition complex (ORC) requires ATP to bind specifically to DNA, a property shared with SV40 T-antigen and the *E. coli* DnaA protein (Borowiec *et al.*, 1990; Kornberg and Baker, 1991). Although only a short consensus sequence is necessary for ORC binding, the final DNA sequence organized by the ORC is much more extensive (Bell *et al.*, 1993a,b), and resembles the *in vivo* footprint at a yeast origin of replication (Diffley and Cocker, 1992).

In contrast to this tight regulation of replication by sequence-specific proteins in viral systems and *S. cerevisiae*, the organization of chromosomal origins of replication in larger eukaryotic cells is presently unknown, but appears to rely in part on aspects of chromosomal structure and the nuclear envelope (Sections 2.5.9 and 4.3). This is a major gap in our current understanding of the molecular genetics of eukaryotic cells. The definition of orc-dependent and origin-specific initiation of DNA replication at defined foci in isolated yeast nuclei (Pasero *et al.*, 1997) potentially offers the opportunity of linking the cell biological characterization of metazoan origins with the molecular genetic dissection of mechanism possible in yeast.

The isolation and purification of eukaryotic transcription factors has allowed the definition of multiple sites of interaction for these proteins both in viral DNA and in normal cellular promoters. Perhaps the best studied example for this class of protein is the role of the transcription factor SP1 in transcriptional regulation. SP1 was originally defined as a promoter-specific transcription factor required for the efficient recognition by RNA polymerase II of the early and late promoters of SV40. SP1 was found to bind at six sites within the SV40 origin region stimulating transcription from both promoters, directed away from the origin (Dynan and Tjian, 1985). Multiple SP1-binding sites were subsequently defined in many normal cellular genes including four sites in the mouse dihydrofolate reductase promoter.

The significance of these multiple sites is not only the increase in the precision of transcriptional regulation, but also that a potential synergism exists in nucleoprotein complex formation mediated by protein–protein interactions between SP1 molecules. SP1 molecules bound to weak and strong sites help each other to bind to DNA, like the DnaA protein does on interaction with oriC (Courey *et al.*, 1989). This synergistic effect extends to SP1 sites some distance (> 1 kb) away from each other (Section 4.1.2). Therefore, the regulation of the specific initiation of transcription at the early and late promoter of SV40 resembles the control of replication initiation in *E. coli*.

Normal eukaryotic promoters are rather more complex than the regulatory elements of the SV40 early and late promoter. However, similar principles apply in the assembly of a regulatory nucleoprotein complex. The inducible enhancer of the human interferon β (IFN-β) gene has a well characterized series of combinatorial interactions between distinct regulatory elements (Thanos and Maniatis, 1992; Du *et al.*, 1993). The enhancer consists of multiple binding sites for transcription factors NF-κB, IRF1 and ATF2-c-jun. Although each binding site will not activate transcription independently, multi-merization facilitates the activation process. However, this multi-merization leads to a loss of regulation by the normal signals that induce transcription. Thus the activity of the intact enhancer is distinct from that of the individual elements. The context in which the elements are organized within three-dimensional space is essential for transcriptional regulation (Tjian and Maniatis, 1994).

The HMG-I/Y protein (Section 2.5.8) is essential for directing the appropriate folding of the IFN-β gene enhancer. HMG-I/Y binds to DNA within the NF-κB binding site and to two regions flanking the ATF2-c-jun site. HMG-I/Y increases the affinity of these factors for DNA and makes direct protein–protein contacts with the factors. The relative phasing of transcription factor and HMG-I/Y binding sites cannot be changed without interfering with the induction of transcription. Thus the assembly of an extended stereospecific nucleoprotein complex requiring extensive protein–protein and protein–DNA interactions regulates the human interferon β gene (Fig. 4.2) (see also Section 2.5.8)

Summary
The regulation of replication in *E. coli* requires the specific association of a DNA-binding protein with multiple copies of a particular DNA sequence. A large complex of DNA and protein is formed due to co-operative interactions between DNA bound and unbound protein

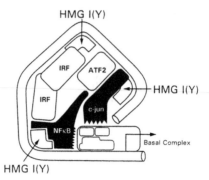

Figure 4.2. Model of nucleoprotein organization of the human interferon β gene enhancer showing the role of HMGI(Y) in directing DNA path.

molecules. Formation of this complex is facilitated by other auxiliary proteins that alter DNA conformation or remove topological constraints. The regulation of replication and transcription initiation in eukaryotic viruses and of transcription initiation in chromosomal promoters show many parallels with the control of *E. coli* replication.

4.1.2 Action at a distance

A common feature of eukaryotic transcriptional regulation is its control by two types of DNA sequences containing clusters of *trans*-acting factor binding sites. One type of sequence is generally located just 5′ to the start site of transcription (proximal promoter elements). The other DNA sequence is generally at some distance (several kilobases) away from the start site of transcription (enhancers). Enhancers have generally been defined through transient transfection assays to act in a position and orientation independent manner to stimulate transcription. It is important to note that physiologically relevant chromatin structures may not be assembled in these assays (Section 3.3). DNA sequences that act in a comparable way to inhibit transcription are known as silencers (Brand *et al.*, 1985). How enhancers work is not known, although they clearly facilitate the binding of *trans*-acting factors to the proximal promoter elements (Mattaj *et al.*, 1985; Weintraub, 1988). It is also possible that enhancers act by causing a local unravelling of the chromatin fibre thereby facilitating transcription factor access to DNA (Martin *et al.*, 1996;

Walters *et al.*, 1996). Another popular potential mechanism is that direct contact occurs between proximal promoter elements and the enhancer through looping-out the intervening DNA (Ptashne, 1986; Grosveld *et al.*, 1993).

Support for the looping-out hypothesis comes from examination of the regulation of transcription in *E. coli* (Gralla, 1991). Most *E. coli* promoters are regulated by the binding of activator or repressor proteins to one or two sites immediately adjacent to the start site of transcription. However, certain promoters, such as the L-arabinose BAD operon, are regulated by protein–DNA interactions over an extensive region requiring the formation of a DNA loop. For example a loop of DNA involving over 210 bp of DNA is mediated by the araC protein leading to the repression of transcription from the adjacent operon (Lobell and Schlief, 1990). The resulting nucleoprotein complex has many similarities to those that regulate replication and recombination (Section 4.1.1). An important limitation to the hypothesis that similar events occur in eukaryotes is that all of the *E. coli* loops are quite small (< 500 bp). In eukaryotes, enhancers can act over kilobases of DNA. It is possible that the removal of eukaryotic enhancers to greater distances is because the intervening DNA can be folded into nucleosomes and the chromatin fibre. Powerful evidence in support of looping in eukaryotes was obtained using the human β-globin locus introduced into transgenic mice (Wijgerde *et al.*, 1995). The locus control region (LCR) of this gene cluster was suggested to associate reversibly (or 'flip-flop') between two different promoters such that active transcription only occurred from the promoter in the presence of the LCR. The DNA between the LCR and the promoters, which is many kilobase pairs in length, was packaged up into a structure compatible with the 30 nm fibre.

Evidence in support of a structural role for chromatin in facilitating communication between enhancers and promoters comes from *in vitro* and *in vivo* approaches. On the *Xenopus* vitellogenin B1 gene, a positioned nucleosome *in vitro* constrains 160 bp between an enhancer and a promoter facilitating communication between these elements and consequently potentiating the transcription process (Fig. 4.3) (Schild *et al.*, 1993; Section 4.2.3). A comparable stimulatory role for a positioned nucleosome has been shown for the human U6 gene, where communication between two 5' regulatory elements separated by 150 bp is facilitated through DNA wrapping around the histones (Stunkel *et al.*, 1997). The capacity of Gal4-VP16, an artificial chimeric protein, to activate transcription *in vitro* when positioned over 1000 bp away from a promoter is facilitated by the assembly of intervening chromatin, in particular by the inclusion of histone H1 into the

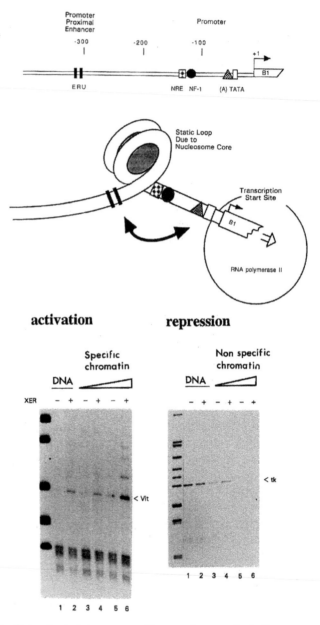

Figure 4.3. Role of a static loop created by a nucleosome including part of the *Xenopus* vitellogenin B1 promoter in potentiating transcription. Transcription from a specific chromatin template that positions nucleosomes increases with increasing numbers of nucleosomes. Transcription from a non-specific chromatin template in which nucleosomes are not positioned decreases with increasing numbers of nucleosomes. Reproduced, with permission from Schild, C. *et al.* (1993) *EMBO J.* **12**, 423–433. Copyright 1993 by IRL Press Ltd.

nucleosomal array (Laybourn and Kadonaga, 1992). This suggests that some form of higher-order structure might be necessary for communication between the Gal4-VP16 binding sites and the promoter. Finally the juxtaposition of regulatory elements within the rat prolactin gene separated by over 1300 bp is facilitated by chromatin assembly *in vivo* (Cullen *et al.*, 1993). Therefore, chromatin proteins have a role in the assembly of specific nucleoprotein architectures including the regulatory elements of eukaryotic genes just as IHF facilitates the assembly of the recombinational intasome (Yang and Nash, 1989).

The capacity of the eukaryotic protein SP1 to activate transcription from distant sites by looping out intervening DNA has been tested. The herpes simplex virus (HSV) thymidine kinase (tk) promoter contains two SP1-binding sites, which are sufficient to provide a measurable stimulation of transcription through the association of SP1. Insertion of six binding sites ~1.8 kb upstream of the tk promoter greatly stimulates transcription (a 90-fold induction occurs; Courey *et al.*, 1989). Direct interaction between these two sites was shown to take place by electron microscopy (Mastrangelo *et al.*, 1991; Su *et al.*, 1991). Such a model of nucleoprotein complex formation involving both enhancer and promoter DNA mediated by interactions with a common DNA-binding protein would be consistent with the capacity of enhancers to act even when not physically linked to a promoter. Other experiments have shown that enhancers can stimulate transcription from a promoter on a separate DNA molecule, when the two distinct DNA molecules containing the enhancer and promoter are intertwined and therefore constrained in space (Muller-Storm *et al.*, 1989).

Summary

Many eukaryotic genes are controlled not only at promoters adjacent to the start site of transcription but also by enhancers that can be several kilobases away. Looping out of intervening DNA to form a common nucleoprotein complex involving both enhancer and promoter DNA sequences might explain the action of enhancers at promoters over great distances. The assembly of chromatin can facilitate communication between enhancers and promoters.

4.1.3 The transcriptional machinery

SP1 is only one example of a plethora of sequence-specific DNA binding proteins that regulate transcription in eukaryotes. These different proteins may be grouped on the basis of structure (Section 4.1.6); however, they all have one thing in common. All of these factors directly or indirectly influence the function of the basal transcriptional machinery, in particular the assembly of a nucleoprotein complex that can be recognized by RNA polymerase (Mitchell and Tjian, 1989).

Roeder and colleagues have defined the basic components required for transcription by RNA polymerase II. These include the TFIID protein that binds to the TATA box (normally 30 bp 5′ to the transcription start site in mammals, as much as 100 bp 5′ in yeast). Binding of TFIID to the TATA box facilitates the stepwise association of the other general transcription factors (TF)II, –B and –F. RNA polymerase II can then associate followed by TFIIE and TFIIH (Fig. 4.4) (Zawel and Reinberg, 1992, 1993; Conaway and Conaway, 1993). Following an ATP-dependent activation step, transcription is initiated. Some repressors interfere directly with TFIID binding (Ohkuma *et al.*, 1990).

The TFIID protein is a remarkably stable complex of a *TATA box binding protein* TBP (Hahn *et al.*, 1988) and at least seven other proteins. These *TBP associated factors* for RNA polymerase II, TAFII

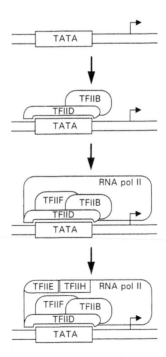

Figure 4.4. Assembly of the pre-initiation complex at a typical promoter containing a TATA box transcribed by RNA polymerase II.

(250, 150, 110, 80, 60, 40 and 30 kDa in size), are required to reconstitute transcriptional activation *in vitro* (Pugh and Tjian, 1990). Systematic analysis by Tjian and colleagues has defined the interaction of TAFIIs with sequence-specific DNA binding proteins (Verrijzer and Tjian, 1996). The glutamine-rich activator domains of bicoid and Sp1 bind to TAFII110, the acidic activation domain such as those in VP16 bind to TAFII40 and TAFII60, proline-rich activation domains such as those in NF1 bind human TAFII55 and the estrogen receptor binds hTAFII30 (Goodrich *et al.*, 1993; Tanese *et al.*, 1991; Hoey *et al.*, 1993; Wong and Tjian, 1994; Gill *et al.*, 1994; Ferreri *et al.*, 1994; Chen *et al.*, 1994b). Recent genetic experiments establish certain TAFIIs as essential for particular regulatory events in *Drosophila* (Sauer *et al.*, 1996), however TAFIIs are not essential for the majority of regulated transcriptional events in *S. cerevisiae* (Apone *et al.*, 1996; Walker *et al.*, 1997). This implies that some redundancy of function exists. The exact role of TAFIIs in transcription remains to be determined (Sauer and Tjian, 1997). It should be noted that connecting a promoter-bound protein to TBP eliminates the need for a transcriptional activation domain (Chatterjee and Struhl, 1995). So it should be remembered that recruiting TBP and some of the associated TAFs will be a key role for many activators (Horikoshi *et al.*, 1988). TBP is highly conserved through evolution and appears required for transcription by all three eukaryotic RNA polymerases (Hernandez, 1993). TBP binding severely distorts DNA through interactions with the minor groove of the TATA box. TBP is shaped like a saddle with two roughly symmetrical halves. The seat of the saddle is the site of many protein–protein interactions both with the TAFs and the general transcription factors (Nikolov and Burley, 1994).

An interesting feature of the TAFs is that at least two share significant sequence identity with archaebacterial histones and with eukaryotic histones H3 and H4 (Kokubo *et al.*, 1993; Section 2.5.1). This raises the interesting evolutionary possibility that components of the eukaryotic transcriptional machinery actually evolved from proteins that had the primary purpose of packaging DNA. This connection between the TAFs and the histones is reinforced by the identification of TAFII250 as a histone acetyltransferase (Mizzen *et al.*, 1996). This interesting protein which serves a central architectural role in the assembly of TFIID also has the capacity to acetylate TFIIE and TFIIF (Imhof *et al.*, 1997) (Fig. 4.5) and to phosphorylate TFIIF (Dikstein *et al.*, 1996). The acetyltransferase activity of TAFII250 is shared with other transcriptional coactivators such as p300/CBP and PCAF (Section 2.5.4). The function of this modification is presently unknown (Imhof *et al.*, 1997).

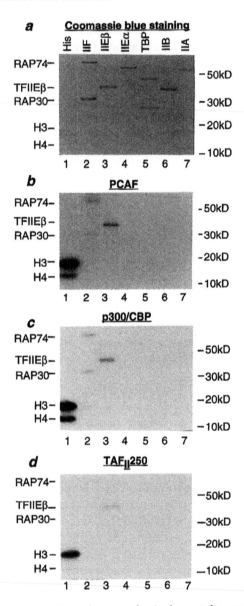

Figure 4.5. Histone acetyltransferases selectively acetylate general transcription factors.

Core histones and purified recombinant general transcription factors for RNA polymerase II were normalized by SDS PAGE and visualized by Coomassie blue staining (a), or acetylated in the presence of [³H]acetyl CoA by PCAF (b), by p300/CBP (c) or by TAF$_{II}$250 (d). Lane 1: core histones; lane 2: RAP74 and RAP30 subunits of TFIIF; lane 3: TFIIEβ(P34); lane 4: TFIIEα(p56); lane 5: TBP; lane 6: TFIIB; lane 7: TFIIA(p55+p12). Protein molecular weight standards in kilodaltons (kD) are indicated on the right. The smaller peptide (~28 kD) detected from TBP fraction is an *E. coli* protein copurified from Ni²⁺ agarose column.

TAFs also appear to make some contact with general transcription factors (Choy and Green, 1993). For example, TAFII40 appears to associate with TFIIB and TAFII250 with TFIIE and TFIIF. Since both TAFs and general transcription factors make contact with the same transcriptional activators, multiple independent contacts could stabilize the assembly of a large functional transcription complex.

TFIIB is the first general transcription factor to associate stably with the TFIID-TATA box complex (Buratowski, 1994). TFIIB interacts with TBP and with DNA sequences flanking both sides of the TATA box. The N-terminus of TFIIB is orientated towards the transcription start site. TFIIB has two domains, an N-terminal zinc finger domain and a carboxyl terminus that recognizes a complex of TBP with DNA (Ha *et al.*, 1993). The C-terminal domain of TFIIB stabilizes TBP-TATA interactions. TFIIB can bind simultaneously with TFIIA. TFIIA provides another stabilizing influence on the TBP-TATA complex (Roeder, 1996; see later). The zinc finger domain of TFIIB recruits the complex of RNA polymerase II and TFIIF to the complex. TFIIF is a two subunit protein and associates with RNA polymerase II even in the absence of DNA. It has functions in both transcriptional initiation and elongation (Price *et al.*, 1989; Flores *et al.*, 1992). TFIIF helps target RNA polymerase II to the preinitiation complex, it also reduces the non-specific association of the enzyme with DNA. The two subunits of TFIIF, RAP74 and RAP30 interact with TAFII250. RAD74 is a target for phosphorylation and acetylation by TAFII250 (see above). Both subunits function in the stabilization of the preinitiation complex and in transcription initiation (Tan *et al.*, 1994). TFIIF together with RNA polymerase II contacts DNA between the TFIIB contact sites (−23 to −14) and position +17 (relative to the start site of transcription at +1) (Roeder, 1996). The minimal complex of TFIID, TFIIB, TFIIF and RNA polymerase II can initiate transcription. Other activities regulate the efficiency with which this minimal complex is assembled and the efficiency with which RNA polymerase is released to elongate along the gene.

General transcription factor TFIIA acts at an early step in the assembly of the minimal transcription complex, probably facilitating the stable interaction of TFIID with DNA (Buratowski *et al.*, 1989). TFIIA appears to interfere with the action of several negative regulators of transcription (Auble and Hahn, 1993). The interaction of TFIIA with TFIID has been the focus of several recent studies concerned with transcriptional repression. Both Ada/Mot 1 and the negative co-factors NC1 and NC2 interfere with transcription complex assembly. Ada/Mot 1 is a member of the SWI family of ATPases that actively releases TBP from DNA (Auble *et al.*, 1994). NC1 was found to be HMG1 (Ge and Roeder, 1994a). HMG1 binds to TBP, such that TBP can no longer bind to TFIIA. NC2 represses transcription by competing with TFIIA and TFIIB for association with the TBP–DNA complex (Meisterernst and Roeder, 1991). NC2 consists of two subunits which contain histone fold domains similar to those of H2A and H2B (Goppelt *et al.*, 1996; Mermelstein *et al.*, 1996). NC2 wraps DNA and may act as a negative regulator by preventing exposure of the recognition sites for TFIIA and TFIIB. Thus the basal transcriptional machinery has two distinct components with histone-like structure: NC2 and TAFII40/TAFII60. These appear to exert opposing functions, since TAFII40 may help recruit TFIIB. Other studies on the TFIIA interaction with TFIID indicate that the complex undergoes an isomerization step that can be promoted by certain regulatory proteins, and that this can contribute to the overall transcriptional activation process (Chi and Carey, 1996). The complex of TFIID and the promoter, which can be remarkably stable under *in vitro* conditions, appears to be potentially destabilized in a regulated fashion within the cell (Buratowski, 1994; see also Wolffe and Brown, 1988).

Transcriptional activators such as SP1 that bind to DNA, and those like VP16 that are naturally tethered to DNA through interactions with other proteins, might target general transcription factors, TAFs and/or RNA polymerase to influence the transcription process (Smale *et al.*, 1990). Direct physical interactions were shown to occur between the acidic activation domain of VP16 and TBP (Stringer *et al.*, 1990). Mutations of the acidic activation domain in which single amino acids are changed and which prevent transcriptional activation by VP16 *in vivo*, also prevent stable interactions with TBP *in vitro* (Ingles *et al.*, 1991). The same mutations also influence VP16 interactions with TFIIB (Lin and Green, 1991). Importantly, TFIIB mutations that are defective in responding to transcriptional activators also fail to bind VP16 (Roberts *et al.*, 1993). Thus TBP (and by implication TFIID) and TFIIB represent targets for transcriptional activators.

More recent work has introduced additional complexity and interest into the targets for transcription activation domains. Genetic screens in yeast revealed that the ADA2p/ADA3p/GCN5p complex interacted with acidic transcription activation domains (Section 2.5.4). Biochemical fractionation revealed a complex of proteins described as 'upstream stimulatory activity' or USA (Meisterernst *et al.*, 1991). The USA fraction contained numerous proteins including PC4 (Ge and Roeder, 1994b; Kretzschmar *et al.*, 1994). PC4 interacts both with the acidic activation domain of VP16 and with TFIIA. This interaction helps recruit TFIIA to stabilize TFIID association with the TATA box.

In *S. cerevisiae* the PC4 homolog SUB1 plays a role in the release of TFIIB from the transcription complex during transcription initiation (Kraus *et al.*, 1996).

RNA polymerase II is recruited to the pre-initiation complex by contacts with TFIIB, TFIIF and potentially with transcriptional activators. The RNA polymerase II enzyme contains at least 12 subunits (Young, 1991). The most studied large subunit contains a repetitive carboxyl-terminal domain that appears to have a regulatory function. This repetitive domain is the site of extensive phosphorylation. This modification blocks incorporation of the polymerase into the initiation complex (Lu *et al.*, 1991; Chesnut *et al.*, 1992). It has been proposed that phosphorylation of the carboxyl-terminal tail domain is necessary to allow the polymerase to leave the initiation complex and begin elongation along the gene (Peterson and Tjian, 1992). Young, Kornberg and colleagues have demonstrated the existence of a 20 polypeptide complex known as the mediator that interacts with the C-terminal tail domain of the large subunit of RNA polymerase II (Thompson *et al.*, 1993; Kim *et al.*, 1994). This complex of the mediator and RNA polymerase II has become known as the holoenzyme (Bjorklund and Kim, 1996). A variety of genetic studies have established the mediator as essential for transcription of most, if not all promoters. The biochemical mechanisms by which the mediator functions remain to be established. There are marked differences between holoenzyme preparations in the literature: Young and colleagues report that TFIIB, TFIIH and components of the SWI/SNF complex are present in the complex (Wilson *et al.*, 1996), this is not the case for Kornberg and colleagues (Cairns *et al.*, 1996). A major limitation in characterizing components of the holoenzyme is the absence of genetic or structural foundation in the definition of biochemical purity. The human holoenzyme complex has been reported to contain numerous basal transcription factors, DNA repair proteins and transcriptional activators (Maldonado *et al.*, 1996).

Two general transcription factors, TFIIE and TFIIH, appear to regulate the transition from RNA polymerase II binding to polymerase elongation. TFIIE contains two subunits, the large subunit is a zinc finger protein (Maxon and Tjian, 1994). TFIIE recruits TFIIH to the initiation complex (Flores *et al.*, 1992). Genetic analysis reveals that *in vivo* TFIIE is essential for growth in yeast (Kuldell and Buratowski, 1997) TFIIH has a kinase activity that will phosphorylate the carboxyl-terminal domain of RNA polymerase. TFIIE stimulates this kinase activity (Ohkuma *et al.*, 1995). The TFIIH protein also contains DNA-dependent ATPase and DNA helicase activities (Serizawa *et al.*, 1993). TFIIE and TFIIH are not required for unwinding of DNA at the start site of transcription, but facilitate the conversion of the initiation complex into an elongation complex (Goodrich and Tjian, 1994). In the process they dissociate (Buratowski, 1994; Svejstrup *et al.*, 1996).

TFIIH is the most complex of all the basal components of the RNA polymerase II transcription factors. The protein contains at least nine polypeptides including the protein kinase CDK7 and cyclin H together with a 5' to 3' helicase, a 3' to 5' helicase, two zinc finger proteins and a ring finger protein. TFIIH functions not only in transcription but also in DNA repair (Svejstrup *et al.*, 1996; Section 4.4). The binding of TFIID itself to the TATA box is necessary to commit a promoter to be transcriptionally active. The TFIID protein remains bound to the promoter following transcription initiation and elongation by RNA polymerase II (van Dyke *et al.*, 1988). The stability of TFIID binding is an important regulatory principle for continued gene activity (Verrijzer *et al.*, 1995). *In vivo* footprinting studies also indicate that proteins remain associated with the TATA region independent of transcriptional activity in at least some genes (Wu, 1984; Mirkovitch and Darnell, 1991). Nevertheless, it is also clear that on active promoters some of the general transcription factors, including TFIIE and TFIIH, must reassociate with each new round of transcription initiation. Thus the activity of a promoter will be determined both by the initial association of TFIID, the retention of other general transcription factors and the reassociation of others (Kingston and Green, 1994). In contrast to the complexity of transcriptional regulation of the class II genes encoding mRNA, the proteins required to transcribe class III and class I genes are relatively simple (White, 1994). The study of the transcription of 5S RNA and tRNA genes (class III) by RNA polymerase III and of ribosomal RNA genes (class I) by RNA polymerase I has been very informative with respect to the influence of chromatin structure on the initiation of transcription and the consequences of transcription for chromatin structure (Sections 4.2 and 4.3). Much of our original insight into the transcriptional regulatory process was established with these genes.

Three proteins are required to assemble a transcription complex on a 5S RNA gene that can be recognized by RNA polymerase III (Segall *et al.*, 1980). Transcription factor (TF) IIIA is a promoter-specific DNA-binding protein that only recognizes the 5S RNA gene specifically, whereas TFIIIC and TFIIIB are proteins required for both 5S RNA and tRNA gene transcription. Unlike class II genes, no enhancer elements are known that can influence the efficiency of transcription complex formation on 5S or tRNA genes. Instead, an interesting hierarchy of transcription factor-DNA and protein–protein interactions between TFIIIA, B and C occurs that serves to regulate differential class III gene transcription.

TFIIIA is a simple protein consisting of an array of zinc-finger domains (Section 4.1.6) that interacts with DNA at a site within the 5S RNA gene called the internal control region. Although TFIIIA associates with DNA at a specific sequence, the binding affinity for DNA is low (K_D 10^{-9} M). However, TFIIIA has to form a complex with the 5S RNA gene before TFIIIC can be sequestered onto the gene. TFIIIC binding can influence the stability with which TFIIIA binds to the 5S RNA gene with important consequences for gene regulation (Hayes *et al.*, 1989; Section 4.1.5). Once TFIIIC is bound, the rate-limiting step in transcription complex formation occurs. This is the binding of TFIIIB to the TFIIIA-C 5S DNA complex. RNA polymerase can only recognize the transcription complex, and thus the 5S RNA gene, after TFIIIB has been sequestered. A similar process occurs on a tRNA gene, except that TFIIIA can be dispensed with. Here TFIIIC binds directly and specifically to the promoter elements of the gene. TFIIIB can then associate with the TFIIIC–tRNA gene complex and RNA polymerase can then recognize the gene (Fig. 4.6).

TFIIIA and TFIIIC can be described as assembly factors, since TFIIIB appears to be the only protein directly recognized by RNA polymerase III (Kassavetis *et al.*, 1989). Using *S. cerevisiae*, Geiduschek and colleagues were able to show that TFIIIA and C could be removed from a 5S RNA gene (or TFIIIC from a tRNA gene) leaving TFIIIB in place. Under these conditions multiple rounds of transcription initiation by RNA polymerase III could still occur. Surprisingly, specific DNA sequences within the 5′ flanking regions of class III genes are not essential for the efficient transcription of these genes, yet it is this region that TFIIIB interacts with. Moreover TFIIIB itself is not a DNA-binding protein in spite of containing TBP as a component (see later). This implies that TFIIIA and TFIIIC are not only essential for bringing TFIIIB to the class III gene, but also for activating its DNA-binding activity and precisely positioning it at the appropriate place to interact with RNA polymerase III. It is remarkable that TFIIIB binds so tightly to DNA that it can only be dissociated by chaotropic agents, yet this binding is activated by protein–protein interactions and is non-specific. TFIIIA and TFIIIC appear to fulfil the same function of directing the association of the protein recognized by RNA polymerase for class III genes that the large number of promoter-specific transcription factors do for class II genes. Gene regulation therefore concerns any process that influences the sequestration at the promoter of the key transcription factor: TFIIIB for class III genes, and TFIIB or TFIID for class II genes (Section 4.1.5).

S. cerevisiae TFIIIB is now known to be composed of three subunits including TBP (Roberts *et al.*, 1996; Ruth *et al.*, 1996; Kassavetis *et al.*,

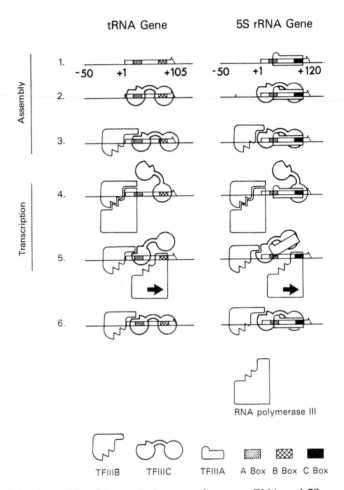

Figure 4.6. Assembly of transcription complexes on tRNA and 5S RNA genes, and a hypothetical mechanism for transcription through the complex without displacement of transcription factors. Transcription factors, their binding sites and RNA polymerase III are shown.

1995; see later). One of these subunits Brf, is regulated to TFIIB. Brf together with the third subunit b″ function together to bing to TFIIIC and recruit RNA polymerase III (Kumar *et al.*, 1997; Kassavetis *et al.*, 1997).

Several analogies exist between the transcription of class III, class II and class I genes (Paule, 1990; Hernandez, 1993). Ribosomal RNA gene transcription requires a single factor (TIF1) in *Acanthamoeba*, which remains in place through multiple transcription initiation events *in vitro* rather like TFIID and TFIIIB. However, in vertebrates two proteins are required called upstream binding factor (UBF) and a protein known as selectivity factor 1, SL1 (Jantzen *et al.*, 1990). SL1 and UBF interact co-operatively with DNA at the promoter element of class I genes (Bell *et al.*, 1988). This interaction requires UBF to provide the correct scaffolding for productive interaction of SL1 molecules with the class I gene promoter (Bazett-Jones *et al.*, 1994; Section 2.5.7). UBF also functions in a similar way to class II transcription factors in that aside from the promoters, it also binds to repetitive sequences 5' to the ribosomal RNA gene promoter that function as enhancer elements. Ribosomal RNA gene transcription is therefore an excellent example of promoter specificity and regulation being assisted by multiple protein–DNA interactions involving a limited number of proteins and repeated recognition sites (Echols, 1986, 1990).

The most remarkable feature in common between the transcription of all three classes of eukaryotic polymerase is the common utilization of TBP (Hernandez, 1993). The SL1 protein consists of TBP and three TAFs (Comai *et al.*, 1992; Heix *et al.*, 1997) and TFIIIB contains TBP and two TAFs, one of which is similar to TFIIB (White, 1994). The reason for this conserved role for TBP is unknown. Class I and most class III promoters do not contain TATA boxes, thus TBP is tethered in the vicinity of the start site of transcription by other proteins. The simplest possible explanation would be that TBP makes contact with some other common component of the transcriptional machinery, perhaps one of the polymerase subunits.

In contrast to the multiple promoter-specific DNA-binding proteins the complex multisubunit RNA polymerases have not been shown to have a role in facilitating the assembly of transcription complexes. However, the fact that they have high affinity for proteins associated with transcription complexes implies that certain transcription factors may associate with RNA polymerase in the absence of DNA. This has been shown to be the case for the general class III transcription factors TFIIIB and TFIIIC (Wingender *et al.*, 1986) and for the basal class II transcription factor TFIIF (Sopta *et al.*, 1989). Some of these proteins may always be associated with RNA polymerase *in vivo*, so the boundary between being a component of the transcription complex or of the polymerase becomes artificial. These arguments are carried to extremes with the multicomponent mammalian RNA polymerase holoenzyme (Maldonado *et al.*, 1996). However, the reader should always examine the definition of biochemical purity for each enzyme preparation. After transcription initiation and promoter clearance by RNA polymerase some of the non-DNA-binding transcription factors such as TFIIE and TFIIH dissociate, preventing a second transcription event by the enzyme until they have rebound (Van Dyke *et al.*, 1988;

Hai *et al.*, 1988). The DNA-binding proteins remain associated with the promoter and may be responsible for the phenomenon of template commitment, a function of a nucleoprotein complex unique to eukaryotes. This is the capacity of a nucleoprotein complex, once assembled to maintain the potential for a particular function such as transcription indefinitely (Section 4.1.4).

Summary
Certain general rules emerge concerning the transcription process relevant to all eukaryotic promoters. RNA polymerase (I, II or III) does not recognize naked DNA itself but a complex multiprotein structure including the promoter DNA sequence and the start site of transcription. Multiprotein transcription complexes undergo a highly ordered assembly process initiated by sequence-specific DNA-binding proteins. Non-DNA-binding proteins are sequestered by virtue of protein–protein contacts. The association of these additional proteins alters pre-existing DNA–protein interactions both qualitatively (extent or site of contacts) and quantitatively (the affinity of the whole complex for the promoter exceeds that of the DNA-binding proteins alone). Over 100 bp of DNA sequence is complexed with multiple proteins on each class (I, II and III) of eukaryotic promoter. Interactions between transcription factors over this length of DNA imply either a precise stereospecific orientation of individual proteins on the surface of the double-helix or that considerable flexibility exists in the structure of the bound proteins. One potential function of the extensive protein–DNA and protein–protein interactions at the promoters is to increase the fidelity of transcription initiation. Each specific protein–DNA and protein–protein interaction is a prerequisite for efficient transcription and provides a reference point for RNA polymerase in aligning the initiation site of transcription.

4.1.4 Stable and unstable transcription complexes

The stable sequestration of transcription factors on to promoter elements responsible for template commitment (Section 4.1.3) has been described for representatives of all classes (I, II and III) of eukaryotic gene (Brown, 1984). One assay for stable transcription complex formation involves the sequential addition of two genes to an *in vitro* transcription extract. If the first gene added binds all the limiting transcription factors in the extract, then the second gene will not be transcribed. If this is true *in vivo*, then stable transcription complexes might explain the maintenance of distinct patterns of gene activity in a terminally differentiated cell. Experiments involving the injection of genes into a living *Xenopus* oocyte nucleus support this idea. Transcription complexes assembled on a 5S RNA gene (*Xenopus* somatic type) are stable for several days, longer than the lifetime of some cells. Moreover, transcription complexes can be found in chromatin isolated from erythrocyte cells in which there are no longer free transcription factors (Darby *et al.*, 1988; Chipev and Wolffe, 1992). This suggests that transcription complexes, at least on 5S RNA genes, can under certain circumstances be stable for several weeks.

Regulatory nucleoprotein complexes are, however, reorganized as a consequence of oocyte maturation into an egg (Landsberger and Wolffe, 1997). This remodelling leads to transcriptional repression under normal physiological conditions. A regulated enhancement in the efficiency of chromatin assembly concomitant with meiotic maturation directs the displacement of the transcriptional machinery and an increase in the incorporation of regulatory DNA into nucleosomes. Either an increase in the abundance of transcriptional activators or a decrease in the efficiency of chromatin assembly can interfere with the repression of transcription normally seen on meiotic maturation. Addition either of repressive components of chromatin (histones) or of transcriptional activators can reversibly influence transcription even after oocyte maturation is complete. These results provide evidence for the existence of a dynamic competition driving the restructuring of the chromosomal environment towards the repression or activation of transcription (see Fig. 3.12).

The events of meiotic maturation on the hsp70 promoter can be explained by nucleosome-directed displacement of transcription factors in a process that is both competitive and reversible. The general conclusion is that the chromosomal environment is not necessarily passive, but in fact has a continual active role in transcription regulation. Other examples are consistent with this alternate vision of regulatory nucleoprotein complexes as dynamic entities.

Transcription complexes are not necessarily stable through the cell cycle (Segil *et al.*, 1996; Section 2.5.1); they can be erased by replication (Wolffe and Brown, 1986 see later) and at mitosis (Shermoen and O'Farrell, 1991; Herskovitz and Riggs, 1995; Martinez-Balbas *et al.*, 1995). The intrinsic instability of individual transcription complexes plays an important role in gene regulation (Wolffe and Brown, 1987, 1988; Kadonaga, 1990; Auble and Hahn, 1993; Auble *et al.*, 1994). Likewise nucleosomes are unstable (Meersseman *et al.*, 1992; Studitsky *et al.*, 1994; Ura *et al.*, 1995), components such as H2A/H2B and H1 exchange in and out of chromatin (Louters and Chalkley, 1985) and molecular machines exist that might facilitate chromatin remodelling (Cote *et al.*, 1994; Tsukiyama *et al.*, 1994; Cairns *et al.*, 1996; Kingston *et al.*, 1996). In light of the dynamic nature of regulatory nucleoprotein complexes it is not surprising that chromatin proteins influence transcriptional efficiency through their association with a promoter in competition with transcription factors. For the oocyte 5S rRNA genes in *Xenopus*, transcriptional efficiency during development depends on the competitive binding of TFIIIA versus histone H1 to a nucleosomal template (Bouvet *et al.*, 1994; Kandolf, 1994). A developmentally regulated increase in the abundance of histone H1 can drive the specific and dominant repression of the oocyte 5S rRNA genes (Section 3.4).

The stability of a transcription complex depends on the multiple protein–protein and protein–DNA interactions involved in its assembly (Section 4.2.3). This stability is distinct from the transition in DNA structure, required during transcription in prokaryotic systems, when the unstable closed complex is transformed into the stable transcriptionally active open complex in which DNA is unwound at the promoter (Hawley and McClure, 1982; Kassavetis *et al.*, 1989). At the eukaryotic transcription complex DNA appears to remain in the double helical form until RNA polymerase initiates transcription. The requirement for many interactions to generate a stable complex affords multiple opportunities for regulating these interactions and thereby modulating gene activity (Section 4.1.5).

An immediate problem in transcribing a class III gene is that all of the essential promoter elements are within the gene sequence (Ciliberto *et al.*, 1983). On a somatic 5S RNA gene, transcription factors associated with these sequences remain stably bound in spite of hundreds of transits by RNA polymerase III (Wolffe *et al.*, 1986). Experiments with bacteriophage SP6 RNA polymerase and RNA polymerase III revealed that the presence of multiple DNA binding proteins coupled together by protein–protein contacts allows individual proteins to anchor the complex to DNA (Wolffe *et al.*, 1986; Bardeleben *et al.*, 1994). Transient dissociation of any one contact need not lead to dissociation of the whole complex (see Fig. 4.6). A similar array of contacts might influence the stability of nucleosome structures during transcription (Section 4.3.4).

A very different result is seen when only a single transcription factor is associated with a promoter, as in the case of the *Acanthamoeba* class I ribosomal RNA gene. RNA polymerase passage through this promoter leads to the displacement of the single transcription factor

(Bateman and Paule, 1988). This result graphically demonstrates the disadvantage of only a limited number of protein–DNA contacts mediating transcription. Similar results have been obtained when TFIIIA alone is bound to a 5S RNA gene (Campbell and Setzer, 1991). This may also explain the significance of transcription termination sites being placed upstream of the promoters of genes arranged in tandem arrays (e.g. ribosomal RNA genes; McStay and Reeder, 1986). Inhibition of transcription from promoters in the path of a transcribing RNA polymerase is a well-known phenomenon for prokaryotic genes (Adhya and Gottesman, 1982; Horowitz and Platt, 1982).

The capacity to maintain protein–DNA interactions in place during transcription may be an important element of transcription complex structure contributing to the regulation of many genes. Some class II genes are known to have regulatory elements and protein–DNA complexes within either exons or introns (Banerji *et al.*, 1983; La Flamme *et al.*, 1987; Theulaz *et al.*, 1988). The capacity to have stable complexes assembled downstream of the promoter contributes another dimension of flexibility to eukaryotic gene regulation (Schaffner *et al.*, 1988). For example, the adenovirus major late promoter directs RNA polymerase to transcribe a gene which contains five other active promoters. Stable complexes are assembled on both the class III VA genes and class II promoters, and these appear to remain in place in spite of the RNA polymerase initiated at the major late promoter moving through them (Berk, 1986). The maintenance of a transcription complex in spite of transcription through it (Wolffe *et al.*, 1986) means that we should not perhaps be too surprised to find overlapping transcription units in the eukaryotic genome.

A related problem of maintaining specific protein–DNA inter-actions associated with a transcription complex occurs when a replication fork passes along a gene. What happens to the transcrip-tion factors comprising the complex may have important implications for the inheritance of patterns of gene activity in eukaryotic cells (Brown, 1984). The stable association of *trans*-acting factors with DNA through the replication event would be a simple way of imprinting a particular expression pattern on a promoter through development. Experiments that attempt to test the maintenance of transcription complexes through replication *in vivo* have generally made use of 'enhancer dependent' promoters. Enhancers facilitate the assembly of transcription complexes at promoters, and in some cases stimulate transcription initiation by over 100-fold (Section 4.1.2). In one experiment, Calame and colleagues established competition for simian virus 40 enhancer factors, after enhancer dependent transcription had been initiated on a gene (Wang and Calame, 1986). Transcription from the promoter of the gene continued in spite of the competition, indicating that transcription factors were stably sequestered at the promoter. Moreover, replication of the transcriptionally active gene did not inhibit transcription even in the presence of the competitor DNA. This shows that either enhancer action was not required to re-establish the transcription complex once it had been formed, or that the transcription complex on the promoter was stable to replication fork passage.

The problem of template commitment *in vivo* has often been investigated and discussed using immunoglobulin genes as examples. These genes alter their utilization of regulatory elements during lymphoid cell differentiation. The immunoglobulin heavy chain (IgH) enhancer is required to activate transcription from IgH promoters early in B-cell differentiation (Banerji *et al.*, 1983). However, several differentiated B-lymphoid cell lines exist that have deleted the IgH enhancer, but retain normal levels of IgH transcription (Wabl and Burrows, 1984; Klein *et al.*, 1985). There are several possible explanations for this result. One of these is that the IgH enhancer is required only for the establishment of the IgH promoter transcription complex early in B-cell differentiation, and later on the enhancer can be deleted and the transcription complex will remain in place in spite of cell division. Alternatively, the enhancer is required for maintenance of the IgH gene transcription complex during cell division, but when deleted can be replaced by other regulatory elements (Grosschedl and Marx, 1988). A third explanation might be that some other modification of active chromatin such as demethylation, which can be propagated at the replication fork, might explain continued gene activity (Kelley *et al.*, 1988; Section 2.5.7). Distinguishing between these possibilities is difficult and has not yet been achieved. The maintenance of the transcription complex through replication remains an attractive possibility. However, definitive proof of the stability of a transcription complex through cell division requires the physical structure of transcription complexes assembled on a particular promoter to be analysed before and after DNA replication.

Evidence that argues against the general maintenance of transcription complexes on all genes during replication comes from *in vitro* experiments in which the physical structure of a 5S RNA gene transcription complex was analysed before and after replication (Wolffe and Brown, 1986). In contrast to the stability of the transcription complex to transcription, replication fork progression disrupted the complex and displaced transcription factors. No selective advantage existed for rebinding factors to the daughter 5S RNA genes, that had initially had a transcription complex, compared to naked 5S DNA. Constitutively expressed genes such as the 5S RNA gene may not require stability to replication. Instead this property may be restricted to the complexes of tissue-specific genes. This experimental approach has recently been extended to the replication of 5S RNA genes and the tissue-specific β-globin gene in nuclei assembled in the *Xenopus* egg extract system (Wolffe, 1993; Barton and Emerson, 1994). In these nuclei, passage of a replication fork through active or inactive chromatin templates was without effect on the transcriptional activity of the template. These results suggest that molecular mechanisms exist to facilitate transcriptional reprogramming of genes in an appropriate chromosomal environment. As we shall discuss, chromatin assembly mechanisms and nucleosome positioning co-operate to allow the reprogramming of the 5S RNA gene with transcription factors in spite of the assembly of chromatin (Section 4.2.3).

A role for extensive co-operative protein–protein and protein–DNA interactions in maintaining transcription complex structure following replication therefore remains to be proven. Similar arguments have been made for the maintenance of specific chromatin structures, once again without proof (Section 4.3.2). It would be particularly attractive if some aspect of the nucleoprotein structure of a regulatory element might be maintained, thereby providing a molecular explanation for the establishment and maintenance of stable states of gene expression during embryonic development (Brown, 1984; Weintraub, 1985). As we have discussed, examples such as X-chromosome inactivation suggest that chromatin structure can imprint a state of gene expression on a chromosome that can be maintained through DNA replication and cell division (Section 2.5.6). The alternative to this type of imprinting is a continual regulation of a state of differentiation that is quite plastic and easily influenced by changes in the abundance of individual *trans*-acting factors that activate or repress genes (Sections 3.1.1 and 3.1.2).

Summary
Transcription complex structure may contribute to several important features characteristic of eukaryotic gene expression. The stable sequestration of transcription factors onto a gene by virtue of co-operative interactions between individual factors can explain the terminal differentiation of a cell type and the stability of a pattern of gene activity over long periods of time. This may be helped by the stability of a complex to processive enzyme complexes such as DNA and RNA polymerases and the compaction of a gene in chromatin. In

particular, the commitment of a gene to a continued state of activity in a given cell lineage might be explained by co-operative interactions between components of a transcription complex and maintenance of the structure through DNA replication and cell division. However, the limited existing experimental evidence supports the disruption of such complexes and the reassembly of chromatin structure *de novo* after every replication event. Other cellular processes such as meiosis or mitosis can also contribute to the remodelling of regulatory nucleoprotien complexes. The molecular mechanism by which the remodelling is achieved has not yet been defined.

4.1.5 Regulation of gene activity

Stable transcription complexes may allow a gene to be active indefinitely; however, many gene systems are regulated. For example, a gene that needs to be inactivated during development might make use of a transcription complex that is unstable. The gene would be active when transcription factors were present at high concentrations but inactive when levels fell below a certain threshold. This type of regulation is seen with the 5S RNA genes of *Xenopus laevis*. During embryogenesis the oocyte 5S RNA genes are turned off, whereas the somatic 5S RNA genes remain active (Wormington and Brown, 1983; Wakefield and Gurdon, 1983). Unlike the situation in *Saccharomyces*, the *Xenopus* transcription complexes depend on the interactions of TFIIIA with TFIIIC with the 5S RNA gene for their stability. Moreover, transcription complexes appear to retain TFIIIA and TFIIIC *in vitro* and *in vivo* (Wolffe and Morse, 1990; Chipev and Wolffe, 1992). Interestingly, TFIIIA and TFIIIC bind rapidly to the 5S RNA gene and are therefore the proteins that might have to compete with histones for access to DNA regulatory elements (Section 4.3.1). A reduction in transcription factor (TFIIIA and C) concentration during embryogenesis leads to the selective dissociation of oocyte 5S RNA gene transcription complexes. Furthermore, chromatin assembly both prevents transcription factors reassociating with the oocyte 5S DNA and directs the dissociation of transcription factors from genes leading to repression (Schlissel and Brown, 1984; Wolffe, 1989b; Bouvet *et al.*, 1994; Kandolf, 1994; Section 4.2.3). Changes in chromosomal composition, including specific transitions in the type of linker histone variant present within chromatin, can exploit differential transcription complex stability on the oocyte and somatic 5S RNA genes to direct the dominant and selective repression of the oocyte 5S RNA genes

(Bouvet *et al.*, 1994; Kandolf, 1994). Higher order chromatin structure is not required to maintain the repressed state (Gurdon *et al.*, 1982).

The DNA sequence differences within the 5S RNA genes responsible for this differential regulation appear to consist of only three base pairs. However, two transcription factors, TFIIIA and C, bind to this region of the gene (Pieler *et al.*, 1987; Wolffe, 1988). Changes in the binding affinity of each of the two proteins amplify their individual effect on complex stability. A major contribution to differential gene activity can be attributed to differences in the stability of protein–protein and protein–DNA interactions in oocyte or somatic 5S RNA gene transcription complexes (Fig. 4.7). Differences in the quality of association of the histone proteins also contribute to the final difference in gene expression between the oocyte and somatic 5S RNA genes of over 1000-fold (Wolffe, 1994c).

The linker histone present in the *Xenopus* oocyte when the oocyte 5S RNA genes are active is histone B4 (Smith *et al.*, 1988; Dimitrov *et al.*, 1993). Histone B4 binds much more weakly to nucleosomal DNA (K_D = 45 nM) than histone H1 (K_D = 2–7 nM) (Ura *et al.*, 1996; Nightingale *et al.*, 1996). Therefore, the stability with which histone H1 binds to chromatin is similar to that of the oocyte 5S RNA gene transcription complex. Histone H1, but not histone B4, restricts the mobility of histone octamers on the 5S RNA genes (Ura *et al.*, 1995, 1996) and selectively

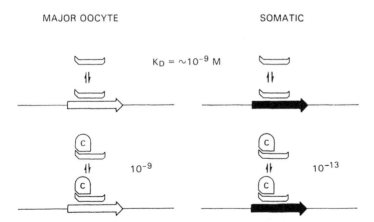

Figure 4.7. Combinatorial interaction of TFIIIC and TFIIIA with oocyte and somatic 5S RNA genes discriminates between them. TFIIIA (open box) binds with equivalent affinity (K_D = ~10^{-9} M) to both oocyte (major variant) and somatic 5S RNA genes. Oocyte genes are shown as open horizontal arrows whereas somatic genes are shown as solid horizontal arrows. In the presence of TFIIIC (C), TFIIIA binding to the oocyte 5S RNA gene is unchanged, whereas a very stable complex (K_D = 10^{-13} M) is formed with the somatic 5S RNA gene.

positions nucleosomes to occlude the TFIIIA binding site on the *Xenopus* oocytes 5S RNA genes (Chiper and Wolffe, 1992; Tomaszewski and Jerzmanowski, 1997).

The interaction between the basal transcriptional machinery and chromatin structure as encapsulated for the *Xenopus* somatic 5S RNA genes provides many opportunities for transcriptional regulation. A similar competition has been shown to exist on the yeast U6 gene where chromatin disruption induced by H4 depletion stimulated transcription of templates in which the promoter had been compromised by mutation, but not from the wild type gene. This indicates that normally transcription factors can efficiently overcome the repressive effects of chromatin assembly on the wild type gene (Marsolier *et al.*, 1995). All of this competition occurs within the context of positioned arrays of nucleosome *in vivo*. For genes transcribed by RNA polymerase II, the recruitment of various coactivators and/or corepressor complexes by sequence-specific DNA binding proteins will influence both the assembly and stability of the basal transcriptional machinery on the promoter. This influence will be exerted directly through protein–protein interactions and indirectly through the modification of chromatin (Section 2.5.4). These transitions in chromatin structure and function are dynamic and offer the opportunity for continual variation in transcriptional activity (Section 4.2.3). Not all transcriptional regulation will occur at the level of the assembly or disassembly of the pre-initiation complex. The transcriptional coactivators and corepressors also have the capacity to modify RNA polymerase or chromatin templates to regulate escape of RNA polymerase II from the pre-initiation complex and transcriptional elongation (Spencer and Groudine, 1990; Brown *et al.*, 1996b).

Many eukaryotic genes require enhancers for maximal gene activity and are regulated through changes in protein–DNA interactions at these sequences (Serfling *et al.*, 1985). Although the precise mechanism of enhancer action is not understood there are many similarities in the assembly of the nucleoprotein structures at both eukaryotic promoters and enhancers. Both complexes are made up of multiple sequence elements, each of which binds a cognate transcription factor (Zenke *et al.*, 1986; Wildeman *et al.*, 1986; Fromental *et al.*, 1988). Both may require stability to either transcription or replication (Schaffner *et al.*, 1988; Wang and Calame, 1986). Stable enhancer complexes may be important in maintaining tissue specificity, even though the activity of particular genes may change (Choi and Engel, 1988). Unstable enhancer complexes are important in regulating gene activity. For example, the mouse mammary tumor virus enhancer is only active when glucocorticoid receptor is bound (Yamamoto, 1985). Removal of the steroid hormone results in transcriptional inactivation, indicating that the glucocorticoid receptor is required for both establishment and maintenance of enhancer-mediated effects. The glucocorticoid receptor is known to exert its enhancer effect through specific chromatin structures (Section 4.2.3).

Many DNA-binding proteins are shared between enhancer sequences and promoters (Falkner and Zachau, 1984; Bienz and Pelham, 1986; Evans *et al.*, 1988), and conceivably non-DNA-binding proteins will also be shared. If the DNA-binding protein itself does not bind co-operatively like SP1, the interaction of a non-DNA-binding protein with a DNA-binding protein at two sites may provide a simple explanation for the possible looping between enhancer and promoter elements (Section 4.1.2). The distinction between transcription complexes and enhancer complexes may in fact be artificial. A single structure combining both elements affords much greater possibilities for each of the potential functions discussed above. For instance, one reason for the separation of enhancers and promoters on DNA over extensive distances may be that any one structure might be disrupted by DNA replication, whereas the other would remain intact (Fig. 4.8). If protein binding to one sequence element influences the binding of proteins to the other, then the intact nucleoprotein complex might facilitate the reformation of the disrupted one (Wolffe, 1990a).

Summary
Eukaryotic transcription complexes have the essential role of directing the accurate and efficient initiation of transcription by RNA polymerase on a particular gene. The focus of much current research in molecular biology lies in understanding the regulation of the frequency with which RNA polymerase initiates transcription. The multiplicity of proteins and protein–DNA interactions involved in assembling transcription complexes of different stabilities affords many opportunities for regulating complex structure and therefore transcription initiation itself. For example, this could occur through combinatorial effects of multiple proteins binding to particular DNA sequences or by transiently associating and dissociating a particular transcription factor. It is also increasingly apparent that both general and targeted modifications of chromatin structure have a regulatory contribution.

4.1.6 Sequence-specific DNA-binding proteins

It has been recognized that proteins interacting specifically, or non-specifically, with nucleic acids fall into several distinct structural

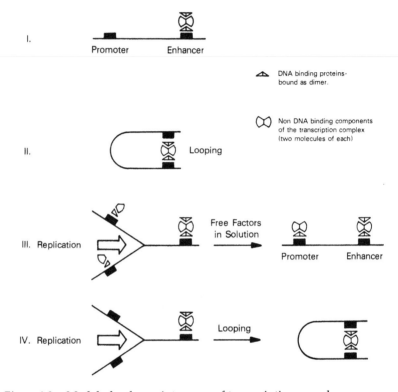

Figure 4.8. Models for the maintenance of transcription complexes through replication.

The regulatory regions of a gene are shown as a promoter and an enhancer (bars). Open boxes are DNA binding or non-DNA binding components of a transcription complex. (I) In this case similar factors are shared between enhancers and promoters. (II) The sequestration of transcription factors onto the promoter (and its activation) is facilitated by looping of the intervening DNA between enhancer and promoter. Common DNA binding proteins associate with both the enhancer and promoter. The bound proteins interact with a common non-DNA binding protein. (III) DNA replication disrupts the transcription complex on the promoter splitting it in half. The remaining transcription factors can sequester free factors from solution generating a complete transcription complex on each daughter chromatid. In this case the enhancer is required for establishment, but not maintenance of the transcription complex. (IV) Alternatively DNA replication again disrupts the complex on the promoter, displacing transcription factors, however because of the distance between enhancer and promoter, the enhancer complex remains intact. DNA looping can establish a new transcription complex on the promoter. Here, the enhancer is required for both establishment and maintenance of the transcription complex through cell division.

classes dependent on what motifs are present in their amino acid sequences (Johnson and McKnight, 1989; Harrison, 1991). In addition, the motifs that can be discerned in the peptide sequence often represent distinct domains or modules of structure (Frankel and Kim, 1991). As such they can be interchanged using molecular genetics (or evolution) to create new proteins with new functions.

In several instances the motifs present in eukaryotic transcription factors have been shown to adopt highly ordered conformations. In general these either represent DNA-binding or multimerization domains. Well characterized DNA-binding domains include the helix-turn-helix found in homeodomain proteins. This is a unit of three α-helices that binds DNA primarily through contacts in the major groove (Otting *et al.*, 1990; Kissinger *et al.*, 1990; Fig. 4.9). An important variation on the conventional helix-turn-helix motif is the 'winged helix' found in transcription factors such as HNF3 and forkhead. This structure consists of three α-helices within the amino terminal half of the protein which adopt a compact structure that presents one of the α-helices to the major groove of DNA (Clark *et al.*, 1993b). The remainder of the nucleic acid binding domain consists of a twisted, antiparallel β-structure and random coil that interacts with the minor groove. This DNA-binding domain has a striking similarity to the structured domain of the linker histone H5 (Ramakrishnan *et al.*, 1993) (Fig. 4.10). Linker histones are known to influence nucleosome positioning (Meerssemann *et al.*, 1991; Chipev and Wolffe, 1992), and it is of great interest that HNF3 will direct the positioning of histone–DNA contacts on the enhancer of the serum albumin gene (McPherson

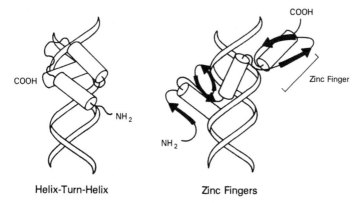

Figure 4.9. Binding of helix-turn-helix and a three zinc finger protein to DNA via contacts in the major groove.
α-helical regions of the protein are shown as cylinders and β sheet regions as opposed arrows.

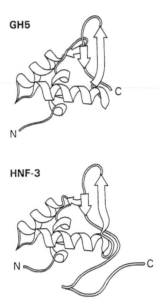

Figure 4.10. A structural comparison between the structured domain of H5 (Ramakrishnan *et al.*, 1993) and transcription factor HNF-3 (Clark *et al.*, 1993b).

et al., 1993; Fig. 2.45). It seems probable that both nucleosome positioning events involve similar interactions by the linker histone and HNF3 with DNA and core histones in the nucleosome.

The similarity in the structures of HNF3 and histone H5 echoes those between TAFII40, TAFII60, histone H3 and histone H4 and those between HMG1 and 2 and LEF1, SRY and UBF (Sections 2.5.8 and 4.1.3). There are two interpretations of these structural identities: either they are fortuitous, representing a limited number of structures with which a DNA-binding protein can be built, or more interestingly, they represent evolutionary and potentially functional relationships of the histones and HMG proteins with the transcription factors. Transcription factors function very effectively within a chromatin environment and it is difficult to imagine a more efficient way of facilitating their function than to have them resemble the normal structural components of chromatin.

The first DNA-binding domain recognized in eukaryotic proteins was the zinc finger. This structure consists of a two-stranded β-sheet and an α-helix in which a single zinc ion is tetrahedrally co-ordinated to two cysteine and two histidine side chains (Lee *et al.*, 1989; Klevit *et al.*, 1990; Pavletich and Pabo, 1991). Like the helix-turn-helix proteins, the zinc finger α-helix recognizes specific base pairs in the major groove (Fig. 4.9). A modification of the basic zinc finger domain is found in the glucocorticoid and oestrogen receptors. This domain contains two α-helices that tetrahedrally co-ordinate two zinc ions through cysteine side chains. One α-helix from each monomer of the receptor dimer is believed to interact with the major groove of DNA (Hard *et al.*, 1990; Schwabe *et al.*, 1990). The different DNA-binding motifs also have distinct requirements for sequence specific recognition of B-form DNA in solution. These requirements may change when DNA is distorted through interaction with histones. In contrast to our rather detailed knowledge of protein–DNA interactions, relatively little is known about protein–protein contacts. One dimerization domain that is well characterized structurally is the leucine zipper. This is a two-stranded, parallel coiled-coil structure (Rasmussen *et al.*, 1991). The multiplicity of domains found in different proteins presumably reflects the opportunity for multiple independent protein–DNA and protein–protein contacts, each of which might be altered with important consequences for gene regulation (Section 4.1.5). It is also possible that nature has found multiple solutions to a single problem.

Several other multimerization and DNA-binding domains have been characterized, including the helix-loop-helix motif and the cold-shock domain of the Y-box proteins (Murre *et al.*, 1989; Tafuri and Wolffe, 1990). The Y-box proteins contain the most evolutionarily conserved structure yet defined, a block of 70 amino acids that has an identity of 42% between *E. coli* and humans. The structure of the cold-shock domain is a five-stranded β-barrel (Schindelin *et al.*, 1993; Schnuckel *et al.*, 1993).

Transcription factors also contain regions of primary amino acid sequence that are involved in their stimulatory action on the transcription process. These include regions rich in acidic amino acids, proline or glutamine (Mitchell and Tjian, 1989). The structure of these regions remains unknown; in fact NMR and circular dichroism (CD) studies reveal no evidence of secondary structure. Mutagenesis studies have suggested that the most important amino acids involved in transcriptional activation may not be the predominant residues such as glutamines or acidic amino acids (Cress and Triezenberg, 1991; Gill *et al.*, 1994). Large hydrophobic amino acids found in the glutamine or acidic activation domains are found to be important for transcriptional activation. This suggests that specificity might follow from hydrophobic interactions (Tjian and Maniatis, 1994). It has also been suggested that multiple weak interactions made between transcription factors might induce structure (induced fit). This phenomenon would potentially allow a number of different stable protein–protein interactions to be generated from a relatively plastic starting conformation. Such conformational transitions have some potential advantages. For example, specificity would be increased if the induced fit only occurred after other components in an extended nucleoprotein complex were incorporated. Two of the few examples of a structurally defined activation domain are the ligand-binding domains of the retinoic acid and thyroid hormone receptors (Renauld *et al.*, 1995; Wagner *et al.*, 1995). In these cases the binding of hormone directs a large conformational transition that repositions an entire amphipathic α helix to generate a functional AF-2 activation domain.

Activation domains have been proven to interact with general transcription factors such as TFIIB and with TFIID, via contacts with TBP and/or the TAFs (Section 4.1.3). Although the activation domains are believed to function through contacts with other non-histone proteins they might also function through recruiting enzymes or molecular machines (Section 2.5.4) that direct the modification of chromatin or nucleosome conformation (Section 4.2.2) or by directing the compartmentalization of promoters to transcriptionally competent regions of the nucleus (Section 2.5.9).

Summary
Trans-acting factors generally have a modular structure with distinct DNA-binding and protein–protein interaction domains. DNA-binding is generally through sequence-specific contacts in the major groove. The modular structure facilitates multiple independent interactions with DNA and protein, each of which might be regulated to control a process.

4.1.7 Problems for nuclear processes in chromatin

A consideration of the complexity of nuclear processes suggests that the formation of the large nucleoprotein complexes required to control these events might be incompatible with the folding of DNA into chromatin and the chromosome. How can several regions of 100–500 bp exist, each requiring the association of multiple DNA-binding proteins to facilitate DNA replication, recombination, repair and transcription; when this DNA may also be wrapped around the core histones and folded into the chromatin fibre? Moreover, how can DNA polymerase and RNA polymerase progress through arrays of nucleosomes and the chromatin fibre even if access to the DNA duplex is achieved? Methodologies to approach these questions have only recently become available. Potential molecular mechanisms to explain the access of *trans*-acting factors to DNA and the progression of polymerases through the chromatin fibre are beginning to be uncovered (Sections 4.2 and 4.3).

Since the packaging of DNA into nucleosomes and the chromatin fibre had been thought to remove DNA from any process of interest in the nucleus, experimental analysis through much of the 1980s focused either on the regulation of naked DNA templates *in vitro* or on templates uncharacterized with respect to chromatin structure through transient transfection of cells (Section 3.3). Results obtained through these analyses are increasingly seen to be oversimplifications of the subtlety and complexity with which genes are regulated in their natural environment – the chromosome (Section 4.2). Progress in several experimental systems has clearly shown that promoter elements are specifically organized within and between nucleosomes, and that the regulation of a gene depends on the organization of DNA in a chromatin template (Simpson, 1991). It has also been demonstrated conclusively that nucleosomes, including histone H1, are present on the majority of transcribed genes (Morse, 1992). Replication obviously has to duplicate both DNA and nucleoprotein complexes in order to create two chromosomes out of one. Understanding how these transcription and replication events occur in the context of chromatin will be seen to have regulatory significance important for all nuclear processes.

Summary

Trans-acting factors form large complexes with DNA in spite of the many apparent obstacles due to the concomitant assembly of chromatin. Likewise, processive enzyme complexes function effectively in a chromatin environment. How these events occur in chromatin structures is only beginning to be understood. Understanding the molecular processes that overcome the many apparent impediments to function should uncover regulatory mechanisms unique to eukaryotes.

4.2 INTERACTION OF *TRANS*-ACTING FACTORS WITH CHROMATIN

The difficulties inherent in having non-histone proteins gain access to a histone-covered template were recognized even before the nucleosome model was developed. Experimental approaches to this problem have continually been refined as the first non-histone proteins were purified and their binding sites on DNA defined. More recently, methodologies for determining specific chromatin structures have been developed. There has been a gradual trend from studying non-specific DNA–protein interactions towards recognition of the role of specific chromatin structures in mediating the function of *trans*-acting factors.

4.2.1 Non-specific interactions

Our knowledge of the accessibility of non-histone proteins to DNA in chromatin has progressed slowly. It has long been known that RNA synthesis using bacterial or bacteriophage RNA polymerase is more efficient from naked DNA than from chromatin, and that the histones are responsible for this inhibition (Georgiev, 1969). These observations led to the idea that DNA was uniformly coated with histones that prevented RNA polymerase from reaching the template. The first experiments to suggest that DNA was not uniformly covered with histones were those of Felsenfeld and colleagues (Clark and Felsenfeld, 1971). Polylysine precipitation of DNA, naked or as chromatin, revealed that as much as 50% of the DNA in chromatin was accessible to the polycation and therefore presumably naked. This number was very similar to the amount of DNA that could be made acid soluble by nucleases. Clearly some DNA sequences in chromatin were more accessible than others.

The development of the nucleosome concept (Section 2.2) led investigators to explore the relative accessibility of linker DNA to DNA-binding proteins compared to DNA tightly associated with the core histones (core DNA). From the initial definition of the nucleosome through the action of endogenous nucleases, linker DNA is by definition more readily cleaved by these enzymes. Bacterial and bacteriophage RNA polymerases do not transcribe eukaryotic genes with any specificity; however, they will initiate transcription at AT-rich sequences resembling natural prokaryotic promoter elements (Maryanka *et al.*, 1979; Pays *et al.*, 1979). These enzymes have been very useful in assessing the relative accessibility of core versus linker DNA in chromatin. Early studies suggested that DNA in the nucleosome is not accessible to *E. coli* RNA polymerase (Cedar and Felsenfeld, 1973; Williamson and Felsenfeld, 1978; Wasylyk and Chambon, 1979). A detailed analysis by Gould and colleagues revealed that linker DNA was more accessible than core DNA to *E. coli* RNA polymerase (Hannon *et al.*, 1984). Titration of linker DNA availability through the addition of histone H1 revealed a rapid decline in accessibility, probably reflecting not only occlusion of linker DNA but also folding of the chromatin fibre. This extensive occlusion of DNA through relatively small changes in linker histone concentration may have significant consequences for the access of other *trans*-acting factors (Section 4.2).

These studies were extended to the problem of how a DNA-binding protein (*E. coli* RNA polymerase) might search for its binding sites in a nucleosomal array (Hannon *et al.*, 1986). Surprisingly, this search occurred with equivalent efficiency in both naked DNA and chromatin that had been depleted of histone H1. Two mechanisms have been envisaged for such a search, either 'sliding' of the DNA-binding protein from site to site or 'hopping' between sites (Berg *et al.*, 1981). As *E. coli* RNA polymerase was known not to be able to slide or progress efficiently through nucleosomes (Section 4.3), it was concluded that the enzyme was able to hop between sites efficiently in a chromatin template. These sites are the regions of relatively accessible linker DNA. Removal of histones H2A/H2B from chromatin increases the accessibility of DNA to RNA polymerase even more (Baer and Rhodes, 1983; Gonzalez and Palacian, 1989). In the chromosomal context the search by RNA polymerase for binding sites is probably an accurate reflection of the search of *trans*-acting factors for recognition sequences. As we discuss later, eukaryotic RNA polymerases recognize transcription complexes, not naked DNA (Sections 4.2.3 and 4.2.4).

Several interesting biological examples exist of changes in RNA polymerase accessibility to chromatin through development. Brown and colleagues were able to document that the normal somatic form of histone H1 was responsible for maintaining the repression of certain types of class III genes in *Xenopus* somatic cells (Schlissel and Brown, 1984). This repressed state is established gradually during development as the amount of somatic histone H1 increases in chromatin (Wolffe, 1989b; Bouvet *et al.*, 1994; Section 3.4). The accumulation of histone H1 causes a general decline in the accessibility of RNA polymerase III to DNA (Andrews *et al.*, 1991; Wolffe, 1991a; Bouvet *et al.*, 1994; Fig. 4.11). This decline in access to *trans*-acting factors is due to changes in chromatin structure. This change also correlates with the cessation of rapid cell division events and the imposition of a normal cell cycle.

Experiments with 'non-specific' prokaryotic RNA polymerases were responsible for the first demonstration that chromatin structure

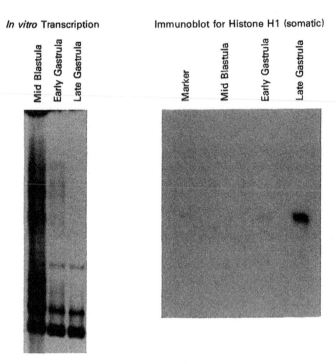

Figure 4.11. The decrease in transcription by RNA polymerase III correlates with the increase of somatic histone H1 in *Xenopus* embryonic chromatin.

Transcription *in vitro* of embryonic chromatin isolated from different developmental stages. Radioactive transcripts were resolved on a denaturing acrylamide gel (specific transcripts are indicated). An immunoblot of *Xenopus* somatic histone H1 in these different chromatin preparations is also shown (see Wolffe, 1989b for details).

over a gene isolated from tissues in which the gene was active, differed from that in which it was repressed. These results followed from the relatively easy access of these polymerases to the DNA of transcriptionally active chromatin. Similar results were later obtained using nucleases (Section 4.2.4). Unfortunately, the specificity of transcription was not improved upon using purified eukaryotic RNA polymerases (I, II and III). No eukaryotic RNA polymerase faithfully transcribes specific genes using purified DNA templates. Roeder and colleagues were responsible for the seminal demonstration that either a natural chromatin template isolated from a cell nucleus or a template reconstituted with transcription factors was necessary for recognition of a gene by RNA polymerase (Parker and Roeder, 1977). At this point the focus of research on gene regulation shifted from the properties of the chromatin template to the properties of the promoter-specific transcription factors (Section 4.1.3).

Summary

Prokaryotic polymerases have been very useful in defining the accessibility of DNA in chromatin to other DNA-binding proteins such as *trans*-acting factors. Like nucleases they preferentially recognize accessible linker DNA between nucleosomes rather than DNA wrapped around the histone core. They also associate selectively with chromatin that is transcriptionally active *in vivo*. Unlike nucleases, they do not destroy the template. Chromatin prevents access of RNA polymerase to DNA; nevertheless, the protein can effectively search out binding sites within exposed linker DNA by 'hopping' between sites.

4.2.2 Specific *trans*-acting factors and non-specific chromatin

The availability of *in vitro* transcription systems employing both specific *cis*-acting elements and *trans*-acting factors for the initiation of transcription by RNA polymerase has led to an increasing number of experiments in which the influence of chromatin structure on transcription has been investigated. A popular experiment with a long history has been to mix a DNA template with histones or a nucleosome assembly system and then to ask whether transcription could still occur. This experiment is responsible for the general belief in the repressive nature of histone–DNA interactions, since the usual result is that the addition of histones inhibits the given process. Although some investigators have undertaken numerous experimental controls to eliminate artefacts, there are often several possible explanations for the observed inhibitory effects that must be excluded.

DNA can precipitate or aggregate following a non-specific association with histones. For example, linker histones (histones H1 or H5) are notorious for forming aggregates on DNA sometimes causing precipitation (Jerzmanowski and Cole, 1990). Most investigators attempt to exclude this possibility by examining the supercoiling or micrococcal nuclease cleavage patterns of their DNA template after nucleosome assembly. Each nucleosome should introduce one negative superhelical turn in the presence of topoisomerase into a closed circular DNA molecule (Fig. 4.12), and protect approximately

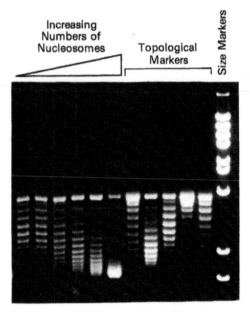

Figure 4.12.   Introduction of supercoils into DNA with increasing numbers of nucleosomes.

An agarose gel resolving topoisomers of a small closed circular plasmid is shown. DNA in the far left lane has very few nucleosomes, more are added (by salt-urea dialysis) from left to right as indicated. The DNA is relaxed with topoisomerase I before deproteinization and resolution on the gel. Markers are also resolved so that the number of nucleosomes can be accurately counted.

146 bp of DNA from micrococcal nuclease (Section 2.2). If these events occur, some fraction of the template must be in solution and contain nucleosomes. However, subnucleosomal particles and proteolysed nucleosomes will also supercoil DNA and protect it from nucleases, therefore these assays give no guarantee of nucleosome integrity (Section 2.2.4). Unfortunately, it is all too easy to detect a few superhelical turns or to detect a single nucleosome length fragment of DNA after micrococcal nuclease, but more difficult to prove that the DNA molecule is efficiently (> 50%) assembled with nucleosomes. Nucleosomes and chromatin are also intrinsically unstable at low concentrations ($\sim 1\,\mu g/\mu l$), so where transcription factors or molecular machines are reported to disrupt histone–DNA interactions, controls should be shown to demonstrate that this disruption is dependent on targeting, such as the presence of a DNA-binding site for the transcription factor (Lilley *et al.*, 1979; Cotton and Hamkalo, 1981; Ausio *et al.*, 1984; Godde and Wolffe, 1995).

Another difficulty with these reconstitution experiments is excluding the possibility that the template is also associated with non-specific DNA-binding proteins. These proteins are often present in crude nucleosome assembly extracts and might also occlude *cis*-acting sequences (Croston *et al.*, 1991). Even if efficient nucleosome assembly does occur, the various systems do not always position nucleosomes as found in the chromosome (Section 4.2.3), nor do they always correctly space nucleosomes as would be found *in vivo* (Section 3.2). These discrepancies might be explained by the fact that *in vivo*, chromatin assembly is coupled to the replication of DNA, special chromatin assembly factors are employed, the histones are post-translationally modified, and nucleosome assembly is staged (Section 3.2). Thus, it is not surprising that the prior association of unmodified histones with DNA under artificial conditions often leads to repressive effects. The physiological significance of the repression may be questionable.

An additional problem is that the unusual composition of chromatin assembled in *Xenopus* and *Drosophila* oocyte, egg or embryo extracts might lead to significant differences in the transcriptional properties of a promoter to those found with the chromatin of normal somatic nuclei. The chromatin assembled in the extracts contains unusual core histone and linker histone variants and much larger amounts of proteins like HMG2 than are normally found. Proteins such as HMG2 might repress transcription independent of chromatin assembly (Ge and Roeder, 1994a). With these reservations in mind it is possible to evaluate critically the large body of data on specific transcription factors and non-specific chromatin.

The pioneering studies of Brown, Roeder and colleagues in developing *in vitro* transcription systems led class III genes to be most intensively studied by this type of analysis (Section 4.1.3). Several experiments showed that mixing histones with class III genes would prevent their transcription (Bogenhagen *et al.*, 1982; Gargiulo *et al.*, 1985); however, the organization of the chromatin template was not characterized. In contrast, it was found that *in vivo*, correctly spaced nucleosomes actually correlated with efficient 5S RNA gene transcription (Weisbrod *et al.*, 1982; Gargiulo and Worcel, 1983). An important conclusion was that the transcription factors had to gain access to DNA before the histones if transcription was to occur (Gottesfeld and Bloomer, 1982; Bogenhagen *et al.*, 1982; Gargiulo *et al.*, 1985). Subsequent studies have shown that a complete transcription complex (TFIIIA, B and C) assembled onto a 5S RNA gene is more resistant to chromatin-mediated repression than the TFIIIA–5S RNA complex alone (Felts *et al.*, 1990; Tremethick *et al.*, 1990). The experiments discussed to this point use chromatin assembly systems that are deficient in histone H1. Reconstitution of histone H1 into chromatin such that the interaction alters nucleosome spacing (with questionable physiological relevance; Section 3.2) allows inhibition of 5S RNA gene transcription at a lower density of nucleosomes per length of DNA sequence than is otherwise required. In the absence of histone H1, very high nucleosome densities (one every 160–180 bp) or removal of free DNA with restriction endonuclease are required to repress transcription (Shimamura *et al.*, 1988, 1989; Morse, 1989; Clark and Wolffe, 1991; Fig. 4.13). Chromatin appears to be repressive, yet if transcription factors gain access to DNA first, RNA polymerase III has no problems finding the transcription complex in chromatin.

A similar series of experiments has examined the effect of nucleosome assembly on the transcription of class II genes (Lorch

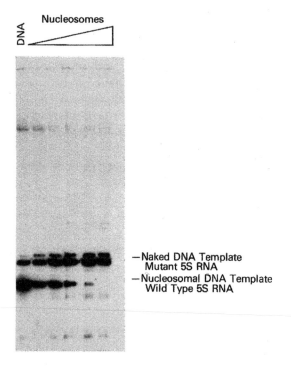

Figure 4.13. The degree of repression of 5S RNA gene transcription depends on the number of nucleosomes reconstituted on the plasmid.

Radioactive transcripts are shown from a mixture of wild type 5S DNA reconstituted with increasing numbers of nucleosomes (nucleosomal DNA) and naked mutant 5S DNA which generates a longer transcript (naked DNA), for details see Clark and Wolffe (1991).

et al., 1992; Paranjape *et al.*, 1994). Luse, Matsui and colleagues showed that prior chromatin assembly restricted transcription from the adenovirus type 2 major late promoter. A major problem of this type of experiment is that only a small percentage (<5%) of templates are actually transcribed. In contrast, class III gene transcription can be very efficient with over 50% of genes in transcription complexes (Clark and Wolffe, 1991). As long as the promoter elements were free it was possible to demonstrate that the small number of active class II genes were in fact assembled into chromatin, suggesting that nucleosomes did not completely inhibit RNA polymerase from transcribing once it had initiated the process (Knezetic and Luse, 1986; Matsui, 1987; Knezetic *et al.*, 1988). As with class III genes, transcription complexes formed prior to chromatin assembly resisted repression. Roeder and colleagues extended this analysis to suggest that TFIID binding alone was sufficient to relieve the inhibition of transcription due to nucleosome assembly. These experiments have, however, used crude assembly extracts supplemented with mixtures of histones including histone H1, and analysis of the resulting chromatin assembly has not been extensive (Workman and Roeder, 1987; Meisterernst *et al.*, 1990). An important point established from these experiments is that promoter-specific factors that stimulate TFIID binding to promoters facilitate transcription of the promoter in the face of whatever is inhibiting transcription in the extract, including chromatin assembly (Workman *et al.*, 1988, 1990). In contrast to this simple race for binding to the promoter, Wu and colleagues have shown that TFIID binding is insufficient for transcriptional activation of the *Drosophila* hsp 70 promoter (Becker *et al.*, 1991). However, a 'potentiated' chromatin template is assembled that can respond to the presence of an 'activated' heat-shock transcription factor (HSTF). It has been suggested that HSTF requires TFIID in order to bind to nucleosomal templates (Taylor *et al.*, 1991).

Kingston and colleagues examined the influence of particular activation domains, especially regions rich in acidic amino acids, in regulating the transcription process in chromatin (Section 4.1.6). It was suggested that the presence of a transcription factor containing these regions could perturb nucleosome structure in the local region (<100 bp) around the factor binding site (Workman *et al.*, 1991). This approach has been extended to replication with the comparable result that a transcription factor perturbs chromatin structure and facilitates the replication process (Cheng and Kelly, 1989; Cheng *et al.*, 1993).

Kingston, Workman and my own group have demonstrated that under appropriate conditions a variety of transcription factors have the potential to associate with nucleosomal DNA even in the absence of activation domains. These experiments initially used short DNA fragments reconstituted with histone octamers, and where the DNA is not always rotationally positioned with respect to the histones. Thus a wide range of DNA conformations within the nucleosome is exposed to the transcription factors. The factors used include Gal4, SP1, USF, TBP and Myc/Max (Workman and Kingston, 1992; Chen *et al.*, 1994a; Li *et al.*, 1994; Wechsler *et al.*, 1994; Coté *et al.*, 1994; Kwon *et al.*, 1994; Imbalzano *et al.*, 1994; Godde *et al.*, 1995). In general the binding of *trans*-acting factors to nucleosomal DNA occurs at an excess of protein approximately 100–1000-fold greater than necessary to saturate the same binding site when present as naked DNA. Several parameters facilitate *trans*-acting factor association within the nucleosome. These include multiple binding sites for the factor being present within the nucleosomal DNA (see also Chang and Gralla, 1994), positioning of the recognition site at the edge of the nucleosome, proteolysis and acetylation of the core histones (Vettesse-Dadey *et al.*, 1994, 1996; Godde *et al.*, 1995), the presence of histone-binding proteins like nucleoplasmin to facilitate histone exchange from DNA (Chen *et al.*, 1994; Walter *et al.*, 1995) and the SWI/SNF general activator complex (Coté *et al.*, 1994; Kwon *et al.*, 1994; Owen-Hughes and Workman, 1995). More recently this analysis has been extended to target nucleosomes placed within long arrays with comparable results (Owen-Hughes and Workman, 1996; Owen-Hughes *et al.*, 1996; Steger and Workman, 1997). These experiments provide a useful illustration of the difficulty of binding *trans*-acting factors to DNA in which substantially random histone–DNA interactions occur. These nucleosomal templates also provide insight into how histone–DNA interactions might prevent inappropriate *trans*-acting factor access within regulatory positioned nucleosomes (Section 4.2.3).

In contrast to the synthetic nucleosomes used in these studies many natural promoters contain intrinsic DNA structure or binding sites for proteins that direct the selective association of histones (see Section 4.2.3). These positioned nucleosomes facilitate *trans*-acting factor access to regulatory elements within chromatin (Section 4.2.3). Under these conditions the affinity of *trans*-acting factors for nucleosomal DNA appears similar to that of naked DNA (Schild *et al.*, 1993; Lee *et al.*, 1993; Wong *et al.*, 1995, 1997a,b).

Histone H1 is a contaminant of many crude *in vitro* transcription extracts. Kadonaga and colleagues have shown that the action of several promoter-specific DNA-binding proteins is to relieve this non-specific inhibitory process. Thus antirepression of transcription in *in vitro* transcription reactions is as important as the actual activation process (Croston *et al.*, 1991; Dusserre and Mermod, 1992). These experiments were extended to the reconstitution of close-packed arrays of nucleosomes with histone H1 (Laybourn and Kadonaga, 1991) and using *Drosophila in vitro* chromatin assembly extract supplemented with exogenous H1 (Kamakaka *et al.*, 1993). There is an increase in repressive character as the type of chromatin assembled more closely achieves a physiologically correct H1 association. High affinity histone H1 binding to chromatin requires both core histones and linker DNA (Hayes and Wolffe, 1993). Histone H1 associated with naked DNA has little resemblance to a native chromatin template.

Emerson and Jones have made effective use of the *in vitro* chromatin assembly systems to establish the roles of individual DNA binding transcription factors in relieving chromatin mediated transcriptional repression (Barton *et al.*, 1993; Sheridan *et al.*, 1995). ATP-dependent chromatin remodelling activities have been shown to be required for transcriptional activation under certain conditions (Pazin *et al.*, 1994). These activities promote nucleosome mobility (Pazin *et al.*, 1997) and may also be involved in the chromatin assembly process itself (Ito *et al.*, 1997).

It is important to note that under the *in vitro* transcription conditions used by many investigators, particularly those that include high concentrations of divalent cations, normal nucleosomal arrays are insoluble (Clark and Kimura, 1990). Aggregation and precipitation effects can inhibit both transcription initiation and elongation within chromatin (Hansen and Wolffe, 1992, 1994). Since histone H1 has a major influence on chromatin aggregation and compaction, it is important to note that *in vivo* the repression of genes by H1 is highly selective (Bouvet *et al.*, 1994; Shen and Gorovsky, 1996). Moreover, it is possible to reconstitute histone H1 into chromatin bound to solid-state matrices without inhibiting the transcription process (Sandaltzopoulos *et al.*, 1994). Chromatin aggregation is likely to be prevented on this type of template, since chromatin is no longer freely diffusible within the matrix.

Summary
Any results involving non-specific chromatin structures and specific *trans*-acting factors should be treated with caution, especially if the organization of the chromatin template is not documented. General conclusions from a large number of experiments are that prior assembly of a template with nucleosomes inhibits *trans*-acting factor access to DNA whereas if the factors bind first, subsequent chromatin assembly will not be inhibitory. RNA polymerase can seek out a transcription complex in a chromatin template without any problem.

Presumably the complex is a highly visible landmark in an invisible chromatin background.

4.2.3 Specific *trans*-acting factors and specific chromatin

Although the association of *trans*-acting factors with specific DNA sequences was readily accepted, the significance of the 'sequence-specific' organization of DNA into nucleosomes has taken longer to be acknowledged. Most investigators accept that there is no logical necessity to organize the vast majority of DNA into chromatin structures that have specific DNA sequences organized in a precise way (Simpson, 1991; Lowary and Widom, 1997). However, it is also clear that nucleosomes are positioned around DNA sequences with important functional consequences (Becker, 1994; Li *et al.*, 1998; Wallrath *et al.*, 1994; Wolffe, 1994d). Formation of such specific chromatin structures is true for the vast majority of genes for which the appropriate assays have been carried out to assess nucleosome positioning. Incorporation of *cis*-acting elements into a positioned nucleosome has important consequences for its accessibility of DNA to *trans*-acting factors. DNA in the nucleosome is highly bent and the helical periodicity of the double helix changes from an average of 10.5 bp/turn to 10.2 bp/turn. Thus, not only is one face of the DNA helix occluded by the histone core, but DNA has an entirely different structure from that in solution (Section 2.2.3).

Martinson and colleagues were the first to examine the issue of *trans*-acting factor access to a specific DNA sequence incorporated into a nucleosome. Their experiments made use of a prokaryotic repressor (the *lac* repressor) and a 144 bp DNA fragment containing binding sites for the repressor. This short DNA fragment is able to assemble a single nucleosome structure following reconstitution with equimolar amounts of the four core histones. The lac repressor is a helix-turn-helix protein that binds to B-form DNA on one side of the double helix in the major groove. Based on sedimentation studies, the authors concluded that both *lac* repressor and the histone octamer could simultaneously occupy the same DNA fragment. This implied that a triple complex of the DNA, histone proteins and *trans*-acting factor formed (Chao *et al.*, 1980a,b). Unfortunately, these early studies lack the resolution necessary to be absolutely sure of either nucleosome positioning or specific association of the DNA-binding proteins; however, they clearly demonstrate the correct approach to this problem.

Transcription factor (TF) IIIA was the first sequence-specific eukaryotic DNA-binding protein to be purified. This protein (38 kDa) consists of a chain of zinc fingers which bind in clusters over the 50 bp internal control region of the 5S RNA gene (Section 4.1.3). Simpson demonstrated that 5S RNA genes contain strong nucleosome positioning sequences (Simpson and Stafford, 1983; FitzGerald and Simpson, 1985; Section 2.2.5). Rhodes examined the interaction of TFIIIA with a *Xenopus* 5S RNA gene that had been incorporated into a nucleosome. It was found that TFIIIA could form a triple complex with the 5S RNA gene and histones (Rhodes, 1985). Subsequent experiments have shown that the efficiency of triple complex formation depends on the stoichiometry and post-transcriptional modification of the histone proteins (Hayes and Wolffe, 1992; Lee *et al.*, 1993). If the nucleosome is deficient in histones H2A/H2B, more DNA is free at the edge of the nucleosome for interaction with TFIIIA (Fig. 4.14). Unlike the *lac* repressor, which can interact with DNA actually within a nucleosome, TFIIIA binds primarily to free DNA at the edge of the nucleosome, not to DNA actually contacting the histones. Interestingly, acetylation of the core histones facilitates the binding of TFIIIA to the 5S RNA gene within the nucleosome (Fig. 4.15; Lee *et al.*, 1993). Since acetylation of the core histones does not apparently alter DNA conformation in the nucleosome directly, or the position of the nucleosome relative to the 5S RNA gene, it is likely that a small change in the conformation of the nucleosome or in the stability of histone–DNA contacts allows TFIIIA to bind (Bauer *et al.*, 1994b) (Section 2.5.2).

The failure of TFIIIA to bind to a 5S RNA gene associated with a complete unmodified octamer of histones within a positioned nucleosome that is constrained with respect to translational position is consistent with functional studies *in vitro* (Shimamura *et al.*, 1988; Morse, 1989; Clark and Wolffe, 1991). Using the yeast minichromosome system, it has been shown that even yeast nucleosomes will position on a 5S RNA gene *in vivo* (Section 3.3). This sequence-directed positioning is not surprising since the 5S RNA gene positioning sequence is one of the strongest natural ones known, probably because of inherent DNA curvature (Shrader and Crothers, 1989; Hayes *et al.*, 1990; Section 2.2.5). In the natural chromosomal context neither the X. *borealis* nor X. *laevis* 5S RNA genes (somatic type) position nucleosomes (Chipev and Wolffe, 1992). However, it is possible that as transcription factors compete for association with a 5S RNA gene, the positioning of a nucleosome or subnucleosomal particle is only transiently important. The positioning of a nucleosome away from the key promoter elements might occur immediately following DNA replication, when competition with transcription factors for

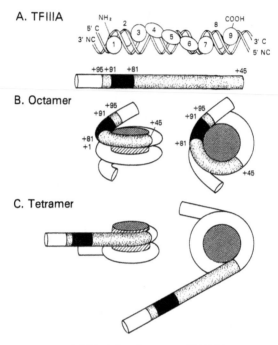

Figure 4.14.   Histone and TFIIIA binding to a 5S RNA gene.
A. A representation of TFIIIA bound to a 5S RNA gene. TFIIIA is believed to consist of nine domains, or zinc fingers. The binding site for TFIIIA is also represented as a cylinder, the speckled region is protected from DNase I cleavage. The region from +81 to +91 is an essential contact for TFIIIA (solid cylinder). B. When associated with a complete octamer of core histones (H2A/H2B/H3/H4)$_2$ the key contacts from +81 to +91 are in contact with the histone core, and TFIIIA cannot bind to the gene. C. When associated with a tetramer of histones (H3/H4)$_2$ the key contacts from +81 to +91 are accessible to TFIIIA which forms a triple complex with the gene.

association with the gene occurs (Section 4.3). Interestingly, repression of the oocyte 5S RNA genes, which bind transcription factors weakly (Section 4.1.5), depends on the formation of a positioned array of nucleosomes, including one containing the 5S RNA gene. In this case nucleosome positioning appears to be not only sequence directed, but also dependent on histone H1 (Chipev and Wolffe, 1992; Sera and Wolffe, 1998; Fig. 4.16).

In vitro experiments attempt to reconstruct phenomena believed to occur *in vivo*, thereby offering mechanistic insights into a process. However, many of our clearest insights into true gene regulation events in chromosomes come directly from the documentation of *in vivo* events (Wolffe, 1990b). The general observation is that promoters organized into chromatin as seen in the living cell are often accessible

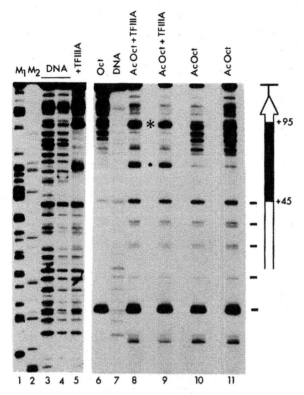

Figure 4.15. Histone acetylation allows the coexistence of TFIIIA and the acetylated octamer on the 5S RNA gene.
Lanes 1 and 2 show markers, lanes 3, 4 and 7, DNase I cleavage of naked DNA, lane 5 cleavage of TFIIIA bound to DNA, lane 6 of the octamer bound to DNA, lanes 8 and 9 of TFIIIA and the acetylated octamer and lanes 10 and 11 of the acetylated octamer alone. The asterisk and dot between lanes 8 and 9 indicate DNase I hypersensitive sites induced by TFIIIA.

to *trans*-acting factors, even when regulatory elements are adjacent to nucleosomes or actually incorporated into them. This conclusion might be contrasted with the vast majority of *in vitro* experiments (Section 4.2.2).

One of the best examples of *in vivo* control of gene expression in a chromosomal context concerns the regulation of the PHO5 gene of *S. cerevisiae* (Svaren and Horz, 1993, 1995, 1997). The yeast PHO5 gene encodes an acid phosphatase that is induced by a reduction in inorganic phosphate concentration (Rudolph and Hinnen, 1987). Two nucleosomes are positioned to either side of an essential promoter element recognized by the *trans*-acting factor PHO4. A second binding site for PHO4 and a site for another *trans*-acting factor PHO2 are

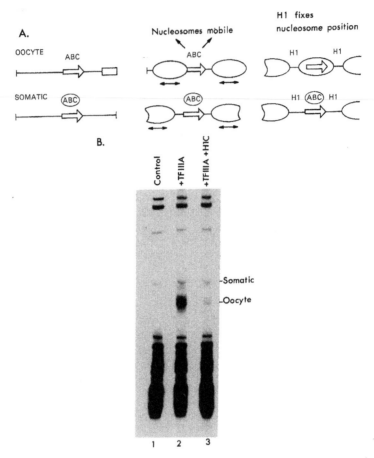

Figure 4.16. Histone H1 and the regulation of *Xenopus* 5S rRNA genes.

A. Model for the regulation of *Xenopus* oocyte and somatic 5S RNA genes. Oocyte and somatic 5S RNA genes are shown associated with transcription factors (TF) IIIA, B, and C forming a transcription complex. Transcription complexes on the oocyte 5S RNA gene are unstable (arrows pointing away from complex), whereas complexes on the somatic 5S RNA gene are stable (circle around complex). In the absence of histone H1 nucleosomes can slide (horizontal double-headed arrows) over the oocyte and somatic 5S RNA genes, especially if transcription factors are not bound to them as in the case of the oocyte 5S RNA genes. When somatic histone H1 accumulates during embryogenesis, nucleosomes are locked into position and the oocyte 5S RNA genes are stably repressed. Since transcription factors are always associated with the somatic 5S RNA genes these genes cannot be incorporated into nucleosomes. B. Histone H1 exerts a dominant repressive effect on oocyte 5S RNA gene transcription. *In vivo* transcription of oocyte 5S RNA genes is normally very low (lane 1); in the presence of excess TFIIIA the oocyte 5S RNA genes are activated (lane 2); however, even in the presence of excess TFIIIA precocious accumulation of histone H1 drives the dominant and specific repression of oocyte 5S RNA gene transcription. Each lane shows radioactive RNA synthesized *in vivo* resolved on a denaturing polyacrylamide gel (see Bouvet *et al.*, 1994, for a detailed description of similar experiments).

incorporated into one of the positioned nucleosomes. PHO2 is a homeodomain protein and PHO4 is a basic helix-loop-helix protein which bind cooperatively to the PHO5 promoter (Barbaris *et al.*, 1990). On induction, all four positioned nucleosomes are disrupted. Nucleosome disruption is independent of replication or transcription events (Schmid *et al.*, 1992; Fascher *et al.*, 1993). However, tethering of the *S. cerevisiae* RNA polymerase II holoenzyme is sufficient to remodel chromatin (Gaudreau *et al.*, 1997). Analysis of mutants reveals that PHO4 is essential for both the transcriptional activation of the PHO5 gene and the rearrangement of chromatin structure. PHO4 activity is regulated by phosphorylation. Under conditions of transcriptional repression PHO4 is phosphorylated (Kaffman *et al.*, 1994). The DNA-binding affinity of PHO4 is substantially increased by phosphate starvation. PHO4 binds to the site between the two nucleosomes and facilitates the disruption of the nucleosome containing the other PHO4-binding site (Venter *et al.*, 1994). The transactivation domain of PHO4 is necessary to mediate this nucleosome disruption process (Svaren *et al.*, 1994a). Changes in DNA sequence in the adjacent nucleosomes can influence transcriptional activation and chromatin rearrangement, suggesting that the chromosomal organization of the whole promoter region is essential for correct regulation (Almer and Horz, 1986; Almer *et al.*, 1986; Fascher *et al.*, 1990; Straka and Horz, 1991). The precise placement of regulatory elements between or within nucleosomes is clearly important for the regulation of this promoter.

An important point is that it is difficult to distinguish between nucleosome displacement, i.e. removal of a complete histone octamer from DNA, and nucleosome disruption, i.e. a conformational change or displacement of a H2A/H2B dimer, on PHO5 induction (Pavlovic and Horz, 1988). The only certain way is to use protein–DNA cross-linking reagents and to examine the association of specific histones with particular DNA fragments containing the promoter element of interest. Using this technique Varshavsky, Mirzebekov and colleagues have presented evidence for continued histone–DNA contacts in actively transcribed genes (Solomon *et al.*, 1988; Nacheva *et al.*, 1989) (Section 4.3.2). Related to this issue are the experiments of Grunstein and colleagues who have shown through genetic manipulation of histone stoichiometry that disruption of the chromatin structure of the PHO5 promoter, even in the absence of induction, can significantly activate transcription (Han *et al.*, 1988; Han and Grunstein, 1988). This further establishes that the repression of PHO5 transcription is related to the chromosomal structure of the gene, and that correct PHO5 gene regulation requires a chromatin template, as is true for many other yeast genes (Grunstein, 1990). The yeast histone deacetylase RPD3 also contributes to the regulation of the PHO5 gene (Vidal and Gaber, 1991). Mutations in RPD3 reduce the extent of both gene activation and repression of the RPD3 gene.

Although experiments with yeast are useful for establishing the major players in the transcriptional regulation of chromatin templates, significant differences exist between the chromosomal architecture of yeast and vertebrates. The *S. cerevisiae* core histones are relatively divergent from those of vertebrates and the protein resembling the histone H1 of vertebrate somatic cells is significantly different in structure, as are potential HMG1 homologs (Section 2.1.2). Yeast histones are more heavily acetylated than those found in normal vertebrate cells and nucleosomes are relatively tightly packed together. A yeast cell has to regulate most of its genes rapidly and continually through the cell cycle in a very different way to a differentiated cell from a metazoan. The only truly stable state of gene expression in a yeast cell analogous to those found in larger eukaryotes is that controlling mating-type. Here multiple chromatin-dependent controls regulate gene activity (Section 2.5.6). It is therefore important to examine the general applicability of any models established in yeast. Pre-eminent among the systems exploited to this end is the regulation by glucocorticoids of transcription of the mouse mammary tumor virus (MMTV) long terminal repeat (LTR) (Zaret and Yamamoto, 1984; Cordingley *et al.*, 1987).

Hager and colleagues established that the MMTV LTR is incorporated into six positioned nucleosomes in both episomes and within a mouse chromosome (Fig. 4.17). The positioned nucleosomes serve to prevent the basal transcriptional machinery associating with the promoter under normal circumstances, that is in the absence of glucocorticoids (Archer *et al.*, 1992). Induction of transcription by glucocorticoids requires binding of the glucocorticoid receptor (GR) to the LTR, disruption of the local chromatin structure initiated by the GR binding to recognition sequences within nucleosomes, and the assembly of a transcription complex over the TATA box (Archer *et al.*, 1989, 1992). Thus comparable events occur on both the PHO5 and MMTV LTR promoters: an inducible transcription factor binds, chromatin structure is rearranged, a transcription complex is assembled and transcription is activated.

Vigorous attempts have been made to reconstruct the transcriptional regulation and concomitant chromatin structural changes of the MMTV LTR *in vitro*. The GR, which is a zinc finger protein, appears to bind nucleosomal DNA with only a slight reduction in affinity relative to naked DNA. This interaction is dependent on the precise position of

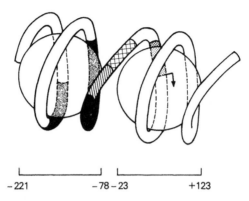

Figure 4.17. The organization of the MMTV LTR into a specific chromatin structure.

Nucleosome positions and key recognition elements for *trans*-acting factors are indicated. The hooked arrow indicates the start site of transcription. Numbers indicate base pair positions of DNA that associated with histones relative to this site.

the nucleosome and hence the translational position of the GR binding site within the nucleosome. GR binding is reported to occur when the nucleosome is at –188 to –45 (Pina *et al.*, 1990a,b) and at –219 to –76 (Perlmann and Wrange, 1988) or –221 to –78 (Archer *et al.*, 1991) relative to the start site of transcription (+1). In these instances the rotational orientation of the individual GR binding sites on the surface of the histone octamer will be similar due to the separation of nucleosome boundaries by almost exactly three helical turns of DNA. The latter *in vitro* nucleosome positions compare favourably with those determined *in vivo* (Richard-Foy and Hager, 1987). Recent work suggests that although a predominant *in vivo* translational position exists for this nucleosome, a number of distinct translational positions can be detected (Fragoso *et al.*, 1995). Detailed *in vitro* analysis suggests that these variant positions are dependent on the DNA sequence (Roberts *et al.*, 1995). This heterogeneity in translational position appears to reflect nucleosome mobility (Meersseman *et al.*, 1992). The DNA sequence containing the GR-binding sites has regions of intrinsic flexibility and curvature that direct the histones to bind it in a particular way (Pina *et al.*, 1990a,b; Roberts *et al.*, 1995). Four GR recognition elements (GREs) are within this DNA segment at –175,

–119, –98 and –83, and have different rotational and translational positions within the nucleosome. The two elements at –119 and –98 face towards the core histones and remain unbound in the presence of GR, whereas the elements at –175 and –83, which face towards solution, are bound by the GR. These sites are separated by 92 bp which places them together on one side of the nucleosome. This proximity might facilitate both the binding and subsequent activity of GR (Perlmann and Wrange, 1991). The display of DNA binding sites on the surface of the nucleosome might actually promote the formation of a functional transcription complex (Chavez and Beato, 1997; Truss *et al.*, 1995).

Experiments that compare the affinity of the GR for a recognition element as free DNA compared to one facing towards solution in the nucleosome at different translational positions, show that binding affinity is reduced 3–11-fold in the nucleosome (Li and Wrange, 1993). This is a remarkably small reduction compared to the complete absence of binding when the recognition element is facing towards the histones (Li and Wrange, 1995). This suggests that the key variable determining the accessibility of the GR to its recognition elements within the nucleosome is not the translational position of the recognition element, but the rotational position of the DNA sequence with respect to the surface of the histones.

The GR is well suited to interact specifically with nucleosomal DNA. GR binds to DNA using a domain containing two zinc fingers: an α-helix in one of the two fingers interacts with a short 6 bp region in the major groove of the double helix, while the other finger is involved in protein–protein interaction (Luisi *et al.*, 1991). The GR associates with DNA as a dimer. The second molecule of the receptor has similar interactions on the same side of the double helix, one helical turn away. Thus GR can bind to DNA on the one side exposed towards solution in the nucleosome, thereby circumventing steric interference by the histone core. Moreover, the GR dimer can bind specifically to DNA containing only one GRE half site, presumably by making both specific and non-specific contacts to DNA in each half of the dimer. Thus highly bent nucleosomal DNA might still provide enough precisely aligned contacts for at least one specific half-site interaction which could then be supplemented by non-specific contacts. This type of non-specific interaction could account in part for the reduction in affinity of the GR for nucleosomal recognition elements (Li and Wrange, 1993). Surprisingly, association of the GR with the nucleosome containing its binding site appears to have no effect on the integrity of the structure *in vitro*, unlike the apparent consequence *in vivo*. Binding of the other promoter-specific transcription factors (NF1), which is facilitated by the GR *in vivo*, does not occur *in vitro* on nucleosomal templates (Candau *et al.*, 1996; Blomquist *et al.*, 1996). Certain nuclear components that presumably facilitate chromatin structural changes *in vivo* have so far been lacking in the *in vitro* system.

What molecular process causes chromatin structure to be disrupted? Some of the possible ways in which a *cis*-acting element could be incorporated into a nucleosome, and the requisite disruption of the nucleosome required for transcription factor access, are shown in Fig. 4.18. Various mechanistic possibilities for disrupting chromatin structure aside from targeting histone modification (Section 2.5.4) are described in Fig. 4.19. DNA replication is the one event certain to disrupt chromatin and provide access of transcription factors to their cognate sequences (Section 4.3.1). However, DNA replication is not required for chromatin disruption and transcriptional activation of the MMTV LTR (Archer *et al.*, 1989). Schutz, Richard-Foy and colleagues have extended these studies by examining not only the initial disruption of chromatin structure over the enhancer of the rat tyrosine amino transferase gene (TAT) following induction of GR binding, but also the reformation of normal chromatin structure on hormone withdrawal. Both of these changes occur within a few hours, implying that neither disruption nor reassembly of nucleosomes is dependent on DNA replication (Carr and Richard-Foy, 1990; Reik *et al.*, 1991; see also Schmid *et al.*, 1992).

If replication is not involved in chromatin rearrangement, nucleosomes must be disrupted in some alternative way. *Trans*-acting factors might displace histones from nucleosomes directly. Alternatively they might wait either for histones to exchange passively out of chromatin or for DNA to spontaneously unravel from association with the histones before binding to their recognition sequences (Polach and Widom, 1995, 1996). The characterization of chromatin remodelling

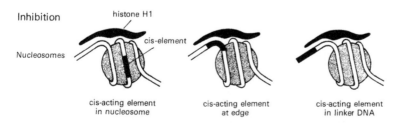

Figure 4.18. Position of a *cis*-acting element in the nucleosome will influence the contacts of the element with histones.
In the nucleosome, the element will contact the core histones, at the edge it will contact the core histones and histone H1, in the linker DNA it will contact predominantly histone H1.

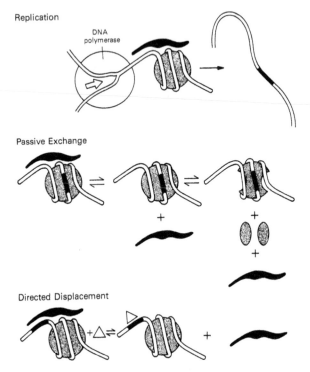

Figure 4.19. Means of disrupting a nucleosome.

Replication may disrupt a nucleosome as DNA duplication requires each strand of DNA to be a template for second strand synthesis, thus one new DNA duplex will be naked. It is also possible that DNA polymerase might displace the histones. Passive exchange requires that an equilibrium exists between histone bound to DNA and free histones in solution. Such exchange is known to occur for histone H1 and histones H2A/H2B. Directed displacement requires that a trans-acting factor (Δ) disrupt histone–DNA contacts.

enzymes and machines at the genetic and biochemical level would make these passive mechanisms appear unlikely. Although they clearly explain some *in vitro* results (Adams and Workman, 1995). In most assays, nucleases are used to examine the incorporation of specific promoter elements into nucleosomes. It is possible that nucleosomes that have altered their position or composition would lose sharp boundaries to nuclease protection even though DNA could remain associated with histones. Protein–DNA cross-linking reagents and antibodies against histones are being used to explore this possibility. Histone H1 and histones H2A/H2B are known to exchange readily in and out of chromatin under physiological conditions (Caron and Thomas, 1981; Louters and Chalkley, 1985) (Sections 3.1.1 and 3.1.2). A certain amount of DNA folding would occur even in the absence of these proteins. The capacity to reassemble a complete nucleosome would also remain in the residual interaction of histones H3/H4 with DNA (Hayes *et al.*, 1991b). Histones H1, H2A and H2B are not able to bind correctly in chromatin unless histones H3/H4 are already sequestered. It is also possible that complete histone octamers might exchange from DNA *in vivo*. However, this seems unlikely, as solutions of high ionic strength (>1 M NaCl) or high temperatures (>80°C) are required to completely disrupt a nucleosome *in vitro* (Section 2.2.1). Only very basic arginine-rich proteins, which are unlike known transcription factors, are known to out-compete histones for association with DNA in a natural context (Section 2.5.5). This integrity of the nucleosome is almost entirely due to the stability of the interaction of the arginine-rich histones H3/H4 with DNA (Section 2.2.4). A partial disruption of the nucleosome in addition to histone acetylation seems by far the most likely mechanism by which *trans*-acting factors might gain access to DNA in chromatin (Fig. 4.20). Removal of histone H1 from the nucleosome might facilitate some mobility of the contacts made by the histone octamer with DNA (Pennings *et al.*, 1991). This local nucleosome sliding could facilitate transcription factor access to recognition sites otherwise constrained within the nucleosome (Chipev and Wolffe, 1992; Ura *et al.*, 1995; Varga-Weisz *et al.*, 1995; Pazin *et al.*, 1997).

The proteins that GR recruits to the MMTV chromatin include the mammalian homologs of the *SWI* (Switch) 1, 2, 3 proteins of *S. cerevisiae* (Yoshinaga *et al.*, 1992; Muchardt and Yaniv, 1993). The exact mechanism whereby they disrupt chromatin is unknown (Section 2.5.4), but they initiate a chain of events that cause the removal of histone H1 from the linker DNA in MMTV chromatin (Bresnick *et al.*, 1992) and a substantial increase in the accessibility of the DNA that is within the positioned nucleosomes to nucleases (Zaret and Yamamoto,

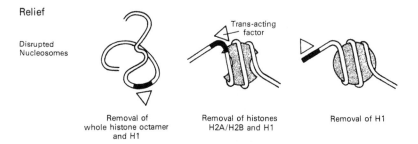

Relief

Disrupted
Nucleosomes

Trans-acting
factor

Removal of
whole histone octamer
and H1

Removal of histones
H2A/H2B and H1

Removal of H1

Figure 4.20. Extent of disruption of nucleosomes shown in Figure 4.18 required to release the *cis*-acting element depends on where the element is within the nucleosome.

1984). The positioning of the nucleosomes is likely to have an important role in the displacement of histone H1, since the linker region between the nucleosomes contains the binding sites for histone H1, NF1 and the octamer factor. It is possible that histone H1 and the transcription factors might compete for binding to this linker region. In any event the transcription factors NF1, the octamer factor and TFIID that lie in the linker DNA and at the periphery of the positioned nucleosomes are recruited to their binding sites in this disrupted chromatin (Archer *et al.*, 1992) and assemble an active transcription complex. Transcriptional activation by GR is only transient and after a few hours the basal transcription complex, NF1, octamer factor and GR are displaced from the MMTV LTR and their binding sites are reincorporated into the positioned nucleosomes and the promoter repressed (Lee and Archer, 1994). The molecular mechanisms responsible for displacement of the transcription factors have not been determined; however, these results indicate that both transcription factor and histone complexes with promoters are likely to be dynamic.

Although targeted histone acetylation is yet to be documented for MMTV chromatin in response to the presence of hormone-bound GR, it provides an attractive mechanism to locally disrupt chromatin and facilitate the function of the basal transcriptional machinery (Section 2.5.4). Consistent with such a mechanism, Beato and colleagues have found that moderate increases in histone acetylation activate the mouse mammary tumor virus promoter and remodel its chromatin structure (Bartsch *et al.*, 1996).

As our information concerning the molecular basis of sequence-directed nucleosome positioning grows (Section 2.2.5) it becomes possible to manipulate specific chromatin structure around the recognition site for a sequence-specific DNA-binding protein. Wolffe and Drew made use of synthetic DNA curves and the selective affinity of histone octamers for DNA in this conformation, to manipulate nucleosome position relative to a promoter for bacteriophage T7 RNA polymerase (Wolffe and Drew, 1989). They found that small changes of less than 10 bp in the association of the promoter (~20 bp in length) with the nucleosome could affect the repression of transcription by 10–20-fold. Here, even interactions of DNA-binding proteins with the periphery of a nucleosome severely inhibit function *in vitro*.

Simpson has carried out similar experiments *in vivo*. Normally in the *TRP1ARS1* minichromosome (Section 3.3) a nucleosome is not positioned so as to include the ARS1 sequence element required for DNA replication. However, using a sequence-directed nucleosome positioning element to move the nucleosome, it was possible to show that incorporation of the 11 bp ARS1 core sequence domain essential for replication into the centre of the nucleosomal DNA, but not the peripheral region, inhibited replication of TRP1ARS1 DNA (Simpson, 1990). Subsequent work from the same laboratory has systematically defined a hierarchy of DNA sequence and protein directed contributions to the assembly of specific chromatin structures (Simpson *et al.*, 1993; Sections 3.3 and 4.2.5). Although certain proteins such as those associated with class III gene transcription can function effectively in a variety of chromatin structures (Morse *et al.*, 1992; Morse, 1993), others cannot (Roth *et al.*, 1990, 1992; Cooper *et al.*, 1994; Patterton and Simpson, 1994). Using similar approaches, the heat-shock transcription factor which appears to bind to nucleosomes weakly in non-specific assays (Taylor *et al.*, 1991) associates efficiently with rotationally positioned DNA within a nucleosome *in vivo* (Pederson and Fidrych, 1994). Alteration of the exact position of the regulatory elements within a heat-shock gene promoter can result in a complete restriction in heat-shock factor access to DNA in a nucleosome (Gross *et al.*, 1993). It is clear that the context in which a DNA recognition element is organized within chromatin can have major consequences for the regulation of gene activity.

Summary
Several studies with positioned nucleosomes have shown that specific *trans*-acting factors can recognize their binding sites when wrapped around the histone core. The histone core is positioned such that the binding sites are accessible to *trans*-acting factors. This suggests that specific chromatin structures assemble in order to fulfil the contrasting needs of DNA compaction and accessibility. Of course, in agreement with experiments using non-specific chromatin structures, certain positioned nucleosomes will inhibit *trans*-acting factor access to DNA. The hypothesis that specific chromatin structures are compatible with nuclear processes gains strength through observations *in vivo*. In yeast and mammalian cells excellent examples exist of promoter elements being regulated through *trans*-acting factors functioning in a chromatin environment and modifying chromatin structure.

4.2.4 *Trans*-acting factors, DNase I sensitivity, DNase I hypersensitive sites and chromosomal architecture

After the early experiments demonstrating the selective association of non-specific DNA-binding proteins (prokaryotic RNA polymerases) with transcriptionally active chromatin, Weintraub, Felsenfeld and colleagues were able to show a comparable general accessibility to nucleases (Weintraub and Groudine, 1976; Wood and Felsenfeld, 1982). This general sensitivity to nucleases includes the coding region of a gene and may extend several kilobases to either side of it. DNase I normally introduces double strand breaks in this transcriptionally active chromatin over 100 times more frequently than in inactive chromatin. However, the exact structural basis of this generalized sensitivity is unknown. Later we discuss structural changes in the transcribed regions of genes (Section 4.3.4); however, the dispersed nature of the generalized sensitivity implies other contributory factors. Careful analysis reveals the non-transcribed regions are just as sensitive to DNase I digestion as are the transcribed regions, provided the last-cut approach to the measurement of DNase I sensitivity is used. This is defined as digestion by DNase I to fragment sizes so small that DNA no longer hybridizes to complementary strands after denaturation (Jantzen *et al.*, 1986). A certain length of DNA is necessary to allow specific recognition (hybridization) of two separated single-stranded regions. An important question that is not yet completely resolved is whether transcription is required to generate generalized nuclease sensitivity in certain instances or whether sensitivity always precedes transcription.

Experiments on the action of mitogens on quiescent cells reveal that a subset of genes, called the immediate-early genes, is rapidly induced (Lau and Nathans, 1987). The most studied examples of such genes are *c-myc* and *c-fos*. These two proto-oncogenes are transcriptionally activated within minutes. Coincident with transcription, the chromatin structure of the proto-oncogenes becomes more accessible to nucleases. Once proto-oncogene transcription ceases, preferential nuclease accessibility is lost (Chen and Allfrey, 1987; Chen *et al.*, 1990; Feng and Villeponteau, 1990). Possible conformational changes in nucleosome structure might account for such effects. It has been proposed that histone H3 sulphydryl residues might become accessible in nuclease-sensitive chromatin; however, transcriptionally active *S. cerevisiae* chromatin which has no cysteine and hence no sulphydryls in the core histones is also retained on the organomercurial agarose columns used to assess nucleosome conformational changes. This suggests that perhaps RNA polymerase or HMGs provide the sulphydryls that bind to the organomercurial agarose columns retaining active chromatin (Walker *et al.*, 1990). However, the rapidity of the changes in nuclease sensitivity ($<90\,$s) and their propagation in both directions 5′ and 3′ to the promoter means that transcription and hence RNA polymerase or HMGs cannot account for all of the observed changes consistent with some retention due to histone H3 in higher eukaryotes (Feng and Villeponteau, 1990; Walker *et al.*, 1990; Section 4.3.4). In fact the speed of the response suggests that changes in nuclease sensitivity precede transcription, and may play a role in regulating *c-fos* expression.

It is interesting that one of the earliest mitogen-induced nuclear signalling events coincident with proto-oncogene induction is the rapid phosphorylation of histone H3 on serine residues within its highly charged, basic N-terminal domain (Section 2.2.4). Whether these changes are localized to chromatin regions containing either *c-fos*, *c-myc* or the other immediate early genes has not yet been determined (Mahadevan *et al.*, 1991). An additional component contributing to the prior sensitization of the proto-oncogenes to nucleases may come from the existence of *trans*-acting factors already associated with the promoter (Herrera *et al.*, 1989). Such interactions are responsible for the second landmark in chromatin: DNase I hypersensitive sites. These sites are the first place DNase I introduces a double-strand break in chromatin. They usually involve small segments of DNA sequences (100–200 bp) and are two or more orders of magnitude more accessible to cleavage than in inactive chromatin (Wu *et al.*, 1979; Wu and Gilbert, 1981; McGhee *et al.*, 1981; Burch and Weintraub, 1983). DNase I hypersensitive sites are often flanked by positioned nucleosomes. DNase I hypersensitive sites are not necessarily nucleosome free and can reflect the stable association of a transcription factor on the surface of a nucleosome (Figs. 4.21, 4.22 and 4.23; Wong *et al.*, 1995, 1997). DNase I hypersensitive sites and nucleosome arrays are usually detected by DNase I or micrococcal nuclease digestion and indirect end-labelling methodologies (Almer *et al.*, 1986; Wu, 1984; see Fig. 3.9). As the most accessible regions of chromatin to non-histone DNA-binding proteins, DNase I hypersensitive sites generally denote DNA sequences with important functions in the nucleus.

DNase I hypersensitive sites were first detected in the SV40 minichromosome (at the ORI region) and in *Drosophila* chromatin (Elgin, 1988). In general, these sites appear to be accessible to all enzymes or reagents that cut duplex DNA. Higher resolution studies have shown these sites often represent clusters of recognition sites for promoter-specific DNA-binding proteins (Emerson *et al.*, 1985). These sites have been mapped to a large number of functional segments of DNA, including promoters, enhancers, locus control elements, transcriptional silencers, origins of replication, recombination elements and structural sites within or around telomeres (Gross and Garrard, 1988).

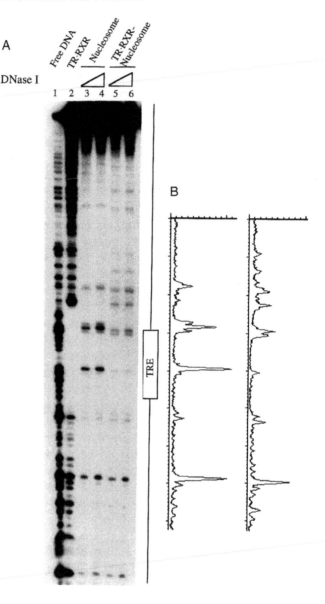

Figure 4.21. DNase I footprinting reveals that TR/RXR hetero-dimer bound to a rotationally positioned nucleosome *in vitro*.
A. TR/RXR was bound to a nucleosome treated with DNase I and the complexes resolved on a non-denaturing polyacrylamide gel. The complexes correspond to nucleosome (lanes 3, 4) and TR/RXR-nucleosome (lanes 5, 6) were then analysed with a 6% sequencing gel. As controls, free probes incubated either with $2\,\mu l$ of control oocyte extract (lane 1) or with $2\,\mu l$ of oocyte extract with TR/RXR (lane 2) in $25\,\mu l$ of binding reaction were digested with DNase I and analysed on the same gel. B. PhosphorImager scanning of lanes 4 and 6, demonstrating strong protection of the TRE region by TR/RXR in the nucleosome.

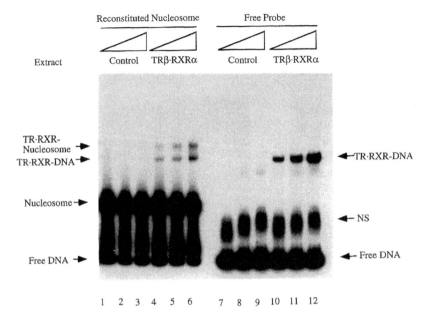

Figure 4.22. TR/RXR heterodimer binds to the thyroid response element reconstituted into nucleosome *in vitro*.
A 160 bp end-labelled DNA fragment from TBβA promoter (from +163 to +322) containing the wild-type TRE was generated by PCR amplification with one of the two primers end-labelled with ^{32}P (position +322), purified and reconstituted into nucleosome *in vitro* with histone octamers purified from chicken erythrocytes. The reconstituted nucleosome was then incubated with extract from oocytes with or without (control) overproduction of TBβ/RXRα (lanes 4–6). For a comparison, the binding experiment was also conducted with the end-labelled naked DNA (lanes 10–12). in a 20 μl binding reaction, ~0.5 ng of free probe or 1 ng of end-labelled probe reconstituted into nucleosome plus 200 ng or carrier nucleosomes plus 1 or 2 μl of control oocyte extract or oocyte extract with TR/RXR was incubated at room temperature for 20 min and then resolved by a 4% native polyacrylamide gel in 0.5 × TBE. (NS) A TR/RXR independent non-specific complex.

Around regulated genes these sites often fall into a hierarchy of patterns. In the chicken β-globin locus containing four globin genes (5′σ – $β^H$ – $β^A$ – ε – 3′) covering over 65 kb, 12 DNase I hypersensitive sites are found (Fig. 4.24). One site is present in all cells independent of whether the genes are transcriptionally active or not. Three sites, upstream of the σ-globin gene, were present only in erythroid cells destined to express the globin genes; these sites were initially without clear functional significance (but see later). However, a similar site was found between the $β^A$ and ε genes that corresponded to an enhancer element. Four sites were found over the promoters of each gene

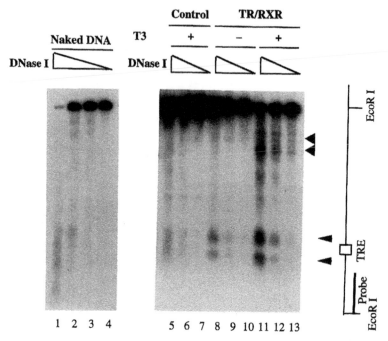

Figure 4.23. DNase I hypersensitive site generated *in vivo* from the binding of the TR/RXR to a TRE within nucleosomal DNA in the absence or presence of hormone as indicated.

Horizontal arrows indicate the DNase I hypersensitivity. The triangles indicate DNase I cleavage over the promoter in the presence of hormone-bound receptor.

depending on whether the gene was transcriptionally active, and three sites were found downstream of the genes corresponding to transcription termination elements (the β^A gene excluded). It is important to note that the formation of DNase I hypersensitive sites at the promoters of the globin genes is a relatively late step in the commitment of these genes to become transcriptionally active. However, it is clear that the formation of such sites precedes the actual initiation of transcription by RNA polymerase; indeed, the generation of these sites may account for a component of the general nuclease sensitivity of a gene (Weintraub *et al.*, 1982; Reitman and Felsenfeld, 1990).

One of the most thorough dissections of a DNase I hypersensitive site has been carried out by Elgin and colleagues (Elgin, 1988). The *Drosophila* heat-shock protein (*hsp*) *26* gene is very rapidly transcriptionally activated by raising the temperature of a fly to a stressful level

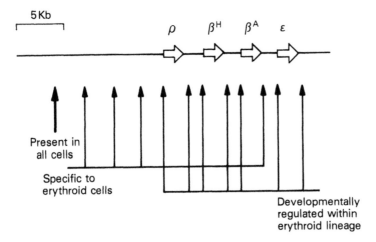

Figure 4.24. DNaseI hypersensitive sites of the chicken β-globin locus.
Genes are shown as open arrows, the three different types of DNaseI hypersensitive sites found are indicated (see Reitman and Felsenfeld, 1990 for details).

(a heat shock of 34°C). Two DNase I hypersensitive sites exist at the promoter of the *hsp 26* gene including recognition sequences for the promoter-specific heat-shock transcription factor (HSTF, a leucine zipper protein; Section 4.1.6) (Fig. 4.25). Following heat shock, HSTF binds to these sites. In contrast, TFIID is bound to the TATA box both before and after heat shock (Section 4.1.3). TFIID alone is insufficient to cause the *hsp 26* gene to be transcribed, the specific association of the HSTF protein is also required. High resolution analysis has revealed that a nucleosome is positioned between the proximal and distal binding sites for HSTF, i.e. between the two DNase I hypersensitive sites. In this case the exact position of histone–DNA contacts within this nucleosome depends not only on the DNA sequence to which the histones bind (from –300 to –140) but also on adjacent DNA sequences. These are repeats of the type (CT)n.(GA)n which bind a specific *trans*-acting factor, the GAGA protein (Lu *et al.*, 1992, 1993; Kerrigan *et al.*, 1991). The (CT)n.(GA)n repeat regions are located to either side of the positioned nucleosome at –347 to –341 and at –135 to –85. The GAGA factor bound to these repeats may function as a 'bookend' to determine exactly where the nucleosome will be positioned (Fedor *et al.*, 1988). Recent evidence suggests that the GAGA factor might function through ATP-dependent mechanisms to actively direct the assembly of a particular chromatin architecture

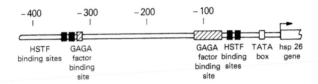

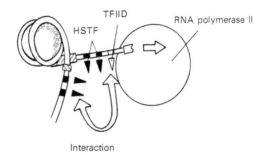

Interaction

Figure 4.25.   The specific chromatin organization of the *Drosophila* hsp 26 promoter.
Key *cis*-acting elements are indicated relative to the start site of transcription (hooked arrow). The organization of these sites on a specific nucleosomal scaffold is indicated together with the interactions necessary to prevent or activate transcription. A tethered RNA polymerase II molecule is released through events initiated by the binding of HSTF (heat shock transcription factor).

(Tsukiyama *et al.*, 1994). Transcription of the gene is regulated through the association of the heat-shock transcription factor with recognition elements at –51, –170, –269 and –340. The sites at –170 and –269 will be wrapped around the core histones in rotational frames that prevent heat-shock transcription factor association. However, the wrapping of DNA around the nucleosome will bring the HSTF molecules bound to the sites at –340 and –51 into juxtaposition, and the histones may potentially facilitate transcription through causing a clustering of HSTF activation domains (Thomas and Elgin, 1988). Direct evidence for transcriptional activation mediated by this nucleosome is yet to be established; however, it is clear that the positioning of a nucleosome in this particular way allows key transcription factors to obtain access to essential regulatory elements in spite of the assembly of the gene into chromatin.

Lis and colleagues have shown that RNA polymerase II is also bound to the promoter next to the TFIID protein. Heat shock and the binding of HSTF allows RNA polymerase II to begin transcriptional elongation by an unknown process (Gilmour and Lis, 1986; Rougvie and Lis, 1988).

Elgin (1988) draws several important conclusions from this analysis. The proximal hypersensitive site exists because there is a region of naked DNA between the specific protein complex at the TATA box and the positioned nucleosome. This region is too small for an additional nucleosome, but contains the binding sites for a promoter-specific factor (HSTF). Thus, HSTF can bind to DNA without hindrance from chromatin structure; in fact the presence of the nucleosome may create a constrained loop which facilitates interactions between HSTF proteins and TFIID (Section 4.1.1). This is another excellent *in vivo* example of specific proteins regulating gene expression through a specific chromatin structure (Section 4.2.3). This role of chromatin proteins in facilitating a complex regulatory process is similar to the role of the histone-like protein HU or IHF in facilitating replication or recombination in prokaryotic systems (Section 4.1.1).

Similar positioned nucleosomes to those on the hsp 26 promoter have been determined to exist for the human U6 RNA gene, *Drosophila* alcohol dehydrogenase gene and the *Xenopus* vitellogenin genes, where a nucleosome is positioned between the enhancer and the promoter (Schild *et al.*, 1993; Jackson and Benyajati, 1993; Stunkel *et al.*, 1997). For the vitellogenin genes, nucleosome positioning directed by DNA sequence occurs between −300 and −140 (see Fig. 4.3). The binding sites for the stimulatory transcription factors, the oestrogen receptor and nuclear factor 1 lie outside the region of DNA that is wrapped around the histones at −300 to −330 and at −120 to −110, respectively. When these sites are brought together either by positioning a nucleosome in between them (or artificially by deleting the intervening DNA), transcription is enhanced about 5–10-fold (see Fig. 4.3). This moderate stimulatory effect is much more significant than it might appear since the assembly of a non-specific chromatin structure would normally lead to a >20-fold repression of transcription as the binding sites for transcription factors are occluded by the histones. Thus in the vitellogenin gene example nucleosome positioning has two roles: (1) to provide a scaffold that allows transcription factors to communicate more effectively; and (2) to prevent the formation of repressive histone–DNA interactions that may prevent any transcription factor from gaining access to a chromatin template.

DNase I hypersensitive sites have been useful in defining important DNA sequences that have no apparent function (at least when first discovered). Among these were four strongly nuclease-sensitive sites located 10–20 kbp upstream of the cluster of human β-globin genes

(Tuan *et al.*, 1985; Forrester *et al.*, 1987; Grosveld *et al.*, 1987). These sites, which have come to be known as locus control regions (LCRs), represent *cis*-acting elements that allow genes that are integrated into a chromosome to be expressed in a way that is independent of chromosomal position, i.e. position effects are abolished (Sections 2.5.5 and 3.2.2). A consequence of this is that LCRs allow each copy of a gene integrated in multiple copies to be expressed equivalently, so that gene expression is copy-number dependent.

When all four DNase I hypersensitive sites comprising the LCR are placed adjacent to reporter genes, they appear to function like enhancers. However, three of the four sites do not function as enhancers in transient expression sites, but will only do so after incorporation into the chromosome. This suggests that the LCRs may play a special role in the stabilization of an accessible chromatin structure distinct from the function of a normal enhancer element (Ryan *et al.*, 1989; Talbot *et al.*, 1990; Philipsen *et al.*, 1990; Talbot and Grosveld, 1991; Fig. 4.26). Nevertheless in certain instances enhancers can relieve position effect variegation in the chromosome (Section 2.5.6). Certain enhancers/LCRs have to interact with promoter sequences in order to confer general DNase I sensitivity on chromatin (Reitman *et al.*, 1993), whereas others can facilitate local factor access within chromatin independent of the presence of a normal eukaryotic promoter (Jenuwein *et al.*, 1993). It is possible that the enhancer-like function of LCRs can be separated from position independence and copy number dependence, since the region between the chicken β^A and ϵ genes that contains the hypersensitive site acts as an enhancer in transient assays and confers position independence and copy number dependence, but does not confer high levels of gene activity in transgenic mice (Reitman *et al.*, 1990). Alternatively, it could be that chicken sequences do not function well to stimulate transcription in mice because they do not bind the appropriate mammalian promoter-specific proteins.

Each LCR hypersensitive site contains multiple binding sites for promoter specific proteins. In the case of the human β-globin genes, one of these sites contains four recognition elements for an erythroid tissue sequence-specific DNA-binding protein known as GATA-1. In spite of this information, how these sites function is unknown. Several models have been proposed including: unravelling of the chromatin fibre, functions similar to normal enhancers (Martin *et al.*, 1996; Section 2.5.6) and stabilization of specific nucleoprotein complexes that are functional for transcription only in the presence of the LCR (Wijgerde *et al.*, 1995). Evidence consistent with the propagation of an altered chromatin structure comes from transgenic experiments in which

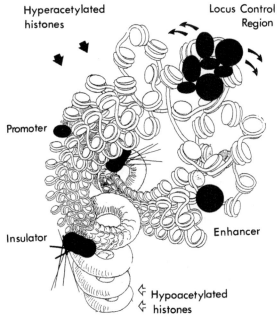

Hyperacetylated histones

Locus Control Region

Promoter

Insulator

Enhancer

Hypoacetylated histones

Figure 4.26. Model for functional chromatin domain.
The LCR disrupts local chromatin structure, as do nucleoprotein complexes associated with promoters and enhancers. The LCR has the capacity to facilitate directly or indirectly chromatin structural changes such as histone acetylation. In certain instances, the entire chromatin domain accumulates hyperacetylated histones as indicated by a looser structure. Outside of the domain, histones are hypoacetylated, and the chromatin fibre is more condensed. The LCR can also make physical contact with promoter and enhancer complexes. The boundaries of LCR function are defined by insulator elements that may make contact with the nuclear matrix.

LCRs confer hypersensitive sites and general DNase I sensitivity whereas promoter elements do not. This suggests that LCRs may be able to overcome the inhibitory influences of localized regions of heterochromatin (Festenstein *et al.*, 1996).

It has been suggested that LCRs or regions adjacent to them might define the boundaries of an active chromatin domain through attachment to a nuclear matrix or scaffold (Section 2.4.2). Such matrix attachment regions (MARs) or scaffold attachment regions (SARs) have been biochemically defined as those AT-rich DNA sequences remaining associated with the nuclear matrix or scaffold during its preparation (Section 2.4.2) (Mirkovitch *et al.*, 1984; Cockerill and Garrard, 1986). We have discussed reservations concerning the specificity of such sites as defined *in vitro* (Section 2.4.2). However, a subfamily of sites determined by such procedures (A elements) can be found at the boundaries of a 24 kbp region of DNase I-sensitive chromatin containing the chicken lysozyme gene (Phi-Van and Stratling, 1988). Sippel and colleagues have demonstrated that the A elements insulate a gene from chromosomal effects in stable transformants, but are not required in transient assays for high levels of gene activity (Stief *et al.*, 1989). These A elements also give high level position-independent, copy-number dependent expression of a transgene in transgenic mice (Bonifer *et al.*, 1990). It should be noted that A elements differ from LCRs in that they function well only when they flank both sides of the gene and its regulatory elements. They also lack intrinsic enhancer activity. The A elements are good candidates for representing the boundaries of domains of chromatin. How they create these boundaries has not been determined.

Schedl and colleagues have used transposable elements (P-elements) that can introduce stable transformants into the germ line of *Drosophila melanogaster*. Using the *hsp 70* gene, these investigators defined DNA sequence elements that contained DNase I hypersensitive sites to either side of the *hsp 70* genes. These specialized chromatin structure (scs) elements conferred position-independent, copy-number dependent transcription from the *white* promoter, with the exception of one insertion into heterochromatin within the *Drosophila* X chromosome. Importantly, the scs elements do not behave as scaffold attachment regions establishing a functional separation between the two sequences (Kellum and Schedl, 1991). The *Drosophila* suppressor of hairy-wing protein and the complex it assembles with the gypsy transposable element has been recognized as having comparable 'insulating' properties (Geyer and Corces, 1992; Section 2.5.6).

An important observation concerning the functions of scs, gypsy or A elements is that they block enhancer function if the element is between the enhancer and the promoter (Eissenberg and Elgin, 1991). This capacity to block activation or silencing effects has led to the current definition of these elements as insulators (Section 2.5.6). Enhancers normally act over very long distances (many kilobases) (Section 4.1.2). It is hard to imagine how a short DNA sequence (100–200 bp) could inhibit DNA looping, suggesting that enhancers in this case are functioning by an alternate mechanism perhaps related to chromatin folding or nuclear compartmentalization (Section 2.5.6). Additional insight comes from experiments suggesting that gene regulation can occur through progressively altering the structure of domains of chromatin that are continuous in the chromosome. In the bithorax complex three genes (*Ubx*, *abd-A* and *Abd-B*) are aligned 5' to 3' in the order in which they are expressed during development and remarkably in the structures whose origin they control in an anterior to posterior direction in the fly. It has been suggested that these three large (> 20 kb) segments of DNA are each activated in succession (Peifer *et al.*, 1987). Several mutations in the bithorax complex are consistent with this hypothesis since by removing a boundary between two domains of chromatin the anterior–posterior boundary between different structures is also removed (Boulet and Scott, 1988). In this model, boundary elements prevent the propagation of a particular modification of chromatin by functioning as a barrier. We have also discussed the propagation of repressive changes in chromatin or chromosome structure on the bithorax complex via the Polycomb group of proteins (Orlando and Paro, 1993) (Section 2.5.6). Once again strong evidence for chromosomal domain boundaries exists.

Summary
Discontinuities in the chromatin fibre are detected as sites that are hypersensitive to DNase I cleavage. DNase I hypersensitive sites can depend on both the binding sites for *trans*-acting factors and also on the formation of positioned nucleosomes. They contain histone-free DNA sequences which often have been found to have important regulatory functions in the nucleus. Definition of such sites led to the discovery of elements that appear to control the function of domains of chromatin-locus control regions, and of elements that appear to define the boundaries between domains of chromatin known as insulators (Section 2.5.6).

DNase I hypersensitive sites exist before a gene is transcribed and they may contribute to establishing domains of general sensitivity to nucleases. Such domains may represent unravelled or destabilized higher-order chromatin structures and their formation may be an important prerequisite for transcription.

4.2.5 *Trans*-acting factors and the local organization of chromatin structure

The presence of promoter-specific factors or basal transcription complexes can prevent nucleosome assembly on these regions (Section 4.2.2). These non-histone protein–DNA interactions contribute to generating nuclease hypersensitive sites in the chromatin fibre (Section 4.2.4). Often such hypersensitive sites are found in the midst of ordered arrays of nucleosomes (Almer *et al.*, 1986; Almer and Horz,

1986; Benezra *et al.*, 1986; Richard-Foy and Hager, 1987; Szent-Gyorgi *et al.*, 1987). In these genes the nucleosome arrays exist even within the coding region when transcription is repressed; the arrays are altered or lost when transcription is activated (Section 4.3.2). Similar arrays can be found in centromeres and telomeres (Bloom and Carbon, 1982; Gottschling and Cech, 1984; Budarf and Blackburn, 1986); these arrays contain not hypersensitive sites but regions that are refractory to nuclease cleavage.

Three components clearly contribute to the ordering of nucleosomes relative to nuclease hypersensitive or refractory sites. One is the sequence-directed positioning of nucleosomes (Section 2.2.5), the second is statistical positioning of nucleosomes (Kornberg, 1981), which relies on the generation of boundaries to nucleosome arrays, and the third is the positioning of nucleosomes through the direct or indirect interaction between *trans*-acting factors and nucleosomes (Roth *et al.*, 1990). Statistical positioning depends on nucleosomes packing adjacent to a boundary and then being phased to either side of that boundary due to spacing constraints. The influence of the boundary should decay with distance from the boundaries. Kornberg and colleagues examined nucleosome positioning with respect to the GAL1-GAL10 intergenic region inserted into *Saccharomyces cerevisiae* minichromosomes. Alterations of the DNA sequences flanking DNase I hypersensitive sites left the nucleosomal array unchanged, showing that nucleosome positioning was not a consequence of sequence-directed histone–DNA interactions but depended on proximity to the boundary hypersensitive site. A particular promoter-specific DNA-binding protein was found to function as a boundary (Fedor *et al.*, 1988). These experiments are interpreted as representing a passive role of sequence-specific DNA-binding proteins in organizing chromatin structure. However, more recent experiments suggest that highly selective interactions between histones and sequence-specific DNA-binding proteins can also occur.

In *S. cerevisiae*, activation of cell-type specific genes depends on expression of the MAT locus (mating type; Dranginis, 1986; Herskowitz, 1989). Two of the proteins regulating gene expression, α2 and α1, have helix-turn-helix domains (homeodomain proteins, Section 4.1.6). α2 represses *a*-cell-type specific gene expression through co-operative interactions with another promoter-specific protein, MCM1. The α2–MCM1 complex associates with a 32 bp sequence located approximately 100 bp upstream of the TATA sequence of five *a*-cell-type specific genes. Repression can also be effected from a comparable distance downstream of the TATA box. In the presence of the α2–MCM1 complex, nucleosomes are positioned right over the

TATA box of at least two of these six genes (Shimizu *et al.*, 1991). The placement of nucleosomes over very different sequences in similar positions suggested that the α2–MCM1 repressor complex might direct this position. The inclusion of the TATA box into a nucleosome does not independently repress transcription (Section 2.1.2), but the assembly of an extended nucleosomal array within a repressive chromosome domain might account for transcriptional repression (Simpson *et al.*, 1993).

Interactions between nucleosomal histones and the α2–MCM1 complex have been suggested by studies with mutants. Although mutations of the amino terminal tails of histones H2A and H2B have little effect on growth or mating ability, deletions of the amino terminal tail of histone H4 leads to a lengthening of the cell cycle and sterility due to aberrant expression of the silent mating loci (Kayne *et al.*, 1988). Activation of these mating loci is specific as a consequence of deletion of the histone H4 tail, since repressed genes such as PHO5 are not transcribed under these conditions. Individual point mutations of amino acids 16–19 in the amino terminal tail of histone H4 have similar effects to the deletions (Johnson *et al.*, 1990; Megee *et al.*, 1990; Park and Szostak, 1990). One of these point mutations alters a lysine residue that could potentially be acetylated (Section 2.5.2). The effect of the point mutations can be suppressed by mutations in a known regulator of the silent mating loci, SIR3, suggesting direct interaction of the H4 histone tail with *trans*-acting factors (Johnson *et al.*, 1990). Aside from the influence of the histone H4 tail on repression of the silent mating type loci, an intact amino terminus is required for the efficient activation of a number of inducible promoters (Durrin *et al.*, 1991).

Simpson and colleagues have extended these primarily genetic observations to ask direct questions about how a *trans*-acting factor, the α2–MCM1 repressor, influences chromatin organization. Re-examining the positioning of nucleosomes adjacent to the α2–MCM1 complex (Shimizu *et al.*, 1991) it was found that deletions of the histone H4 amino terminal tail prevented formation of a stably positioned nucleosome next to the α2–MCM1 complex. Individual point mutations in the tail region had a comparable effect. An *a*-cell specific promoter was derepressed in the mutant strains, suggesting the positioning of nucleosomes directed by α2 is essential for stringent gene regulation (Roth *et al.*, 1992). The general transcriptional repressors of yeast Tup1 and SSN6 are also essential for directing the positioning of nucleosomes next to the α2–MCM1 complex. Mutations in these proteins also disrupt positioning and activate transcription, further cementing the correlation between specific chromatin structure and gene silencing (Cooper *et al.*, 1994; see Section 3.3). Roth and colleagues have been able to establish that the Tup1 repressor makes direct contacts with the amino-terminal tails of histones H3 and H4 (Edmondson *et al.*, 1996). The Tup1 histone-binding domain coincides with the previously defined domain of Tup1 that is essential for repression. Using their binding asssay, these investigators were able to show directly that histone acetylation interfered with Tup1 association. These studies establish biochemical correlations with the existing genetic analysis and offer the prospect of a structural characterization of the repression mechanism directed by α2–MCM1.

Other examples of sequence-specific DNA-binding proteins that appear to contribute actively to the assembly of specific chromatin structures include the role of the GAGA factor at the *Drosophila* hsp70 and hsp26 promoters (Tsukijama *et al.*, 1994; Becker, 1994). On the hsp70 promoter the GAGA factor directs the repositioning of nucleosomes in a process dependent on ATP hydrolysis. On the hsp26 promoter, interaction of the GAGA factor with its binding sites in chromatin reconstituted in a *Drosophila* embryo chromatin assembly extract resulted in a structure closely resembling that seen *in vivo* (Section 4.2.4). Regulatory elements are kept accessible to factors and a positioned nucleosome is assembled between these sites (Becker, 1994). The active process indicated by the requirement for ATP hydrolysis has not yet been defined. These remodelling experiments might reflect a role for molecular machines that contain the *Drosophila* homologs of the SWI/SNF general activator complex such as NURF or CHRAC (Tsukiyama and Wu, 1995; Tsukiyama *et al.*, 1995; Varga-Weisz *et al.*, 1997) (Section 2.5.8) or they might reflect residual association of chromatin assembly proteins, e.g. CAF1 (Section 3.4.3). These molecular chaperones might retain the capacity to cause a dynamic exchange of histone from assembled nucleosomes.

On the vitellogenin B1 promoter, NF1 may contribute to the exact positioning of the regulator nucleosome that exists between the enhancer and the promoter (Fig. 4.3). Screening for protein–protein interactions using the yeast two-hybrid method, and mutational analysis, have revealed that the histone-fold domain of histone H3 interacts with the proline-rich transcriptional activation domain of NF1. These studies were confirmed by biochemical analysis demonstrating that NF1 binds H3 but not H4, H2A or H2B in isolation. NF1 also binds to mixtures of H3 and H4, presumably as a consequence of the specific heterodimerization of H3 and H4. Importantly, the transcriptional activation domain of NF1 interacts with H3 in a nucleosomal context, and therefore provides positional information for the adjacent nucleosome leading the the specific organization of histone–DNA interactions (Alevizopoulos *et al.*, 1995). There is an excellent correlation between the capacity of mutant forms of NF1 to activate transcription and their capacity to interact with histone H3. These results lead to a model in which the primary role of the transcription factor NF1 is to direct the assembly of a positioned nucleosome that stimulates transcription from the vitellogenin B1 promoter as part of an extended regulatory nucleoprotein complex.

As we have discussed, the HNF3 protein has also been shown to direct the assembly of positioned nucleosomes on the serum albumin enhancer (McPherson *et al.*, 1993). This observation is especially interesting because the structural identity between linker histones and transcription factors such as HNF3 provides a potential mechanism for how nucleosome positioning might be achieved (Section 4.1.6). Clearly HNF3 might substitute for the linker histones in the nucleosome and still recognize DNA specifically.

It appears that in some cases histones and *trans*-acting factors do interact to generate specific chromatin structures *in vivo*. Nucleosome positioning and hence chromatin structure is not a passive filling in of the gaps, but is directed by the appropriate *trans*-acting factor.

Summary
Aside from sequence-directed nucleosome positioning, nucleosomes can be ordered into a specific organization by *trans*-acting factors. This can occur through the generation of boundaries by non-histone proteins that cause nucleosomal arrays to phase themselves relative to these boundaries – statistical positioning. Alternatively, *trans*-acting factors may not passively constrain where nucleosomes form but direct nucleosome positioning through interactions with the core histones. *Trans*-acting factors in some cases have an active role in organizing specific chromatin structures.

4.3 PROCESSIVE ENZYME COMPLEXES AND CHROMATIN STRUCTURE

The compaction of DNA into chromatin provides an obstacle course for any enzyme complex that has to progress along the double helix (Fig. 4.27). DNA and RNA polymerases are large multisubunit enzyme complexes that at least transiently unwind the double helix. In eukaryotes these enzymes are over seven times larger than the

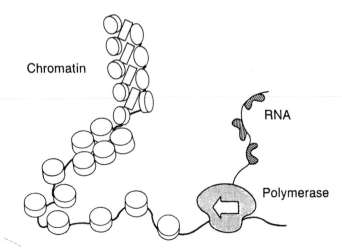

Figure 4.27. Chromatin structure poses many impediments to the progression of RNA polymerase molecules along the DNA molecule. Presumably the chromatin fibre has to be unwound to allow transcription to occur.

combined mass of the histones in a nucleosome. The coexistence of a nucleosome and a DNA segment on which a polymerase is attached seems impossible. How these enzymes work in chromatin and the consequences for chromatin structure of their progression through it are important issues.

4.3.1 Replication and the access of transcription factors to DNA

The period in the eukaryotic cell cycle when the genome is duplicated (S-phase) is crucially important both for establishing and maintaining programs of gene activity. The majority of genes in a proliferating cell of a defined type or line continually retain the same states of transcriptional activity through cell division. This reflects the commitment of that cell type to a particular state of determination (Section 4.1.4). How this commitment is established and maintained is not yet resolved; however, several experiments have suggested a solution to this problem (Wolffe, 1991c). Furthermore, many examples exist in which states of gene activity change following replication events: genes may be either transcriptionally activated or repressed. These events have major consequences for the differentiation of a particular cell type during development. A consideration of the processes occurring at the eukaryotic replication fork again suggests molecular mechanisms that might explain these phenomena.

We have discussed the basic requirements for establishing a eukaryotic gene in a transcriptionally active state (Section 4.1.4). The initial direct binding of transcription factors to DNA is rapid; the sequestration of non-DNA-binding factors is relatively slow. *In vitro,* the process of assembling a complete transcription complex takes several minutes. The mutually exclusive binding of the histone octamer or transcription factors to promoter elements in the absence of specific nucleosome positioning is well established (Section 4.2.2). Prior assembly of nucleosomes prevents transcription factors from binding to DNA *in vitro* and conversely, the prior assembly of a transcription complex prevents nucleosome formation from repressing transcription. Although these results provide a plausible molecular basis for the maintenance of stable states of gene expression in a terminally differentiated non-dividing cell, they do not explain why either transcription complexes or nucleosomes are assembled onto DNA in the first place. Clearly, as both nucleoprotein structures can incorporate the same DNA molecule, the possibility exists of a competition occurring between the assembly of the two structures. As discussed later, this competition in fact occurs; however, chromatin assembly is staged (Section 3.2). The opportunity of a *trans*-acting factor to interact with immature chromatin structures facilitates transcription complex formation. Such a general mechanism for *trans*-acting factor access to DNA could, of course, be assisted by the assembly of specific chromatin structures (Section 4.2.3).

Following replication, nascent DNA is assembled into a chromatin structure that is more sensitive to nucleases than mature chromatin (Section 3.3). The maturation of this nascent chromatin from a nuclease-sensitive conformation takes approximately 10–20 min in a mammalian cell. Fractionation of nascent chromatin at various times following replication allowed the demonstration of a two-stage maturation process for chromatin in that histones H3 and H4 are sequestered onto DNA before histones H2A and H2B. Histone H4 is diacetylated and phosphorylated as it is deposited onto DNA. Histone H1 is the last histone to be stably incorporated into chromatin. This observation is related to the structure of a nucleosome since histones H3 and H4 form the core of the structure, whereas histones H2A and H2B bind at the periphery of the nucleosome and histone H1 can only associate in its proper place once two turns of DNA are wrapped around the core histones (Section 2.2.4).

The development of *in vitro* replication systems has permitted a further dissection of the assembly of chromatin on replicating DNA

(Section 3.2). The replication of DNA *in vitro* clearly facilitates chromatin assembly in comparison to non-replicating DNA incubated in the same extracts. The exact mechanisms underlying this increase in the efficiency of chromatin assembly are unknown, but appear to involve proteins such as CAF1 associated with the replication fork (Section 3.2). What is clear is that an intermediate in chromatin assembly forms on replicating DNA, and that this structure has properties consistent with it being a tetramer of histones H3 and H4. The subsequent addition of histones H2A and H2B to this complex generates chromatin with a nuclease sensitivity similar to that expected for the mature state. The nuclease-sensitive stage presumably reflects the accessibility of DNA to the DNA-binding nucleases when complexed with only histones H3 and H4. Other DNA-binding proteins, such as transcription factors, might also gain access to DNA at this stage in chromatin assembly. Only later when histones H2A, H2B and H1 are sequestered would access to DNA in the nucleosome be restricted.

The presence of a transcription complex (Section 4.1.3) or of a repressive chromatin structure over the promoter elements of a gene would establish a stable state of gene activity in a cell that would not be perturbed until the next round of DNA replication. Which nucleoprotein complex forms on DNA will depend on many variables. For example, the rate of transcription complex formation will principally depend on the concentration of transcription factors and the stability of their interaction with specific DNA sequences (Section 4.1.5). Likewise, nucleosome formation will depend on histone concentrations and the relative stability of a nucleosome containing a particular DNA sequence. Experiments using non-replicating DNA templates for transcription have suggested that the rate of transcription complex formation in the face of chromatin assembly is an important variable in determining gene activity (Section 4.2.2). A second variable follows from the observation that transcription complexes can have different stabilities depending on the promoter elements within the complex. Therefore the final equilibrium binding of transcription factors during chromatin assembly can also affect gene transcription (Section 4.1.5). These observations do not explain how genes might be effectively programmed in spite of nucleosome assembly; however, they do indicate that transcription factors must be abundant if a transcription complex is to be assembled on to the promoter.

The first insight as to how the efficient programming of genes is accomplished following replication came from experiments using class III genes and non-replicating DNA. The assembly of *Xenopus* 5S

RNA genes into complete nucleosomes has been shown to inhibit transcription complex formation (Sections 4.2.2 and 4.2.3). Importantly, the assembly of only histones H3 and H4 onto 5S RNA genes did not inhibit transcription (Tremethick *et al.*, 1990). Similarly, a chromatin template isolated from *Xenopus* sperm, which is naturally deficient in histones H2A, H2B and histone H1 especially after incubation in egg or oocyte extracts, was as active as naked DNA in an *in vitro* transcription reaction (Wolffe, 1989b). These results suggest that a key intermediate in nucleosome assembly, the complex of histones H3 and H4 with DNA, is accessible to transcription factors. This would be consistent with the hypothesis that the newly replicated DNA in the first stage of chromatin maturation could be programmed with transcription complexes. An additional contributory factor to the efficient programing of genes at the replication fork is the requirement for a certain length of double-stranded DNA (113 bp) to be synthesized before histones can be stably assembled on to templates (Svaren *et al.*, 1994b). This is presumably because of the fact that the first histone complex assembled following replication – the $(H3, H4)_2$ tetramer – requires almost 120 bp of DNA to interact with in forming a stable complex (Hayes *et al.*, 1991b). Sequence-specific transcription factors can stably associate with much shorter double-stranded DNA fragments.

Xenopus egg extracts have the capacity to carry out complex biological processes such as the complete replication of sperm nuclei (Section 3.1). Simplified systems derived from these extracts retain the ability to replicate single-stranded DNA templates with efficiencies of DNA synthesis approaching those observed *in vivo*. The enzymatic processes involved closely resemble those occurring on the lagging strand of the chromosomal replication fork. These extracts also assemble chromatin efficiently on the newly replicated DNA and transcribe class III genes (Wolffe and Brown, 1987; Almouzni and Méchali, 1988a). Using single-stranded DNA as starting material, it was demonstrated that a competition existed between transcription complex assembly on a 5S RNA gene and chromatin assembly on replicating DNA (Almouzni *et al.*, 1990b). This competition depended on the presence of all four core histones. Subsequent work dissected the components of chromatin assembly responsible for the competition between transcription factors and histones for association with the 5S RNA gene. Conditions were established under which the association of histones H2A and H2B with DNA was made limiting (Almouzni *et al.*, 1991). This involved not supplementing the replication reaction with exogenous Mg^{2+}/ATP. Under these circumstances histones H3 and H4 were incorporated into chromatin normally, but complete nucleosomes did not form until either exogenous Mg^{2+}/ATP or exogenous histones H2A and H2B were added. The histone H3/H4 complex was transcriptionally active and apparently accessible to transcription factors. Addition of histones H2A and H2B resulted in the competitive effects on transcription observed in earlier work (Almouzni *et al.*, 1990b, 1991). These competitive effects were shown to be due to the displacement of a non-specific DNA-binding transcription factor (TFIIIB) whose association with DNA depends on specific interactions made by other proteins (TFIIIA and TFIIIC; Section 4.1.4). This provides further evidence for the absence of an immediate competition between transcription factors and histones H3 and H4 for binding to the 5S RNA gene. Perhaps the *trans*-acting factors interacting with other regulatory elements will gain access to DNA in a similar manner.

Summary
Experiments using replicating DNA as a substrate demonstrate that a complex of DNA with histones H3 and H4 is initially assembled and that DNA in this complex is accessible to transcription factors (Fig. 4.28). These observations are consistent with experiments on the accessibility of chromatin formed on non-replicating DNA to transcription factors. Most importantly they provide confirmation that the molecular mechanisms adopted by the cell to assemble chromatin can accommodate access of essential regulatory elements in DNA to *trans*-acting factors. It is only the subsequent sequestration of histones H2A and H2B followed by histone H1 that renders chromatin progressively less accessible to the transcriptional machinery.

4.3.2 The fate of nucleosomes and transcription complexes during replication

The above discussion (Section 4.3.1) has focused on how genes are programmed *de novo*. The experiments described how a gene can be rendered active in spite of on-going chromatin assembly if transcription factors are available. The presumption implicit in this description is that pre-existing nucleoprotein complexes are disrupted as a consequence of DNA replication. Whether or not this disruption actually occurs is the focus of much current research.

Disruption of pre-existing nucleoprotein complexes is not an inevitable consequence of replication. In the bacteriophage T4, the

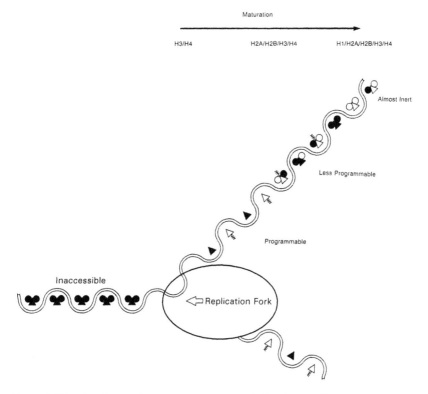

Figure 4.28. As chromatin structure matures following replication it becomes much less accessible to *trans*-acting factors.
Triangles are (H3, H4)$_2$ tetramers, circles H2A, H2B dimers, acetylation is indicated by the zig-zag lines.

DNA replication machinery can cope very effectively with a head-on collision with a transcriptionally engaged RNA polymerase (Liu and Alberts, 1995). When the DNA polymerase collides with an *E. coli* RNA polymerase heading in the opposite direction, the DNA polymerase pauses and dependent on the activity of the T4 gene 41 DNA helicase, it passes the RNA polymerase after a few seconds. When both polymerases are moving in the same direction, the pause is even shorter. In both instances the RNA polymerase remains transcriptionally engaged and resumes RNA chain elongation once the DNA polymerase has passed.

It has long been thought that, once established, the nucleoprotein complexes determining gene activity or repression might be capable of reproducing these structures during replication (Tsanev and Sendov, 1971; Brown, 1984; Weintraub, 1985). Under certain conditions, nucleosomes that exist prior to replication can be shown to associate with the daughter DNA molecules (Cusick *et al.*, 1984; Sogo *et al.*, 1986; Bonne-Andrea *et al.*, 1990). However, there is no apparent preference for pre-existing nucleosomes to be reformed on either the leading or lagging strand of the replication fork. This observation allows strong arguments to be made against any imprinting mediated by the arrangement of histones on DNA. Of course, pre-existing histone modification states and associated proteins might remain concentrated in the vicinity of the replication fork for reassembly on nascent DNA (Perry *et al.*, 1993). The dispersive segregation of nucleosomes on daughter chromatids is consistent with evidence from electron microscopy that histones are displaced from DNA by the replication fork (Sogo *et al.*, 1986). Furthermore, disruption of nucleosomes during replication would explain the presence of both newly synthesized histones and old histones within the same nucleosome in newly assembled chromatin (Jackson, 1987, 1988, 1990). This disruption of chromatin structure during replication suggests a simple mechanism for facilitating gene activation by replication (Svaren and Chalkley, 1990).

As described previously, if transcription factors are not available, the formation of complete nucleosomes following replication will render promoter elements inaccessible unless specific chromatin structures form that mediate access. This repressed state would not be altered by the appearance of transcription factors later in the cell cycle. However, replication events offer the opportunity to disrupt repressive chromatin structures. If transcription factors are present at the instant of replication, the staged assembly of chromatin will facilitate their association with DNA and subsequent gene activation. This hypothesis has recently been tested *in vivo* in an insightful series of experiments using *S. cerevisiae* (Aparicio and Gottschling, 1994).

Aparicio and Gottschling made use of the repressive influence on gene expression at the telomeres due to position effect variegation. They investigated the molecular mechanism by which a silent repressed gene switched to a state of active transcription. They found that they could increase the efficiency of switching to active transcription by increasing the amount of a transcription factor in the cell. Importantly, this effect was dependent on the period within the cell cycle that the transcription factor was expressed. If the cell cycle was arrested towards the end of S-phase the transcription factor could activate transcription, but it could not reverse repression before the DNA was replicated. These experiments suggest that the disassembly of repressed chromatin structures during replication is necessary to facilitate transcription factor mediated activation of gene expression.

Although most investigators believe in the dispersive segregation of nucleosomes at the replication fork, some researchers have supported the direct templating or reproduction of pre-existing chromatin structures. This type of conclusion generally emerged from studies in which histone synthesis was artificially inhibited. Under these conditions replication proceeds, but no new nucleosome assembly occurs (Riley and Weintraub, 1979; Seidman *et al.*, 1979; Handeli *et al.*, 1989). Nucleosomes were observed to segregate to the leading strand at the replication fork. This result appears to be an artefact, since De Pamphilis and colleagues demonstrated that ongoing protein synthesis is required for DNA synthesis on the lagging strand of the replication fork. Thus duplex DNA, which can be assembled into nucleosomes, is only present on the leading strand (Burhans *et al.*, 1991). Single-stranded DNA, which is not assembled efficiently into nucleosomal structures, remains on the lagging strand (Almouzni *et al.*, 1990b). Thus, the mistaken conclusion that nucleosomes segregated conservatively at the replication fork was reached.

A similar series of experiments addressed the fate of transcription complexes on DNA during replication (Section 4.1.4). A proposal for how transcription complexes might confer a stable state of gene activity through replication came from experiments using *Xenopus* 5S RNA genes (Brown, 1984). Brown proposed the existence of strong co-operative interactions between transcription factors bound to DNA. If just one component of a complex remained associated with a daughter gene after replication, then it could nucleate formation of a complete complex by attracting excess transcription factors from the nucleoplasm. This model was tested by assembling a transcription complex on a 5S RNA gene and replicating through this complex using a viral *in vitro* replication system (Wolffe and Brown, 1986). Replication was dominant to transcription and a direct consequence of replication fork progression through the active 5S RNA gene was the displacement of specific transcription factors. Several correlations from *in vivo* work support this observation. There is a clear antagonism between transcription and replication on efficiently replicating SV40 molecules (Lebkowski *et al.*, 1985; Lewis and Manley, 1985). Replication forks invade the transcriptionally active ribosomal RNA genes in yeast (Saffer and Miller, 1986). In mammalian cells a RNA polymerase I specific termination factor TTF-1 blocks progression of the replication machinery into the rDNA transcription unit preventing a head-on collision with RNA polymerase (Gerber *et al.*, 1997). The disruption of transcription complexes by replication would provide a simple means of inactivating genes. A replication event through a gene that occurred when transcription factors were not available to bind to the promoter would cause inactivation of the gene through nucleosome assembly unless specific chromatin structures are assembled that might allow subsequent access (Solomon and Varshavsky, 1987; Barton *et al.*, 1993).

A final issue relevant to this discussion is the significance of transcription factors for the initiation of replication and the timing of this initiation in S-phase. If replication disrupts both active and repressed chromatin structures, then the entire nucleus has to be remodelled after each replication event. The accessibility of immature chromatin on newly replicated DNA provides a means to accomplish this remodelling; however, the reformation of nuclear structures has other implications. If there are limiting transcription factors available in a cell then a gene that is replicated early in S-phase has more opportunity to assemble an active transcription complex than a gene that replicates late. This is simply because the gene that replicates early is available for transcription factors to bind before all of the early replicating portion of the genome has sequestered these factors. A late-replicating gene will therefore experience a relative deficiency in factor availability (Gottesfeld and Bloomer, 1982; Wormington *et al.*, 1983). Transcriptionally active genes replicate early in S-phase (Goldman *et al.*, 1984; Gilbert, 1986; Guinta and Korn, 1986). The reason for this early replication is unknown, but possibilities include the local disruption of chromatin structure by transcription complexes making that DNA more accessible to the replication machinery (Wolffe and Brown, 1988) and the observation that many transcription factors are in fact also replication factors (De Pamphilis, 1988). An attractive variation of this model is that the type of chromatin assembled early in S-phase is more accessible to transcription factors than chromatin assembled late in S-phase. Early replicating chromatin may sequester histones that are more highly acetylated, and consequently more accessible to the transcription factors that maintain continued transcriptional activity. The CENP-A protein that replaces histone H3 in the mammalian centromere is synthesized at the very end of S-phase, providing an example of how the compartmentalization of protein synthesis might contribute to the assembly of specialized chromatin structure (Shelby *et al.*, 1997). Although a general test of the significance of this model has not been made, it remains an attractive mechanism for explaining the maintenance of specific patterns of gene expression in a proliferating cell type.

Our current understanding of the developmental regulation of differential gene expression has followed not only from the definition of *cis*-acting DNA sequence elements and *trans*-acting factors, but also from knowledge of how these function in a chromatin environment in the context of DNA replication events (Wolffe and Brown, 1988). There are many examples that further suggest an integration of these various aspects of nuclear function. A few of these are illustrated below.

The simplest examples of a requirement for replication in the developmental regulation of transcription involve the rapid and precisely programmed life cycle of eukaryotic viruses. Work with adenoviruses, herpes viruses and vaccinia indicates that replication events are required for the activation of certain genes. In these instances DNA replication is proposed to remove or dilute inhibitors that deny access of newly synthesized transcription factors to the viral genome (Thomas and Mathews, 1980; Crossland and Raskas, 1983; Gaynor and Berk, 1983; Mavromara-Nazos and Roizman, 1987; Keck *et al.*, 1990). An instance of a replication event being associated with the repression of genes (as opposed to activation) occurs in yeast where a single round of replication is associated with the inactivation of the silent mating type loci (Miller and Nasmyth, 1984). The replication fork in this case is proposed to remove transcriptional activators.

Early *Drosophila* and *Xenopus* development is characterized by rapid rounds of cell division in the absence of transcription. Each of these embryos contains huge stores of nuclear components, including transcription factors and histones (Laskey *et al.*, 1983; Kadonaga, 1990; Almouzni and Wolffe, 1993b). DNA replication appears to play a major role in suppressing efficient transcription complex formation, presumably through the repeated disruption of transcription complexes that do form. Procedures that inhibit replication in these embryos lead to the premature activation of transcription (Edgar and Schubiger, 1986; Kimelman *et al.*, 1987).

In *Caenorhabditis elegans* and the sea urchin, replication events are correlated with changes in the commitment of cells to a particular developmental fate (Mita-Miyazawa *et al.*, 1985; Edgar and McGhee, 1988). Similar changes can occur in differentiated cells that express one set of specialized genes, and that can switch to another program of gene expression only after one or more cell divisions (e.g. Wolffian regeneration of the lens, Takata *et al.*, 1964). However, replication events are not necessarily essential to change gene expression within a particular cell (Sections 3.1.1 and 3.1.2). This is not surprising since chromatin structure is not completely inert *in vivo*. We have also discussed how the formation of specific chromatin structures and the use of targeted chromatin remodelling machines can alleviate the requirement for DNA replication in order to activate genes (Sections 2.5.4 and 4.2.3).

Summary
DNA replication has an important role in the regulation of eukaryotic gene expression. The replication process transiently disrupts both active and repressed chromatin structures. This provides a key opportunity for mediating changes in the programming of genes. In those cases that have been tested experimentally, imprinting of gene activity mediated by the prior formation of transcription complexes or nucleosomes, followed by their conservative segregation to daughter DNA molecules, has not been observed. However, the mere fact that a gene is active or inactive may influence the time during S-phase when the gene is replicated. By determining the availability of transcription factors at the instant of replication, this may confer continued gene activity or repression.

4.3.3 Chromatin structure and DNA replication

The initiation of DNA replication in eukaryotic chromosomes is perhaps the most highly regulated event that occurs within the nucleus. All DNA must be replicated once per cell cycle. An origin must be activated a single time and not multiple times within a given S-phase. From our earlier discussion of replication origin utilization in prokaryotic systems (Section 4.1.1), it might be anticipated that the nucleoprotein complexes regulating replication in eukaryotic chromosomes will be extensive and will rely on manifold protein–DNA and protein–protein interactions for their assembly and regulation. Most of our best current information about chromosomal origins derives from experiments in *S. cerevisiae*. Within *S. cerevisiae* chromosomes, origins are spaced every 40 kb and have a defined temporal order of initiating replication (Newlon, 1988; Fangman and Brewer, 1991). DNA sequences have been defined (autonomously replicating sequences, ARS) that allow extrachromosomal plasmid DNA molecules to promote high-frequency transformation of yeast cells by allowing the plasmids to replicate autonomously. Not all ARS elements within a chromosome are used; some can be repressed by position effects such as those found at the telomere (Section 2.5.5) (Newlon *et al.*, 1993). It appears that adjacent ARS elements can determine the frequency with which they individually serve as origins implying that each origin might define a chromatin domain of influence (Brewer *et al.*, 1993).

We have discussed how a large nucleoprotein origin recognition complex (ORC) has been defined that associates with the ARS consensus sequence (Section 4.1.1). Genetic studies on mutations in components of the ORC clearly establish that this nucleoprotein complex is required for replication initiation (Fox *et al.*, 1993). *In vivo* footprinting also suggests that components of the ORC remain intact and associated with DNA within the chromosome throughout the cell cycle (Diffley and Cocker, 1992). However, the context in which the regulatory elements directing ORC assembly are presented within the chromosome is important, since incorporating part of the ORC-binding site into a nucleosome can inhibit origin utilization (Simpson, 1990).

The multiple ARS elements present in the yeast chromosome all have the potential to associate with the ORC, but are not all used as origins. This led to the concept that the ORC also fulfils other essential functions within the chromosome. The ORC complex has a role in transcriptional silencing at one of the silent mating type loci, HMR (Fox *et al.*, 1993). Thus ORC may have a key regulatory role in controlling the transcriptional activity of chromatin domains (Section 2.5.5).

The utilization of origins of replication within the chromosomes of metazoans appears much more complex than that of *S. cerevisiae* (Hamlin *et al.*, 1993; Gilbert *et al.*, 1993). Although it is possible to define limited segments of DNA as small as 500 bp where bidirectional replication is initiated (Burhans *et al.*, 1990), no consensus sequences that independently direct replication initiation have been defined. It appears that extended chromosomal regions as large as several kilobase pairs in length can facilitate replication initiation. A model was proposed to explain these results reflecting the Jesuit maxim that 'many are called, but few are chosen' (De Pamphilis, 1993). This suggests that although there are many detectable sites at which replication might begin, the assembly of DNA into nucleosomal arrays followed by the subsequent assembly of higher-order chromatin structures represses many potential origins, while potentiating the activity of others. This might also contribute to the selective utilization of ARS elements as origins of replication in yeast. This model does not offer any explanation of why and how a replication-competent nucleoprotein complex is assembled at an origin.

Arguments for nuclear organization influencing the initiation of DNA replication derive from *in vitro* experiments in *Xenopus* egg extracts and *in vivo* experiments in the chromosomes of *Drosophila* and *S. cerevisiae*. Chromatin assembly and nuclear structures are both necessary for replication to occur in *Xenopus* egg extracts (Sheehan *et al.*, 1988; Blow and Sleeman, 1990; Newport *et al.*, 1990; Meier *et al.*, 1991). Additional evidence for chromatin structure influencing origin utilization comes from the suppression of origin utilization in regions of *S. cerevisiae* and *Drosophila* chromosomes that contain heterochromatin (Forrester *et al.*, 1990; Ferguson and Fangman, 1992; Karpen and Spradling, 1990). Nuclear scaffold attachment regions have also been proposed to facilitate aspects of the replication process (Bode *et al.*, 1992). Chromosomal replication within the nucleus clearly occurs within morphologically defined factories attached to a nuclear scaffold (Nakamura *et al.*, 1986; Jackson and Cook, 1986; Mills *et al.*, 1989; Hozak *et al.*, 1993). Finally, the nuclear envelope has a regulatory role in determining the activity of the 'licensing' process that capacitates replication initiation (Blow and Laskey, 1988; Leno *et al.*, 1992).

Summary
Chromatin structure has an established role in suppressing the utilization of origins of replication in *S. cerevisiae* chromosomes. The origin recognition complex (ORC) of yeast is required for the initiation of replication, and it also has other functions in regulating transcription. No clear picture regarding the local chromosomal requirements for replication in the chromosomes of metazoans has been established. Nuclear assembly is important for replication to occur. Replication occurs at factories. This is a powerful example of selective compartmentalization of proteins necessary for a specialized function within the eukaryotic nucleus.

4.3.4 Transcription and chromatin integrity *in vivo*

Some of the most compelling photographs in biology are those made by Miller and colleagues in examining transcription units in *Drosophila* embryos (McKnight *et al.*, 1978; McKnight and Miller, 1979). Active ribosomal RNA genes (class I genes) were found to be densely packed with RNA polymerase I with few, if any, nucleosomes present. During transcription of ribosomal RNA genes (Conconi *et al.*, 1989) biochemical analysis using psoralen cross-linking (see below) also indicates that nucleosomes disappear from the transcribed sequences. In contrast, non-ribosomal transcription units (class II genes) in which RNA polymerase II molecules are more widely dispersed, are clearly assembled into nucleosomes. Measurements of the length of DNA within a gene and its compaction into nucleosomal structures reveal that compaction is inversely proportional to the number of RNA polymerase molecules, suggesting that RNA polymerase disrupts nucleosomes at least over the DNA it is bound to.

Most investigators detect nucleosomes on actively transcribed class II genes using nucleases, although the micrococcal nuclease cleavage pattern may become less defined during transcription (Pavlovic *et al.*, 1989). Chemical cross-linking studies also support the continued association of nucleosomes with transcribed class II genes. Formaldehyde cross-linking of histones to DNA in whole cells followed by immunoprecipitation demonstrated a quantitatively similar association of these proteins with transcribed and non-transcribed sequences (Solomon *et al.*, 1988). Slightly different cross-linking methodologies demonstrated that fewer core histone DNA contacts mediated by the globular domains overall were actually established on active genes in comparison to inactive ones. In contrast, contacts by the tails remained unchanged. This suggests that some form of altered chromatin structure exists on transcribed sequences (Karpov *et al.*, 1984; Nacheva *et al.*, 1989). These observations further indicate that the presence of nucleosomes is not incompatible with the efficient elongation by RNA polymerase and that RNA polymerase molecules may displace nucleosomes from DNA. Displacement of nucleosomes could be a consequence of some aspect of the transcription process or simply due to steric occlusion of histone–DNA contacts. At the opposite extreme, Tata and colleagues have shown that the formation of heterochromatin in erythrocyte nuclei (Section 2.5.6) actually inhibits the elongation of engaged RNA polymerase II molecules (Hentschel and Tata, 1978). Clearly the elongation of RNA polymerase can influence chromatin structure and the compaction of chromatin can influence the processivity of RNA polymerases.

Studies on transcriptionally active chromatin have examined the structural basis for its increased accessibility to non-specific DNA-binding proteins and nucleases (Section 4.2.1). Among the most noticeable differences between actively transcribed and inactive chromatin is the level of histone acetylation (Csordas, 1990). Nucleosomes released from nuclei at early times of nuclease digestion are enriched in actively transcribed DNA sequences and acetylated histones (Section 2.5.2). Studies using organomercurial agarose column chromatography have shown actively transcribed genes to copurify with highly acetylated histones H4, H3 and H2B (Sterner *et al.*, 1987). Crane-Robinson and colleagues established a more direct linkage between core-histone acetylation and transcribed gene sequences using antibodies against the acetylated lysines of the basic tails, to immunoprecipitate acetylated chromatin. The acetylated chromatin was found to be enriched in actively transcribed sequences (Hebbes *et al.*, 1988). Although these observations do not establish a causal role for histone acetylation in facilitating elongation through chromatin by RNA polymerase II, they do suggest that acetylation of histones will be at least a consequence of transcribing chromatin.

Evidence consistent with a possible conformational change in the nucleosome is that the histone H3 sulphydryls of nucleosomes on active genes may be accessible to organomercurial columns, although sulphydryls of other non-histone proteins might contribute to these results (Chen and Allfrey, 1987; Chen *et al.*, 1990; Section 2.5.2). In transcribed ribosomal RNA genes from *Physarum polycephalum*, the histone H3 sulphydryls are accessible to chemicals such as iodacetamide. This accessibility would imply that the nucleosome would have to be disrupted at least transiently during transcription. Furthermore, electron microscopy reveals extended particulate structures on these genes that did not appear to be the same as normal nucleosomes (Prior *et al.*, 1983). It has been proposed that conformational changes in nucleosomes might occur during transcription (Weintraub *et al.*, 1976), although the vast majority of physicochemical studies indicate that a splitting of a nucleosome is very unlikely. It is possible that a dimer of histones H2A and H2B might readily dissociate (Sections 2.2.3 and 2.2.4). This might account for the increase in histone H3 sulphydryl accessibility.

Nuclease accessibility studies have clearly shown a regular change in a canonical nucleosome repeat in the *S. cerevisiae* heat-shock protein (*hsp*) *82* gene (Szent-Gyorgi *et al.*, 1987). When the promoter of this gene is crippled by mutation, transcription ceases and the coding sequence is packaged into a positioned array of nucleosomes. The nucleosomal repeat as seen with DNase I is approximately 165 bp. When transcription of the *hsp 82* gene is induced by heat shock, the nucleosomal repeat disappears, and DNase I cuts chromatin approximately every 80–100 bp. It is possible that these new repeats represent an ordered array of non-histone proteins such as RNA polymerases (Lee and Garrard, 1991a,b) although a change in nucleosome conformation is possible. It was originally suggested that torsional stress might be necessary both for transcription and for chromatin disruption (Lee and Garrard, 1991a,b). However, recent experiments clearly establish that this is not the case (Liang and Garrard, 1997). Jackson and colleagues have shown that exchange of histones H2A/H2B out of chromatin *in vivo* is facilitated by transcription, whereas little effect is seen for histones H3/H4 (Jackson, 1990). The transcriptionally active fraction of chromatin, enriched in acetylated histones, is deficient in histone H1 and has a particulate structure with a mass consistent with a loss of a dimer of histones H2A/H2B (Locklear *et al.*, 1990). All of these data are consistent with changes in nucleosome structure caused by the transcription process *in vivo*.

Early studies that fractionated chromatin based on its solubility at different ionic strengths (Section 2.3.1), suggested that histone H1 was deficient in transcriptionally active chromatin (Rose and Garrard, 1984; Rocha *et al.*, 1984; Xu *et al.*, 1986). Different approaches have suggested that histone H1 might still be present in chromatin, albeit interacting differently with DNA. Based on nuclease accessibility studies, Weintraub suggested that the histone H1 in active chromatin could not mediate chromatin folding (Weintraub, 1984). Thomas, Mirzabekov and colleagues have made use of cross-linking methodologies to suggest that in transcribed chromatin, reduced amounts of histone H1 remain associated with DNA. This histone H1 no longer associates through the globular domain, but does so through the basic carboxyl and amino terminal tails (Nacheva *et al.*, 1989; Kamakaka and Thomas, 1990). Furthermore, Daneholt and colleagues have used immunoelectron microscopy to show the presence of histone H1 on transcribed Balbiani ring chromatin (Ericsson *et al.*, 1990; Grossbach *et al.*, 1990). UV cross-linking studies suggest that actively transcribed chromatin is slightly deficient in histone H1 and that this deficiency might account for subsequent difficulty in folding the chromatin fibre *in vitro*. Thus, although histone H1 seems to remain in transcribed chromatin to a certain extent, its deficiency might account for the increased acetylation of the core histones, through greater access of the histone acetylase enzyme to its substrate in the chromatin fibre.

Although nucleosome structure may be changed as a consequence of transcription, are nucleosomes actually displaced at the instant of polymerase progression along the DNA in the nucleosome? The evidence discussed earlier suggests that this is indeed the case for class I genes. For RNA polymerase II genes, similar events appear to occur. Psoralen cross-linking has been used to study the consequences for nucleosomes during transcription by RNA polymerase II of SV40 minichromosomes in the living cell. UV light will induce psoralen to cross-link duplex DNA together; however, DNA in the nucleosome reacts less well with psoralen. This is because psoralen has to intercalate into DNA to exert its effects. This is more difficult when DNA is wrapped around the core histones. Thus on denaturation a nucleosome will appear as a single-stranded bubble in the electron microscope. It is also possible to cross-link nascent mRNA to DNA, and to observe such structures adjacent to nucleosome-size single-stranded bubbles. Surprisingly, these results suggest that the RNA polymerase must either coexist with nucleosomes, or that histones must rapidly reassemble to form a nucleosome after RNA polymerase progression through the nucleosome (De Bernardin *et al.*, 1986). These observations are consistent with many electron microscopic studies showing nucleosomes immediately behind an elongating RNA polymerase molecule (McKnight and Miller, 1979; Bjorkroth *et al.*, 1988). Pederson and Morse have examined the topological consequences of transcription in yeast minichromosomes. If nucleosomes unfold or if histones are lost during transcription of a closed circular plasmid *in vivo* a change in supercoiling would be expected. No such changes are observed (Pederson and Morse, 1990). These results demonstrate that nucleosomes must be able to rapidly reform following RNA polymerase II passage.

An interesting regulatory possibility is that nucleosomes might provide regulated blocks to transcriptional elongation. The HIV 5′ long terminal repeat contains a positioned nucleosome immediately downstream of the start of transcription initiation (Verdin *et al.*, 1993). In response to normal inducers of transcriptional activity this nucleosome is disrupted. Nucleosome disruption is independent of replication and transcription. It is attractive to speculate that the positioned nucleosome on the HIV LTR blocks transcriptional elongation until it is removed (Fig. 4.29). Incubation of several cell

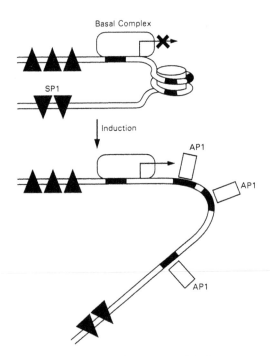

Figure 4.29. Model for nucleosomal-mediated block to transcription in the HIV 5′LTR.
Induced binding of AP-1 disrupts the nucleosome and allows transcriptional elongation to proceed.

lines latently infected with HIV-1 with the specific inhibitors of histone deacetylase, Trapoxin and Trichostatin A leads to the transcriptional activation of the HIV-1 promoter and an increase in virus production (Van Lint *et al.*, 1996). The nucleosome that is located at the transcription start site is disrupted in the presence of Trapoxin and Trichostatin A. This disruption is independent of the transcription process itself. Therefore, targeted histone acetylation might have a significant role in HIV-1 promoter regulation in chromatin. The human heat shock transcription factor has been reported to recruit human SWI/SNF activity to relieve a nucleosomal block to RNA polymerase elongation on the hsp70 promoter (Brown *et al.*, 1996b). Many eukaryotic promoters contain template engaged RNA polymerase (Gilmour and Lis, 1986). This suggests that chromatin structure might have a general role in the regulation of transcriptional elongation.

Summary

RNA polymerases progress through a chromatin template *in vivo*. Nucleosomes appear to be displaced by polymerase, but can rapidly reform after polymerase progression. Histone H1 and a full complement of the core histones are present on transcribed chromatin, but their mode of interaction with DNA may differ from that in the nucleosome. Histone H1 in transcribed regions may be slightly deficient relative to non-transcribed sequences. The core histones are acetylated on transcribed regions with unknown consequences for the processivity of polymerases.

4.3.5 Transcription and chromatin integrity *in vitro*

Although interesting correlations can be made between transcription and chromatin structure *in vivo*, mechanistic studies have to be made *in vitro* in order to understand how RNA polymerase can progress through nucleosomal DNA and what the direct consequences are for the integrity of the nucleosome or chromatin fibre (Kornberg and Lorch, 1991). Felsenfeld and colleagues quantitated both the decrease in accessibility of prokaryotic RNA polymerase to chromatin (Section 4.2.1) and the decrease (66%) in the rate of elongation of the polymerase in chromatin relative to naked DNA (Cedar and Felsenfeld, 1973). Subsequent experiments used a specific template, bacteriophage T7 DNA reconstituted into nucleosomes (Williamson and Felsenfeld, 1978). An important observation was that the rate of elongation by *E. coli* RNA polymerase was reduced at low ionic strength ($< 0.1\,M$) compared to naked DNA, but approached that seen with naked DNA as a template at higher ionic strengths ($0.5\,M$). Nucleosomes do not dissociate at these salt concentrations although their interactions with DNA are weakened (Section 3.3). These results for the first time suggested that nucleosomes might not present a major impediment to the progress of RNA polymerase in an *in vitro* system.

The next step in resolving how polymerases progress through chromatin came from experiments using defined chromatin templates and purified prokaryotic RNA polymerase. Kornberg and collaborators found that transcription through a single nucleosome by SP6 RNA polymerase occurred without impediment at low ionic strength ($40\,mM$ Tris, $6\,mM$ $MgCl_2$). The nucleosome was disrupted by the transcription process (Lorch *et al.*, 1987). An almost identical experiment by Losa and Brown (1987) reached the same conclusion with respect to little impediment existing to SP6 RNA polymerase transcription, but the opposite conclusion with respect to the integrity of the nucleosome. In this case the nucleosome remained intact rather like the 5S RNA gene transcription complex (Section 4.1.4). Subsequent experiments suggested that the explanation for this discrepancy was an unusual stability of the nucleosome positioned on the 5S RNA gene (Lorch *et al.*, 1988) due to the intrinsic DNA curvature in 5S DNA (Section 2.2.5).

Felsenfeld and colleagues resolved this issue with a series of biochemical experiments which are a true tour-de-force (Clark and Felsenfeld, 1992; Studitsky *et al.*, 1994, 1995). These investigators presented evidence in favour both of displacement of a histone octamer during transcription and the reformation of nucleosomes on the same DNA fragment. They used the same approach as the Kornberg and Brown labs of assembling a nucleosomal structure on both long closed circular DNA and on short DNA fragments next to a promoter for SP6 RNA polymerase. The initial experiments on the circular DNA demonstrated that the histone octamer moved during transcription to sites about 1000 bp upstream of the transcription start site. This demonstrated that the octamer was displaced from its original position, but that a nucleosome reformed elsewhere on the template. Comparable results were obtained by Prunell and colleagues (O'Donohue *et al.*, 1994). The experiments on the short DNA fragments clearly demonstrated that the histone octamer could step around the transcribing RNA polymerase over a distance of less than 100 bp without leaving the template (Fig. 4.30). A careful series of experiments demonstrated that this movement occurred once the SP6

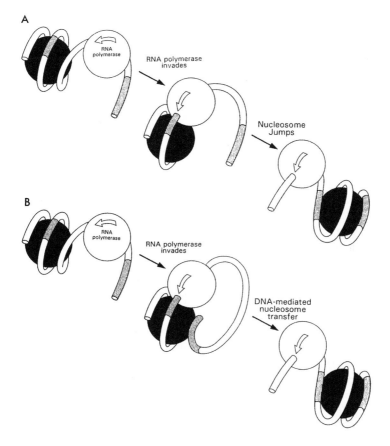

Figure 4.30. Models for how histone octamer steps around an elongating RNA polymerase.
The DNA sequence is marked for reference such that the octamer moves to a position 5′ of its original, upstream of polymerase. A. The RNA polymerase invades the nucleosome disrupting key histone–DNA contacts. Once sufficient contacts are broken the histone octamer jumps to free DNA downstream of the RNA polymerase. B. The RNA polymerase invades the nucleosome unwinding DNA from the surface of the histone octamer. Once sufficient DNA is unwound from the octamer surface it can loop back and mediate the transfer of the octamer to a site upstream of the RNA polymerase.

RNA polymerase had moved across the dyad axis of the nucleosome (Studitsky *et al.*, 1995). Hence the polymerase has to disrupt these key histone–DNA contacts to facilitate movement of the whole octamer. This model has been recently used to generate very similar results for RNA polymerases III (Studitsky *et al.*, 1995).

Bradbury and colleagues examined the elongation of T7 RNA polymerase through an array of nucleosomes finding that nucleosomes remained in place during transcription and that elongation was inhibited over 80% dependent on nucleosome density. Each nucleosome inhibited 15% of RNA polymerases from progressing through it (O'Neill *et al.*, 1992). In contrast to the above experiments Jackson and colleagues found that when nucleosomal templates were transcribed with T7 RNA polymerase in the presence of high levels of topoisomerase I, nucleosomes were disrupted. This demonstrates that under certain conditions nucleosomes cannot reform after transient disruption at the instant of polymerase passage (Pfaffle *et al.*, 1990). A displaced histone octamer will spontaneously disrupt (Gallego *et al.*, 1995; Protacio and Widom, 1996). Reconciling these disparate experimental results is difficult; however, it is clear that in most circumstances the nucleosome can stay in place or rapidly reform after transcription.

Comparable experiments have been carried out for eukaryotic RNA polymerases. Chambon and colleagues have shown that the rate of transcription elongation for RNA polymerases I and II is greatly decreased for nucleosomal templates. Likewise several investigators have suggested that RNA polymerase III has difficulty progressing through arrays of nucleosomes (Morse, 1989; Felts *et al.*, 1990). Some of these inhibitory effects could be explained by aggregation of nucleosomal arrays under the ionic conditions used to investigate elongation efficiencies (Hansen and Wolffe, 1992, 1994). Free divalent cation (Mg^{2+}) causes the compaction of spaced nucleosomes (Section 2.3.1). Thus the folding of chromatin might further impede RNA polymerase or represent the major constraint to transcription as RNA polymerase II proceeds through a single nucleosome without hindrance (Lorch *et al.*, 1987) whereas multiple nucleosomes cause the polymerase to stall (Izban and Luse, 1991). Displacement of histones (H2A, H2B) from nucleosomes facilitates the progression of RNA polymerase III through nucleosomal arrays (Puerta *et al.*, 1993; Hansen and Wolffe, 1994). Arrays of histone tetramers can only be very inefficiently compacted under transcription conditions.

An interesting study (Heggeler-Bordier *et al.*, 1995) compared the fate of nucleosomes during *in vitro* transcription of long linear and supercoiled minichromosomes by either T7 RNA polymerase or RNA polymerase II. Under the reaction conditions used, T7 RNA polymerase disrupted nucleosomes on both linear and supercoiled templates. In contrast RNA polymerase II transcription in the presence of a rat liver total nuclear extract preserved the nucleosomal architecture. Addition of the extract to the T7 polymerase transcription reaction allowed RNA synthesis with the retention of nucleosomes.

The authors suggested that an activity in the extract stimulated the reformation of nucleosomes following RNA polymerase transit.

How might an RNA polymerase molecule progress through nucleosomal DNA? The nucleosome must be at \least transiently disrupted as RNA polymerase progresses through chromatin. This could occur either by the histones individually dissociating from DNA, by the complete octamer of core histones being released and transferred to other DNA regions as a unit (Studitsky *et al.*, 1994), or by the core histones remaining bound but only releasing one or two key contacts at a time to allow the nucleosome to retain its integrity (Thoma, 1991).

Several interesting possibilities follow from the suggestion of Liu and Wang (1987) that processive enzyme complexes, i.e. DNA and RNA polymerases, might transiently introduce positive superhelical stress ahead of the complex as it unwinds duplex DNA, and negative superhelical stress behind it. This phenomenon has been shown to occur *in vivo* (Brill and Sternglanz, 1988). Although nucleosomes appear stable to both positive and negative superhelical stress (Clark and Felsenfeld, 1991; Clark and Wolffe, 1991), since DNA is overwound in each nucleosome (Section 2.2.3), core histones prefer to associate with DNA that contains negative superhelical turns. It is, therefore, possible that nucleosome transfer might be favoured from the DNA sequences ahead of the RNA polymerase to those behind the enzyme. *In vitro*, the transfer of the histone octamer means that any topological stress effects must be very local in action (Studitsky *et al.*, 1994). *In vivo*, it is possible that other cellular components, perhaps those involved in chromatin assembly, might facilitate nucleosome dissolution and reformation.

Summary
In vitro experiments indicate that prokaryotic RNA and eukaryotic polymerases are not impeded during elongation through the DNA in a single nucleosome. In contrast, arrays of nucleosomes impede both prokaryotic and eukaryotic RNA polymerases. The consequences of transcription are controversial, although it is clear that in some instances the nucleosome remains associated with the template and reforms rapidly once RNA polymerase has elongated through it.

4.4 CHROMATIN STRUCTURE AND DNA REPAIR

DNA repair offers a unique opportunity to examine the influence of the packaging of DNA within chromatin on the access of DNA damaging agents to the double helix and the mechanism of chromatin disruption for the DNA repair machinery to target and then correct the damaged chromosome.

4.4.1 Influence of chromatin structure on DNA damage

There are numerous ways in which DNA can be damaged by ultraviolet (UV) light and by various chemicals. The major UV photoproduct in DNA (*cis-syn* cyclobutane pyrimidine dimer, CPD) is found to have a random distribution between linker DNA and nucleosome core DNA following the irradiation of cells (Niggli and Cerutti, 1982; Suquet and Smerdon, 1995). The second most significant form of UV damage to DNA is the formation of pyrimidine (6-4) pyrimidone dimers. These are found preferentially in the linker regions between nucleosome cores (Mitchell *et al.*, 1990; Suquet and Smerdon, 1995). More detailed analysis has focused on sites of UV photoproducts within the 146 bp of DNA within the nucleosome core (Gale *et al.*, 1987; Gale and Smerdon, 1988a,b, 1990). The efficiency of CPD formation in nucleosome cores is modulated with an average periodicity 10.3 bp, such that the sites of maximal CPD formation are farthest away from the histone surface. The modulation in CPD formation appears due to changes in DNA chemistry induced by the bending of DNA around the histone octamer. Structural analysis of a specific decamer sequence containing a single CPD photoproduct shows that DNA is bent by 9° at that site (Kim *et al.*, 1995). The formation of a CPD photoproduct after photon absorption is more probable if the appropriate double-bonds in the pyrimidine base are aligned. This occurs when the minor groove is on the outside of a curved double helix. This sensitivity to DNA distortion was most definitively shown by looping a 50 bp DNA sequence between two binding sites for the λ repressor, and examining the sites of CPD formation (Pehrson and Cohen, 1992). Maximal sites of damage were on the outside for the DNA curve created by the loop.

CPD photoproduct distribution provides a useful probe of nucleosomal structure: nucleosome cores unfold when they are exposed to very low ionic strengths. This transition in chromatin structure can be seen from the progressive loss of CPD modulation following UV treatment (Brown *et al.*, 1993). The presence of CPDs in DNA before nucleosome assembly also leads to informative results, because both the yield and modulation of CPDs is reduced for the three central turns of DNA in the nucleosome core (Suquet and Smerdon, 1993). These three turns of DNA have an unusual structure when the double helix is underwound (Hayes *et al.*, 1991b). The exclusion of CPDs indicates that the histones find it difficult to accommodate DNA with the induced structural alterations in that particular region of the nucleosome (Kim *et al.*, 1995). These studies make use of native DNA that is not treated with a photosensitizer to enhance the formation of photoproducts. One concern with the use of photosensitizers is that they might alter DNA structure directly. With this caveat, Pehrson was able to demonstrate using a photosensitizer that histone H1 was able to alter thymine dimer formation around the dyad axis of the nucleosome core (Pehrson, 1989). Subsequent work indicated that linker DNA was less tightly coiled in chromatin than that in the nucleosome core (Pehrson, 1995).

Pyrimidine (6-4) pyrimidone dimers (6-4 PDs) severely distort the path of the double helix, each photoproduct induces a bend of 44° in DNA (Kim *et al.*, 1995). This bend probably explains why they are found in the more deformable linker DNA in chromatin, rather than the tightly constrained DNA in the nucleosome core. With respect to chemical damage of DNA it might be expected that linker DNA would be more accessible to bulky drugs than DNA in the nucleosome core. The outer surface of nucleosomal DNA is likely to be more accessible than the inner surface. These expectations are in general confirmed for the distribution of benzo(a)pyrene diol epoxide (BPDE) adducts in nucleosome containing 5S DNA from *Xenopus borealis* (Thrall *et al.*, 1994) and *Xenopus laevis* (Smith and McLeod, 1993). The formation of adducts is reduced two fold in the nucleosome core compared to naked DNA. Comparable results are obtained with bleomycin and neocarzinostatin (Smith *et al.*, 1994).

In general it appears that nucleosomes can tolerate most of the common forms of DNA damage without major distortions of structure. Nevertheless these sites of DNA damage are subsequently targeted for DNA and chromosomal repair.

Summary
UV photoproducts provide information about DNA structure in the nucleosome and about the capacity of the histones to assemble nucleosomes using damaged DNA.

4.4.2 Repairing DNA and chromatin

A major unresolved question in the DNA repair field is the nature of the targeting of the repair machinery to sites of damage. Some insights have emerged for transcribed DNA sequences. The molecular machinery involved in DNA repair includes components of the eukaryotic transcriptional machinery. Human cells exist that are defective in excision repair; seven complementation groups have been characterized that help define genes involved in the process. These genes are designated excision repair cross complement (ERCC) genes (Drapin *et al.*, 1994). The largest subunit of the transcription factor TFIIH, p89, is identical to the ERCC3 gene (Schaeffer *et al.*, 1993). ERCC3 has a DNA unwinding activity that depends on ATP hydrolysis. The yeast homolog RAD25, previously defined as a gene involved in DNA repair, is present in yeast TFIIH. RAD25 has a helicase activity and is required for transcription by RNA polymerase II (Qui *et al.*, 1993). Other components of yeast TFIIH are both involved in excision repair and are required for transcription. Thus TFIIH has a dual function in DNA repair and transcription.

It has been proposed that TFIIH travels along with RNA polymerase until it reaches the site of DNA damage. The model then requires TFIIH to dissociate and act as a nucleation site for recruiting other ERCC proteins. A variation of this hypothesis is that TFIIH does not travel with RNA polymerase, but instead is recruited to the site of DNA damage by a stalled RNA polymerase. An alternative proposal is that TFIIH might be recruited to the site of DNA damage by other proteins that recognize damage. This model would then have TFIIH act as a helicase to unwind DNA within a chromatin environment much as it might do following the initiation of transcription. Such helicase activity is likely to locally disrupt chromatin structure. Such a relationship between the transcriptional machinery and DNA accessibility within chromatin has been previously proposed for the recombination process (Schlissel and Baltimore, 1989). How TFIIH or other components of the repair machinery would be recruited to nontranscribed segments of the genome such as centromeres and telomeres is unclear.

DNA repair occurs first in regions of DNA not associated with the core histones, and only later within the core DNA (Jensen and Smerdon, 1990; Sidik and Smerdon, 1990). This is in agreement with the general accessibility of DNA-binding proteins to linker DNA and core DNA (Section 4.2.1). Nucleosomes have to be rearranged for this second phase of DNA repair to occur. The repair of UV damage is biphasic: at early times after damage (3–6 h) repair synthesis occurs preferentially at the edges of the nucleosome core (Lan and Smerdon, 1985), at later times repair synthesis is much more random. 6-4 PDs are removed much more rapidly from chromatin than CPDs (Mitchell and Tjian, 1989). A significant rearrangement of nucleosome structure occurs during the repair of UV-induced photoproducts (Smerdon and Lieberman, 1978). There is an initial rapid increase in nuclease sensitivity over repair patches coupled to a loss of the 10-11 bp modulation of DNase I cleavage (Smerdon and Lieberman, 1980). These alterations in DNase I cleavage are diagnostic of a release of DNA from a normal nucleosomal organization. Nucleosomes are then slowly recovered on the repaired DNA (Smerdon, 1986; Nissen *et al.*, 1986).

Several investigators have examined the potential role of histone modification in facilitating access of the DNA repair machinery to chromatin. Acetylation of the core histones correlates with enhanced excision repair of UV photoproducts (Smerdon *et al.*, 1982). Most of the enhanced repair synthesis is found in regions of chromatin containing hyperacetylated histones (Ramanathan and Smerdon, 1989). Poly (ADP) ribosylation of histones has been proposed to facilitate access of the repair machinery to chromatin (Mathis and Althaus, 1990). The efficiency of DNA repair correlates with the activity of poly(ADPribose) polymerase (Bhatia *et al.*, 1990; Mathis and Althaus, 1990). This enzyme contains zinc fingers which bind to breaks in the double helix (Gradwohl *et al.*, 1990). Synthesis of long chains of poly(ADP-ribose) has been proposed to provide a competing polyanion which might facilitate the displacement of histones from DNA (Huletsky *et al.*, 1989; Section 2.5.2). Surprisingly mice lacking poly(ADP-ribose) polymerase develop normally (Wang *et al.*, 1995), suggesting that the enzyme is not essential for DNA repair.

Chromatin reassembly once DNA repair is complete utilizes CAF1 (Section 3.2) and is very efficiently targeted presumably by features of repaired DNA or by components of the repair machinery (Gaillard *et al.*, 1996). Using the yeast minichromosomes, Smerdon and Thoma (1990) examined the rate and efficiency of DNA repair in actively transcribed and inactive regions of chromatin (see also Smerdon *et al.*, 1990). Transcription of chromatin facilitates DNA repair, suggesting that the passage of RNA polymerase further facilitates access of the repair enzymes to DNA. This may be a consequence of the delivery of TFIIH to the site of DNA damage by RNA polymerase (Drapin *et al.*, 1994).

Summary
DNA repair enzymes rely on components of the basal transcription machinery and on the post-translational modification of histones to locally disrupt chromatin structure. Disruption of chromatin structure mediated by the transcription process can facilitate DNA repair.

CHAPTER FIVE

Future Prospects

The importance of understanding the structure of chromatin and chromosomes in order to understand function is now well established. Nevertheless there are still large gaps in our knowledge. We know very little about chromatin structure beyond the nucleosome itself. Even less is known about the dynamics of higher order structures and how they might be compartmentalized within the chromosome and the nucleus. These features of nuclear organization have been implicated in numerous epigenetic effects in plants, animals and humans. At this time most of these observations are phenomenological with little foundation in molecular mechanism. Their importance in connecting basic research with problems of biotechnological and medical relevance provides a wealth of opportunity for a new generation of scientists.

5.1 LOCAL CHROMATIN STRUCTURE

The detailed information available concerning the structure of the nucleosome allows predictions to be made and experimentally tested concerning the organization of a particular DNA sequence on the histone core. How non-histone proteins recognize DNA after deformation of the double-helix in the nucleosome is unknown. Manipulation of the position of specific DNA sequences within the nucleosome will allow this problem to be addressed. It might be expected that several aspects of gene regulation will be changed by virtue of differences in the recognition of specific chromatin structures rather than of naked DNA by regulatory proteins. Of particular relevance to this issue is the role of unstable repeat sequences in human disease genes. Here is an example of a naturally occurring variation in chromatin structure that may contribute to the generation of a disease. Alterations in nucleosome stability could directly influence both the expansion of the repeats and the transcription of genes in which these repeats are found. There may be many such architectural punctuation marks in chromatin and chromosomal structure.

The organization of the core histones within the nucleosome is substantially understood. This information provides the basis for understanding the significance of core histone modification and mutation. These alterations in the core histones associated with promoter regions have largely unknown consequences for the organization of DNA in the nucleosome and for the subsequent association of non-histone proteins. The effect of such changes on the integrity of the nucleosome is an active area of investigation. Considerable insight into the nucleosome as a structure whose dynamics are regulated by post-translational modification of histones may well be forthcoming. Deficiencies in our basic understanding of nucleosomal dynamics are highlighted by the discovery of targeted histone modification by components of the transcriptional machinery. The nucleosome is not an invariant structure. One of the most exciting emerging issues in chromatin function is the evidence for nucleosomal allostery. Post-translational modifications such as acetylation of the core histones influence histone–DNA contacts and the properties of the nucleosome in general. The next step is to understand how these modifications contribute to increasing or decreasing the activities of the general transcriptional machinery. It might be anticipated that significant regulatory events will occur not only at the level of transcription complex assembly, but also subsequent to the recruitment of RNA polymerase. Histone acetylation is not a stable modification and regulation of the turnover of acetylated lysines in the nucleosome is likely to be of significance. The targeting of other histone modifications, such as phosphorylation, ubiquitination and methylation is also probable. Diverse modifications may be used to illuminate distinct regions of the chromosome for recognition by regulatory complexes.

Linker histone phosphorylation is correlated with almost as many biological phenomena as core histone acetylation. The molecular basis of targeting and the direct functional consequences need to be determined. Some nucleosomes will contain linker histones, others will not. Inclusion of linker histones alters core histone–DNA interactions. Likewise, the content of linker histone and core histone variants will differ in a way which may reflect the function of the DNA within the nucleosome. For example, if a nucleosomal array is transcriptionally active, histone variants synthesized outside of S-phase, especially of H2A and H2B, are likely to accumulate in the transcribed chromatin. Similar considerations apply to the inclusion of HMG proteins into nucleosomal structures. Thus, there are potentially many different nucleosomes that will presumably reflect distinct structural and functional requirements. The discovery of histone variants such as CENP-A and macro H2A clearly suggests the adaptation of chromatin structures to carry out specialized functions. The isomorphous nature of some transcription factors and histones indicates that the division between packaging protein and regulatory molecule can be difficult to distinguish. All of these results imply that specific chromatin structures exist. How they are targeted, assembled and regulated is only beginning to be discovered.

Major deficiencies exist with respect to our knowledge of the molecular machines such as SWI/SNF that might disrupt chromatin structure. Genetic evidence suggests that this process is important for gene regulation, but we have as yet little information as to how the proteins involved are targeted to chromosomal domains. Presumably the assembly of large regulatory complexes involving many *trans*-acting factors will effect a local compartmentalization within the nucleus that will allow the further recruitment of the enzymes or molecular machines.

Many advances in elucidating the specific role of histones in gene regulation have followed from molecular genetic experiments in yeast. The combination of mutational analysis of the yeast histones with functional studies at specific promoters has proven the driving force behind much of the conceptual progress in the field. More elaborate genetic screens are uncovering *trans*-acting factors that appear to interact specifically with both DNA and the histones. Some of these direct chromatin assembly and others specialized chromosomal structures. Although the final resolution of molecular mechanisms will require the establishment of *in vitro* systems, these genetic approaches are providing a rich store of interesting histone-specific interactions to explore. This is most clearly seen in recent progress concerning the phenomenon of position effect variegation. Although yeast is the *Escherichia coli* of eukaryotic genetics, and the broad principles of nuclear architecture and function are the same as for higher eukaryotes, sufficient differences exist in the structural proteins of chromatin to make parallel studies using viral episomes extremely important. Moreover, with the exception of mating-type regulation, yeast has no need for the establishment and maintenance of stable patterns of gene expression such as those seen in a differentiated cell of a metazoan. However, there is no doubt that the major players in the field will be defined through work with *Saccharomyces*. A particularly attractive avenue for future research is the development of biochemical systems competent for chromatin and chromosome assembly in yeast (Schultz *et al.*, 1997). The possibility of coupling structural, biochemical and genetic analysis offers unrivalled opportunities for advancing our basic knowledge of chromosomal biology.

Research on the influence of chromosomal structure on gene expression has undergone a conceptual revolution. The long-standing dogma that chromatin merely packages DNA away from any significant role is seen to be as much of an oversimplification as gene regulation occurring on naked DNA. Understanding how *trans*-acting factors and chromatin structural components interact to regulate transcription presents molecular biologists with important questions, the answers to which will have general applicability for other processes such as DNA replication, recombination and repair. It is clear that eukaryotic *trans*-acting factors have evolved to operate in a chromatin environment and that histones have evolved to let them function. The nucleus itself compartmentalizes functional enzymatic complexes that utilize chromatin as a substrate. The future offers considerable promise for reconstructing, and thus understanding, the correct regulation of genes within an environment that recapitulates that found within the nucleus.

Summary

We need to know much more about the conformational transitions available to the nucleosome and how these are influenced by post-translational modification and the various nucleosome remodelling machines. The precise role of histone modification at a regulated eukaryotic promoter needs to be determined. In particular we need to know where in the transcription process chromatin modification exerts an effect.

5.2 LONG-RANGE CHROMATIN AND CHROMOSOMAL STRUCTURE

Although the nucleosome is perhaps the best-defined large nucleo-protein complex yet analysed, the further folding of arrays of nucleosomes is poorly understood. Viewpoints concerning chromatin fibre structure range from a rigid solenoid to the postulate that no such structure exists *in vivo*. Fortunately, systems are now available that allow pure components to be reconstituted into spaced arrays of nucleosomes. Such systems will contribute to our understanding of the next step in chromatin assembly, the correct incorporation of linker histones. Assays for this event are now available through the development of techniques to map histone–DNA contacts in terms of DNA and protein sequence.

These studies erode established dogmas concerning the necessity of linker histones for the assembly of the chromatin fibre, the exact determinants of linker histone association in the nucleosome and the role of interactions between linker histones in the assembly of higher-order chromatin structures. The structural role of linker histones is very much unknown. Likewise functional studies in which linker histones are manipulated *in vivo* indicate that linker histones do not have a uniformly repressive role, but are highly selective in their action. The consequences of the assembly of higher-order structures for the access of *trans*-acting factors and RNA polymerase are only beginning to be defined. The implications of core histone modification, linker histone modification and the incorporation of non-histone structural proteins such as HMGs for the structure of the chromatin fibre ought to be accessible through these studies.

The further folding of the chromatin fibre through association with other specialized chromosomal proteins is becoming clearer. How this folding is regulated and its significance for nuclear function beyond DNA compaction are subjects of intense interest. Genetic evidence from *Caenorhabditis elegans* and *Drosophila* clearly indicates that long-range chromosome organization can selectively influence gene activity. The targeting of individual chromosomes for these effects raises many interesting questions for future investigation. Pre-eminent among the genetically defined chromosomal proteins that need biochemical definition of their functions are the chromodomain proteins and the SMC proteins. Polycomb represents an important chromodomain protein that functions as a regulator with known target loci. However, we do not know how polycomb arrives at a particular locus or what it does once it is there. Likewise the SMC family of proteins have both general and specific roles in modulating chromosomal structure and function. How do they work?

The relevance of specific DNA sequences, such as insulators and locus control regions, to establishing domains of chromatin activity is a topic of particular importance. Transgenic mice and *Drosophila* offer a means of defining the DNA sequences involved in mediating domain-specific effects; however, in order to establish the molecular mechanisms responsible for these phenomena such domains of chromatin will have to be reconstructured *in vitro* together with their functional properties. *Xenopus* and *Drosophila* cell-free systems offer many avenues for future progress towards this type of reconstruction. Defining the determinants of domains of influence and the constraints on insulators and locus control regions should provide us with valuable information about how DNA is packaged into high-order chromatin structures and how such structures move within the nucleus.

Most issues concerning nuclear compartmentalization itself are open for investigation. The existence of a scaffold, matrix and skeleton is established, however the true functional significance is yet to be resolved. Perhaps the definition of the large molecular machines that control chromatin organization will eventually lead to a definition of their structural roles within the nucleus. Many of these chromatin remodelling complexes are abundant nuclear constituents. The existence of chromosomal territories, splicing and replication factories are attractive cell biological concepts that await biochemical reconstruction and genetic definition.

Summary

We need to know how the chromosome organizing complexes and molecular machines that utilize chromatin as a substrate are compartmentalized in the nucleus. There is little known about the dynamics and movement of chromatin domains within the nucleus and how this might reflect different functional states.

References

Aasland, R. and Stewart, A.F. (1995). The chromo shadow domain, a second chromo domain in heterochromatin-binding protein-1, HP1. *Nucleic Acids Res.* **23**, 3168–73.

Abé, S.I. and Hiyoshi, H. (1991). Synthesis of sperm-specific basic nuclear proteins (SPs) in cultured spermatids from *Xenopus laevis*. *Exp. Cell Res.* **194**, 397–414.

Adachi, Y. and Laemmli, U.K. (1992). Identification of nuclear pre-replication centers poised for DNA synthesis in *Xenopus* egg extracts: immunolocalization study of replication protein A. *J. Cell Biol.* **119**, 1–15.

Adachi, Y., Kas, E. and Laemmli, U.K. (1989). Preferential, cooperative binding of topoisomerase II to scaffold associated regions. *EMBO J.* **8**, 3997–4006.

Adachi, Y., Luke, M. and Laemmli, U.K. (1991). Chromosome assembly *in vitro*: topoisomerase II is required for condensation. *Cell* **64**, 137–48.

Adams, C.C and Workman, J.L. (1995). The binding of disparate transcriptional activators to nucleosomal DNA is inherently cooperative. *Mol. Cell. Biol.* **15**, 1405–21.

Adamson, E.D. and Woodland, H.R. (1974). Histone synthesis in early amphibian development. Histone and DNA syntheses are not coordinated. *J. Mol. Biol.* **88**, 263–85.

Adhya, S. and Gottesman, M. (1982). Promoter ooclusion: transcription through a promoter may inhibit its activity. *Cell* **29** 939–44.

Alevizopoulos, A., Dusserre, Y., Tsai-Pflulgfelderm, Von Der Weid, T., Wahli, W. and Mermod, N. (1995). A proline-rich TGF-β-responsive transcriptional activator interacts with histone H3. *Genes Dev.* **9**, 3051–66.

Alfonso, P.J., Crippa, M.J., Hayes, J.J. and Bustin, M. (1994). The footprint of chromosomal proteins HMG-14 and HMG-17 on chromatin subunits. *J. Mol. Biol.* **236**, 189–98.

Allan, J., Hartman, P.G., Crane-Robinson, C. and Aviles, F.X. (1980). The structure of histone H1 and its location in chromatin. *Nature* **288**, 675–9.

Allan, J., Cowling, G.J., Harborne, N., Cattani, P., Craigie, R. and Gould, H. (1981). Regulation of the higher-order structure of chromatin by histones H1 and H5. *J. Cell Biol.* **90**, 279–88.

Allan, J., Mitchell, T., Harborne, N., Bohm, L. and Crane-Robinson, C. (1986). Roles of H1 domains in determining higher order chromatin structure and H1 location. *J. Mol. Biol.* **187**, 591–601.

Alland, L., Muhle, R., Hou Jr., H., Potes, J., Chin, L., Schreiber-Agus, N. and De Pinho, R.A. (1997). Role of NCoR and histone deacetylase in Sin3-mediated transcriptional and oncogenic repression. *Nature* **387**, 49–55.

Allen, G.C., Hall, G.E. Jr., Childs, L.C., Wissinger, A.K., Spiker, S. and Thompson, W.F. (1993). Scaffold attachment regions increase reporter gene expression in stably transformed plant cells. *Plant Cell* **5**, 603–13.

Allen, G.C., Hall, G. Jr., Michalowski, S., Newman, W., Spiker, S. Weissinger, A.K. and Thompson, W.F. (1996). High level transgene expression in plant cells: effects of a strong scaffold attachment region from tobacco. *Plant Cell* **8**, 899–913.

Allfrey, V., Faulkner, R.M. and Mirsky, A.E. (1964). Acetylation and methylation of histones and their possible role in the regulation of RNA synthesis. *Proc. Natl Acad. Sci. USA* **51**, 786–94.

Allis, C.D. and Gorovsky, M.A. (1981). Histone phosphorylation in macro- and micronuclei of *Tetrahymena thermophila*. *Biochemistry* **20**, 3828–33.

Allis, C.D., Colavito-Shepanski, M. and Gorovsky, M.A. (1987). Scheduled and unscheduled DNA synthesis during development in conjugating *Tetrahymena*. *Dev. Biol.* **124**, 469–80.

Allshire, R.C., Javerzat, J.P., Redhead, N.J. and Cranston, G. (1994). Position effect variegation at fission yeast centromeres. *Cell* **76**, 157–69.

Allshire, R.C., Nimmo, E.R., Ekwall, K., Javerzat, J.P. and Cranston, G. (1995). Mutations derepressing silent centromeric domains in fission yeast disrupt chromosome segregation. *Genes Dev.* **9**, 218–33.

Almer, A. and Horz, W. (1986). Nuclease hypersensitive regions with adjacent positioned nucleosomes mark the gene boundaries of the PHO5/PHO3 locus in yeast. *EMBO J.* **5**, 2681–7.

Almer, A., Rudolph, H., Hinnen, A. and Horz, W. (1986). Removal of positioned nucleosomes from the yeast PHO5 promoter upon PHO5 induction releases additional activating DNA elements. *EMBO J.* **5**, 2689–96.

Almouzni, G. (1994). The origin replication complex (ORC): the stone that kills two birds. *BioEssays* **16**, 233–5.

Almouzni, G. and Méchali, M. (1988a). Assembly of spaced chromatin involvement of ATP and DNA topoisomerase activity. *EMBO J.* **7**, 4355–65.

Almouzni, G. and Méchali, M. (1988b). Assembly of spaced chromatin by DNA synthesis in extracts from *Xenopus* eggs. *EMBO J.* **7**, 664–72.

Almouzni, G. and Wolffe, A.P. (1993a). Replication coupled chromatin assembly is required for the repression of basal transcription *in vivo*. *Genes Dev.* **7**, 2033–47.

Almouzni, G. and Wolffe, A.P. (1993b). Nuclear assembly, structure and function: the use of *Xenopus in vitro* systems. *Exp. Cell Res.* **205**, 1–15.

Almouzni, G. and Wolffe, A.P. (1995). Constraints on transcriptional activator function contribute to transcriptional quiescence during early *Xenopus* embryogenesis. *EMBO J.* **14**, 1752–65.

Almouzni, G., Clark, D.J., Méchali, M. and Wolffe, A.P. (1990a). Chromatin assembly on replicating DNA *in vitro*. *Nucleic. Acids Res.* **18**, 5767–74.

Almouzni, G., Méchali, M. and Wolffe, A.P. (1990b). Competition between transcription complex assembly and chromatin assembly on replicating DNA. *EMBO J.* **9**, 573–82.

Almouzni, G., Méchali, M. and Wolffe, A.P. (1991). Transcription complex disruption caused by a transition in chromatin structure. *Mol. Cell. Biol.* **11**, 655–65.

Almouzni, G., Khochbin, S., Dimitrov, S. and Wolffe, A.P. (1994). Histone acetylation influences both gene expression and development of Xenopus laevis. *Dev. Biol.* **165**, 654–69.

Amati, B.B. and Gasser, S.M. (1988). Chromosomal ARS and CEN elements bind specifically to the yeast nuclear scaffold. *Cell* **54**, 967–78.

Ambrose, C., Rajadhyaksha, A., Lowman, H. and Bina, M. (1989). Locations of nucleosomes on the regulatory region of simian virus 40 chromatin. *J. Mol. Biol.* **209**, 255–63.

An, W., Leuba, S.H., van Holde, K.E. and Zlatanova, J. (1998). Linker histone protects linker DNA on only one side of the core particle and in a sequence-dependent manner. *Proc. Natl Acad. Sci. USA* **95**, 3396–3401.

Anderson, D.M. and Smith, L.D. (1978). Patterns of synthesis and accumulation of heterogenous RNA in lampbrush stage oocytes of *Xenopus laevis* (Daudin). *Dev. Biol.* **67**, 274–86.

Andrews, M.T., Loo, S. and Wilson, L.R. (1991). Coordinate inactivation of class III genes during the gastrula–neurula transition in *Xenopus. Dev. Biol.* **146**, 250–4.

Annunziato, A.T. (1989). Inhibitors of topoisomerases I and II arrest DNA replication, but do not prevent nucleosome assembly *in vivo. J. Cell. Sci.* **93**, 593–603.

Annunziato, A.T., Frado, L.L., Seale, R.L. and Woodcock, C.L. (1988). Treatment with sodium butyrate inhibits the complete condensation of interphase chromatin. *Chromosoma* **96**, 132–8.

Annunziato, A.T., Eason, M.B. and Perry, C.A. (1995). Relationship between methylation and acetylation of arginine-rich histones in cycling and arrested HeLa cells. *Biochemistry* **34**, 2916–24.

Antequera, F., Macleod, D. and Bird, A. (1989). Specific protection of methylated CpGs in mammalian nuclei. *Cell* **58**, 509–17.

Antequera, F., Boyes, J. and Bird, A. (1990). High levels of *de novo* methylation and altered chromatin structure at CpG islands in cell lines. *Cell* **62**, 503–14.

Aparicio, O.M. and Gottschling, D.E. (1994). Overcoming telomeric silencing: a *trans*-activator competes to establish gene expression in a cell cycle dependent way. *Genes Dev.* **8**, 1133–46.

Aparicio, O.M., Billington, B.L. and Gottschling, D.E. (1991). Modifiers of position effect are shared between telomeric and silent mating-type loci in *S. cerevisiae. Cell* **66**, 1279–87.

Apone, L.M., Virbasius, C.A., Reese, J.C. and Green, M.R. (1996) Yeast TAFII90 is required for cell cycle progression through G2/M but not for general transcription activation. *Genes Dev.* **10**, 2368–79.

Appels, R. and Wells, J.R.E. (1972). Synthesis and turnover of DNA-bound histone during maturation of avian blood cells. *J. Mol. Biol.* **70**, 425–34.

Archer, T.K., Cordingley, M.G., Marsaud, V., Richard-Foy, H. and Hager, G.L. (1989). Steroid transactivation at a promoter organized in a specifically positioned array of nucleosomes. In *Proceedings: Second International CBT Symposium on the Steroid/Thyroid Receptor Family and Gene Regulation*, Springer Verlag, Berlin, pp. 221–38.

Archer, T.K., Cordingley, M.G., Wolford, R.G. and Hager, G.L. (1991). Transcription factor access is mediated by accurately positioned nucleosomes on the mouse mammary tumor virus promoter. *Mol. Cell. Biol.* **11**, 688–98.

Archer, T.K., Lefebvre, P., Wolford, R.G. and Hager, G.L. (1992). Transcription factor loading on the MMTV promoter: a bimodal mechanism for promoter activation. *Science* **255**, 1573–6.

Arents, G. and Moudrianakis, E.N. (1993). Topography of the histone octamer surface: repeating structural motifs utilized in the docking of nucleosomal DNA. *Proc. Natl Acad. Sci. USA* **90**, 10489–93.

Arents, G., Burlingame, R.W., Wang, B.W., Love, W.E. and Moudrianakis, E.N. (1991). The nucleosomal core histone octamer at 3.1 Å resolution: a tripartite protein assembly and a left-handed superhelix. *Proc. Natl Acad. Sci. USA* **88**, 10148–52.

Ariel, M., Selig, S., Brandeis, M., Kitsberg, D., Kafri, T., Weiss, A., Keshet I., Razin, A. and Cedar, H. (1993). Allele specific structures in the mouse Igf2-H19 domain. *Cold Spring Harb. Symp. Quant. Biol.* **58**, 307–14.

Arndt, K.T., Styles, C.A. and Fink, G.R. (1989). A supressor of a HIS4 transcriptional defect encodes a protein with homology to the catalytic subunit of protein phosphatases. *Cell* **56**, 527–37.

Athey, B.D., Smith, M.F., Rankert, D.A., Williams, S.P. and Langmore, J.P. (1990). The diameters of frozen-hydrated chromatin fibres increase with DNA linker length: evidence in support of variable diameter models for chromatin. *J. Cell Biol.* **111**, 795–806.

Aubert, D., Garcia, M., Benchaibi, M., Poncet, D., Chebloune, Y., Verdier, G., Nigon, V., Samarut, J. and Mura, C.V. (1991). Inhibition of proliferation of primary avian fibroblasts through expression of histone H5 depends on the degree of phosphorylation of the protein. *J. Cell Biol.* **11**, 497–506.

Auble, D.T. and Hahn, S. (1993). An ATP-dependent inhibitor of TBP binding to DNA. *Genes Dev.* **7**, 844–56.

Auble, D.T., Hansen, K.E., Mueller, C.G.F., Lane, W.S., Thorner, J. and Hahn, S. (1994). Mot1, A global repressor of RNA polymerase II transcription, inhibits TBP binding to DNA by an ATP-dependent mechanism. *Genes Dev.* **8**, 1920–34.

Ausio, J., Serger, D. and Eisenberg, H. (1984). Nucleosome core particle stability and conformational change. *J. Mol. Biol.* **176**, 77–104.

Ausio, J., Dong, F. and van Holde, K.E. (1989). Use of selectively trypsinized nucleosome core particles to analyze the role of the histone tails in the stabilization of the nucleosome. *J. Mol. Biol.* **206**, 451–63.

Avery, O.T., MacLeod, C.M. and McCarty, M. (1944). Studies on the chemical nature of the substance inducing transformation of pneumococcal types. *J. Exp. Med.* **79**, 137–58.

Ayer, D.E., Kretzner, L. and Eisenman, R.N. (1993). Mad: a heterodimeric partner for Max that antagonizes Myc transcriptional activity. *Cell* **72**, 211–22.

Ayer, D.E., Lawrence, Q.A. and Eisenman, R.N. (1995). Mad–Max transcriptional repression is mediated by ternary complex formation with mammalian homologs of yeast repressor Sin3. *Cell* **80**, 767–76.

Bachellerie, J.P., Puvion, E. and Zalta, J.P. (1975). Ultrastructural organization and biochemical characterization of chromatin–RNA–protein complexes isolated from mammalian cell nuclei. *Eur. J. Biochem.* **58**, 327–37.

Baer, B.W. and Rhodes, D. (1983). Eukaryotic RNA polymerase II binds to nucleosome cores from transcribed genes. *Nature (London)* **301**, 482–8.

Ball, D.J., Gross, D.S. and Garrard, W.T. (1983). 5-methylcytosine is localised in nucleosomes that contain H1. *Proc. Natl Acad. Sci. USA* **80**, 5490–4.

Banerjee, S. and Cantor, C.R. (1990). Nucleosome assembly of simian virus 40 DNA in a mammalian cell extract. *Mol. Cell. Biol.* **10**, 2863–73.

Banerjee, S., Bennion, G.R., Goldberg, M.W. and Allen, T.D. (1991). ATP dependent histone phosphorylation and nucleosome assembly in a human cell free extract. *Nucleic. Acids Res.* **19**, 5999–6006.

Banerjee, S., Smallwood, A. and Hulten, M. (1995). ATP-dependent reorganization of human sperm nuclear chromatin. *J. Cell Sci.* **108**, 755–65.

Banerji, J., Olson, L. and Schaffner, W. (1983). A lymphocyte-specific cellular enhancer is located downstream of the joining region in immunoglobulin heavy chain genes. *Cell* **33**, 729–40.

Barbaris, S., Munsterkotter, M., Svaren, J. and Horz, W. (1996). The homeodomain protein Pho2 and the basic-helix-loop-helix protein Pho4 bind DNA cooperatively at the yeast PHO5 promoter. *Nucleic. Acids Res.* **24**, 4479–86.

Bardeleben, C., Kassavetis, G.A. and Geiduschek, E.P. (1994). Encounters of *Saccharomyces cerevisiae* RNA polymerase III with its transcription factors during RNA chain elongation. *J. Mol. Biol.* **235**, 1193–205.

Barlev, N.A., Candau, R., Wang, L., Darpino, P., Silverman, N. and Berger, S.L. (1995). Characterization of physical interactions of the putative transcriptional adaptor ADA2 with acidic activation domains and TATA-binding-protein. *J. Biol. Chem.* **270**, 19337–43.

Baron, M.H. (1993). Reversibility of the differentiated state in somatic cells. *Curr. Opin. Cell Biol.* **5**, 1050–6.

Baron, M.H. and Farrington, S.M. (1994). Positive regulators of the lineage specific transcription factor GATA-1 in differentiating erythroid cells. *Mol. Cell. Biol.* **14**, 3108–14.

Baron, M.H. and Maniatis, T. (1986). Rapid reprogramming of globin gene expression in transient heterokaryons. *Cell* **46**, 591–602.

Baron, M.H. and Maniatis, T. (1991). Regulated expression of human α and β-globin genes in transient heterokaryons. *Mol. Cell Biol.* **11**, 1239–47.

Barry, J.M. and Merriam, R.W. (1972). Swelling of hen erythrocyte nuclei in cytoplasm from *Xenopus* eggs. *Exp. Cell. Res.* **71**, 90–6.

Bartolomé, S., Bermudez, A. and Daban, J.-R. (1994). Internal structure of the 30 nm chromatin fiber. *J. Cell Sci.* **107**, 2983–92.

Barton, M.C. and Emerson, B.M. (1994). Regulated expression of the β-globin gene locus in synthetic nuclei. *Genes Dev.* **8**, 2453–65.

Barton, M.C., Madani, N. and Emerson, B.M. (1993). The erythroid protein cGATA-1 functions with a stage-specific factor to activate transcription of chromatin-assembled β globin genes. *Genes Dev.* **7**, 1796–1809.

Bartsch, J., Truss, M., Bode, J. and Beato, M. (1996). Moderate increase in histone acetylation activates the mouse mammary tumour virus promoter and remodels its nucleosome structure. *Proc. Natl Acad. Sci. USA* **93**, 10741–6.

Bashkin, J., Hayes, J.J., Tullius, T.D. and Wolffe, A.P. (1993). Structure of DNA in a nucleosome core at high salt concentration and at high temperature. *Biochemistry* **32**, 1895–8.

Bateman, E. and Paule, M.R. (1988). Promoter occlusion during ribosomal RNA transcription. *Cell* **54**, 985–92.

Bates, D.L. and Thomas, J.O. (1981). Histones H1 and H5: one or two molecules per nucleosome. *Nucleic. Acids Res.* **9**, 5883–94.

Bates, G. and Lehrach, H. (1994). Trinucleotide repeat expansions and human genetic disease. *BioEssays* **16**, 277–84.

Bauer, D.W., Murphy, C., Wu, Z., Wu, C.H. and Gall, J.G. (1994a). *In vitro* assembly of coiled bodies in *Xenopus* egg extract. *Mol. Biol. Cell.* **5**, 633–44.

Bauer, W.R., Hayes, J.J., White, J.H. and Wolffe, A.P. (1994b). Nucleosome structural changes due to acetylation. *J. Mol. Biol.* **236**, 685–90.

Baumann, H., Knapp, S., Lundback, T., Ladenstein, R. and Hard, T. (1994). Solution structure and DNA-binding properties of a thermostable protein from the archaeon *Solfolobus solfataricus*. *Nature Struct. Biol.* **1**, 808–19.

Bavykin, S.G., Usachenko, S.I., Zalensky, A.O. and Mirzabekov, A.D. (1990). Structure of nucleosomes and organization of internucleosomal DNA in chromatin. *J. Mol. Biol.* **212**, 495–511.

Baxevanis, A.D., Bryant, S.H. and Landsman, D. (1995). Homology model building of the HMG-1 box structural domain. *Nucleic. Acids Res.* **23**, 1604–13.

Bazett-Jones, D.P., LeBlanc, B., Herfort, M. and Moss, T. (1994). Short range DNA looping by the *Xenopus* HMG- box transcription factor, xUBF. *Science* **264**, 1134–6.

Beachy, P.A., Helfand, S.L. and Hogness, D.S. (1985). Segmental distribution of bithorax complex proteins during *Drosophila* development. *Nature* **313**, 545–51.

Becker, P.B. (1994). The establishment of active promoters in chromatin. *BioEssays* **16**, 541–7.

Becker, P.B. and Wu, C. (1992). Cell-free system for assembly of transcriptionally repressed chromatin from *Drosophila* embryos. *Mol. Cell. Biol.* **12**, 2241–9.

Becker, P.B., Rabindran, S.K. and Wu, C. (1991). Heat shock-regulated transcription *in vitro* from a reconstituted chromatin template. *Proc. Natl Acad. Sci. USA* **88**, 4109–13.

Bednar, J., Horowitz, R.A., Dubochet, J. and Woodcock, C.L. (1995). Chromatin conformation and salt induced compaction: three dimensional structural information from cryoelectron microscopy. *J. Cell Biol.* **131**, 1365–76.

Behrens, J., von Kries, J.P., Kuhl, M., Bruhn, L., Wedlich, D., Grosschedl, R. and Birchmeier, W. (1996). Functional interaction of β-catenin with the transcription factor LEF-1. *Nature* **382**, 638–42.

Bell, S.P. and Stillman, B. (1992). ATP-dependent recognition of eucaryotic origins of DNA replication by a multiprotein complex. *Nature* **357**, 128–35.

Bell, S.P., Learned, R.M., Jantzen, H-M. and Tjian, R. (1988). Functional cooperativity between transcription factors UBF1 and SL1 mediates human ribosomal RNA synthesis. *Science* **241**, 1192–7.

Bell, S.P., Kobayashi, R. and Stillman, B. (1993a). Yeast origin recognition complex functions in transcription silencing and DNA replication. *Science* **262**, 1844–9.

Bell, S.P., Marahrens, Y., Rao, H. and Stillman, B. (1993b). The replicon model and eukaryotic chromosomes. *Cold Spring Harbor Symp. Quant. Biol.* **58**, 435–42.

Belmont, A.S. and Bruce, K. (1994). Visualization of G1 chromosomes: a folded, twisted, supercoiled chromonema model of interphase chromatid structure. *J. Cell Biol.* **127**, 287–302.

Belmont, A.S., Sedat, J.W. and Agard, D.A. (1987). A three dimensional approach to mitotic chromosome structure: evidence for a complex hierarchical organization. *J. Cell Biol.* **105**, 77–92.

Belmont, A.S., Braunfeld, M.B., Sedat, J.W. and Agard, D.A. (1989). Large scale chromatin structural domains within mitotic and interphase chromosomes *in vivo* and *in vitro*. *Chromosoma* **98**, 129–43.

Bendig, M.M. (1981). Persistence and expression of histone genes injected into *Xenopus laevis* eggs in early development. *Nature (Lond.)* **292**, 65–7.

Benezra, R., Cantor, C.R. and Axel, R. (1986). Nucleosomes are phased along the mouse β-major globin gene in erythroid and non-erythroid cells. *Cell* **44**, 697–704.

Benyajati, C. and Worcel, A. (1976). Isolation, characterization and structure of the folded interphase genome of *Drosophila melanogaster*. *Cell* **9**, 393–407.

Berg, O.G., Winter, R.B. and von Hippel, P.H. (1981). Diffusion-driven mechanisms of protein translocation on nucleic acids. *Biochemistry* **20**, 6929–77.

Berger, S.L., Pina, B., Silverman, N., Marcus, G.A., Agapite, J., Regier, J.L., Triezenberg, S.J. and Guarente, L. (1992). Genetic isolation of ADA2: a potential transcriptional adaptor required for function of certain acidic activation domains. *Cell* **70**, 251–65.

Berk, A.J. (1986). Adenovirus promoters and E1A transactivation. *Annu. Rev. Genet.* **20**, 45–79.

Berman, H.M. (1991). Hydration of DNA. *Curr. Opin. Struct. Biol.* **1**, 423–7.

Berrios, M. and Avilion, A.A. (1990). Nuclear formation in a *Drosophila* cell-free system. *Exp. Cell Res.* **191**, 64–70.

Bestor, T.H. and Verdine, G.L. (1994). DNA methyltransferases. *Curr. Opin. Cell Biol.* **6**, 380–9.

Beyer, A. L. and Osheim, Y. N. (1988). Splice site selection, rate of splicing, and alternative plicing on nascent transcripts. *Genes. Dev.* **2**, 754–65.

Bhatia, K., Pommier, Y., Giri, C., Fornace, J., Imaizumi, M., Breitman, T.R., Cherney, B.W. and Smulson, M.E. (1990). Expression of the poly(ADP-ribose) polymerase gene following natural and induced DNA strand breakage and effect of hyperexpression on DNA repair. *Carcinogenesis* **11**, 123–8.

Bianchi, M.E., Beltrame, M. and Paonessa, G. (1989). Specific recognition of cruciform DNA by nuclear protein HMG1. *Science* **243**, 1056–9.

Bieker, J.J., Martin, P.L. and Roeder, R.G. (1985). Formation of a rate limiting intermediate in 5S RNA gene transcription. *Cell* **40**, 119–27.

Bienz, M. (1986). A CCAAT box confers cell-type-specific regulation of the *Xenopus* hsp70 gene in oocytes. *Cell* **46**, 1037–42.

Bienz, M. and Pelham, H.R.B. (1986). Heat shock regulatory elements function as an inducible enhancer in the *Xenopus* hsp 70 gene and when linked to a heterologous promoter. *Cell* **45**, 753–60.

Bird, A. (1995). Gene number, noise reduction and biological complexity. *Trends Genet.* **11**, 94–100.

Bird, A.P. (1986). CpG-rich islands and the function of DNA methylation. *Nature* **321**, 209–13.

Bischoff, F.R. and Ponstingl, H. (1991). Catalysis of guanine nucleotide exchange on Ran by the mitotic regulator RCC1. *Nature (London)* **354**, 80–2.

Bjorklund, S. and Kim, Y.-J. (1996). Mediator of transcriptional regulation. *Trends Biochem. Sci.* **21** 335–7.

Bjorkroth, B., Ericsson, C., Lamb, M.M. and Daneholt, B. (1988). Structure of the chromatin axis during transcription. *Chromosoma* **96**, 333–40.

Blasquez, V.C., Xu, M., Moses, S.C. and Garrard, W.T. (1989). Immunoglobulin K gene expression after stable integration. 1. Role of the intronic MAR and enhancer in plasmacytoma cells. *J. Biol. Chem.* **264**, 21183–9.

Blau, H.M. (1992) Differentiation requires continuous active control. *Annu. Rev. Biochem.* **61**, 1213–30.

Blau, H.M. and Baltimore, D. (1991). Differentiation requires continuous regulation. *J. Cell Biol.* **112**, 781–3.

Blau, H.M., Chiu, C-P. and Webster, C. (1983). Cytoplasmic activation of human nuclear genes in stable heterokaryons. *Cell* **32**, 1171–80.

Blau, H.M., Parlath, G.K., Hardeman, E.C., Chiu, C-P, Silberstein, L., Webster, S.F., Miller, S.C. and Webster, C. (1985). Plasticity of the differentiated state. *Science* **230**, 758–66.

Blobel, G. (1985). Gene gating: a hypothesis. *Proc. Natl Acad. Sci. USA* **82**, 8527–9.

Blomquist, P., Li, Q. and Wrange, O. (1996). The affinity of nuclear factor 1 for its DNA site is drastically reduced by nucleosome organization irrespective of its rotational and translational position. *J. Biol. Chem.* **271**, 154–9.

Bloom, K.S. and Carbon, J. (1982). Yeast centromere DNA is a unique and highly ordered structure in chromosomes and small circular minichromosomes. *Cell* **29**, 305–17.

Blow, J.J. and Laskey, R.A. (1986). Initiation of DNA replication in nuclei and purified DNA by a cell-free extract of *Xenopus* eggs. *Cell* **47**, 577–87.

Blow, J.J. and Laskey, R.A. (1988). A role for the nuclear envelope in controlling DNA replication within the cell cycle. *Nature* **332**, 546–8.

Blow, J.J. and Sleeman, A.M. (1990). Replication of purified DNA in *Xenopus* egg extract is dependent on nuclear assembly. *J. Cell Sci.* **95**, 383–91.

Bode, J., Gomez-Lira, M.J. and Schroter, H. (1983). Nucleosomal particles open as the histone core becomes hyperacetylated. *Eur. J. Biochem.* **130**, 437–45.

Bode, J., Kohwi, Y., Dickinson, L., Joh, T., Klehr, D., Mielke, C. and Kohwi-Shigematsu, T. (1992). Biological significance of unwinding capability of nuclear matrix-associating DNAs. *Science* **155**, 195–8.

Bogenhagen, D.F., Wormington, W.M. and Brown, D.D. (1982). Stable transcription complexes of *Xenopus* 5S RNA genes: a means to maintain the differentiated state. *Cell* **28**, 413–21.

Bohmann, K., Ferreira, J., Santama, N., Weis, K. and Lamond, A.I. (1995). Molecular analysis of the coiled body. *J. Cell Sci.* **19**, 107–13.

Bone, J.R., Lavender, J., Richman, R., Palmer, M.J., Turner, M.B. and Kuroda, M.L. (1994). Acetylated histone H4 on the male X chromosome is associated with dosage compensation in *Drosophila*. *Genes Dev.* **8**, 96–104.

Bonifer, C., Vidal, M., Grosveld, F. and Sippel, A.E. (1990). Tissue specific and position independent expression of the complete gene domain for chicken lysozyme in transgenic mice. *EMBO J.* **9**, 2843–8.

Bonne-Andrea, C., Harper, F., Sobczak, J. and De Recondo, A-M (1984). Rat liver HMG1: a physiological nucleosome assembly factor. *EMBO J.* **3**, 1193–9.

Bonne-Andrea, C., Wong, M.L. and Alberts, B.M. (1990). *In vitro* replication through nucleosomes without histone displacement. *Nature* **343**, 719–26.

Bonner, W.M., Wu, R.S., Panusz, H.T. and Muneses, C. (1988). Kinetics of accumulation and depletion of soluble newly synthesized histone is the reciprocal regulation of histone and DNA synthesis. *Biochemistry* **27**, 6542–50.

Borowiec, J.A., Dean, F.B., Bullock, P.A. and Hurwitz, J. (1990). Binding and unwinding – how T antigen engages the SV40 origin of DNA replication. *Cell* **60**, 181–91.

Borrow, J., Stanton, V.P. Jr, Andresen, J.M., Becher, R., Behm, F.G., Chaganti, R.S., Civin, C.I., Disteche, C., Dube, I., Frischauf, A.M., Horsman, D., Mitelman, F., Volinia, S., Watmore, A.E. and Housman, D.E. (1996). The translocation t(8;16)(p11;p13) of acute myeloid leukaemia fuses a putative acetyltransferase to the CREB-binding protein.*Nat. Genet.* **14**, 33–41.

Boulet, A.M. and Scott, M.P. (1988). Control elements of the P2 promoter of the *Antennapedia* gene. *Genes Dev.* **2**, 1600–14.

Boulikas, T., Wiseman, J.M. and Garrard, W.T. (1980). Points of contact between histone H1 and the histone octamer. *Proc. Natl Acad. Sci. USA* **77**, 127–31.

Bouvet, P. and Wolffe, A.P. (1994). A role for transcription and FRGY2 in masking maternal mRNA in *Xenopus* oocytes. *Cell* **77**, 931–41.

Bouvet, P., Dimitrov, S. and Wolffe, A.P. (1994). Specific regulation of chromosomal 5S rRNA gene transcription *in vivo* by histone H1. *Genes Dev.* **8**, 1147–59.

Boy de la Tour, E. and Laemmli, U.K. (1988). The metaphase scaffold is helically folded: sister chromatids have predominantly opposite helical handedness. *Cell* **55**, 937–44.

Boyes, J. and Bird, A. (1991). DNA methylation inhibits transcription indirectly via a methyl-CpG binding protein. *Cell* **64**, 1123–34.

Boyes, J. and Bird, A. (1992). Repression of genes by DNA methylation depends on CpG density and promoter strength: evidence for involvement of a methyl-CpG binding protein. *EMBO J.* **11**, 327–33.

Bradbury, E.M., Inglis, R.J. and Matthews, H.R. (1974). Control of cell division by very lysine rich histone (F1) phosphorylation. *Nature (London)* **247**, 257–61.

Brand, A.H., Breeden, L., Abraham, J., Sternglanz, R. and Nasmyth, K. (1985). Characterization of a 'silencer' in yeast: a DNA sequence with properties opposite to those of a transcriptional enhancer. *Cell* **41**, 41–8.

Brandhorst, B.P. (1980). Simultaneous synthesis, translation and storage of mRNA including histone mRNA in sea urchin eggs. *Dev. Biol.* **52**, 310–11.

Braunstein, M., Rose, A.B., Holmes, S.G., Allis, C.D. and Broach, J.R. (1993). Transcriptional silencing in yeast is associated with reduced histone acetylation. *Genes Dev.* **7**, 592–604.

Braunstein, M., Sobel, R.E., Allis, C.D., Turner, B.M. and Broach, J.R. (1996). Efficient transcriptional silencing in *Saccharomyces cerevisiae* requires a heterochromatin acetylation pattern. *Mol. Cell. Biol.* **16**, 4349–56.

Bresnick, E.H., Bustin, M., Marsaud, V., Richard-Foy, H. and Hager, G.L. (1992). The transcriptionally-active MMTV promoter is depleted of histone H1. *Nucleic. Acids Res.* **20**, 273–8.

Brewer, B.J., Diller, J.D., Friedman, K.L., Kolor, K.M., Raghuraman, M.K. and Fangman, W.L. (1993). The topography of chromosome replication in yeast. *Cold Spring Harbor Symp. Quant. Biol.* **58**, 425–34.

Brill, S.J. and Sternglanz, R. (1988). Transcription-dependent DNA supercoiling in yeast topoisomerase mutants. *Cell* **54**, 403–11.

Brizuela, B.J., Elfring, L., Ballard, J., Tamkun, J.W. and Kennison, J.A. (1994). Genetic analysis of the *brahma* gene of *Drosophila melanogaster* and polytene subdivisions 72AB. *Genetics* **137**, 803–13.

Brockdorff, N., Ashworth, A., Kay, G.F., Cooper, P., Smith, S., McCabe, V., Norris, D., Penny, G., Patel, D. and Rastan, S. (1991). Conservation of position and exclusive expression of mouse Xist from the inactive X chromosome. *Nature* **351**, 329–31.

Brockdorff, N., Ashworth, A., Kay, G., McCabe, V., Norris, D., Cooper, D., Swift, S. and Rastan, S. (1992). The product of the mouse Xist gene is a 15 kb inactive X-specific transcript containing no conserved ORF and located in the nucleus. *Cell* **71**, 515–26.

Brown, C.-J., Ballabio, A., Rupert, J., Lafreiniere, R., Grompe, M., Tonlorenzi, R. and Willard, H.F. (1991). A gene from the region of the human X inactivation centre is expressed exclusively from the inactive X chromosome. *Nature* **349**, 38–42.

Brown, C.-J., Hendrich, B., Rupert, J., Lafrieniere, R., Xing, Y., Lawrence, J. and Willard, H.F. (1992). The human XIST gene: analysis of a 17 Kb inactive X-specific RNA that contains conserved repeats and is highly localized within the nucleus. *Cell* 71, 527–38.

Brown, D.D. (1981). Gene expression in eukaryotes. *Science* 211, 667–74.

Brown, D.D. (1984). The role of stable complexes that repress and activate eukaryotic genes. *Cell* 37, 359–65.

Brown, D.D. and Gurdon, J.B. (1977). High fidelity transcription of 5S DNA injected into *Xenopus* oocytes. *Proc. Natl Acad. Sci. USA* 74, 2064–8.

Brown, D.D. and Littna, E. (1966). Synthesis and accumulation of low molecular weight RNA during embryogenesis of *Xenopus laevis. J. Mol. Biol.* 20, 95–112.

Brown, D.W., Libertini, L.J., Suquet, C., Small, E.W. and Smerdon, M.J. (1993). Unfolding of nucleosome cores dramatically changes the distribution of UV photoproducts. *Biochemistry* 32, 10527–31.

Brown, K.D., Coulson, R.M., Yen, T.J. and Cleveland, D.W. (1994). Cyclin-like accumulation and loss of the putative kinetochore motor CENP-E results from coupling continuous synthesis with specific degradation at the end of mitosis. *J. Cell Biol.* 125, 1303–12.

Brown, K.D., Wood, K.W. and Cleveland, D.W. (1996a). The kinesin-like protein CENP-E is kinetochore associated throughout poleward chromosome segregation during anaphase A. *J. Cell Sci.* 109, 961–9.

Brown, S.A., Imbalzano, A.N. and Kingston, R.E. (1996). Activator-dependent regulation of transcriptional pausing on nucleosomal templates. *Genes Dev.* 10, 1479–90.

Brownell, J.E., Zhou, J., Ranalli, T., Kobayashi, R., Edmondson, D.G., Roth, S.Y. and Allis, C.D. (1996b). *Tetrahymena* histone acetyltransferase A: a homolog to yeast Gcn5p linking histone acetylation to gene activation. *Cell* 84, 843–51.

Broyles, S.S. and Pettijohn, D.E. (1986). Interaction of the *Escherichia coli* HU protein with DNA. Evidence for formation of nucleosome-like structures with altered DNA helical pitch. *J. Mol. Biol.* 187, 47–58.

Bryk, M., Banerjee, M., Murphy, M., Kundsen, K.E., Garfinkel, D.J. and Curcio, M.J. (1997). Transcriptional silencing of Ty1 elements in the RDN1 locus of yeast. *Genes Dev.* 11, 255–69.

Budarf, M.L. and Blackburn, E.H. (1986). Chromatin structure of the telomeric region and 3'-nontranscribed spacer of *Tetrahymena* ribosomal RNA genes. *J. Biol. Chem.* 261, 363–9.

Buonigiorno-Nardelli, M., Micheli, G., Carri, M.T. and Marilley, M. (1982). A relationship between replicon size and supercoiled loops domains in the eukaryotic genome. *Nature* 298, 100–2.

Buratowski, S. (1994). The basics of basal transcription by RNA polymerase II. *Cell* 77, 1–3.

Buratowski, S., Hahn, S., Guarente, L. and Sharp, P.A. (1989). Five intermediate complexes in transcription initiation by RNA polymerase. *Cell* 56, 549–61.

Burch, J.B.E. and Weintraub, H. (1983). Temporal order of chromatin structural changes associated with activation of the major chicken vitellogenin gene. *Cell* 33, 65–76.

Burhans, W.C., Vassilev, L.T., Caddle, M.S., Heintz, N.H. and De Pamphilis, M.L. (1990). Identification of an origin of bidirectional DNA replication in mammalian chromosomes. *Cell* 62, 955–65.

Burhans, W.C., Vassilev, L.T., Wu, J., Sogo, J.M., Nallaseth, F.S. and De Pamphilis, M.L. (1991). Emetine allows identification of origins of mammalian DNA replication by imbalanced DNA synthesis, not through conservative nucleosome segregation. *EMBO J.* **10**, 3419–28.

Buschhausen, G., Wittig, B., Graessmann, M. and Graessman, A. (1987). Chromatin structure is required to block transcription of the methylated herpes simplx virus thymidine kinase gene. *Proc. Natl Acad. Sci. USA* **84**, 1177–81.

Busslinger, M., Hurst, J. and Flavell, R.A. (1983). DNA methylation and the regulation of globin gene expression. *Cell* **34**, 197–206.

Bustin, M. and Reeves, R. (1996). High-mobility-group chromosomal proteins: architectureal components that facilitate chormatin function. *Prog. Nucleic. Acid Res. Mol. Biol.* **54**, 35–100.

Cairns, B.R., Kim, Y.-J., Sayre, M.H., Laurent, B.C. and Kornberg, R.D. (1994). A multisubunit complex containing the SWI1/ADR6, SWI2/SNF2, SWI3, SNF5, and SNF6 gene products isolated from yeast. *Proc. Natl Acad. Sci. USA* **91**, 1950–4.

Cairns, B.R., Lorch, Y., Li, Y., Zhang, M. Lacomis, L., Evdjument-Bromage, H., Tempst, P., Du, J., Laurent, B. and Kornberg, R.D. (1996). RSC, an essential abundant chromatin-remodeling complex. *Cell* **87**, 1249–60.

Calladine, C.R. and Drew, H.R. (1997). *Understanding DNA: The Molecule and How it Works.* 2nd ed. Academic Press, London.

Callan, H.G. (1986). *Lampbrush Chromosomes.* Springer Verlag, Berlin.

Callan, H.G., Gall, J.G. and Berg, C.A. (1987). The lampbrush chromosomes of *Xenopus laevis*: preparation, identification and distribution of 5S DNA sequences. *Chromosoma* **95**, 236–50.

Camerini-Otero, R.D. and Zasloff, M.A. (1980). Nucleosomal packaging of the thymidine kinase gene of herpes simplex virus transferred into mouse cells: an actively expressed single copy gene. *Proc. Natl Acad. Sci. USA* **77**, 5079–83.

Camerini-Otero, R.D., Sollner-Webb, B. and Felsenfeld, G. (1976). The organization of histones and DNA in chromatin: evidence for an arginine-rich histone kernel. *Cell* **8**, 333–47.

Campbell, F.E. and Setzer, D.R. (1991). Displacement of *Xenopus* transcription factor IIIA from a 5S rRNA gene by a transcribing RNA polymerase. *Mol. Cell. Biol.* **11**, 3978–86.

Campoy, F.J., Meehan, R.R., McKay, S., Nixon, J. and Bird, A. (1995). Binding of histone H1 to DNA is indifferent to methylation at CpG sequences. *J. Biol. Chem.* **270**, 26473–81.

Candau, R., Chavez, S. and Beato, M. (1996). The hormone responsive region of mouse mammary tumor virus positions a nucleosome and precludes access of nuclear factor 1 to the promoter. *J. Steroid Biochem. Mol. Biol.* **57**, 19–31.

Candido, E.P.M., Reeves, R. and Davie, J.R. (1978). Sodium butyrate inhibits histone deacetylation in cultured cells. *Cell* **14**, 105–15.

Cardoso, M. C., Leonhardt, H. and Nadal-Ginard, B. (1993). Reversal of terminal differentiation and control of DNA replication: cyclin A and Cdk2 specifically localize at subnuclear sites of DNA replication. *Cell* **74**, 979–92.

Carlson, M., Osmond, B.C., Neigeborn, L. and Botstein, D. (1984). A suppressor of Snf1 mutations causes constitutive high level invertase synthesis in yeast. *Genetics* **107**, 19–32.

Carmen, A.A., Rundlett, S.E. and Grunstein, M. (1996). HDA1 and HDA3 are components of a yeast histone deacetylase (HDA) complex. *J. Biol. Chem.* **271**, 15837–44.

Carmo-Fonseca, M., Ferreira, J. and Lamond, A. I. (1993). Assembly of snRNP-containing coiled bodies is regulated in interphase and mitosis–evidence that the coiled body is a kinetic nuclear structure. *J. Cell Biol.* **120**, 841–52.

Caron, F. and Thomas, J.O. (1981). Exchange of histone H1 between segments of chromatin. *J. Mol. Biol.* **146**, 513–37.

Carr, K.D. and Richard-Foy, H. (1990). Glucocorticoids locally disrupt an array of positioned nucleosomes on the rat tyrosine aminotransferase promoter in hepatoma cells. *Proc. Natl Acad. Sci. USA* **87**, 9300–4.

Carroll, S.B., Layman, R.A., McCutcheon, M.A. and Riley, P.D. (1986). The localization and regulation of *Antennapedia* protein expression in *Drosophila* embryos. *Cell* **47**, 113–22.

Carter, K. C., Taneja, K. L. and Lawrence, J. B. (1991). Discrete nuclear domains of poly(A) RNA and their relationship to the functional organization of the nucleus. *J. Cell Biol.* **115**, 1191–202.

Carter, K. C., Bowman, D., Carrington, W., Fogarty, K., McNeil, J. A., Fay, F. S. and Lawrence, J. B. (1993). A three-dimensional view of precursor messenger RNA metabolism within the mammalian nucleus [see comments]. *Science* **259**, 1330–5.

Cary, P.D., Crane-Robinson, C., Bradbury, E.M. and Dixon, G.H. (1982). Effect of acetylation on the binding of N-terminal peptides of histone H4 to DNA. *Eur. J. Biochem.* **127**, 137–43.

Cedar, H. and Felsenfeld, G. (1973). Transcription of chromatin *in vitro*. *J. Mol. Biol.* **77**, 237–54.

Celnikev, S.E., Sharma, S., Keelan, D.J. and Lewis, E. (1990). The molecular genetics of the *bithorax* complex of *Drosophila*: *Cis* regulation of the Abdominal-B domain. *EMBO J.* **9**, 4277–86.

Cereghini, S. and Yaniv, M. (1984). Assembly of transfected DNA into chromatin: structural changes in the origin–promoter–enhancer region upon replication. *EMBO J.* **3**, 1243–53.

Challberg, M.D. and Kelly, T.J. (1989). Animal virus replication. *Annu. Rev. Biochem.* **58**, 671–717.

Chambers, S.A.M. and Shaw, B.R. (1984). Levels of histone H4 diacetylation decrease dramatically during sea urchin embryonic development and correlate with cell doubling rate. *J. Biol. Chem.* **259**, 13458–63.

Chan, C.S., Rastelli, L. and Pirrotta, V. (1994). A Polycomb response element in the ubx gene that determines an epigenetically inherited state of repression. *EMBO J.* **13**, 2553–64.

Chang, C. and Gralla, J.D. (1994). A critical role for chromatin in mounting a synergistic transcriptional response to GAL4-VP16. *Mol. Cell Biol.* **14**, 5175–81.

Chang, L., Loranger, S.S., Mizzen, C., Ernst, S.G., Allis, C.D. and Annunziato, A.T. (1997). Histones in transit: cytosolic histone complexes and diacetylation of H4 during nucleosome assembly in human cells. *Biochemistry* **36**, 469–80.

Chao, M.V., Gralla, J.D. and Martinson, H.G. (1980a). *Lac* operator nucleosomes I. Repressor binds specifically to operators within the nucleosome core. *Biochemistry* **19**, 3254–60.

Chao, M.V., Martinson, H.G. and Gralla, J.D. (1980b). *Lac* operator nucleosomes can change conformation to strengthen binding by *Lac* repressor. *Biochemistry* **19**, 3260–9.

Charlieu, J.-P., Larsson, S., Miyagawa, K., van Heyningen, V. and Hastie, N.D. (1995). Does the Wilms' tumor suppressor gene, WT1, play roles in both splicing and transcription? *J. Cell Sci.* **19**, 95–9.

Chastain, P.D., Eichler, E.E., Kang, S., Nelson, D.L., Levene, S.D. and Siden, R.R. (1995). Anomalous rapid electrophoretic mobility of DNA containing triplet repeats associated with human disease genes. *Biochemistry* **34**, 16125–31.

Chatterjee, S. and Struhl, K. (1995). Connecting a promoter-bound protein to TBP by passes the need for a transcriptional activation domain. *Nature* **374**, 820–3.

Chavez, S. and Beato, M. (1997). Nucleosome-mediated synergism between transcription factors on the mouse mammary tumor virus promoter. *Proc. Natl Acad. Sci. USA* **94**, 2885–90.

Chen, G.L., Yang, L., Rowe, T.C. , Halligan, B.D., Tewey, K.M. and Liu, L.F. (1984). Nonintercalative antitumor drugs interfere with the breakage reunion reaction of mammalian DNA topoisomerase II. *J. Biol. Chem.* **259**, 13560–6.

Chen, H., Li, B. and Workman, J.L. (1994). A histone-binding protein nucleoplasmin stimulates transcription factor binding to nucleosomes and factor-induced nucleosome disassembly. *EMBO J.* **13**, 380–90.

Chen, H., Lin, R.J., Schiltz, R.L., Chakravarti, D., Nash, A., Nagy, L., Privalsky, M.L., Nakatani, Y. and Evans, R.M. (1997). Nuclear receptor coactivator ACTR is a novel histone acetyltransferase and forms a multimeric activation complex with PCAF and CBP/p300. *Cell* **90**, 569–80.

Chen, J., Willingham, T., Margraf, L.R, Schreiber-Agus, N., DePinho, R.A. and Nisen, P.D. (1995). Effects of the MYC oncogene antagonist, MAD, on proliferation, cell cycling and the malignant phenotype of human brain tumor cells. *Nat. Med.* **1**, 638–43.

Chen, J.L., Attardi, L.D., Verrijzer, C.P., Yokomori, K. and Tjian, R. (1994b) Assembly of recombinant TFIID reveals differential coactivator requirements for distinct transcriptional activators. *Cell* **79**, 93–105.

Chen, T.A. and Allfrey, V.G. (1987). Rapid and reversible changes in nucleosome structure accompany the activation, repression and super-induction of the murine proto-oncogenes *c-fos* and *c-myc*. *Proc. Natl Acad. Sci. USA* **84**, 5252–6.

Chen, T.A., Sterner, R., Cozzolino, A. and Allfrey, V.G. (1990). Reversible and irreversible changes in nucleosome structure along the *c-fos* and *c-myc* oncogenes following inhibition of transcription. *J. Mol. Biol.* **212**, 481–93.

Cheng, L. and Kelly, T.J. (1989). The transcriptional activator nuclear factor 1 stimulates the replication of SV40 minichromosomes *in vivo* and *in vitro*. *Cell* **59**, 541–51.

Cheng, L., Workman, J.L., Kingston, R.E. and Kelly, T.J. (1993). Regulation of DNA replication *in vitro* by the transcriptional activation domain of GAL4-VP16. *Proc. Natl Acad. Sci. USA* **89**, 589–93.

Chesnut, J.D., Stephens, J.H. and Dahmus, M.E. (1992). The interaction of RNA polymerase II with the adenovirus-2 major late promoter is precluded by phosphorylation of the C-terminal domain of subunit IIa. *J. Biol. Chem.* **267**, 10500–6.

Chi, T. and Carey, M. (1996). Assembly of the isomerized TFIIA–TFIID–TATA ternary complex is necessary and sufficient for gene activation. *Genes Dev.* **10**, 2540–50.

Chiba, H., Muramatsu, M., Nomoto, A. and Kato, H. (1994). Two human homologs of *Saccharomyces cerevisiae* SWI2/SNF2 and *Drosophila* brahma are transcriptional coactivators cooperating with the estrogen receptor and the retinoic acid receptor. *Nucleic. Acids Res.* **22**, 1815–20.

Chicoine, L.G. and Allis, C.D. (1986). Regulation of histone acetylation during macronuclear differentiation in *Tetrahymena*. Evidence for control at the level of acetylation and deacetylation. *Dev. Biol.* **116**, 477–85.

Chipev, C.C. and Wolffe, A.P. (1992). The chromosomal organization of *Xenopus laevis* 5S ribosomal RNA genes *in vivo*. *Mol. Cell. Biol.* **12**, 45–55.

Chiu, C.-P. and Blau, H.M. (1984). Reprogramming cell differentiation in the absence of DNA synthesis. *Cell* **37**, 879–87.

Cho, H. and Wolffe, A.P. (1994). Characterization of the *Xenopus laevis* B4 gene: an oocyte specific vertebrate linker histone gene containing introns. *Gene* **143**, 233–8.

Choi, O-R.B. and Engel, J.D. (1988). Developmental regulation of ϵ globin gene switching. *Cell* **55**, 17–26.

Choy, B. and Green, M.R. (1993). Eukaryotic activators function during multiple steps of preinitiation complex assembly. *Nature* **366**, 531–6.

Christians, E., Campion, E., Thompson, E.M. and Renard, J.P. (1995). Expression of the HSP 70.1 gene, a landmark of early zygotic activity in the mouse embryo is restricted to the first burst of transcription. *Development* **112**, 113–22.

Chuang, P.T., Albertson, D.G. and Meyer, B.J. (1994). DPY-27: A chromosome condensation protein homolog that regulates *C. elegans* dosage compensation through association with the X chromosome. *Cell* **79**, 459–74.

Chung, J.H., Whiteley, M. and Felsenfeld, G. (1993). A 5′ element of the chicken β-globin domain serves as an insulator in human erythroid cells and protects against position effect in *Drosophila*. *Cell* **74**, 505–14.

Churchill, M.E.A. and Travers, A.A. (1991). Protein motifs that recognize structural features of DNA. *Trends Biochem. Sci.* **16**, 92–7.

Ciliberto, G., Castagnoli, L. and Cortese, R. (1983). Transcription by RNA polymerase III. *Curr. Top. Dev. Bio* **18**, 59–88.

Cirillo, L.A., McPherson, C.E., Bossard, P., Stevens, K., Cherian, S., Shim, E.Y., Clark, K.L., Burley, S.K. and Zaret, K.S. (1998). Binding of the winged helix transcription factor HNF3 to a linker histone site on the nucleosome. *EMBO J.* **17**, 244–254.

Clark, D.J. and Felsenfeld, G. (1991). Formation of nucleosomes on positively supercoiled DNA. *EMBO J.* **10**, 387–95.

Clark, D.J. and Felsenfeld, G. (1992). A nucleosome core is transferred out of the path of a transcribing polymerase. *Cell* **71**, 11–22.

Clark, D.J. and Kimura, T. (1990). Electrostatic mechanism of chromatin folding. *J. Mol. Biol.* **211**, 883–96.

Clark, D.J. and Thomas, J.O. (1986). Salt-dependent co-operative interaction of histone H1 with linker DNA. *J. Mol. Biol.* **187**, 569–80.

Clark, D.J. and Wolffe, A.P. (1991). Superhelical stress and nucleosome mediated repression of 5S RNA gene transcription *in vitro*. *EMBO J.* **10**, 3419–28.

Clark, D.J., Hill, C.S., Martin, S.R. and Thomas, J.O. (1988). α-Helix in the carboxy-terminal domains of histones H1 and H5. *EMBO J.* **7**, 69–75.

Clark, D.J., O'Neill, L.P. and Turner, B.M. (1993a). Selective use of H4 acetylation sites in the yeast *Saccharomyces cerevisiae*. *Biochem. J.* **294**, 557–61.

Clark, K.L., Halay, E.D., Lai, E. and Burley, S.K. (1993b). Co-crystal structure of the HNF-3/fork head DNA recognition motif resembles histone H5. *Nature* **364**, 412–20.

Clark, R.J. and Felsenfeld, G. (1971). Structure of chromatin. *Nature New Biol.* **229**, 101–6.

Clark-Adams, C.D., Norris, D., Osley, M.A., Fassler, J.S. and Winston, F. (1988). Changes in histone gene dosage alter transcription in yeast. *Genes Dev.* **2**, 150–9.

Clarke, H.J., Oblin, C. and Bustin, M. (1992). Developmental regulation of chromatin composition during mouse embryogenesis: somatic histone H1 is first detectable at the 4-cell stage. *Development* **115**, 791–9.

Cockerill, P.N. and Garrard, W.T. (1986). Chromosomal loop anchorage of the kappa immunoglobulin gene occurs next to the enhancer in a region. *Cell* **44**, 273–82.

Collas, P., Courvalin, J.C. and Poccia, D. (1996). Targeting of membranes to sea urchin sperm chromatin is mediated by a Lamin B receptor-like integral membrane protein. *J. Cell Biol.* **135**, 1715–25.

Comai, L., Tauese, N. and Tjian, R. (1992). The TATA-binding protein and associated factors are integral components of the RNA polymerase I transcription factor SL1. *Cell* **68**, 965–76.

Conaway, R.C. and Conaway, J.W. (1993). General initiation factors for RNA polymerase II. *Annu. Rev. Biochem.* **62**, 161–90.

Conconi, A., Widmer, R.M., Koller, T. and Sogo, J.M. (1989). Two different chromatin structures coexist in ribosomal RNA genes throughout the cell cycle. *Cell* **57**, 753–61.

Conrad, M.N., Wright, J.H., Wolf, A.J. and Zakian, V.A. (1990). RAP1 protein interacts with yeast telomeres *in vivo*: over production alters telomere structure and decreases chromosome stability. *Cell* **63**, 739–50.

Cook, P.R. and Brazell, I.A. (1975). Supercoils in human DNA. *J. Cell Sci.* **19**, 261–79.

Cook, P.R. (1988). The nucleoskeleton: artefact, passive framework or active site? *J. Cell Sci.* **90**, 1–6.

Cook, P.R. (1991). The nucleoskeleton and the topology of replication. *Cell* **66**, 627–37.

Cook, P. R. (1994). RNA polymerase: structural determinant of the chromatin loop and the chromosome. *Bioessays* **16**, 425–30.

Cook, C.A., Bernat, R.L. and Earnshaw, W.C. (1990). CENP-B: a major human centromere protein located beneath the kinetochore. *J. Cell Biol.* **110**, 1475–88.

Cooper, J.P., Roth, S.Y. and Simpson, R.T. (1994). The global transcriptional regulators SSN6 and Tup1, play distinct roles in the establishment of a repressive chromatin structure. *Genes Dev.* **8**, 1400–10.

Corces, V.G. and Geyer, P.K. (1991). Interactions of retrotransposons with the host genome: the case of the *gypsy* element of *Drosophila*. *Trends Genet.* **7**, 69–73.

Cordingley, M.G., Riegel, A.T. and Hager, G.L. (1987). Steroid dependent interaction of transcription factors with the inducible promoter of mouse mammary tumor virus *in vivo*. *Cell* **48**, 261–70.

Coté, J., Quinn, J., Workman, J.L. and Peterson, C.L. (1994). Stimulation of GAL4 derivative binding to nucleosomal DNA by the yeast SWI/SNF complex. *Science* **265**, 53–60.

Cotten, M. and Chalkley, R. (1987). Purification of a novel, nucleoplasmin-like protein from somatic nuclei. *EMBO J.* **6**, 3945–54.

Cotton, R.W. and Hamkalo, B.A. (1981). Nucleosome dissociation at physiological ionic strengths. *Nucleic. Acids Res.* **9**, 445–58.

Courey, A.J., Holtzman, D.A., Jackson, S.P. and Tjian, R. (1989). Synergistic activation by the glutamine-rich domains of human transcription factor Sp1. *Cell* **59**, 827–36.

Cox, L.S. and Laskey, R.A. (1991). DNA replication occurs at discrete sites in pseudonuclei assembled from purified DNA *in vitro*. *Cell* **66**, 271–5.

Craig, J.M., Boyle, S., Perry, P. and Bickmore, W.A. (1997). Scaffold attachments within the human genome. *J. Cell Sci.* **110**, 2673–2682.

Cremer, T., Kurz, A., Zirbel, R., Dietzel, S., Rinke, B., Schrock, E., Speicher, M. R., Mathieu, U., Jauch, A. and Emmerich, P. (1993). Role of chromosome territories in the functional compartmentalization of the cell nucleus. *Cold. Spring. Harb. Symp. Quant. Biol.* **58**, 777–92.

Cremisi, C. and Yaniv, M. (1980). Sequential assembly of newly synthesized histones on replicating SV40 DNA. *Biochem. Biophys. Res. Commun.* **92**, 1117–23.

Cress, W.D. and Triezenberg, S.J. (1991). Critical structural elements of the VP16 transcriptional activation domain. *Science* **251**, 87–90.

Crippa, M.P., Alfonso, P.J. and Bustin, M. (1992). Nucleosome core binding region of chromosomal protein HMG-17 acts as an independent functional domain. *J. Mol. Biol.* **228**, 442–9.

Crippa, M.P., Trieschmann, L., Alfonso, P.J., Wolffe, A.P. and Bustin, M. (1993). Deposition of chromosomal protein HMG-17 during replication affects the nucleosomal ladder and transcriptional potential of nascent chromatin. *EMBO J.* **12**, 3855–64.

Crossland, L.D. and Raskas, H.J. (1983). Identification of adenovirus genes that require template replication for expression. *J. Virol.* **46**, 737–48.

Croston, G.E., Kerrigan, L.A., Liva, L.M., Marshak, D.R. and Kadonaga, J.T. (1991). Sequence-specific antirepression of histone H1-mediated inhibition of basal RNA polymerase II transcription. *Science* **251**, 643–9.

Csordas, A. (1990). On the biological role of histone acetylation. *Biochem. J.* **265**, 23–38.

Cullen, K.E., Kladde, M.P. and Seyfred, M.A. (1993). Interaction between transcription regulatory regions of prolactin chromatin. *Science* **261**, 203–6.

Cusick, M.E., Lee, K.-S., DePamphilis, M.L. and Wasserman, P.M. (1983). Structure of chromatin at deoxyribonucleic acid replication forks: Nuclease hypersensitivity results from both prenucleosomal deoxyribonucleic acid and an immature chromatin structure. *Biochemistry* **22**, 3873–84.

Cusick, M.E., DePamphilis, M.L. and Wasserman, P.M. (1984). Dispersive segregation of nucleosomes during replication of simian virus 40 chromosomes. *J. Mol. Biol.* **178**, 249–71.

Dabauvalle, M.-C., Loos, K., Merkert, H. and Scheer, U. (1991). Spontaneous assembly of pore complex-containing membranes (annulate lamellae) in *Xenopus* egg extract in the absence of chromatin. *J. Cell. Biol.* **112**, 1073–82.

Dallas, P.B., Yaciuk, P. and Moran, E. (1997). Characterization of monoclonal antibodies raised against p300: both p300 and CBP are present in intracellular TBP complexes. *J. Virol.* **71**, 1726–31.

Darby, M.K., Andrews, M.T. and Brown, D.D. (1988). Transcription complexes that program *Xenopus* 5S RNA genes are stable *in vivo*. *Proc. Natl Acad. Sci. USA* **85**, 5516–20.

Dasso, M. (1993). RCC1 in the cell cycle: the regulator of chromosome condensation takes on new roles. *Trends Biochem. Sci.* **18**, 96–101.

Dasso, M. and Newport, J.W. (1990). Completion of DNA replication is monitored by a feedback system that controls the initiation of mitosis *in vitro*: studies in *Xenopus*. *Cell* **61**, 811–23.

Dasso, M., Nishitani, H., Korubluth, S., Nishimoto, T. and Newport, J.W. (1992). RCC1, a regulator of mitosis, is essential for DNA replication. *Mol. Cell. Biol.* **12**, 3337–45.

Dasso, M., Dimitrov, S. and Wolffe, A.P. (1994a). Nuclear assembly is independent of linker histones. *Proc. Natl Acad. Sci. USA* **91**, 12477–81.

Dasso, M., Seki, T., Azuma, Y., Ohba, T. and Nishimoto, T. (1994b). A mutant form of the Ran/TC4 protein disrupts nuclear formation in *Xenopus laevis* egg extracts by inhibiting the RCC1 protein, a regulator of chromosome condensation. *EMBO J.* **13**, 5732–44.

Davey, C., Pennings, S., Meersseman, G., Wess, T.J. and Allan, J. (1995). Periodicity of strong nucleosome positioning sites around the chicken adult β-globin gene may encode regularly spaced chromatin. *Proc. Natl Acad. Sci. USA* **92**, 11210–14.

Davidson, E.H. (1986). *Gene activity in Early Development* 3rd edn. Academic Press Orlando.

Dean, A., Pederson, D.S. and Simpson, R.T. (1989). Isolation of yeast plasmid chromatin. *Methods Enzymol.* **170**, 26–40.

De Bernardin, W., Koller, T. and Sogo, J.M. (1986). Structure of *in vivo* transcribing chromatin as studied in simian virus 40 minichromosomes. *J. Mol. Biol.* **191**, 469–82.

De Camillis, M., Cheng, N.S., Pierre, D. and Brock, H. (1992). The polyhomeotic gene of *Drosophila* encodes a chromatin protein that shares polytene chromosome binding sites with Polycomb. *Genes Dev.* **6**, 223–32.

De Lange, R.J., Farnbrough, D.M., Smith, E.L. and Bonner, J. (1969a). Calf and pea histone IV: the complete amino acid sequence of calf thymus histone IV; presence of N-acetyllysine. *J. Biol. Chem.* **244**, 319–34.

De Lange, R.J., Farnbrough, D.M., Smith, E.L. and Bonner, J. (1969b). Calf and pea histone IV: complete amino acid sequence of pea seedling histone IV; comparison with the homologous calf thymus histone. *J. Biol. Chem.* **244**, 5669–79.

De Lange, T. (1992). Human telomeres are attached to the nuclear matrix. *EMBO J.* **11**, 717–24.

De Lange, T., Shine, L., Myers, R.M., Cox, D.R., Naylor, S.L., Killery, A.M. and Varmus, H.E. (1990). Structure and variability of human chromosome ends. *Mol. Cell. Biol.* **10**, 518–27.

De Pamphilis, M.L. (1988). Transcriptional elements as components of eukaryotic origins of DNA replication. *Cell* **52**, 635–8.

De Pamphilis, M.L. (1993). Eukaryotic DNA replication: anatomy of an origin. *Annu. Rev. Biochem.* **62**, 29–59.

De Rubertis, F., Kadosh, D., Henchoz, S., Pauli, D., Reuter, G., Struhl, K. and Spierer, P. (1996). The histone deacetylase RPD3 counteracts genomic silencing in *Drosophila* and yeast. *Nature* **384**, 589–91.

Di Bernardino, M.A. (1987). Genomic potential of differentiated cells analyzed by nuclear transplantation. *Am. Zool.* **27**, 623–44.

DiNardo, S., Voelkel, K. and Sternglanz, R. (1984). DNA topoisomerase II is required for segregation of daughter molecules at the termination of DNA replication. *Proc. Natl Acad. Sci. USA* **81**, 2616–20.

Diffley, J.F.X. and Cocker, J.H. (1992). Protein DNA interactions at a yeast replication origin. *Nature* **357**, 169–71.

Diffley, J.F. and Stillman, B. (1989). Transcriptional silencing and lamins. *Nature* **342**, 24.

Dikstein, R., Ruppert, S. and Tjian, R. (1996). TAF$_{II}$250 is a bipartite protein kinase that phosphorylates the basal transcription factor RAP74. *Cell* **84**, 781–90.

Dilworth, S.M., Black, S.J. and Laskey, R.A. (1987). Two complexes that contain histones are required for nucleosome assembly *in vitro*: role of nucleoplasmin and N1 in *Xenopus* egg extracts. *Cell* **51**, 1009–18.

Dimitrov, S. and Wolffe, A.P. (1996). Remodeling somatic nuclei in *Xenopus laevis* egg extracts: molecular mechanisms for the selective release of histone H1 and H1° from chromatin and the acquisition of transcriptional competence. *EMBO J.* **15**, 5897–906

Dimitrov, S.I., Russanova, V.R. and Pashev, I.G. (1987). The globular domain of histone H5 is internally located in the 30 nm fibre: an immunochemical study. *EMBO J.* **6**, 2387–92.

Dimitrov, S., Almouzni, G., Dasso, M., and Wolffe, A.P. (1993). Chromatin transitions during early *Xenopus* embryogenesis: changes in histone H4 acetylation and in linker histone type. *Dev. Biol.* **160**, 214–27.

Dimitrov, S., Dasso, M.C. and Wolffe, A.P. (1994). Remodeling sperm chromatin in *Xenopus laevis* egg extracts: the role of core histone phosphorylation and linker histone B4 in chromatin assembly. *J. Cell Biol.* **126**, 591–601.

DiNardo, S., Voelkel, K. and Sternglanz, R. (1984). DNA topoisomerase II is required for segregation of daughter molecules at the termination of DNA replication. *Proc. Natl Acad. Sci. USA* **81**, 2616–20.

Ding, H.-F., Rimsky, S., Batson, S.C., Bustin, M. and Hansen, U. (1994). Stimulation of RNA polymerase II elongation by chromosomal protein HMG-4. *Science* **265**, 796–9.

Dingwall, C. and Laskey, R. A. (1986). Protein import into the cell nucleus. *Annu. Rev. Cell Biol.* **2**, 367–90.

Dingwall, C., Dilworth, S.M., Black, S.J., Kearsey, S.E., Cox, L.S. and Laskey, R.A. (1987). Nucleoplasmin cDNA reveals polyglutamic acid tracts and a cluster of sequences homologous to putative nuclear localisation signals. *EMBO J.* **6**, 69–74.

Doenecke, I. and Tonjes, R. (1986). Differential distribution of lysine and arginine residues in the closely related histones H1° and H5: analysis of a human H1° gene. *J. Mol. Biol.* **187**, 461–4.

Dong, F. and van Holde, K.E. (1991). Nucleosome positioning is determined by the (H3-H4)$_2$ tetramer. *Proc. Natl Acad. Sci. USA* **88**, 10596–600.

Dong, F., Hansen, J.C. and van Holde, K.E. (1989). DNA and protein determinants of nucleosome positioning on sea urchin 5S rRNA gene sequences *in vitro*. *Proc. Natl Acad. Sci. USA* **87**, 5724–8.

Doucas, V., Ishov, A. M., Romo, A., Juguilon, H., Weitzman, M. D., Evans, R. M. and Maul, G. G. (1996). Adenovirus replication is coupled with the dynamic properties of the PML nuclear structure. *Genes. Dev.* **10**, 196–207.

Dranginis, A.M. (1986). Regulation of cell type in yeast by the mating type locus. *Trends Biochem. Sci.* **11**, 328–31.

Drapin, R., Sancar, A. and Reinberg, D. (1994). Where transcription meets repair. *Cell* **77**, 9–12.

Draves, P.H., Lowary, P.T. and Widom, J. (1992). Co-operative binding of the globular domain of histone H5 to DNA. *J. Mol. Biol.* **225**, 1105–21.

Drew, H.R. (1984). Structural specificities of five commonly-used DNA nucleases. *J. Mol. Biol.* **176**, 535–57.

Drew, H.R. and Travers, A.A. (1985). DNA bending and its relation to nucleosome positioning. *J. Mol. Biol.* **186**, 773–90.

Drew, H.R., McCall, M.J. and Calladine, C.R. (1988). Recent studies of DNA in the crystal. *Annu. Rev. Cell. Biol.* **4**, 1–20.

Dreyfuss, G., Matunis, M. J., Pinol-Roma, S. and Burd, C. G. (1993). hnRNP proteins and the biogenesis of mRNA. *Annu. Rev. Biochem.* **62**, 289–321.

Du, W., Thanos, D. and Maniatis, T. (1993). Mechanisms of transcriptional synergism between distinct virus-inducible enhancer elements. *Cell* **74**, 887–98.

Dunaway, M. (1989). A transcription factor TFIS, interacts with both the promoter and enhancer of the *Xenopus* rRNA genes. *Genes Dev.* **3**, 1768–78.

Dunphy, W.G. and Newport, J.W. (1988). Unraveling of mitotic control mechanisms. *Cell* **55**, 925–8.

Durrin, L.K., Mann, R.K., Kayne, P.S. and Grunstein, M. (1991). Yeast histone H4 N-terminal sequence is required for promoter activation *in vivo*. *Cell* **65**, 1023–31.

Dusserre, Y. and Mermod, N. (1992). Purified cofactors and histone H1 mediate transcriptional regulation by CTF/NF1. *Mol. Cell. Biol.* **12**, 5228–37.

Dyck, J. A., Maul, G. G., Miller, W. H.,Jr, Chen, J. D., Kakizuka, A. and Evans, R. M. (1994). A novel macromolecular structure is a target of the promyelocyte-retinoic acid receptor oncoprotein. *Cell* **76**, 333–43.

Dynan, W.S. and Tjian, R. (1985). Control of eukaryotic messenger RNA synthesis by sequence-specific DNA-binding proteins. *Nature (London)* **316**, 774–8.

Earnshaw, W.C. (1987). Anionic regions in nuclear proteins. *J. Cell Biol.* **105**, 1479–82.

Earnshaw, W.C. (1988). Mitotic chromosome structure. *Bioessays* **9**, 147–50.

Earnshaw, W.C. (1991). Large scale chromosome structure and organization. *Curr. Opin. Struct. Biol.* **1**, 237–44.

Earnshaw, W.C. and Heck, M.M.S. (1985). Localization of topoisomerase II in mitotic chromosomes. *J. Cell. Biol.* **100**, 1716–25.

Earnshaw, W.C., Honda, B.M., Laskey, R.A. and Thomas, J.O. (1980). Assembly of nucleosomes: the reaction involving X. *laevis* nucleoplasmin. *Cell* **21**, 373–83.

Earnshaw, W.C., Halligan, B., Cooke, C.A., Heck, M.M.S. and Liu, L.F. (1985). Topoisomerase II is a structural component of mitotic chromosome scaffolds. *J. Cell Biol.* **100**, 1706–15.

Ebralidse, K.K., Grachev, S.A. and Mirzabekov, A.D. (1988). A highly basic histone H4 domain bound to the sharply bent region of nucleosomal DNA. *Nature* **331**, 365–7.

Ebralidse, K.K., Hebbes, T.R., Clayton, A.L., Thorne, A.W. and Crane-Robinson, C. (1993). Nucleosomal structure at hyperacetylated loci probed in nuclei by DNA–histone crosslinking. *Nucleic. Acids Res.* **21**, 4734–8.

Echols, H. (1986). Multiple DNA–protein interactions governing high-precision DNA transactions. *Science* **233**, 1050–6.

Echols, H. (1990). Nucleoprotein structures initiating DNA replication, transcription and site-specific recombination. *J. Biol. Chem.* **265**, 14697–700.

Eckley, D.M, Ainsztein, A.M., McKay, A.M., Goldberg, I.G. and Earnshaw, W.C. (1996). Chromosomal proteins and cytokinesis: patterns of cleavage furrow formation and inner centromere protein positioning in mitotic heterokaryons and mid anaphase cells. *J. Cell Biol.* **136**, 1169–83.

Edgar, B.A. and Schubiger, G. (1986). Parameters controlling transcriptional activation during early *Drosophila* development. *Cell* **44**, 871–7.

Edgar, B.A., Kiehle, C.P. and Schubiger, G. (1986). Cell cycle control by the nucleo-cytoplasmic ratio in early *Drosophila* development. *Cell* **44**, 365–72.

Edgar, L.G. and McGhee, J.D. (1988). DNA synthesis and the control of embryonic gene expression in C. *elegans. Cell* **53**, 589–99.

Edmondson, S.P., Qui, L. and Shriver, J.W. (1995). Solution structure of the DNA-binding protein Sac7d from the hyperthermophile *Sulfolobus acidocaldarius. Biochemistry* **34**, 13289–304.

Edmondson, D.G., Smith, M.M. and Roth, S.Y. (1996). Repression domain of the yeast global repressor Tup1 interacts directly with histones H3 and H4. *Genes Dev.* **10**, 1247–59.

Eickbusch, T.H. and Moudrianakis, E.N. (1978). The histone core complex: an octamer assembled by two sets of protein–protein interactions. *Biochemistry* **17**, 4955.

Einck, L. and Bustin, M. (1985). The intracellular distribution and function of the high mobility group chromosomal proteins. *Exp. Cell Res.* **156**, 295–310.

Eisenmann, D.M., Dollard, C. and Winston, F. (1989). SPT15, the gene encoding the yeast TATA binding factor TFIID is required for normal transcription initiation *in vivo*. *Cell* **58**, 1183–91.

Eissenberg, J.C. and Elgin, S.C.R. (1991). Boundary functions in the control of gene expression. *Trends Genet.* **7**, 335–40.

Eissenberg, J.C., James, T.C., Foster, H.D.M., Hartnett, T., Ngan, V. and Elgin, S.C. (1990). Mutation in a heterochromatin-specific chromosomal protein is associated with suppression of position effect variegation in *Drosophila melanogaster*. *Proc. Natl Acad. Sci. USA* **87**, 9923–7.

Ekwall, K., Javerzat, J.P., Lorentz, A., Schmidt, H., Cranston, G. and Allshire, R.C. (1995). The chromodomain protein Swi6: a key component at fission yeast centromeres. *Science* **269**, 1429–31.

Elgin, S.C.R. (1988). The formation and function of DNase I hypersensitive sites in the process of gene activation. *J. Biol. Chem.* **263**, 19259–62.

Elgin, S.C.R. (1990). Chromatin structure and gene activity. *Curr. Opin. Cell. Biol.* **2**, 437–45.

Elliott, D. J., Stutz, F., Lescure, A. and Rosbash, M. (1994). mRNA nuclear export. *Curr. Opin. Genet. Dev.* **4**, 305–9.

Emerson, B.M., Lewis, C.D. and Felsenfeld, G. (1985). Interaction of specific nuclear factors with the nuclease-hypersensitive region of the chicken adult β-globin gene: nature of the binding domain. *Cell* **41**, 21–30.

Engelke, D.R., Ng S-Y., Shastry, B.S. and Roeder, R.G. (1980). Specific interaction of a purified transcription factor with an internal control region of 5S RNA genes. *Cell* **19**, 717–28.

Englander, E.W., Wolffe, A.P. and Howard, B.H. (1993). Nucleosome interactions with a human Alu element. Transcriptional repression and effects of template methylation. *J. Biol. Chem.* **268**, 19565–73.

Englert, C., Vidal, M., Maheswaran, S., Ge, Y., Ezzeli, R.M., Isselbacher, K.J. and Haber, D.A. (1995). Truncated WT1 mutants alter the subnuclear localization of the wild-type protein. *Proc. Natl Acad. Sci. USA* **92**, 11960–4.

Ephrussi, B. (1972). *Hybridisation of Somatic Cells*. Princeton University Press, Princeton, NJ.

Epstein, H., James, T.C. and Singh, P.B. (1992). Cloning and expression of *Drosophila* HP1 homologs from a mealybug, *Planococcus citri*. *J. Cell Sci.* **101**, 463–74.

Erard, M.S., Belenguer, P., Caizergues-Ferrer, M., Pantaloni, A. and Amalric, F. (1988). A major nucleolar protein, nucleolin, induces chromatin decondensation by binding to histone H1. *Eur. J. Biochem.* **175**, 525–30.

Ericsson, C., Mehlin, H., Björkroth, B., Lamb, M.M. and Daneholt, B. (1989). The ultrastructure of upstream and downstream regions of an active Balbiani ring gene. *Cell* **56**, 631–9.

Ericsson, C., Grossbach, U., Björkroth, B. and Daneholt, B. (1990). Presence of histone H1 on an active Balbiani ring gene. *Cell* **60**, 73–83.

Ernst, S.G., Miller, H., Brenner, G.A., Nocenta-McGrath, C., Francis, C. and McIsaac, R. (1987). Characterization of a cDNA clone coding for a sea urchin histone H2A variant related to the H2A.F/Z histone protein of verterbrates. *Nucleic. Acids Res.* **15**, 4629–44.

Evans, T. and Felsenfeld, G. (1989). The erythroid specific transcription factor Eryf1: a new finger protein. *Cell* **58**, 877–85.

Evans, T. and Felsenfeld, G. (1991). Trans-activation of a globin promoter in nonerythroid cells. *Mol. Cell Biol.* **11**, 843–53.

Evans, T., Reitman, M. and Felsenfeld, G. (1988). An erythrocyte specific DNA binding factor recognizes a regulatory sequence common to all chicken globin genes. *Proc. Natl Acad. Sci. USA* **85**, 5976–80.

Fakan, S. (1994). Perichromatin fibrils are in situ forms of nascent transcripts. *Trends Cell Biol.* **4**, 86–90.

Fakan, S., Leser, G. and Martin, T. E. (1984). Ultrastructural distribution of nuclear ribonucleoproteins as visualized by immunocytochemistry on thin sections. *J. Cell Biol.* **98**, 358–63.

Falkner, F.G. and Zachau, H.G. (1984). Correct transcription of an immunoglobulin K gene requires an upstream fragment containing conserved sequence elements. *Nature (London)* **310**, 71–4.

Fangman, W.L. and Brewer, B.J. (1991). Activation of replication origins within yeast chromosomes. *Annu. Rev. Cell. Biol.* **7**, 375–96.

Fascher, K.D., Schmitz, J. and Horz, W. (1990). Role of *trans*-activating proteins in the generation of active chromatin at the PHO 5 promoter in *S. cerevisiae*. *EMBO J.* **9**, 2523–8.

Fascher, K.D., Schmitz, J. and Horz, W. (1993). Structural and functional requirements for the chromatin transition at the PHO5 promoter in *Saccharomyces cerevisiae* upon PHO5 activation. *J. Mol. Biol.* **231**, 658–67.

Fedor, M.J., Lue, N.F. and Kornberg, R.D. (1988). Statistical positioning of nucleosomes by specific protein-binding to an upstream activating sequence in yeast. *J. Mol. Biol.* **204**, 109–27.

Felsenfeld, G. (1992). Chromatin: an essential part of the transcriptional apparatus. *Nature (London)* **355**, 219–24.

Felsenfeld, G. and McGhee, J.D. (1986). Structure of the 30 nm fibre. *Cell* **44**, 375–7.

Felsenfeld, G., Nickol, J., Behe, M., McGhee, J. and Jackson, D. (1983). Methylation and chromatin structure. *Cold Spring Harbor Symp. Quant. Biol.* **47**, 577–84.

Felts, S.J., Weil, P.A. and Chalkley, R. (1990). Transcription factor requirements for *in vitro* formation of transcriptionally competent 5S rRNA gene chromatin. *Mol. Cell. Biol.* **10**, 2390–401.

Feng, J. and Villeponteau, B. (1990). Serum stimulation of the *c-fos* enhancer induces reversible changes in *c-fos* chromatin structure. *Mol. Cell. Biol.* **10**, 1126–32.

Ferguson, B.M. and Fangman, W.L. (1992). A position effect on the time of replication origin activation in yeast. *Cell* **68**, 333–43.

Ferrari, S., Harley, V., Pontiggia, A., Goodfellow, P.N., Lovell-Badge, R. and Bianchi, M.E. (1992). SRY, like HMG1, recognizes sharp angles in DNA. *EMBO J.* **11**, 4497–506.

Ferreri, K., Gill, G. and Montminy, M. (1994). The cAMP regulated transcription factor CREB interacts with a component of the TFIID complex. *Proc. Natl Acad. Sci. USA* **91**, 1210–13.

Festenstein, R., Tolaini, M., Corbella, P., Mamaliki, C., Parrington, J., Fox, M., Millou, A., Jones, M. and Kioussis, D. (1996). Locus control region function and heterochromatin-induced position effect variegation. *Science* **271** 1123–5.

Fiering, S., Epner, E., Robinson, K., Zhuang, Y., Telling, A., Hu, M., Martin, D.I.K., Enver, T., Ley, T. and Groudine, M. (1995). Targeted deletion of 5′ HS2 of the murine β-globin locus reveals that it is not essential for proper regulation of the β-globin locus. *Genes Dev.* **9**, 2203–13.

Filipski, J., Leblanc, J., Youdale, T., Sikorska, M. and Walker, P.R. (1990). Periodicity of DNA folding in higher-order chromatin structures. *EMBO J.* **9**, 1319–27.

Finch, J.T., Lutter, L.C., Rhodes, D., Brown, A.S., Rushton, B., Levitt, M. and Klug, A. (1977). Structure of nucleosome core particles of chromatin. *Nature (London)* **269**, 29–36.

FitzGerald, P.C. and Simpson, R.T. (1985). Effects of sequence alterations in a DNA segment containing the 5S rRNA gene from *Lythechinus variegatus* on positioning of a nucleosome core particle *in vitro*. *J. Biol. Chem.* **260**, 15318–24.

Fitzsimmons, D.W. and Wolstenholme, G.E.W. (1976). The Structure and Function of Chromatin. *CIBA Found. Symp.* **28**, 368.

Fletcher, T.M. and Hansen, J.C. (1995). Core histone tail domains mediate oligonucleome folding and nucleosomal DNA organization through distinct molecular mechnaisms. *J. Biol. Chem.* **270**, 25359–62.

Fletcher, T.M. and Hansen, J.C. (1996). The nucleosomal array: structure/ function relationships. *Crit. Rev. in Euk. Gene Expression* **6**, 149–88.

Flores, O., Lu, H. and Reinberg, D. (1992). Factors involved in specific transcription initiation by RNA polymerase II: identification and characterization of factor IIH. *J. Biol. Chem.* **267**, 2786–93.

Flynn, J.M. and Woodland, H.R. (1980). The synthesis of histone H1 during early amphibian development. *Dev. Biol.* **75**, 222–30.

Foisner, R. and Gerace, L. (1993). Integral membrane proteins of the nuclear envelope interact with lamins and chromosomes, and binding is modulated by mitotic phosphorylation. *Cell* **73**, 1267–79.

Fong, H.K.W., Hurley, J.B., Hopkins, R.S., Miake-Lye, R., Johnson, M.S., Doolittle, R.F. and Simon, M. (1986). Repetitive segmental structure of the transducin β subunit: Homology with the CDC4 gene and identification of related mRNAs. *Proc. Natl Acad. Sci.* **83**, 2162–6.

Forbes, D.J., Kirschner, M.W. and Newport, J.W. (1983). Spontaneous formation of nucleus-like structures around bacteriophage DNA micro-injected into *Xenopus* eggs. *Cell* **34**, 13–23.

Forrester, W., Takagawa, S., Papayannopoulou, T., Stamatoyannopoulos, G. and Groudine, M. (1987). Evidence for a locus activation region: The formation of developmentally stable hypersensitive sites in globin-expressing hybrids. *Nucleic. Acids Res.* **15**, 10159–77.

Forrester, W.C., Epner, E., Driscoll, M.C., Enver, T., Brice, M., Papayannopou-lou, T. and Groudine, M. (1990). A deletion of the human β-globin locus activation region causes a major alteration in chromatin structure and replication across the entire β-globin region. *Genes Dev.* **4**, 1637–49.

Fotedar, R. and Roberts, J.M. (1989). Multistep pathway for replication dependent nucleosome assembly. *Proc. Natl Acad. Sci. USA* **86**, 6459–63.

Fox, C.A., Loo, S., Rivier, D.H., Foss, M.A. and Rine, J. (1993). A transcriptional silencer as a specialized origin of replication that establishes functional domains of chromatin. *Cold Spring Harbor Symp. Quant. Biol.* **58**, 443–55.

Fragoso, G., John, S., Robertis, S. and Hager, G.L. (1995). Nucleosome positioning on the MMTV LTR from the frequently-biased occupancy of multiple frames. *Genes Dev.* **9**, 1933–47.

Franke, A. (1991). The Polycomb protein: genetic and molecular analyses of interactions with other Polycomb-group genes in *Drosophila melanogaster.* Ph.D. Thesis. University of Heidelberg, Germany.

Franke, A., DeCamillis, M., Zink, D., Cheng, N., Brock, H. and Paro, R. (1992). *Polycomb* and polyhomeatic are constituents of a mullimeric protein complex in chromatin of *Drosophila melanogaster. EMBO J.* **11**, 2941–50.

Franke, W.W. (1974). Structure, biochemistry and functions of the nuclear envelope. *Int. Rev. Cytol. Suppl.* **4**, 71–236.

Franke, W.W. (1987). Nuclear lamins and cytoplasmic intermediate filament proteins: a growing multigene family. *Cell* **48**, 3–4.

Frankel, A.D. and Kim, P.S. (1991). Molecular structure of transcription factors: implications for gene regulation. *Cell* **65**, 717–19.

Freeman, L., Kurumizaka, H. and Wolffe, A.P. (1996). Functional assays for assembly of histones H3 and H4 into the chromatin of *Xenopus* embryos. *Proc. Natl Acad. Sci. USA* **93**, 12780–5.

Frey, M. R. and Matera, A. G. (1995). Coiled bodies contain U7 small nuclear RNA and associate with specific DNA sequences in interphase human cells. [published erratum appears in *Proc. Natl Acad. Sci. USA* 1995 **92**; 8532]. *Proc. Natl Acad. Sci. USA* **92**, 5915–19.

Fromental, C., Konoo, M., Nomiyama, H. and Chambon, P. (1988). Cooperativity and hierarchical levels of functional organization in the SV40 enhancer. *Cell* **54**, 943–53.

Fu, X. D. and Maniatis, T. (1990). Factor required for mammalian spliceosome assembly is localized to discrete regions in the nucleus. *Nature* **343**, 437–41.

Funabiki, H., Hagan, I., Uzawa, S. and Yanagida, M. (1993). Cell cycle-dependent specific positioning and clustering of centromeres and telomeres in fission yeast. *J. Cell Biol.* **121**, 961–76.

Fundele, R.H. and Surani, M.A. (1994). Experimental embryological analysis of genetic imprinting in mouse development. *Dev. Genet.* **15**, 515–22.

Gacy, A.M., Goellner, G., Juranic, N., Macura, S. and McMurray, C.T. (1995). Trinucleotide repeats that expand in human disease form hairpin structures in vitro. *Cell* **81**, 533–40.

Gaillard, P.-H.L., Martini, E.M.-D., Kaufman, P.D., Stillman, B., Moustacchi, E. and Almouzni, G. (1996). Chromatin assembly coupled to DNA repair: a new role for chromatin assembly factor I. *Cell* **86**, 887–96.

Gale, J.M. and Smerdon, M.J. (1988a). Photofootprint of nucleosome core DNA in intact chromatin having different structural states. *J. Mol. Biol.* **204**, 949–58.

Gale, J.M. and Smerdon, M.J. (1988b). UV-induced pyrimidine dimers and trimethylpsoralen crosslinks do not alter chromatin folding *in vivo. Biochemistry* **27**, 7197–205.

Gale, J.M. and Smerdon, M.J. (1990). UV induced (6-4) photoproducts are distributed differently than cyclobutane dimers in nucleosomes. *Photochem. Photobiol.* **51**, 411–17.

Gale, J.M., Nissen, K.A. and Smerdon, M.J. (1987). UV induced formation of pyrimidine dimers in nucleosome core DNA is strongly modulated with a period of 10.3 bases. *Proc. Natl Acad. Sci. USA* **84**, 6644–8.

Gall, J. G., Tsvetkov, A., Wu, Z. and Murphy, C. (1995). Is the sphere organelle/coiled body a universal nuclear component? *Dev. Genet.* **16**, 25–35.

Gallego, F., Fernandez-Busquets, X. and Daban, J.-R. (1995). Mechanism of nucleosome dissociation produced by transcription elongation in a short chromatin template. *Biochemistry* **34**, 6711–19.

Garcia-Ramirez, M., Dong, F. and Ausio, J. (1992). Role of the histone 'tails' in the folding of oligonucleosomes depleted of histone H1. *J. Biol. Chem.* **267**, 19587–95.

Garcia-Ramirez, M., Rocchini, C. and Ausio, J. (1995). Modulation of chromatin folding by histone acetylation. *J. Biol. Chem.* **270**, 17923–28.

Gargiulo, G. and Worcel, A. (1983). Analysis of chromatin assembled in germinal vesicles of *Xenopus* oocytes. *J. Mol. Biol.* **170**, 699–722.

Gargiulo, G., Razvi, F., Ruberti, I., Mohr, I. and Worcel, A. (1985). Chromatin-specific hypersensitive sites are assembled on a *Xenopus* histone gene injected into *Xenopus* oocytes. *J. Mol. Biol.* **181**, 333–49.

Garner, M.M. and Felsenfeld, G. (1987). Effect of ZDNA on nucleosome placement. *J. Mol. Biol.* **196**, 581–90.

Gasser, S.M. and Laemmli, U.K. (1986). The organization of chromatin loops: characterization of a scaffold attachment site. *EMBO J.* **5**, 511–18.

Gasser, S.M., Laroche, T., Falquet, J., Boy de la Tour, E. and Laemmli, U.K. (1986). Metaphase chromosome structure involvement of topoisomerase II. *J. Mol. Biol.* **188**, 613–29.

Gasser, S.M., Amati, B.B., Cardenas, M.E. and Hofmann, J.F. (1989). Studies on scaffold attachement sites and their relationship to genome function. *Int. Rev. Cytol.* **119**, 57–96.

Gaudreau, L., Schmid, A., Blaschke, D., Ptashne, M. and Horz, W. (1997). RNA polymerase II holoenzyme recruitment is sufficient to remodel chromatin at the yeast *PHO5* promoter. *Cell* **89**, 55–62.

Gaynor, R.B. and Berk, A.J. (1983). *Cis*-acting induction of adenovirus transcription. *Cell* **33**, 683–93.

Ge, H. and Roeder, R.G. (1994a). The high mobility protein HMG1 can reversibly inhibit class II gene transcription by interaction with the TATA-binding protein. *J. Biol. Chem.* **269**, 17136.

Ge, H. and Roeder, R.G. (1994b). Purification, cloning and characterization of a human coactivator PC4 that mediates transcriptinal activation of class II genes. *Cell* **78**, 515–23.

Georgakopoulos, T. and Thireos, G. (1992). Two distinct yeast transcriptional activators require the function of the GCN5 protein to promote normal levels of transcription. *EMBO J.* **11**, 4145–52.

Georgakopoulos, T., Gounalaki, N. and Thireos, G. (1995). Genetic evidence for the interaction of the yeast transcriptional co-activator proteins GCN5 and ADA2. *Mol. Gen. Genet.* **246**, 723–8.

Georgiev, G.P. (1969). Histones and the control of gene action. *Annu. Rev. Genet.* **3**, 155–80.

Gerace, L. and Burke, B. (1988). Functional organization of the nuclear envelope. *Annu. Rev. Cell Biol.* **4**, 335–74.

Gerber, J.K., Gogel, E., Berger, C., Wallisch, M., Muller, F., Grummt, I. and Grummt, F. (1997). Termination of mammalian rDNA replication: polar arrest of replication fork movement by transcription termination factor TTF-1. *Cell* **90**, 559–67.

Geyer, P.K. and Corces, V.G. (1992). DNA position-effect repression of transcription by a *Drosophila* zinc-finger protein. *Genes Dev.* **6**, 1865–73.

Giese, K., Cox, J. and Grosschedl, R. (1992). The HMG domain of lymphoid enhancer factor 1 bends DNA and facilitates assembly of functional nucleoprotein structures. *Cell* **69**, 185–95.

Gilbert, D.M. (1986). Temporal order of replication of *Xenopus laevis* 5S ribosomal RNA genes in somatic cells. *Proc. Natl Acad. Sci. USA* **83**, 2924–8.

Gilbert, D.M., Miyazawa, H., Nallaseth, F.S., Ortega, J.M., Blow, J.J. and De Pamphilis, M.L. (1993). Site-specific initiation of DNA replication in metazoan chromosomes and the role of nuclear organization. *Cold Spring Harbor Symp. Quant. Biol.* **58**, 475–85.

Gill, G., Pascal, E., Tseng, Z.H. and Tjian, R. (1994). A glutamine-rich hydrophobic patch in transcription factor Sp1 contacts the dTAFII110 component of the *Drosophila* TFIID complex and mediates transcriptional activation. *Proc. Natl Acad. Sci. USA* **91**, 192–6.

Gilmour, D.S. and Lis, J.T. (1986). RNA polymerase II interacts with the promoter region of the non-induced hsp 70 gene in *Drosophila melanogaster* cells. *Mol. Cell. Biol.* **6**, 3984–9.

Glass, J.R. and Gerace, L. (1990). Lamins A and C bind and assemble at the surface of mitotic chromosomes. *J. Cell. Biol.* **111**, 1047–57.

Glass, C.A., Glass, J.R., Taniura, H., Hasel, K.W., Blevitt, J.M. and Gerace, L. (1993). The α-helical rod domain of human lamins A and C contains a chromatin binding site. *EMBO J.* **12**, 4413.

Glikin, G.C., Ruberti, I. and Worcel, A. (1984). Chromatin assembly in *Xenopus* oocytes: *in vitro* studies. *Cell* **37**, 33–41.

Glotov, B.O., Itkes, A.V., Nikolaev, L.G. and Severin, E.S. (1978). Evidence for close proximity between histones H1 and H3 in chromatin of intact nuclei. *FEBS Lett.* **91**, 149–52.

Godde, J.S. and Wolffe, A.P. (1995). Disruption of reconstituted nucleosomes: the effect of particle concentration $MgCl_2$, and KCl concentration, the histone tails and temperature. *J. Biol. Chem.* **270**, 27399–402.

Godde, J.S. and Wolffe, A.P. (1996). Nucleosome assembly of CTG triplet repeats. *J. Biol. Chem.* **271**, 15222–9.

Godde, J.S., Nakatani, Y. and Wolffe, A.P. (1995). The amino-terminal tails of the core histones and the translational position of the TATA box determine TBP/TFIIA association with nucleosomal DNA. *Nucleic. Acids Res.* **23**, 4557–64.

Godde, J.S., Kass, S.U., Hirst, M.C. and Wolffe, A.P. (1996). Nucleosome assembly on methylated CGG triplet repeats in the Fragile X Mental Retardation Gene 1 promoter. *J. Biol. Chem.* **271**, 24325–8.

Goldberg, I.G., Sawhney, H., Pluta, A.F., Warburton, P.E. and Earnshaw, W.C. (1996). Surprising deficiency of CENP-B binding sites in African Green Monkey α-satellite DNA: implications for CENP-B function at centromeres. *Mol. Cell. Biol.* **16**, 5156–8.

Goldman, M.A., Holmquist, G.P., Gray, M.C., Caston, L.A. and Nag, A. (1984). Replication timing of genes and middle repetitive sequences. *Science* **224**, 686–92.

Gonzalez, P.J. and Palacian, E. (1989). Interaction of RNA polymerase II with structurally altered nucleosomal particles. *J. Biol. Chem.* **264**, 18457–62.

Goodrich, J.A. and Tjian, R. (1994). Transcription factor IIE and IIH and ATP hydrolysis direct promoter clearance by RNA polymerase II. *Cell* **77**, 145–56.

Goodrich, J.A., Hoey, T., Thut, C.J., Admon, A. and Tjian, R. (1993). *Drosophila* TAFII40 interacts with both VP16 activation domain and the basal transcription factor TFIIB. *Cell* **75**, 519–30.

Goodwin, G.H., Woodhead, L. and Johns, E.W. (1977). The presence of high mobility group non-histone proteins in isolated nucleosomes. *FEBS* **73**, 85.

Goppelt, A., Stelzer, G., Lottspeich, F. and Meisterernst, M. (1996). A mechanism for repression of class II gene transcription through specific binding of NC2 to TBP–promoter complexes via heterodimeric histone fold domains. *EMBO J.* **15** 3105–16.

Gorman, C.M., Howard, B.H. and Reeves, R. (1983). Expression of recombinant plasmids in mammalian cells is enhanced by sodium butyrate. *Nucleic. Acids Res.* **11**, 7631–48.

Gorovsky, M.A. (1986). *Molecular Biology of Ciliated Protozoa.* Academic Press, New York.

Gorovsky, M.A. (1973). Macro- and micronuclei of *Tetrahymena pyriformis*: a model system for studying the structure and function of eukaryotic nuclei. *J. Protozool.* **20**, 19–25.

Gorovsky, M.A., Pleger, G.L., Keevert, J.B. and Johmann, C.A. (1973). Studies on histone fraction F2A1 in macro and micronuclei of *Tetrahymena pyriformis*. *J. Cell. Biol.* **57**, 773–81.

Gotta, M., La voche, T., Formenton, A., Maillet, L., Scherthan, H. and Gasser, S.M. (1996). The clustering of telomeres and colocalization with Rap1, Sir3, and Sir4 proteins in wild type *Saccharomyces cerevisiae. J. Cell Biol.* **134**, 1349–63.

Gottesfeld, J. and Bloomer, L.S. (1982). Assembly of transcriptionally active 5S RNA gene chromatin *in vitro. Cell* **28**, 781–91.

Gottschling, D.E. (1992). Telomere-proximal DNA in *Saccharomyces cerevisiae* is refractory to methyltransferase activity *in vivo. Proc. Natl Acad. Sci. USA* **89**, 4062–5.

Gottschling, D.E. and Cech, T.R. (1984). Chromatin structure of the molecular ends of Oxytricha macronuclear DNA: phased nucleosomes and a telomeric complex. *Cell* **38**, 501–10.

Gottschling, D.E., Aparicio, O.M., Billington, B.L. and Zakian, V.A. (1990). Position effect at *S. cerevisiae* telomeres: reversible repression of pol II transcription. *Cell* **63**, 751–62.

Goustin, A.S. and Wilt, F.H. (1981). Protein synthesis, polyribosomes, and peptide elongation in early development of *Strongylocentrotus purpuralus. Dev. Biol.* **82**, 32–45.

Goytisolo, F.A., Gerchman, S.E., Yu, X., Rees, C., Graziano, V., Ramakrishnan, V. and Thomas, J.O. (1996). Identification of two DNA-binding sites on the globular domain of histone H5. *EMBO J.* **15**, 3421–9.

Gradwohl, G., De Murcia, J.M., Molinete, M., Simonin, F., Koken, M., Hoeijmakers, J.H.J. and De Murcia, G. (1990). The second zinc-finger domain of poly(ADP-ribose) polymerase determines specificity for single-stranded breaks in DNA. *Proc. Natl Acad. Sci. USA* **87**, 2990–4.

Graham, C.F, Arms, K. and Gurdon, J.B. (1966). The induction of DNA synthesis in frog egg cytoplasm. *Dev. Biol.* **14**, 349–81.

Gralla, J.D. (1991). Transcriptional control-lessons from an *E. coli* promoter data base. *Cell* **66**, 415–18.

Grant, S.G. and Chapman, V.M. (1988). Mechanisms of X-chromosome regulation. *Annu. Rev. Genet.* **22**, 199–233.

Graziano, V., Gerchman, S.E., Schneider, D.K. and Ramakrishnan, V. (1994). Histone H1 is located in the interior of the chromatin 30 nm filament. *Nature* **368**, 351–4.

Green, G.R. and Poccia, D.L. (1985). Phosphorylation of sea urchin sperm H1 and H2B histones precedes chromatin decondensation and H1 exchange during pronuclear formation. *Dev. Biol.* **108**, 235–45.

Grigliatti, T. (1991). Position-effect variegation: an assay for non-histone chromosomal proteins and chromatin assembly and modifying factors. *Methods Cell. Biol.* **35**, 587–627.

Gross, D.S. and Garrard, W.T. (1988). Nuclease hypersensitive sites in chromatin. *Annu. Rev. Biochem.* **57**, 159–97.

Gross, D.S., Adams, C.C., Lee, S. and Stentz, B. (1993). A critical role for heat shock transcription factor in establishing a nucleosome free region over the TATA-initiation site of the yeast HSP82 heat shock gene. *EMBO J.* **12**, 3931–45.

Grossbach, E.R., Bjorkroth, B. and Daneholt, B. (1990). Presence of histone H1 on an active Balbiani ring gene. *Cell* **60**, 78–83.

Grossbach, U. (1995). Selective distribution of histone H1 variants and high mobility group proteins in chromosomes. *Semin. in Cell Biol.* **6**, 237–46.

Grosschedl, R. and Marx, M. (1988). Stable propagation of the active transcriptional state of an immunoglobulin μ gene requires continuous enhancer function. *Cell* **55**, 645–54.

Grosschedl, R., Giese, K. and Pagel, J. (1994). HMG domain proteins: architectural elements in the assembly of nucleoprotein structures. *Trends Genet.* **10**, 94–100.

Grosveld, F., van Assendelft, G.B., Greaves, D.R. and Kollias, G. (1987). Position independent, high level expression of the human β-globin gene in transgenic mice. *Cell* **51**, 975–85.

Grosveld, F., Antoniou, M., Berry, M., de Boer, E., Dillou, N., Ellis, J., Fraser, P., Hurst, J., Iman, A., Meijer, D., Philipsen, S., Pruzina, S., Strouboulis, J. and Whyatt, D. (1993). Regulation of human globin gene switching. *Cold Spring Harbor Symp. Quant. Biol.* **58**, 7–15.

Groudine, M. and Conkin, K.F. (1985). Chromatin structure and *de novo* methylation of sperm DNA: implications for activation of the paternal genome. *Science* **228**, 1061–8.

Gruenbaum, Y., Szyf, M., Cedar, H. and Razin, A. (1983). Methylation of replicating and post replicated mouse L-cell DNA. *Proc. Natl Acad. Sci. USA* **80**, 4919–21.

Grunstein, M. (1990). Histone function in transcription. *Annu. Rev. Cell Biol.* **6**, 643–78.

Grunstein, M., Durrin, L.K., Mann, R.K., Fisher-Adams, G. and Johnson, L.M. (1992). Histones: regulators of transcription in yeast. In *Transcriptional Regulation*, ed. S. McKnight and K. Yamamoto. Cold Spring Harbor Press, New York, pp. 1295–315.

Grunstein, M., Hecht, A., Fisher-Adams, G., Wan, J., Mann, R.K., Strahl-Bolsinger, S., Laroche, T. and Gasser, S. (1995). The regulation of euchromatin and heterochromatin by histones in yeast. *J. Cell. Sci.* **519**, 29–36.

Gruss, C., Gutierrez, C., Burhans, W.C., DePamphilis, M.L., Koller, T. and Sogo, J.M. (1990). Nucleosome assembly in mammalian cell extracts before and after DNA replication. *EMBO J.* **9**, 2911–22.

Gruss, C., Wu, J., Koller, T. and Sogo, J.M. (1993). Disruption of nucleosomes at replication forks. *EMBO J.* **12**, 4533–45.

Guarente, L. (1995). Transcriptional coactivators in yeast and beyond. *Trends Biochem. Sci.* **20**, 517–27.

Guinta, D.E. and Korn, L.J. (1986). Differential order of replication of *Xenopus laevis* 5S RNA genes. *Mol. Cell. Biol.* **6**, 2537–42.

Guldner, H. H., Szostecki, C., Grotzinger, T. and Will, H. (1992). IFN enhance expression of Sp100, an autoantigen in primary biliary cirrhosis. *J. Immunol.* **149**, 4067–73.

Gurdon, J.B. (1962). Adult frogs derived from the nuclei of single somatic cells. *Dev. Biol.* **4**, 256–76.

Gurdon, J.B. (1963). Nuclear transplantation in amphibia and the importance of stable nuclear changes in promoting cellular differentiation. *Q. Rev. Biol.* **38**, 54–78.

Gurdon, J.B. (1968). Changes in somatic cell nuclei inserted into growing and maturing amphibian oocytes. *J. Embryol. Exp. Morphol.* **20**, 401–14.

Gurdon, J.B. (1974). *The Control of Gene Expression in Animal Development.* Oxford University Press, Oxford.

Gurdon, J.B. (1976). Injected nuclei in frog eggs: fate, enlargement and chromatin dispersal. *J. Embryol. Exp. Morphol.* **36**, 523–40.

Gurdon, J.B. and Brown, D.D. (1965). Cytoplasmic regulation of RNA synthesis and nucleolus formation in developing embryos of *Xenopus laevis. J. Mol. Biol.* **12**, 27–35.

Gurdon, J.B., Partington, G.A. and De Robertis, E.M. (1976). Injected nuclei in frog oocytes: RNA synthesis and protein exchange. *J. Embryol. Exp. Morphol.* **36**, 541–53.

Gurdon, J.B., Dingwall, C., Laskey, R.A. and Korn, L.J. (1982). Developmental inactivity of 5S RNA genes persists when chromosomes are cut between genes. *Nature (London)* **299**, 652–3.

Gurley, L.R., D'Anna, J.A., Barham, S.S., Deavan, L.L. and Tobey, R.A. (1978). Histone phosphorylation and chromatin structure during mitosis in Chinese hamster cells. *Eur. J. Biochem.* **84**, 1–15.

Gushchin, D.Y., Ebralidse, K.K. and Mirzabekov, A.D. (1991). Structure of the nucleosome: localization of the segments of the H2A histones that interact with DNA by DNA–protein crosslinking. *Molek. Biol.* **25**, 1400–11.

Guschin, D., Chandler, S. and Wolffe, A.P. (1998). Asymmetric linker histone association directs the asymmetric rearrangement of core histone interactions in a positioned nucleosome containing a thyroid hormone response element. *Biochemistry* (in press).

Ha, I., Roberts, S., Maldonado, E., Sun, X., Kim, L.U., Green, M. and Reinberg, D. (1993). Multiple functional domains of human transcription factor IIB: distinct interactions with two general transcription factors and RNA polymerase II. *Genes Dev.* **7**, 1021–32.

Hahn, S., Buratowski, S., Sharp, P.A. and Guarente, L. (1988). Isolation of the gene encoding yeast TATA binding protein TFIID: a gene identical to the SPT15 supressor of Ty element insertions. *Cell* **58**, 1173–81.

Hai T., Horikoshi, M., Roeder, R.G. and Green, M.R. (1988). Analysis of the role of the transcription factor ATF in the assembly of a functional preinitiation complex. *Cell* **54**, 1043–51.

Hamlin, J.L., Mosca, P.J., Dijkwel, P.A. and Lin, H.B. (1993). Initiation of replication at a mammalian chromosomal origin. *Cold Spring Harbor Symp. Quant. Biol.* **58**, 467–74.

Han, J., Hsu, C., Zhu, Z., Longshore, J.W. and Finley, W.H. (1994). Over-representation of the disease associated (CAG) and (CGG) repeats in the human genome. *Nucleic. Acids Res.* **22**, 1735–40.

Han, M. and Grunstein, M. (1988). Nucleosome loss activates yeast downstream promoters *in vivo. Cell* **55**, 1137–45.

Han, M., Chang, M., Kim, U.-J. and Grunstein, M. (1987). Histone H2B repression causes cell-cycle-specific arrest in yeast: effects on chromosomal segregation, replication and transcription. *Cell* **48**, 589–97.

Han, M., Kim, U.-J., Kayne, P. and Grunstein, M. (1988). Depletion of histone H4 and nucleosomes activates the PHO5 gene in *Saccharomyces cerevisiae. EMBO J.* **7**, 2221–8.

Handeli, S., Klar, A., Meuth, M. and Cedar, H. (1989). Mapping replication units in animal cells. *Cell* **57**, 909–20.

Hannon, R., Bateman, E., Allan, J., Harborne, N. and Gould, H. (1984). Control of RNA polymerase binding to chromatin by variations in linker histone composition. *J. Mol. Biol.* **180**, 131–49.

Hannon, R., Richards, E.G. and Gould, H. (1986). Facilitated diffusion of a DNA binding protein on chromatin. *EMBO J.* **5**, 3313–19.

Hansen, J.C. and Ausio, J. (1992). Chromatin dynamics and the modulation of genetic activity. *Trends Biochem. Sci.* **17**, 187–91.

Hansen, J.C. and Wolffe, A.P. (1992). The influence of chromatin folding on transcription initiation and elongation by RNA polymerase III. *Biochemistry* **31**, 7977–88.

Hansen, J.C. and Wolffe, A.P. (1994). A role for histones H2A/H2B in chromatin folding and transcriptional repression. *Proc. Natl Acad. Sci. USA* **91**, 2339–43.

Hansen, J.C., Ausio, J., Stanik, V.H. and van Holde, K.E. (1989). Homogenous reconstituted oligonucleosomes, evidence for salt-dependent folding in the absence of histone H1. *Biochemistry* **28**, 9129–36.

Hansen, J.C., van Holde, K.E. and Lohr, D. (1991). The mechanism of nucleosome assembly onto oligomers of the sea urchin 5S DNA positioning sequence. *J. Biol. Chem.* **266**, 4276–82.

Hard, T., Kellenbach, E., Boelens, R., Maler, B.A., Dahlman, K., Freedman, L.P., Carlstedt-Duke, J., Yamamoto, K.R., Gustafsson, J.A. and Kaptein, R. (1990). Solution structure of the glucocorticoid receptor DNA-binding domain. *Science* **249**, 157–60.

Hardeman, E.C., Chiu, C.P., Minty, A. and Blau, H.M. (1986). The pattern of actin expression in human fibroblast x mouse muscle heterokaryons suggests that human muscle regulatory factors are produced. *Cell* **47**, 123–30.

Harding, K. and Levine, M. (1988). Gap genes define the limits of *Antennapedia* and *bithorax* gene expression during early development in *Drosophila*. *EMBO J.* **7**, 205–14.

Harland, R.M. (1982). Inheritance of DNA methylation in microinjected eggs of Xenopus laevis. *Proc. Natl Acad. Sci. USA* **79**, 2323–7.

Harland, R.M., Weintraub, H. and McKnight, S.L. (1983). Transcription of DNA injected into *Xenopus* oocytes is influenced by template topology. *Nature (London)* **302**, 38–43.

Harrison, S.C. (1991). A structural taxonomy of DNA-binding domains. *Nature (London)* **353**, 715–19.

Hartl, P., Gottesfeld, J. and Forbes, D.J. (1993). Mitotic repression of transcription *in vitro*. *J. Cell. Biol.* **120**, 613–24.

Harvey, R.P., Whiting, J.A., Coles, L.S., Krieg, P.A. and Wells, J.R.E. (1983). H2A.F: an extremely variant histone H2A sequence expressed in the chicken embryo. *Proc. Natl Acad. Sci. USA* **80**, 2819–23.

Hastie, N. D. (1994). The genetics of Wilms' tumor – a case of disrupted development. *Annu. Rev. Genet.* **28**, 523–58.

Hatanaka, M. (1990). Discovery of the nucleolar targeting signal. *BioEssays* **12**, 143–8.

Hatch, C.L. and Bonner, W.M. (1988). Sequence of cDNAs for mammalian H2A.Z, an evolutionarily diverged but highly conserved basal histone H2A isoprotein species. *Nucleic. Acids Res.* **16**, 1113–24.

Hawley, D.K. and McClure, W.R. (1982). Mechanism of activation of transcription initiation from the λPrm promoter. *J. Mol. Biol.* **157**, 493–525.

Hayes, J.J. (1996). Site-directed cleavage of DNA by a linker histone-Fe(II) EDTA coinugate: localization of a globular domain binding site within a nucleosome. *Biochemistry* **35**, 11931–7.

Hayes, J.J. and Wolffe, A.P. (1992). Histones H2A/H2B inhibit the interaction of TFIIIA with 5S DNA in a nucleosome. *Proc. Natl Acad. Sci. USA* **89**, 1229–33.

Hayes, J.J. and Wolffe, A.P. (1993). Preferential and asymmetric interaction of linker histones with 5S DNA in the nucleosome. *Proc. Natl Acad. Sci. USA* **90**, 6415–19.

Hayes, J., Tullius, T.D. and Wolffe, A.P. (1989). A protein–protein interaction is essential for stable complex formation on a 5S RNA gene. *J. Biol. Chem.* **264**, 6009–12.

Hayes, J.J., Tullius, T.D. and Wolffe, A.P. (1990). The structure of DNA in a nucleosome. *Proc. Natl Acad. Sci. USA* **87**, 7405–9.

Hayes, J.J., Bashkin, J., Tullius, T.D. and Wolffe, A.P. (1991a). The histone core exerts a dominant constraint on the structure of DNA in a nucleosome. *Biochemistry* **30**, 8434–40.

Hayes, J.J., Clark, D.J. and Wolffe, A.P. (1991b). Histone contribution to the structure of DNA in a nucleosome. *Proc. Natl Acad. Sci. USA* **88**, 6829–33.

Hayes, J.J., Pruss, D. and Wolffe, A.P. (1994). Histone domains required to assemble a chromatosome including the *Xenopus borealis* somatic 5S rRNA gene. *Proc. Natl Acad. Sci. USA* **91**, 7817–21.

Hayes, J.J., Kaplan, R., Ura, K., Pruss, D. and Wolffe, A.P. (1996). A putative DNA binding surface in the globular domain of a linker histone is not essential for specific binding to the nucleosome. *J. Biol. Chem.* **271**, 25817–22.

Hebbes, T.R., Thorne, A.W. and Crane-Robinson, C. (1988). A direct link between core histone acetylation and transcriptionally active chromatin. *EMBO J.* **7**, 1395–402.

Hebbes, T.R., Thorne, A.W., Clayton, A.L. and Crane-Robinson, C. (1992). Histone acetylation and globin gene switching. *Nucleic. Acids Res.* **20**, 1017–22.

Hebbes, T.R., Clayton, A.L., Thorne, A.W. and Crane-Robinson, C. (1994). Core histone hyperacetylation co-maps with generalized DNase I sensitivity in the chicken β-globin chromosomal domain. *EMBO J.* **13**, 1823–30.

Hecht, A., Laroche, T., Strahl-Bolsinger, S., Gasser, S.M. and Grunstein, M. (1995). Histone H3 and H4 N-termini interact with SIR3 and SIR4 proteins: a molecular model for the formation of heterochromatin in yeast. *Cell* **80**, 583–92.

Hecht, A., Strahl-Bolsinger, S. and Grunstein, M. (1996). Spreading of the transcriptional repressor SIR3 from telomeric chromatin. *Nature* **383**, 92–6.

Heck, M.M.S. and Earnshaw, W.C. (1986). Topoisomerase II: a specific marker for cell proliferation. *J. Cell Biol.* **103**, 2569–81.

Heggeler-Bordier, B., Schild-Oulter, C., Chapel, S. and Wahli, W. (1995). Fate of linear and supercoiled multinucleosomic templates during transcription. *EMBO J.* **14** 2561–9.

Heinzel, T., Laviusky, R.M., Mullen, T.M., Soderstrom, M., Laherty, C.D., Torchia, J.T., Yang, W.-M., Brard, C., Ngo, S.G., Davie, J.R., Seto, E., Eisenman, R.M., Rose, D.W., Glass, C.K. and Rosenfeld, M.G. (1997). N-CoR, mSIN3, and histone deacetylase in a complex required for repression by nuclear receptors and Mad. *Nature* **387**, 43–8.

Heitz, E. (1928). Das Heterochromatin der Moose. I. *Jahrb. Wiss. Bot.* **69**, 762–819.

Heix, J., Zomerdijk, J.C., Ravanpay, A., Tjian, R. and Grummt, I. (1997). Cloning of murine RNA polymerase I – specific TAF factors: conserved interactions between the subunits of the species specific transcription initiation factor TIF-1B/SL-1. *Proc. Natl Acad. Sci. USA* **94**, 1733–8.

Henikoff, S. (1990). Position effect variegation after 60 years. *Trends Genet.* **6**, 422–6.

Henikoff, S. (1994). A reconsideration of the mechanisms of position effect. *Genetics* **138**, 1–5.

Hentschel, C.C. and Tata, J.R. (1978). Template-engaged and free RNA polymerases during *Xenopus* erythroid cell maturation. *Dev. Biol.* **65**, 496–507.

Hernandez, N. (1993). TBP, a universal eukaryotic transcription factor? *Genes Dev.* **7**, 1291–308.

Herrera, R.E., Shaw, P.E. and Nordheim, A. (1989). Occupation of the *c-fos* serum response element *in vivo* by a multi-protein complex is unaltered by growth factor induction. *Nature (London)* **340**, 68–70.

Herschbach, B.M., Arnaud, M.B. and Johnson, A.D. (1994). Transcriptional repression directed by the yeast α2 protein *in vitro*. *Nature* **370**, 309–11.

Herskovitz, M. and Riggs, A.D. (1995). Metaphase chromosome analysis by ligation-mediated PCR: heritable chromatin structure and a comparison of active and inactive X chromosomes. *Proc. Natl Acad. Sci. USA* **92**, 2379–83.

Herskowitz, I. (1989). A regulatory hierarchy for cell specialization in yeast. *Nature (London)* **342**, 749–57.

Herskowitz, I., Andrews, B., Kruger, W., Ogas, J., Sil, A., Coburn, C. and Peterson, C. (1992). Integration of multiple regulatory inputs in the control of HO expression in yeast. In *Transcriptional Regulation*, Vol. 2, eds S. McKnight and K. Yamamoto. Cold Spring Harbor Press, New York, pp. 949–74.

Heslop-Harrison, J.-S. and Bennett, M.D. (1990). Nuclear architecture in plants. *Trends Genet.* **6**, 401.

Hewish, D.R. and Burgoyne, L.A. (1973). Chromatin sub-structure: the digestion of chromatin DNA at regularly spaced sites by a nuclear deoxyribonuclease. *Biochem. Biophys. Res. Commun.* **52**, 504–10.

Hill, C.S. and Thomas, J.O. (1990). Core histone–DNA interactions in sea urchin sperm chromatin. The N-terminal tail of H2B interacts with linker DNA. *Eur. J. Biochem.* **187**, 145–53.

Hill, C.S., Rimmer, J.M., Green, B.N., Finch, J.T. and Thomas, J.O. (1991). Histone–DNA interactions and their modulation by phosphorylation of Ser-Pro-X-Lys/Arg-motifs. *EMBO J.* **10**, 1939–48.

Hirano, T. and Mitchison, T.J. (1991). Cell cycle control of higher-order chromatin assembly around naked DNA *in vitro*. *J. Cell Biol.* **115**, 1479–89.

Hirano, T. and Mitchison, T.J. (1993). Topoisomerase II does not play a scaffolding role in the organization of mitotic chromosomes assembled in *Xenopus* egg extracts. *J. Cell Biol.* **120**, 601–12.

Hirano, T. and Mitchison, T.J. (1994). A heterodimeric coiled-coil protein required for mitotic chromosome condensation *in vitro*. *Cell* **79**, 449–58.

Hirschhorn, J.N., Brown, S.A., Clark, C.D. and Winston, F. (1992). Evidence that SNF2/SWI2 and SNF activate transcription in yeast by altering chromatin structure. *Genes Dev.* **6**, 2288–98.

Hiyoshi, H., Yokata, T., Katagari, C., Nishida, H., Takai, M., Agata, K., Eguchi, G. and Abe, S. (1991). Isolation of cDNA for a *Xenopus* sperm-specific basic nuclear protein (SP4) and evidence for expression of SP4 mRNA in primary spermatocytes. *Exp. Cell Res.* **194**, 95–9.

Hochstrasser, M., Mathog, D., Gruenbaum, Y., Saumweber, H. and Sedat, J. W. (1986). Spatial organization of chromosomes in the salivary gland nuclei of *Drosophila melanogaster*. *J. Cell Biol.* **102**, 112–23.

Hock, R., Moorman, A., Fischer, D. and Scheer, U. (1993). Absence of somatic histone H1 in oocytes and preblastula embryos of *Xenopus laevis*. *Dev. Biol.* **158**, 510–22.

Hoeller, M., Westin, G., Jiricny, J. and Schaffner, W. (1988). Sp1 transcription factor binds DNA and activates transcription even when the binding site is CpG methylated. *Genes Dev.* **2**, 1127–35.

Hoey, T., Weinzieri, R.O.J., Gill, G., Chen, J., Dynlacht, B.D. and Tjian, R. (1993). Molecular cloning and functional analysis of *Drosophila* TAF110 reveal properties expected of coactivators. *Cell* **72**, 247–60.

Hogan, M.E., Rooney, T.F. and Austin, R.H. (1987). Evidence for kinks in DNA folding in the nucleosome. *Nature (London)* **328**, 554–7.

Holdridge, C. and Dorsett, D. (1991). Repression of *hsp70* heat shock gene transcription by the suppressor of hairy-wing protein of *Drosophila melanogaster*. *Mol. Cell. Biol.* **11**, 1894–900.

Holliday, R. (1987). Inheritance of epigenetic defects. *Science* **238**, 163–70.

Holm, C., Goto, T., Wang, J.C. and Botstein, D. (1985). DNA topoisomerase II is required at the time of mitosis in yeast. *Cell* **41**, 553–63.

Holmes, S.G. and Broach, J.R. (1996). Silencers are required for inheritance of the repressed state in yeast. *Genes Dev.* **10**, 1021–32.

Hood, L., Kronenberg, M. and Hunkapiller (1985). T cell antigen receptors and the immunoglobulin supergene family. *Cell* **40**, 225–9.

Horikoshi, M., Carey, M.F., Kakidani, H. and Roeder, R.G. (1988). Mechanism of action of a yeast activator: direct effect of GAL4 derivatives on mammalian TFIID promoter interactions. *Cell* **54**, 665–9.

Horiuchi, J., Silverman, N., Marcus, G.A. and Guarente, L. (1995). ADA3, a putative transcriptional adaptor, consists of two separable domains and interacts with ADA2 and GCN5 in a trimeric complex. *Mol. Cell Biol.* **15**, 1203–9.

Horlein, A.J., Naar, A.M., Heinzel, T., Torchia, J., Gloss, B., Kurokawa, R., Ryan, A., Kamei, Y., Soderstrom, M., Glass, C.K. and Rosenfeld, M.G. (1995). Ligand-independent repression by the thyroid hormone receptor mediated by a nuclear receptor co-repressor. *Nature* **377**, 397–404.

Horowitz, D.S. and Wang, J.C. (1984). Torsional rigidity of DNA and length dependence of the free energy of DNA supercoiling. *J. Mol. Biol.* **173**, 75–91.

Horowitz, H. and Platt, T. (1982). Regulation of transcription from tandem and convergent promoters. *Nucleic. Acids Res.* **10**, 5447–65.

Horowitz, R.A., Agard, D.A., Sedat, J.W. and Woodcock, C.L. (1994). The three dimensional architecture of chromatin *in situ*: electron tomography reveals fibers composed of a continuously variable zig-zag nucleosomal ribbon. *J. Cell Biol.* **125**, 1–10.

Horsley, D., Hutchings, A., Butcher, G.W. and Singh, P.B. (1996). M32, a murine homologue of *Drosophila* heterochromatin protein 1 (HP1), localises to enchromatin within interphase nuclei and is largely excluded from constitutive heterochromatin cytogenet. *Cell Genet.* **73**, 308–11.

Howlett, S.K., Barton, S.C. and Surani, M.A. (1987). Nuclear cytoplasmic interactions following nuclear transplantation in mouse embryos. *Development* **101**, 915–23.

Hozak, P., Hassan, A.B., Jackson, D.A. and Cook, P.R. (1993). Visualization of replication factories attached to a nucleoskeleton. *Cell* **73**, 361–73.

Hsieh, CH. and Griffith, J.D. (1988). The terminus of SV40 DNA replication and transcription contains a sharp sequence-directed curve. *Cell* **52**, 535–44.

Hsieh, C.-L. (1994). Dependence of transcriptional repression on CpG methylation density. *Mol. Cell. Biol.* **14**, 5487–94.

Hu, C.H., McStay, B., Jeong, S.-Y. and Reeder, R.H. (1994). xUBF, an RNA polymerase I transcription factor, binds crossover DNA with low sequence specificity. *Mol. Cell. Biol.* **14**, 2871–82.

Huang, S. and Spector, D. L. (1991). Nascent pre-mRNA transcripts are associated with nuclear regions enriched in splicing factors. *Genes. Dev.* **5**, 2288–302.

Huang, S., Deerinck, T. J., Ellisman, M. H. and Spector, D. L. (1994). *In vivo* analysis of the stability and transport of nuclear poly(A)+ RNA. *J. Cell Biol.* **126**, 877–99.

Hughes, T.A., Pombo, A., McManus, J., Hozak, P., Jackson, D.A. and Cook, P.R. (1995). On the structure of replication and transcription factories. *J. Cell Sci.* **19**, 59–65.

Huletsky, A., De Murcia, G., Muller, S., Hengartner, M., Menard, L., Lamarre, D. and Poirier, G.G. (1989). The effect of poly(ADP-ribosyl)ation on native and H1-depleted chromatin. A role of poly(ADP-ribosyl)ation on core nucleosome structure. *J. Biol. Chem.* **264**, 8878–86.

Hurlin, P.J., Queva, C., Koskinen, P.J., Steingrimsson, E., Ayer, D.E., Copeland, NG., Jenkins, N.A. and Eisenman, R.N. (1995). Mad3 and Mad4: novel Max-interacting transcriptional repressors that suppress c-Myc-dependent transformation and are expressed during neural and epidermal differentiation. *EMBO J.* **14**, 5646–59.

Hutchinson, C.J., Cox, R. and Ford, C.C. (1988). The control of DNA replication in a cell-free extract that recapitulates a basic cell cycle *in vitro*. *Development* **103**, 553–66.

Hutchison, N. and Weintraub, H. (1985). Localization of DNAase I-sensitive sequences to specific regions of interphase nuclei. *Cell* **43**, 471–82.

Hyrien, O., Maric, C. and Mechali, M. (1995). Transition in specification of embryonic metazoan DNA replication origins. *Science* **270**, 994–7.

Ichimura, S., Mita, K. and Zama, M. (1982). Essential role of arginine residues in the folding of deoxyribonucleic acid into nucleosome cores. *Biochemistry* **21**, 5329–34.

Iguchi-Ariga, S.M.M. and Schaffner, W. (1989). CpG methylation of the cAMP responsive enhancer/promoter sequence TGACGTCA abolishes specific factor binding as well as transcriptional activation. *Genes Dev.* **3**, 612–19.

Ikegami, S., Ooe, Y., Shimizu, T., Kasahara, T., Tsuruta, T., Kijima, M., Yoshida, M. and Beppu, T. (1993). Accumulation of multiacetylated forms of histones by trichostatin A and its developmental consequences in early starfish embryos. *Roux's Arch. Dev. Biol.* **202**, 144–51.

Illmensee, K. (1978). *Drosophila* chimeras and the problem of determination. In *Genetic Mosaics and Cell Differentiation*, vol. 9, ed. W.J. Gehring, Springer Verlag, Berlin, 51–69.

Imbalzano, A.M., Kwon, H., Green, M.R. and Kingston, R.E. (1994). Facilitated binding of TATA-binding protein to nucleosomal DNA. *Nature* **370**, 481–5.

Imhof, A., Yang, X.J., Ogryzko, V.V., Nakatani, Y., Wolffe, A.P. and Ge, H. (1997). Acetylation of general transcription factors by histone acetyltransferases: identification of a major site of acetylation in TFIIEβ. *Curr. Biol.* **7**, 689–92.

Ingles, C.J., Shales, M., Cress, W.D., Triezenberg, S.J. and Greenblatt, J. (1991). Reduced binding of TFIID to transcriptionally compromised mutants of VP16. *Nature* **351**, 588–90.

Irish, V.F., Martinez-Arias, A. and Akam, M. (1989). Spatial regulation of the *Antennapedia* and *Ultrabithorax* homeotic genes during *Drosophila* early development. *EMBO J.* **8**, 1527–37.

Isackson, P.J., Debold, W.A. and Reeck, G.R. (1980). Isolation and separation of chicken erythrocyte high mobility group non-histone chromatin proteins by chromatography on phosphocellulose. *FEBS Lett.* **119**, 337.

Ito, T., Bulger, M., Kobayashi, R. and Kadonaga, J.T. (1996a). Drosophila NAP-1 is a core histone chaperone that functions in ATP-facilitated assembly of regulatory spaced nucleosmal arrays. *Mol. Cell Biol.* **16**, 3112–24.

Ito, T., Tyler, J.K., Bulger, M., Kobayashi, R. and Kadonaga, J.T. (1996b). ATP facilitated chromatin assembly with a nucleoplasmin like protein from *Drosophila melanogaster. J. Biol. Chem.* **271**, 25041–8.

Ito, T., Bulger, M., Pazin, M.J., Kobayashi, R. and Kadonaga, J.T. (1997). ACF, an ISWI-containing and ATP utilizing chromatin assembly and remodeling factor. *Cell* **90**, 145–55.

Izaurralde, E., Kas, E. and Laemmli, U.K. (1989). Highly preferential nucleation of histone H1 assembly on scaffold associated regions. *J. Mol. Biol.* **210**, 573–85.

Izaurralde, E., Lewis, J., McGuigan, C., Jankowska, M., Darzynkiewicz, E. and Mattaj, I. W. (1994). A nuclear cap binding protein complex involved in pre-mRNA splicing. *Cell* **78**, 657–68.

Izban, M.G. and Luse, D.S. (1991). Transcription of nucleosomal templates by RNA polymerase II *in vitro*: inhibition of elongation with enhancement of sequence-specific pausing. *Genes Dev.* **5**, 683–96.

Jack, J., Dorsett, D., Delotto, Y. and Liu, S. (1991). Expression of the *cut* locus in the *Drosophila* wing margin is required for cell type specification and is regulated by a distal enhancer. *Development* **113**, 735–48.

Jackson, D.A. and Cook, P.R. (1986). Replication occurs at a nucleoskeleton. *EMBO J.* **5**, 1403–11.

Jackson, D.A., Yuan, J. and Cook, P.R. (1988). A gentle method for preparing cyto- and nucleoskeletons and associated chromatin. *J. Cell Sci.* **90**, 365–78.

Jackson, D.A., Dickinson, P. and Cook, P.R. (1990). The size of chromatin loops in HeLa cells. *EMBO J.* **9**, 567–71.

Jackson, D.A., Hassan, A.B., Errington, R.J. and Cook, P.R. (1993). Visualization of focal sites of transcription within human nuclei. *EMBO J.* **12**, 1059–65.

Jackson, J.B., Pollock Jr, J.M. and Rill, R.L. (1979). Chromatin fractionation procedure that yields nucleosomes containing near-stoichiometric amounts of high mobility group non histone chromosomal proteins. *Biochemistry* **18**, 3739.

Jackson, J.R. and Benyajati, C. (1993). Histone interactions are sufficient to position a single nucleosome juxtaposing *Drosophila Adh* adult enhancer and distal promoter. *Nucleic. Acids Res.* **21**, 957–67.

Jackson, V. (1987). Deposition of newly synthesized histones: new histones H2A and H2B do not deposit in the same nucleosome with new histones H3 and H4. *Biochemistry* **26**, 2315–25.

Jackson, V. (1988). Deposition of newly synthesized histones: hybrid nucleosomes are not tandemly arranged on daughter DNA strands. *Biochemistry* **27**, 2109–20.

Jackson, V. (1990). *In vivo* studies on the dynamics of histone–DNA interaction: evidence for nucleosome dissolution during replication and transcription and a low level of dissolution independent of both. *Biochemistry* **29**, 719–31.

Jackson, V., Shires, A., Tanphaichitr, N. and Chalkley, R. (1976). Modification to histones immediately after synthesis. *J. Mol. Biol.* **104**, 471–83.

Jacobovits, E.B., Bratosin, S. and Aloni, Y. (1980). A nucleosome free region in SV40 minichromosomes. *Nature (London)* **285**, 263–5.

Jaehning, J.A. and Roeder, R.G. (1977). Transcription of specific adenoviral genes in isolated nuclei by exogenous RNA polymerases. *J. Biol. Chem.* **252**, 8753–61.

Jaenisch, R. (1997). DNA methylation and imprinting: who needs it anyway? *Trends Genet.* **13**, 323–9.

James, T.C. and Elgin, S.C.R. (1986). Identification of a non histone chromosomal protein associated with heterochromatin in *Drosophila melanogaster* and its gene. *Mol. Cell. Biol.* **6**, 3862–72.

Janknecht, R. and Hunter, T. (1996). A growing coactivator network. *Nature* **383**, 22–3.

Jantzen, H.M., Admon, A., Bell, S.P. and Tjian, R. (1990). Nuclear transcription factor hUBF contains a DNA-binding motif with homology to HMG proteins. *Nature (London)* **344**, 830–6.

Jantzen, K., Fritton, H.P. and Igo-Kemenes, T. (1986). The DNase I sensitive domain of the chicken lysozyme gene span 24 kb. *Nucleic. Acids Res.* **14**, 6085–99.

Jensen, K.A. and Smerdon, M.J. (1990). DNA repair within nucleosome cores of UV-irradiated human cells. *Biochemistry* **29**, 4773–82.

Jenuwein, T., Forrester, W.C., Qui, R.-G. and Grosschedl, R. (1993). The immunoglobulin μ enhancer core establishes local factor access in nuclear chromatin independent of transcriptional stimulation. *Genes Dev.* **7**, 2016–32.

Jeppesen, P. and Turner, B.M. (1993). The inactive X chromosome in female mammals is distinguished by a lack of histone H4 acetylation, a cytogenetic marker for gene expression. *Cell* **74**, 281–91.

Jerzmanowski, A. and Cole, R.D. (1990). Flanking sequences of *Xenopus* 5S RNA genes determine differential inhibition of transcription by H1 histone *in vitro*. *J. Biol. Chem.* **265**, 10726–32.

Jiang, Y.W. and Stillman, D.J. (1992). Involvement of the SIN4 global transcriptional regulatory in the chromatin structure of *Saccharomyces cerevisiae*. *Mol. Cell Biol.* **12**, 4503–14.

Jiang, Y.W. and Stillman, D.J. (1995). Regulation of HIS4 expression by the *Saccharomyces cerevisiae* SIN transcription regulatory. *Genetics* **140**, 103–14.

Jimenez-Garcia, L. F., Segura-Valdez, M. L., Ochs, R. L., Rothblum, L. I., Hannan, R. and Spector, D. L. (1994). Nucleologenesis: U3 snRNA-containing prenucleolar bodies move to sites of active pre-rRNA transcription after mitosis. *Mol. Biol. Cell* **5**, 955–66.

Johns, E.W. (1982). *The HMG Chromosomal Proteins*. Academic Press, New York.

Johnson, L.M., Kayne, P.S., Kahn, E.S. and Grunstein, M. (1990). Genetic evidence for an interaction between SIR3 and histone H4 in the repression of silent mating loci in *Saccharomyces cerevisiae*. *Proc. Natl Acad. Sci. USA* **87**, 6286–90.

Johnson, P. and McKnight, S. (1989). Eucaryotic transcriptional regulatory proteins. *Annu. Rev. Biochem.* **58**, 799–839.

Johnson, T.C. and Holland, J.J. (1965). Ribonucleic acid and protein synthesis in mitotic HeLa cells. *J. Cell Biol.* **27**, 565–74.

Jones, P.A. (1985). Altering gene expression with 5-Azacytidine. *Cell* **40**, 485–6.

Jones, P.L., Veenstra, G.J.C., Wade, P.A. Vermaak, D., Kass, S.U., Landsberger, N., Strouboulis, J. and Wolffe, A.P. (1998). Methylated DNA and MeCP2 recruit histone deacetylase to repress transcription. *Nature Genet.* (in press).

Jost, J.-P. and Hofsteenge, J. (1992). The repressor MDBP-2 is a member of the histone H1 family that binds preferentially *in vitro* and *in vivo* to methylated non specific DNA sequences. *Proc. Natl Acad. Sci. USA* **89**, 9499–503.

Jutglar, L., Borell, J.I. and Ausio, J. (1991). Primary, secondary and tertiary structure of the core of a histone H1-like protein from the sperm of *Mytilus. J. Biol. Chem.* **266**, 8184–91.

Kadonaga, J.T. (1990). Assembly and disassembly of the *Drosophila* RNA polymerase II complex during transcription. *J. Biol. Chem.* **265**, 2624–31.

Kadosh, D., and Struhl, K. (1997). Repression by Ume6 involves recruitment of a complex containing Sin3 corepressor and Rpd3 histone deacetylase to target promoters. *Cell* **89**, 365–71.

Kaffman, A., Herskowitz, I., Tjian, R. and O'Shea, E.K. (1994). Phosphorylation of the transcription factor PHO4 by a cyclinCDK complex, PHO80PHO85. *Science* **263**, 1153–6.

Kamakaka, R.T. and Thomas, J.O. (1990). Chromatin structure of transcriptionally competent and repressed genes. *EMBO J.* **9**, 3997–4006.

Kamakaka, R.T., Bulger, M. and Kadonaga, J.T. (1993). Potentiation of RNA polymerase II transcription by Gal4-VP16 during but not after DNA replication and chromatin assembly. *Genes Dev.* **7**, 1779–95.

Kamakaka, R.T., Bulger, M., Kaufman, P.D., Stillman, B. and Kadonaga, J.T. (1996). Post-replicative chromatin assembly by Drosophila and human chromatin assembly factor-1. *Mol. Cell Biol.* **16**, 810–17.

Kamei, Y., Xu, L., Heinzel, T., Torchia, J., Kurokama, R., Gloss, B., Lin, S.-C., Heyman, R.A., Rose, D.W., Glass, C.K., and Rosenfeld, M.G. (1996). A CBP integrator complex mediates transcriptional activation and AP-1 inhibition by nuclear receptors. *Cell* **85**, 403–14.

Kandolf H. (1994). The H1A histone variant is the *in vivo* repressor of oocyte-type 5S gene transcription in *Xenopus laevis* embryos. *Proc. Natl Acad. Sci. USA* **91**, 7257–60.

Karantza, V., Freire, E. and Moudrianakis, E.N. (1996). Thermodynamic studies of the core histones: pH and ionic strength effects on the stability of the (H3-H4)/(H3-H4)$_2$ system. *Biochemistry* **35**, 2037–46.

Karpen, G.H. and Spradling, A.C. (1990). Reduced DNA polytenization of a minichromosome region undergoing position effect variegation in *Drosophila. Cell* **63**, 97–107.

Karpov, V.L., Preobrazhenskaya, O.V. and Mirzabekov, A.D. (1984). Chromatin structure of hsp 70 genes, activated by heat shock: selective removal of histones from the coding region and their absence from the 5' region. *Cell* **36**, 423–31.

Kass, K.U., Goddard, J.P. and Adams, R.L.P. (1993). Inactive chromatin spreads from a focus of methylation. *Mol. Cell Biol.* **13**, 7372–9.

Kass, S., Tyc, K., Steitz, J. A. and Sollner-Webb, B. (1990). The U3 small nucleolar ribonucleoprotein functions in the first step of preribosomal RNA processing. *Cell* **60**, 897–908.

Kass, S.U., Landsberger, N. and Wolffe, A.P. (1997a). DNA methylation directs a time-dependent repression of transcription initiation. *Curr. Biol.* **7**, 157–65.

Kass, S.U., Pruss, D. and Wolffe, A.P. (1997b) How does DNA methylation repress transcription? *Trends Genet.* **13**, 444–9.

Kassavetis, G.A., Braun, B.R., Nguyen, L.H. and Geiduschek, E.P. (1989). *S. cerevisiae* TFIIIB is the transcription initiation factor proper of RNA polymerase III, while TFIIIA and TFIIIC are assembly factors. *Cell* **60**, 235–45.

Kassavetis, G.A., Nguyen, S.T., Kobayashi, R., Kumar, A., Geiduschek, E.P. and Pisano, M. (1995). Cloning, expression and function of TFC5, the gene encoding the B″ component of the Saccharomyces cerevisiae RNA polymerase III transcription factor TFIIIB. *Proc. Natl Acad. Sci. USA* **92**, 9786–90.

Kassavetis, G.A., Bardeleben, C., Kumar, A., Ramirez, E. and Geiduschek, E.P. (1997). Domains of the Brf component of RNA polymerase III transcription factor IIIB (TFIIIB): functions in assembly of TFIIIB–DNA complexes and recruitment of RNA polymerase to the promoter. *Mol. Cell Biol.* **17**, 5299–306.

Kauffmann, S.A. (1980). Heterotopic transplantation in the syncytial blastoderm of *Drosophila*: evidence for anterior and posterior nuclear commitments. *Roux's Archiv. Dev. Biol.* **189**, 135–45.

Kaufman, P.D. and Botchan, M.R. (1994). Assembly of nucleosomes: do multiple assembly factors mean multiple mechanisms? *Curr. Opin. Genet. Dev.* **4**, 229–35.

Kaufman, P.D., Kobayashi, R., Kessler, N. and Stillman, B. (1995). The p150 and p60 subunits of chromatin assembly factor I: a molecular link between newly synthesized histones and DNA replications. *Cell* **81**, 1105–14.

Kaufman, T.C., Lewis, R. and Wakimoto, B. (1980). Cytogenetic analysis of chromosome 3 in *Drosophila melanogaster*: the momeotic gene complex in polytene chromosome interval 84A-B. *Genetics* **94**, 115–33.

Kayne, P.S., Kim, U.-J., Han, M., Mullen, J.R., Yoshizaki, F. and Grunstein, M. (1988). Extremely conserved histone H4 N terminus is dispensable for growth but essential for repressing the silent mating loci in yeast. *Cell* **55**, 27–39.

Keck, J.G., Baldick, Jr. C.J. and Moss, B. (1990). Role of DNA replication in vaccinia virus gene expression: a naked template is required for transcription of three late *trans*-activator genes. *Cell* **61**, 801–9.

Kedes, L.H. and Gross, P.R. (1969). Synthesis and function of messenger RNA during early embryonic development. *J. Mol. Biol.* **42**, 559–675.

Keleher, C.A., Goutte, C. and Johnson, A.D. (1988). The yeast cell-type-specific repressor alpha 2 acts cooperatively with a non-cell-type-specific protein. *Cell* **53**, 927–36.

Keleher, C.A., Redd, M.J., Schultz, J., Carlson, M. and Johnson, A.D. (1992). Ssn6-Tup1 is a general repressor of transcription in yeast. *Cell* **68**, 709–19.

Kelley, D.E., Pollock, B.A., Atchison, M.L. and Perry, R.T. (1988). The coupling between enhancer activity and hypomethylation of κ immunoglobulin genes is developmentally regulated. *Mol. Cell Biol.* **8**, 930–7.

Kelly, C., van Driel, R. and Wilkinson, G. W. (1995). Disruption of PML-associated nuclear bodies during human cytomegalovirus infection. *J. Gen. Virol.* **76**, 2887–93.

Kellum, R. and Schedl, P. (1991). A position effect assay for boundaries of higher order chromosomal domains. *Cell* **64**, 941–50.

Kellum, R. and Schedl, P. (1992). A group of scs elements function as domain boundaries in an enhancer-blocking assay. *Mol. Cell. Biol.* **12**, 2424–31.

Kempnauer, K.-H., Fanning, E., Otto, B. and Knippers, R. (1980). Maturation of newly replicated chromatin of simian virus 40 and its host cell. *J. Mol. Biol.* **136**, 359–74.

Kennedy, B.K., Gotta, M., Sinclair, D.A., Mills, K., McNabb, D.S., Murthy, M., Dak, S.M., LaRoche, T., Gasser, S.M. and Guarente, L. (1997). Redistribution of silencing proteins from telomeres to the nucleolus is associated with extension of lifespan in *S. cerevisiae*. *Cell* **89**, 381–91.

Kennison, J.A. (1993). Transcriptional activation of *Drosophila* homeotic genes from distant regulatory elements. *Trends Genet.* **9**, 75–9.

Kennison, J.A. and Tamkun, J.W. (1988). Dosage-dependent modifiers of *Polycomb* and *Antennapedia* mutations in *Drosophila*. *Proc. Natl Acad. Sci. USA* **85**, 8136–40.

Kerrigan, L.A., Croston, G.E., Lira, L.M. and Kadonaga, J.T. (1991). Sequence-specific antirepression of the *Drosophila* Kruppel gene by the GAGA factor. *J. Biol. Chem.* **266**, 574–82.

Keshet, I., Lieman-Hurwitz, J. and Cedar, H. (1986). DNA methylation affects the formation of active chromatin. *Cell* **44**, 535–43.

Khavari, P.A., Peterson, C.L., Tamkun, J.W., Mendel, D.B. and Crabtree, G.R. (1993). BRG1 contains a conserved domain of the SWI2/SNF2 family necessary for normal mitotic growth and transcription. *Nature* **366**, 170–4.

Khochbin, S. and Wolffe, A.P. (1993). Developmental regulation and butyrate inducible transcription of the *Xenopus* histone H1° promoter. *Gene* **128**, 173–80.

Khochbin, S. and Wolffe, A.P. (1994). Developmentally regulated expression of linker-histone variants in vertebrates. *Eur. J. Biochem.* **225**, 501–10.

Kidwell, W.R. and Mage, M.G. (1976). Change in poly(adenosine diphosphate ribose) and poly(adenosine diphosphate ribose) polymerase in synchronous HeLa cells. *Biochemistry* **15**, 1213–17.

Kim, J.K., Patel, D. and Choi, B.S. (1995). Constrasting structural impacts induced by *cis*-syn cyclobutane dimer and (6-40) adduct in DNA duplex decamers: implication in mutagenesis and repair activity. *Photochem. Photobiol.* **62(1)**, 44–50.

Kim, Y-J., Bjorklund, S., Li, Y., Sayre, M.H. and Kornberg, R.D. (1994). A multiprotein mediator of transcriptional activation and its interaction with the C-terminal repeat domain of RNA polymerase II. *Cell* **77**, 599–608.

Kimelman, D., Kirschner, M. and Scherson, T. (1987). The events of the midblastula transition in *Xenopus* are regulated by changes in the cell cycle. *Cell* **48**, 399–407.

Kingsley, P.D., Angerer, L.M. and Angerer, R.C. (1993). Major temporal and spatial patterns of gene expression during differentiation of the sea urchin embryo. *Dev. Biol.* **155**, 216–34.

Kingston, R.E. and Green, M.R. (1994). Modeling eukaryotic transcriptional activation. *Curr. Biol.* **4**, 325–32.

Kingston, R.E., Bunker, C.A. and Imbalzano, A. (1996). Repression and activation by multiprotein complex that alter chromatin structure. *Genes Dev.* **10**, 905–20.

Kissinger, C.R., Liu, B., Martin-Blanco, E., Kornberg, T.B. and Pabo, C.O. (1990). Crystal structure of an engrailed homeodomain–DNA complex at 2.8 Å resolution. *Cell* **63**, 579–90.

Klar, A.J.S. and Bonaduce, M.J. (1991). *swi6*, a gene required for mating-type switching, prohibits meiotic recombination in the *mat2-mat3* 'cold spot' of fission yeast. *Genetics* **129**, 1033–42.

Klar, A.J.S. (1992). Developmental choices in mating type interconversion in fission yeast. *Trends Genet.* **8**, 208–13.

Kleene, K.C. and Flynn, J.F. (1987). Characterization of a cDNA clone encoding the basic protein, TP2, involved in chromatin condensation during spermiogenesis in the mouse. *J. Biol. Chem.* **262**, 17272–7.

Klein, F., Laroche, T., Cardenas, M.E., Hoffmann, J.F.X., Schweizer, D. and Gasser, S.M. (1992). Localization of RAP1 and topoisomerase II in nuclei and mieotic chromosomes of yeast. *J. Cell Biol.* **117**, 935–48.

Klein, S., Gerster, T., Picard, D., Radbruch, A. and Schaffner, W. (1985). Evidence for transient requirement of the 1gH enhancer. *Nucleic. Acids Res.* **13**, 8901–12.

Kleinschmidt, A.M. and Martison, H.G. (1981). Structure of nucleosome core particles containing uH2A. *Nucleic Acids Res.* **27**, 565–74.

Kleinschmidt, J.A. and Seiter, A. (1988). Identification of domains involved in nuclear uptake and histone binding of protein N1 of *Xenopus laevis*. *EMBO J.* **7**, 1605–14.

Kleinschmidt, J.A. and Steinbeisser, H. (1991). DNA dependent phosphorylation of histone H2A.X during nucleosome assembly in *Xenopus laevis* oocytes: involvement of protein phosphorylation in nucleosome spacing. *EMBO J.* **10**, 3043–50.

Kleinschmidt, J.A., Scheer, U., Dabauville, M.-C., Bustin, M. and Franke, W.W. (1983). High mobility group proteins of amphibian oocytes: a large storage pool of a soluble high mobility group-1-like protein and involvement in transcriptional events. *J. Cell Biol.* **97**, 838–48.

Kleinschmidt, J.A., Fortkamp, E., Krohne, G., Zentgraf, H. and Franke, W.W. (1985). Co-existence of two different types of soluble histone complexes in nuclei of *Xenopus laevis* oocytes. *J. Biol. Chem.* **260**, 1166–76.

Kleinschmidt, J.A., Dingwall, C., Maier, G. and Franke, W.W. (1986). Molecular characterization of a karyophilic histone-binding protein: cDNA cloning, amino acid sequence and expression of nuclear protein N1/N2 of *Xenopus laevis*. *EMBO J.* **5**, 3547–52.

Kleinschmidt, J.A., Seiter, A. and Zentgraf, H. (1990). Nucleosome assembly *in vitro*: separate histone transfer and synergistic interaction of native histone complexes purified from nuclei of *Xenopus laevis* oocytes. *EMBO J.* **9**, 1309–18.

Klesert, T.R., Otten, A.D., Bird, T.D. and Tappscott, S.J. (1997). Trinucleotide repeat expansion at the myotonic dystrophy locus reduces expression DMAHP. *Nature Genet.* **16**, 402–6.

Klevit, R.E., Hamiot, J.R. and Horvath, S.J. (1990). Solution structure of a zinc finger domain of yeast ADR1. *Proteins* **7**, 215–26.

Klobutcher, L.A., Jahn, C.L. and Prescott, D.M. (1984). Internal sequences are eliminated from genes during macronuclear development in the ciliated protozoan *Oxytricha nova*. *Cell* **36**, 1045–55.

Klug, A. and Lutter, L.C. (1981). The helical periodicity of DNA on the nucleosome. *Nucleic. Acids Res.* **9**, 4267–83.

Knezetic, J.A. and Luse, D.S. (1986). The presence of nucleosomes on a DNA template prevents initiation by RNA polymerase II *in vitro*. *Cell* **45**, 95–104.

Knezetic J.A., Jacob, G.A. and Luse, D.S. (1988). Assembly of RNA polymerase II preinitiation complexes before assembly of nucleosomes allows efficient initiation of transcription on nucleosomal templates. *Mol. Cell Biol.* **8**, 3114–21.

Knowles, J.A. and Childs, G.J. (1984). Temporal expression of late histone messenger RNA in the sea urchin *Lytechinus pictus*. *Proc. Natl Acad. Sci. USA* **81**, 2411–15.

Koken, M. H., Linares-Cruz, G., Quignon, F., Viron, A., Chelbi-Alix, M. K., Sobczak-Thepot, J., Juhlin, L., Degos, L., Calvo, F. and de The, H. (1995). The PML growth-suppressor has an altered expression in human oncogenesis. *Oncogene*. **10**, 1315–24.

Koken, M. H., Puvion-Dutilleul, F., Guillemin, M. C., Viron, A., Linares-Cruz, G., Stuurman, N., de Jong, L., Szostecki, C., Calvo, F. and Chomienne, C. (1994). The t(15;17) translocation alters a nuclear body in a retinoic acid-reversible fashion. *EMBO J.* **13**, 1073–83.

Kokubo, T., Gong, D.-W., Wootton, J.C., Horikoshi, M. and Roeder, R.G. (1993). Molecular cloning, structural relationships and interactions of *Drosophila* TFIID subunits. *Nature* **367**, 484–7.

Komachi, K., Redd, M.J., and Johnson, A.D. (1994). The WD repeats of TUP1 interact with the homeo domain protein α2. *Genes Dev.* **8**, 2857–67.

Koo, H-S., Wu, H.M. and Crothers, D.M. (1986). DNA bending at adenine thymine tracts. *Nature (London)* **320**, 501–6.

Koonin, E.V., Zhou, S.B. and Lucchesi, J.C. (1995). The chromosuperfamily new members, duplication of the chromo domain and possible role in delivering transcription regualtors in chromatin. *Nucleic. Acids Res.* **23**, 4229–33.

Kornberg, A. (1988). DNA replication. *J. Biol. Chem.* **263**, 1–4.

Kornberg, A. and Baker, T.A. (1991). *DNA Replication*, 2nd edn. W.H. Freeman, New York.

Kornberg, R. (1974). Chromatin structure: a repeating unit of histones and DNA. *Science* **184**, 868–71.

Kornberg, R.D. (1981). The location of nucleosomes in chromatin: specific or statistical? *Nature (London)* **292**, 579–80.

Kornberg, R.D. and Lorch, Y. (1991). Irresistible force meets immovable object: transcription and the nucleosome. *Cell* **67**, 833–6.

Kornberg, R. and Thomas, J.O. (1974). Chromatin structure: oligomers of histones. *Science* **184**, 865–8.

Kossel, A. (1928). *The Protamines and Histones*. Longmans, London.

Koshland, D. and Strunnikov, A. (1996). Mitotic chromosome segregation. *Annu. Rev. Cell Dev. Biol.* **12**, 305–33.

Kraus, R., Pollock, R. and Guarente, L. (1996). yeast SUB1 is a suppressor of TFIIB mutations and has homology to the human coactivator PC4. *EMBO J.* **15**, 1933–40.

Kreidberg, J. A., Sariola, H., Loring, J. M., Maeda, M., Pelletier, J., Housman, D. and Jaenisch, R. (1993). WT-1 is required for early kidney development. *Cell* **74**, 679–91.

Kretzschmar, M., Kaiser, K., Lottspeich, F. and Meisterernst, M. (1994). A novel mediator of class II transcription with homology to viral immediate-early transcriptional regulators. *Cell* **78**, 525–34.

Krieg, P.A. and Melton, D. (1985). Developmental regulation of a gastrula-specific gene injected into fertilized *Xenopus* eggs. *EMBO J.* **4**, 3463–71.

Krieg, P.A. and Melton, D. (1987). An enhancer responsible for activating transcription at the mid-blastula transition in Xenopus development. *Proc. Natl Acad. Sci. USA* **84**, 2331–5.

Krohne, G.K. and Franke, W.W. (1980). Immunological identification and localization of the predominant nuclear protein of the amphibian oocyte nucleus. *Proc. Natl Acad. Sci. USA* **77**, 1034–8.

Kroll, K.L. and Gerhart, J.C. (1994). Transgenic *X. laevis* embryos from eggs transplanted with nuclei of transfected cultured cells. *Science* **266**, 650–3.

Krude, T. and Knippers, R. (1993). Nucleosome assembly during complementary DNA strand synthesis in extracts from mammalian cells. *J. Biol. Chem.* **268**, 14432–42.

Krude, T., de Maddalena, C. and Knippers, R. (1993). A nucleosome assembly factor is a constituent of SV40 minichromosomes. *Mol. Cell. Biol.* **13**, 1059–68.

Kruger, W. and Herskowitz, I. (1991). A negative regulator of HO transcription, SIN1 (SPT2), is a non specific DNA binding protein related to HMG1. *Mol. Cell. Biol.* **11**, 4135–46.

Kruger, W., Peterson, C.L., Sil, A., Coburn, C., Arents, G., Moudrianakis, E.N., and Herskowitz, I. (1995). Amino acid substitutions in the structured domains of histones H3 and H4 partially relieve the requirement of the yeast SWI/SNF complex for transcription. *Genes Dev.* **9**, 2770–9.

Kuldell, N.H. and Buratowski, S. (1997). Genetic analysis of the large subunit of yeast transcription factor IIE reveals two regions with distinct function. *Mol. Cell Biol.* **17**, 5288–98.

Kumar, A., Kassavetis, G.A., Geiduschek, E.P., Hambalko, M. and Brent, C.J. (1997). Functional dissection of the B″ component of RNA polymerase III transcription factor IIIB: a scaffolding protien with multiple roles in assembly and initiation of transcription. *Mol. Cell Biol.* **17** 1868–80.

Kuo, M.-H., Brownell, J.E., Sobal, R.E., Ranalli, T.A., Cook, R.G., Edmondson, D.G., Roth, S.Y. and Allis, C.D. (1996). Transcription-linked acetylation by Gcn5p of histones H3 and H4 at specific lysines. *Nature* 383, 269–72.

Kurumizaka, H. and Wolffe, A.P. (1997). Sin mutations of histone H3: influence on nucleosome core structure and function. *Mol. Cell Biol.* **17**, 6953–69.

Kuziora, M.A. and McGinnis, W. (1988). Different transcripts of the *Drosophila Abd-B* gene correlate with distinct genetic sub-functions. *EMBO J.* **7**, 3233–44.

Kwon, H., Imbalzano, A.N., Khavarl, P.A., Kingston, R.E. and Green, M.R. (1994). Nucleosome disruption and enhancement of activator binding by a human SWI/SNF complex. *Nature* **370**, 477–81.

La Flamme, S., Acuto, S., Markowitz, D., Vick, L., Landschultz, W. and Bank, A. (1987). Expression of chimeric human β- and δ-globin genes during erythroid differentiation. *J. Biol. Chem.* **262**, 4819–26.

La Marca, M.J., Strobel-Fidler, M.C., Smith, L.D. and Keem, K. (1975). Hormonal effects on RNA synthesis by stage 6 oocytes of *Xenopus laevis*. *Dev. Biol.* **47**, 384–93.

Laemmli, U.K., Cheng, S.M., Adolph, K.W., Paulson, J.R., Brown, J.A. and Brumbach, W.R. (1978). Metaphase chromosome structure: the role of non histone proteins. *Cold Spring Harb. Symp. Quant. Biol.* **42**, 351–60.

Laherty, C.D., Yang, W.M., Sun, J.M., Davie, J.R., Seto, E., and Eisenman, R.M. (1997). Histone deacetylase associated with the mSin3 corepressor mediate Mad transcriptional repression. *Cell* **89**, 349–56.

Lahoz, E.G., Xu, L., Schreiber-Agus, N. and DePinho, R.A. (1994). Suppression of Myc, but not E1a, transformation activity by Max-associated proteins, Mad and Mxi1. *Proc. Natl Acad. Sci. USA* **91**, 5503–7.

Lambert, S., Muyldermans, S., Baldwin, J., Kilner, J., Ibel, K. and Wijns, L. (1991). Neutron scattering studies of chromatosomes. *Biochem. Biophys. Res. Commun.* **179**, 810–16.

Lamond, A. I. and Carmo-Fonseca, M. (1993). Localisation of splicing snRNPs in mammalian cells. *Mol. Biol. Rep.* **18**, 127–33.

Lan, S.Y. and Smerdon, M.J. (1985). A nonuniform distribution of excision repair synthesis in nucleosome DNA. *Biochemistry* **24**, 7771–83.

Landsman, D. (1996). Histone H1 in *Saccharomyces cerevisiae*: a double mystery solved. *Trends Biochem. Sci.* **21**, 287–8.

Landsberger, N. and Wolffe, A.P. (1995a). Chromatin and transcriptional activity in early *Xenopus* development. *Semin. Cell Biol.* **6**, 191–200.

Landsberger, N. and Wolffe, A.P. (1995b). The role of chromatin and *Xenopus* heat shock transcription factor (XHSF1) in the regulation of the *Xenopus* hsp70 promoter *in vivo*. *Mol. Cell. Biol.* **15**, 6013–24.

Landsberger, N. and Wolffe, A.P. (1997). Remodeling of regulatory nucleoprotein complexes on the Xenopus hsp70 promoter during meiotic maturation of the *Xenopus* oocyte. *EMBO J.* **16**, 4361–73.

Landsberger, N., Ranjan, M., Almouzni, G., Stump, D. and Wolffe, A.P. (1995). The heat shock response in *Xenopus* oocytes, embryos and somatic cells: an essential regulatory role for chromatin. *Dev. Biol.* **170**, 62–74.

Langan, T.A. (1982). Characterization of highly phosphorylated subcomponents of rat thymus H1 histone. *J. Biol. Chem.* **257**, 14835–46.

Langan, T.A., Gautier, J., Lohka, M., Hollingsworth, R., Moreno, S., Nurse, P., Maller, J. and Sclafani, R.A. (1989). Mammalian growth-associated H1 histone kinase: a homolog of cdc^{2+}/CDC28 protein kinases controlling mitotic entry in yeast and frog cells. *Mol. Cell. Biol.* **9**, 3860–8.

Larsson, S. H., Charlieu, J. P., Miyagawa, K., Engelkamp, D., Rassoulzadegan, M., Ross, A., Cuzin, F., van Heyningen, V. and Hastie, N. D. (1995). Subnuclear localization of WT1 in splicing or transcription factor domains is regulated by alternative splicing. *Cell* **81**, 391–401.

Laskey, R.A. and Earnshaw, W.C. (1980). Nucleosome assembly. *Nature (London)* **286**, 763–7.

Laskey, R.A., Honda, B.M., Mills, A.D. and Finch, J.T. (1978). Nucleosomes are assembled by an acidic protein which binds histones and transfers them to DNA. *Nature (London)* **275**, 416–20.

Laskey, R.A., Harland, R.M. and Méchali, M. (1983). Induction of chromosome replication during maturation of amphibian oocytes. *CIBA. Found. Symp.* **98**, 25–36.

Lassar, A., Hamer, D.M. and Roeder, R.G. (1985). Stable transcription complex as a class III gene in a minichromosome. *Mol. Cell. Biol.* **5**, 40–5.

Latham, K.E., Solter, D. and Schultz, R.M. (1992). Acquisition of a transcriptionally permissive state during the 1-cell stage of mouse embryogenesis. *Dev. Biol.* **149**, 457–62.

Lau, L.F. and Nathans, D. (1987). Expression of a set of growth-related immediate early genes in BALBc/3T3 cells: coordinate regulation with c-*fos* and c-*myc*. *Proc. Natl Acad. Sci. USA* **84**, 1182–9.

Laurenson, P. and Rine, J. (1992). Silencers, silencing and heritable transcriptional states. *Microbiol. Rev.* **56**, 543–60.

Laurent, B.C., and Carlson, M. (1992). Yeast SNF2/SWI2, SNF5 and SNF6 proteins function coordinately with the gene-specific transcriptional activators Gal4 and Bicoid. *Genes Dev.* **6**, 1707–15.

Laurent, B.C., Treioch, I. and Carlson, M. (1993). The yeast SNF2/SWI2 protein has DNA stimulated ATPase activity required for transcriptional activation. *Genes Dev.* **7**, 583–91.

Lavender, J.S., Birley, A.J., Palmer, M.J., Kuroda, M.I. and Turner, B.M. (1994). Histone H4 acetylated at lysine 16 and other components of the *Drosophila* dosage compensation pathway colocalize on the male X chromosome through mitosis. *Chromosome Res.* **2**, 398–404.

Lawrence, J. B., Carter, K. C. and Xing, X. (1993). Probing functional organization within the nucleus: is genome structure integrated with RNA metabolism? *Cold Spring Harb. Symp. Quant. Biol.* **58**, 807–18.

Laybourn, P. and Kadonaga, J.T. (1991). Role of nucleosomal cores and histone H1 in the regulation of transcription by RNA polymerase II. *Science* **254**, 238–45.

Laybourn, P.J. and Kadonaga, J.T. (1992). Threshold phenomena and long distance activation of transcription by RNA polymerase II. *Science* **257**, 1682–5.

Learn, B., Karzai, A.W. and McMacken, R. (1993). Transcription stimulates the establishment of bidirectional λ DNA replication *in vitro*. *Cold Spring Harb. Symp. Quant. Biol.* **58**, 389–401.

Lebkowski, J.S., Clancy, S. and Calos, M.P. (1985). Simian virus 40 replication in adenovirus-transformed human cells antagonizes gene expression. *Nature (London)* **317**, 169–71.

Le Blanc, B., Read, C. and Moss, T. (1993). Recognition of the *Xenopus* ribosomal core promoter by the transcription factor xUBF interdomain interaction. *EMBO J.* **12**, 513–25.

LeBowitz, J.H. and McMacken, R. (1986). The *Escherichia coli* dnaB replication protein is a DNA helicase. *J. Biol. Chem.* **261**, 4738–50.

Le Douarin, B., Nielsen, A.L., Garnier, J.M., Ichinose, H., Jeanmougin, G., Losson, R. and Chambon, P. (1996). A possible involvement of TIF1α and TIF1β in the epigenetic control of transcription by nuclear receptors. *EMBO J.* **15**, 6701–15.

Lee, D.Y., Hayes, J.J., Pruss, D. and Wolffe, A.P. (1993). A positive role for histone acetylation in transcription factor binding to nucleosomal DNA. *Cell* **72**, 73–84.

Lee, H.H. and Archer, T.K. (1994). Nucleosome-mediated disruption of transcription factor-chromatin initiation complexes at the mouse mammary tumor virus long terminal repeat *in vivo*. *Mol. Cell. Biol.* **14**, 32–41.

Lee, K.-M. and Hayes, J.J. (1997). The N-terminal tail of histone H2A binds to two distinct sites within the nucleosome core. *Proc. Natl Acad. Sci. USA* **94**, 8959–64.

Lee, K-M. and Hayes, J.J. (1998). Linker DNA and H1-dependent reorganization of histone-DNA interactions within the nucleosome. *Biochemistry* (in press).

Lee, M.S. and Garrard, W.T. (1991a). Transcription-induced nucleosome 'splitting' an underlying structure for DNase I sensitive chromatin. *EMBO J.* **10**, 607–15.

Lee, M.S. and Garrard, W.T. (1991b). Positive supercoiling generates a chromatin conformation characteristic of highly active genes. *Proc. Natl Acad. Sci. USA* **88**, 9675–9.

Lee, M.S., Gippert, G.P., Soman, K.V., Case, D.A. and Wright, P.E. (1989). Three dimensional solution structure of a single zinc finger DNA-binding domain. *Science* **245**, 635–7.

Leffak, I.M. (1984). Conservative segregation of nucleosome core histones. *Nature (London)* **307**, 82–5.

LeMaire, M. F. and Thummel, C. S. (1990). Splicing precedes polyadenylation during Drosophila E74A transcription. *Mol. Cell Biol.* **10**, 6059–63.

Lenfant, F., Mann, R.K., Thomsen, B., Ling, X. and Grunstein, M. (1996). All four core histone N-termini contain sequences required for the repression of basal transcription in yeast. *EMBO J.* **15**, 3974–85.

Leng, M. and Felsenfeld, G. (1966). The preferential interaction of polylysine and polyarginine with specific base sequences in DNA. *Proc. Natl Acad. Sci. USA* **56**, 1325–32.

Lennard, A.C. and Thomas, J.O. (1985). The arrangement of H5 molecules in extended and condensed chicken erythrocyte chromatin. *EMBO J.* **4**, 3455–62.

Leno, G.H. and Laskey, R.A. (1991). The nuclear membrane determines the timing of DNA replication in *Xenopus* egg extracts. *J. Cell. Biol.* **112**, 557–66.

Leno, G.H., Downes, C.S. and Laskey, R.A. (1992). The nuclear membrane prevents replication of human G2 nuclei but not G1 nuclei in *Xenopus* egg extract. *Cell* **69**, 151–61.

Leno, G.H., Mills, A.D., Philpott, A. and Laskey, R.A. (1996). Hyperphosphorylation of nucleoplasmin facilitates *Xenopus* sperm decondensation at fertilization. *J. Biol. Chem.* **271**, 7253–6.

Leonhardt, H., Page, A.W., Weier, H.U. and Bestor, T.H. (1992). A targeting sequence directs DNA methyltransferase to sites of DNA replication in mammalian nuclei. *Cell* **71**, 865–73.

Leuba, S.H., Zlatanova, J. and van Holde, K. (1993). On the location of histones H1 and H5 in the chromatin fibre. *J. Mol. Biol.* **229**, 917–29.

Leuba, S.H., Yang, G., Robert, C., Samori, B., van Holde, K., Zlatanova, J. and Bustamante, C. (1994a). Three-dimensional structure of extended chromatin fibers as revealed by tapping-mode scanning force microscopy. *Proc. Natl Acad. Sci. USA* **91**, 11621–5.

Leuba, S.H., Zlatanova, J. and van Holde, K. (1994b). On the location of linker DNA in the chromatin fibre. *J. Mol. Biol.* **235**, 871–90.

Levine, A., Yeirin, A., Ben-Asher, E., Aloni, Y. and Razin, A. (1993). Histone H1-mediated inhibition of transcription initiation of methylated templates *in vitro*. *J. Biol. Chem.* **268**, 21754–9.

Levinger, L. and Varshavsky, A. (1980). High-resolution fractionation of nucleosomes: minor particles 'whiskers' and separation of nucleosomes containing and lacking A24 semihistone. *Proc. Natl Sci. USA* **77**, 3244–8.

Levitt, M. (1978). How many base-pairs per turn does DNA have in solution and in chromatin? Some theoretical calculation. *Proc. Natl Acad. Sci. USA* **75**, 640–4.

Lewis, C.D. and Laemmli, U.K. (1982). Higher order metaphase chromosome structure: evidence for metalloprotein interactions. *Cell* **29**, 171–81.

Lewis, E.B. (1950). The phenomenon of position effect. *Adv. Genet.* **3**, 73–115.

Lewis, E.B. (1978). A gene complex controlling segmentation. *Nature* **276**, 565–70.

Lewis, E.D. and Manley, J.L. (1985). Repression of simian virus 40 early transcription by viral DNA replication in human 293 cells. *Nature (London)* **317**, 172–5.

Lewis, J.D., Meehan, R.R., Henzel, W.J., Maurer-Fogy, I., Jeppensen, P., Klein, G. and Bird, A. (1992). Purification, sequence and cellular localization of a novel chromosomal protein that binds to methylated DNA. *Cell* **69**, 905–14.

Lewis, J.D., Gunderson, S.I. and Mattaj, I.W. (1995). The influence of 5′ and 3′ end structures on pre-mRNA metabolism. *J. Cell Sci. Suppl.* **19**, 13–19.

Li, B., Adams, C.C. and Workman, J.L. (1994). Nucleosome binding by the constitutive factor Sp1. *J. Biol. Chem.* **269**, 7756–63.

Li, E., Bestor, T.H. and Jaenisch, R. (1992). Targeted mutation of the DNA methytransferase gene results in embryonic lethality. *Cell* **69**, 915–26.

Li, E., Beard, C. and Jaenisch, R. (1993). Role of DNA methylation in genomic imprinting. *Nature* **366**, 362–5.

Li, G., Chandler, S.P., Wolffe, A.P. and Hall, T.C. (1998). Architectural specificity in chromatin structure at the TATA box *in vivo*: transcription dependent remodeling of a seed specific regulatory nucleosome. *Proc. Natl Acad. Sci. USA* (in press).

Li, J.J. and Kelly, T.J. (1984). Simian virus 40 DNA replication *in vitro*. *Proc. Natl Acad. Sci. USA* **81**, 6973–7.

Li, Q. and Wrange, O. (1993). Translational positioning of a nucleosomal glucocorticoid response element modulates glucocorticoid receptor affinity. *Genes Dev.* **7**, 2471–82.

Li, Q. and Wrange, O. (1995). The accessibility of a glucocorticoid response element dependent on its rotational positioning. *Mol. Cell. Biol.* **15**, 4375–84.

Li, Y., Bjorklund, S., Jiang, Y.W., Kim, Y.J., Lane, W.S., Stillman, D.J. and Kornberge, R.D. (1995). Yeast global transcriptional repressors Sin4 and Rgr1 are components of mediator complex/RNA polymerase II holoenzyme. *Proc. Natl Acad. Sci. USA* **92**, 10864–8.

Liang, C.P. and Garrard, W.T. (1997). Template topology and transcription: chromatin templates relaxed by localized linearization are transcriptionally active in yeast. *Mol. Cell Biol.* **17** 2825–34.

Lilley, D.M. (1992). DNA–protein interactions. HMG has DNA wrapped up. *Nature* 357, 282–3.

Lilley, D.M. and Tatchell, K. (1977). Chromatin core particle unfolding induced by tryptic cleavage of histones. *Nucleic. Acids Res.* **4**, 2039–55.

Lilley, D.M.J., Jacobs, M.F. and Houghton, M. (1979). The nature of the interaction of nucleosomes with RNA polymerase II. *Nucleic. Acids Res.* 7, 377–98.

Lima-de Faria, A. (1983). Processes of directing expression, mutation and rearrangements. In *Molecular Evolution and Organization of the Chromosome*, ed. A. Lima de Faria. Elsevier Science Publishers B.V., Amsterdam, The Netherlands, 507–604.

Lin, R., Leone, J.W., Cook, R.G. and Allis, C.D. (1989). Antibodies specific to acetylated histones document the existence of deposition- and transcription-related histone acetylation in *Tetrahymena*. *J. Cell. Biol.* **108**, 1577–88.

Lin, R., Cook, R.G. and Allis, C.D. (1991). Proteolytic removal of core histone amino termini and dephosphorylation of histone H1 correlate with the formation of condensed chromatin and transcriptional silencing during *Tetrahymena* macronuclear development. *Genes Dev.* **5**, 1601–10.

Lin, S-Y and Riggs, A.D. (1975). The general affinity of *lac repressor* for E. coli DNA: implication for gene regulation in procaryotes and eucaryotes. *Cell* **4**, 107–11.

Lin, Y.S. and Green, M.R. (1991). Mechanism of action of an acidic transcriptional activator *in vitro*. *Cell* **64**, 971–82.

Lindsey, G.G., Orgeig, S., Thompson, P., Davies, N. and Maeder, D.L. (1991). Extended C-terminal tail of wheat histone H2A interacts with DNA of the linker region. *J. Mol. Biol.* **218**, 805–13.

Ling, X., Harkness, T.A.A., Schultz, M.C., Fisher-Adams, G. and Grunstein, M. (1996). Yeast histone H3 and H4 amino termini are important for nucleosome assembly *in vivo* and *in vitro*: redundant and position-independent functions in assembly but not in gene regulation. *Genes Dev.* **10**, 686–99.

Liu, B. and Alberts, B.M. (1995). Head-on collision betwen a DNA replication apparatus and RNA polymerase transcription complex. *Science* **267**, 1131–7.

Liu, K., Lauderdale, J.D. and Stein, A. (1993). Signals in chicken β globin DNA influence chromatin assembly *in vitro*. *Mol. Cell Biol.* **13**, 7596–603.

Liu, L.F. and Wang, J.C. (1987). Supercoiling of the DNA template during transcription. *Proc. Natl Acad. Sci. USA* **84**, 7024–7.

Lobell, R.B. and Schlief, R.F. (1990). DNA looping and unlooping by Ara C protein. *Science* **250**, 528–32.

Lock, L.F., Takagi, N. and Martin, G.R. (1987). Methylation of the *hprt* gene on the inactive X occurs after chromosome inactivation. *Cell* **48**, 36–46.

Locke, J., Kotarski, M.A. and Tartof, K.D. (1988). Dosage dependent modifiers of position effect variegation in *Drosophila* and a mass action model that explains their effect. *Genetics* **120**, 181–8.

Locklear, L., Risdale, J.A., Bazett-Jones, D.P. and Davie, J.R. (1990). Ultrastructure of transcriptionally competent chromatin. *Nucleic. Acids Res.* **18**, 7015–24.

Lohka, M.J. and Masui, Y. (1983). Formation *in vitro* of sperm pronuclei and mitotic chromosomes induced by amphibian ooplasmic components. *Science* **220**, 719–21.

Lohka, M.J. and Masui, Y. (1984). Roles of the cytosol and cytoplasmic particles in nuclear envelope assembly and sperm pronuclear formation in cell-free preparations from amphibian eggs. *J. Cell Biol.* **98**, 1222–30.

Longo, F.J. (1972). An ultrastructural analysis of mitosis and cytokinesis in the sea urchin, *Arbacia punctulata. J. Morphol.* **138**, 207–38.

Loo, S. and Rine, J. (1994a). Silencers create domains of generalized repression on eukaryotic chromosomes. *Science* **264**, 5166–8.

Loo, S. and Rine, J. (1994b). Silencers and domains of generalized repression. *Science* **264**, 1768–71.

Lorch, Y., La Pointe, J.W. and Kornberg, R.D. (1987). Nucleosomes inhibit the initiation of transcription but allow chain elongation with the displacement of histones. *Cell* **49**, 203–10.

Lorch, Y., La Pointe, J.W. and Kornberg, R.D. (1988). On the displacement of histones from DNA by transcription. *Cell* **55**, 743–4.

Lorch, Y., La Pointe, J.W. and Kornberg, R.D. (1992). Initiation on chromatin templates in a yeast RNA polymerase II transcription system. *Genes Dev.* **6**, 2282–7.

Lorentz, A., Heim, L. and Schmidt, H. (1992). The switching gene *swi6* affects recombination and gene expression in the mating-type region of *Schizosaccharomyces pombe. Mol. Gen. Genet.* **233**, 436–42.

Lorentz, A., Ostermann, K., Fleck, O. and Schmidt, H. (1994). Switching gene *swi6* involved in the repression of silent mating-type loci in fission yeast, encodes a homolog of chromatin associated proteins from *Drosophila* and mammals. *Gene* **143**, 1–8.

Losa, R. and Brown, D.D. (1987). A bacteriophage RNA polymerase transcribes *in vitro* through a nucleosome core without displacing it. *Cell* **50**, 801–8.

Losa, R., Thoma, F. and Koller, T. (1984). Involvement of the globular domain of histone H1 in the higher order structures of chromatin. *J. Mol. Biol.* **175**, 529–51.

Louters, L. and Chalkley, R. (1985). Exchange of histones H1, H2A and H2B *in vivo. Biochemistry* **24**, 3080–5.

Lovering, R., Hanson, I. M., Borden, K. L., Martin, S., O'Reilly, N. J., Evan, G. I., Rahman, D., Pappin, D. J., Trowsdale, J. and Freemont, P. S. (1993). Identification and preliminary characterization of a protein motif related to the zinc finger. *Proc. Natl Acad. Sci. USA* **90**, 2112–16.

Lowary, P.T. and Widom, J. (1989). Higher-order structure of *Saccharomyces cerevisiae* chromatin. *Proc. Natl Acad. Sci. USA* **86**, 8266–70.

Lowary, P.T. and Widom, J. (1997). Nucleosome packaging and nucleosome positioning of genomic DNA. *Proc. Natl Acad. Sci. USA* **94**, 1183–8.

Lu, H., Flores, O., Weinmann, R. and Reinberg, D. (1991). The non-phosphorylated form of RNA polymerase II preferentially associates with the preinitiation complex. *Proc. Natl Acad. Sci. USA* **88**, 10004–8.

Lu, Q., Wallrath, L.L., Allan, B.D., Glaser, R.L., Lis, J.T. and Elgin, S.C.R. (1992). Promoter sequence containing (CT)n (GA)n repeats is critical for the formation of the DNase I hypersensitive sites in the *Drosophia hsp26* gene. *J. Mol. Biol.* **225**, 985–98.

Lu, Q., Wallrath, L.L., Granok, H. and Elgin, S.C.R. (1993). (CT)n (GA)n repeats and heat shock elements have distinct roles in chromatin structure and transcriptional activation of the *Drosophila hsp26* gene. *Mol. Cell. Biol.* **13**, 2802–14.

Luerssen, H., Hoyer-Fender, S. and Engel, W. (1989). The nucleotide sequence of rat transition protein 2 (TP2) cDNA. *Nucleic. Acids Res.* **17**, 3585.

Luger, K., Mader, A.W., Richmond, R.K., Sargent, D.F. and Richmond, T.J. (1997). Crystal structure of the nucleosome core particle at 2.8A resolution. *Nature* **389**, 251–60.

Luisi, B.F., Xu, W.X., Otwinowski, Z., Freedman, L.P., Yamamoto, K.R. and Sigler, P. (1991). Crystallographic analysis of the interaction of the glucocorticoid receptor with DNA. *Nature* **352**, 497–502.

Luke, M. and Bogenhagen, D. (1989). Quantitation of type II topoisomerase in oocytes and eggs of *Xenopus laevis*. *Develop. Biol.* **136**, 459–68.

Lund, E. and Dahlberg, J.E. (1992). Control of 4-8S RNA transcription at the midblastula transition in *Xenopus laevis* embryos. *Genes Dev.* **6**, 1097–106.

Lutter, L. (1978). Kinetic analysis of deoxyribonuclease I cleavage sites in the nucleosome core: evidence for a DNA superhelix. *J. Mol. Biol.* **124**, 391–420.

Macatee, T., Jiang, Y.W., Stillman, D.J., and Roth, S.Y. (1997). Global alterations in chromatin accessibility associated with loss of *SIN4* function. *Nucleic. Acids Res.* **25**, 1240-7.

Macleod, D., Charlton, J., Mullins, J. and Bird, A.P. (1994). Sp1 sites in the mouse aprt gene promoter are required to prevent methylation of the CpG island. *Genes Dev.* **8**, 2282–92.

Mahadevan, L.C., Willis, A.C. and Barrah, M.J. (1991). Rapid histone H3 phosphorylation in response to growth factors, phorbol esters, okadaic acid and protein synthesis inhibitors. *Cell* **65**, 775–83.

Maillet, L., Boscheron, C., Gotta, M., Marcand, S., Gilson, E. and Gasser, S.M. (1996). Evidence for silencing compartments within the yeast nucleus: a role for telomere proximity and Sir protein concentration in silencer-mediated repression. *Genes Dev.* **10** 1796–811.

Majumder, D., Miranda, M., De Pamphilis, M. (1993). Analysis of gene expression in mouse preimplantation embryos demonstrates that the primary role of enhancers is to relieve repression of promoters, *EMBO J.* **12**, 1131–40.

Makarov, V.L., Lejnine, S., Bedoyan, J. and Langmore, J.P. (1993). Nucleosomal organization of telomere-specific chromatin in rat. *Cell* **73**, 775–87.

Maldonado, E., Shiekhattar, R., Sheldon, M., Cho, H., Drapkin, R., Rickert, P., Lees, E., Anderson, C. and Reinberg, D. (1996). A human RNA polymerase II complex associated with SRB and DNA-repair proteins. *Nature* **381**, 86–9.

Manders, E. M., Stap, J., Brakenhoff, G. J., van Driel, R. and Aten, J. A. (1992). Dynamics of three-dimensional replication patterns during the S-phase, analysed by double labelling of DNA and confocal microscopy. *J. Cell Sci.* **103**, 857–62.

Mandl, B., Brandt, W.F., Superli-Furga, G., Graninger, P.G., Birnstiel, M.L. and Busslinger, M. (1997). The five cleavage-stage (CS) histones of the sea urchin are encoded by a maternally expressed family of replacement histone genes: functional equivalence of the CS H1 and frog H1M (B4) proteins. *Mol. Cell Biol.* **17**, 1189–200.

Mann, R.K. and Grunstein, M. (1992). Histone H3 N-terminal mutations allow hyperactivation of the yeast GAL1 gene *in vivo*. *EMBO J.* **11**, 3297–306.

Mannironi, C., Bonner, W.M. and Hatch, C.L. (1989). H2A.X a histone isoprotein with a conserved C-terminal sequence is encoded by a novel mRNA with both DNA replication type and poly A 3′ processing signals. *Nucleic. Acids Res.* **17**, 9113–26.

Marcus, G.A., Silverman, N., Berger, S.L., Horiuchi, J., and Guarente, L. (1994). Functional similarity and physical association between GCN5 and ADA2: putative transcriptional adaptors. *EMBO J.* **1354**, 4807–15.

Mardian, J.K.W., Paton, A.E., Bunick, G.J. and Olins, D.E. (1980). Nucleosome cores have two specific binding sites for nonhistone chromosomal proteins HMG14 and HMG17. *Science* **209**, 1534–6.

Marsolier, M.C., Tanaka, S., Livinston-Zatchej, M., Grunstein, M., Thoma, F. and Sentenac, A. (1995). Reciprocal interferences between nucleosomal organization and transcriptional activity of the yeast SNR6 gene. *Genes Dev.* **9**, 410–22.

Martin, D.I.K., Fiering, S. and Groudine, M. (1996). Regulation of β-globin gene expression: straightening out the locus. *Curr. Opin. Genet. Dev.* **6**, 488–95.

Martin, L.D., Vesenka, J.P., Henderson, E. and Dobbs, D.L. (1995). Visualization of nucleosomal substructure in native chromatin by atomic force microscopy. *Biochemistry* **34**, 4610–16.

Martindale, D.W., Allis, C.D. and Bruns, P.J. (1982). Conjugation in *Tetrahymena thermophila*: a temporal analysis of cytological stages. *Exp. Cell Res.* **140**, 227–36.

Martinez-Balbas, M.A., Dey, A., Rabindran, S.K., Ozato, K. and Wu, C. (1995). Displacement of sequence-specific transcription factors from mitotic chromatin. *Cell* **83**, 29–38.

Martinez-Salas, E., Linney, E., Hassell, J. and De Pamphilis, M.L. (1989). The need for enhancers in gene expression first appears during mouse development with formation of the zygotic nucleus. *Genes Develop.* **3**, 1493–506.

Marushige, K. and Dixon, G.H. (1969). Developmental changes in chromosomal composition and template activity during spermatogenesis in trout testis. *Dev. Biol.* **19**, 397–414.

Maryanka, D., Cowling, G.J., Allan, J., Fey, S.J., Huvos, P. and Gould, H. (1979). Transcription of globin genes in reticulocyte chromatin. *FEBS Lett.* **105**, 131–6.

Mastrangelo, I.A., Hough, P.V.C., Wilson, V.G., Wall, J.S., Hainfeld, J.F. and Tegtmeyer, P. (1985). Monomers through trimers of large tumor antigen bind in region I and monomers through tetramers bind in region II of simian virus 40 origin of replication DNA as stable structures in solution. *Proc. Natl Acad. Sci. USA* **82**, 3626–30.

Mastrangelo, I.A., Courey, A.J., Wall, J.S., Jackson, S.P. and Hough, P.V.C. (1991). DNA looping and Sp1 multimer links: a mechanism for transcriptional synergism and enhancement. *Proc. Natl Acad. Sci. USA* **88**, 5670–4.

Masumoto, H., Masukata, H., Muro, Y., Nozaki, N. and Okazaki, T. (1989). A human centromere antigen (CENP-B) interacts with a short specific sequence in alphoid DNA, a human centromeric satellite. *J. Cell Biol.* **109**, 1963–73.

Mathis, D.J., Oudet, P., Waslyk, B. and Chambon, P. (1978). Effect of histone acetylation on structure and *in vitro* transcription of chromatin. *Nucleic. Acids Res.* **5**, 3523–47.

Mathis, G. and Althaus, F.R. (1990). Uncoupling of DNA excision repair and nucleosomal unfolding in poly(ADP-ribose) depleted mammalian cells. *Carcinogenesis* **11**, 1237–9.

Matsui, T. (1987). Transcription of adenovirus 2 major late and peptide IX genes under conditions of *in vitro* nucleosome assembly. *Mol. Cell Biol.* **7**, 1401–8.

Matsumoto, K., Anzai, M., Nakagata, N., Takahashi, A., Takahashi, Y. and Miyata, K. (1994). Onset of paternal gene activation in early mouse embryos fertilized with transgenic mouse sperm. *Mol. Reprod. Dev.* **39**, 136–40.

Mattaj, I. W. (1994). RNA processing. Splicing in space [news; comment]. *Nature* **372**, 727–8.

Mattaj, I., Lienhard, S., Jiricny, J. and De Robertis, E. (1985). An enhancer-like sequence within the *Xenopus* U2 gene promoter facilitates the formation of stable transcription complexes. *Nature (London)* **316**, 163–7.

Mattern, K.A., Humbel, B.M., Muijsers, A.O., de Jong, L. and van Driel, R. (1996). hnRNP proteins and B23 are the major proteins of the internal nuclear matrix of HeLa S3 cells. *J. Cell. Biochem.* **62**, 275–89.

Mavromara-Nazos, P. and Roizman, B. (1987). Activation of herpes simplex 1 γ2 genes by viral DNA replication. *Virology* **161**, 593–8.

Maxon, M.E. and Tjian, R. (1994). Transcriptional activity of transcription factor IIE is dependent on zinc binding. *Proc. Natl Acad. Sci. USA* **91**, 9529–33.

Maxson, R., Mohun, T., Gormezano, G., Childs, G. and Kedes, L. (1983). Distinct organization and patterns of expression of early and late histone sets in the sea urchin. *Nature (London)* **301**, 120–7.

McArthur, M. and Thomas, J.O. (1996). A preference of histone H1 for methylated DNA. *EMBO J.* **15**, 1705–14.

McDowall, A.W., Smith, J.M. and Dubochet, J. (1986). Cryo-electron microscopy of vitrified chromosomes *in situ*. *EMBO J.* **5**, 1395–402.

McGhee, J.D. and Felsenfeld, G. (1979). Reaction of nucleosome DNA with dimethyl sulfate. *Proc. Natl Acad. Sci. USA* **76**, 2133–7.

McGhee, J.D. and Felsenfeld, G. (1982). Reconstitution of nucleosome core particles containing glucosylated DNA. *J. Mol. Biol.* **158**, 685–98.

McGhee, J.D., Rau, D.C. and Felsenfeld, G. (1980). Orientation of the nucleosome within the higher order structure of chromatin. *Cell* **22**, 87–96.

McGhee, J.D., Wood, W.I., Dolan, M., Engel, J.D. and Felsenfeld, G. (1981). A 200 base pair region at the 5′ end of the chicken adult β-globin gene is accessible to nuclease digestion. *Cell* **27**, 45–55.

McGhee, J.D., Nickol, J.M., Felsenfeld, G. and Rau, D.C. (1983a). Higher order structure of chromatin: orientation of nucleosomes within the 30 nm chromatin solenoid is independent of species and spacer length. *Cell* **33**, 831–41.

McGhee, J.D., Nickol, J.M., Felsenfeld, G. and Rau, D.C. (1983b). Histone acetylation has little effect on the higher order fold of chromatin. *Nucleic. Acids Res.* **11**, 4065–75.

McGrath, J.D. and Solter, D. (1984). Inability of mouse blastomere nuclei transferred to enucleated zygotes to support development *in vitro*. *Science* **226**, 1317–19.

McKeon, F., Kirschner, M. and Caput, D. (1986). Homologies in both primary and secondary structure between nuclear envelope and intermediate filament proteins. *Nature (London)* **319**, 463–8.

McKnight, S.L. and Miller, O.L. Jr. (1979). Post replicative nonribosomal transcription units in *D. melanogaster* embryos. *Cell* **17**, 551–63.

McKnight, S.L., Bustin, M. and Miller, O.L. Jr. (1978). Electron microscopic analysis of chromosome metabolism in the *Drosophila melanogaster* embryo. *Cold Spring Harbor Symp. Quant. Biol.* **42**, 741–54.

McPherson, C.E., Shim, E.Y., Friedman, D.S. and Zaret, K.S. (1993). An active tissue-specific enhancer and bound transcription factors existing in a precisely positioned nucleosomal array. *Cell* **75**, 387–98.

McStay, B. and Reeder, R.H. (1986). A termination site for *Xenopus* RNA polymerase I also acts as an element of an adjacent promoter. *Cell* **47**, 913–20.

McStay, B., Sullivan, G.J. and Cairns, C. (1997). The *Xenopus* RNA polymerase I transcription factor, UBF, has a role in transcriptional enhancement distinct from that at the promoter. *EMBO J.* **16**, 396–405.

Méchali, M. and Harland, R.M. (1982). DNA synthesis in a cell-free system from *Xenopus* eggs: priming and elogation on single stranded DNA *in vitro*. *Cell* **30**, 93–101.

Meehan, R.R., Lewis, J.D., McKay, S., Kleiner, E.L. and Bird, A.P. (1989). Identification of a mammalian protein that binds specifically to DNA containing methylated CpGs. *Cell* **58**, 499–507.

Meehan, R.R., Lewis, J.D. and Bird, A.P. (1992). Characterization of MeCP2, a vertebrate DNA binding protein with affinity for methylated DNA. *Nucleic. Acids Res.* **20**, 5085–92.

Meeks-Wagner, D. and Hartwell, L.H. (1986). Normal stoichiometry of histone dimer sets is necessary for high fidelity of mitotic chromosome transmission. *Cell* **44**, 53–63.

Meersseman, G., Pennings, S. and Bradbury, E.M. (1991). Chromatosome positioning on assembled long chromatin. Linker histones affect nucleosome placement on 5S DNA. *J. Mol. Biol.* **220**, 89–100.

Meersseman, G., Pennings, S. and Bradbury, E.M. (1992). Mobile nucleosomes-a general behavior. *EMBO J.* **11**, 2951–9.

Megee, P.C., Morgan, B.A., Mittman, B.A. and Smith, M.M. (1990). Genetic analysis of histone H4: essential role of lysines subject to acetylation. *Science* **247**, 4932–4.

Megee, P.C., Morgan, B.A. and Smith, M.M. (1995). Histone H4 and the maintenance of genome integrity. *Genes Dev.* **9**, 1716–27.

Mehlin, H. and Danҽholt, B. (1993). The balbiani ring particle: a model for the assembly and export of RNPs from the nucleus? *Trends Cell Biol.* **3**, 443–7.

Meier, J., Campbell, K.H.S., Ford, C.C., Stick, R. and Hutchison, C.J. (1991). The role of lamin LIII in nuclear assembly and DNA replication in cell free extracts of *Xenopus* eggs. *J. Cell Sci.* **98**, 271–9.

Meier, U. T. and Blobel, G. (1992). Nopp140 shuttles on tracks between nucleolus and cytoplasm. *Cell* **70**, 127–38.

Meisterernst, M. and Roeder, R.G. (1991). Family of proteins that interct with TFIID and regulate promoter activity. *Cell* **67**, 557–67.

Meisterernst, M., Horikoshi, M. and Roeder, R.G. (1990). Recombinant yeast TFIID, a general transcription factor, mediates activation by the gene specific factor USF in a chromatin assembly assay. *Proc. Natl Acad. Sci. USA* **87**, 9153–7.

Meisterernst, M., Roy, A.L., Lieu, H.M. and Roeder, R.G. (1991). Activation of class II gene transcription by regulatory factors is potentiated by a novel activity. *Cell* **66** 981–93.

Mélin, F., Miranda, M., Montreau, N., De Pamphilis, M.L. and Blangy, D. (1993). Transcription enhancer factor-1 (TEF-1) DNA binding sites can specifically enhance gene expression at the beginning of mouse development. *EMBO J.* **12**, 4657–66.

Mensa-Wilmot, K., Caroll, K. and McMacken, R. (1989). Transcriptional activation of bacteriophage λ DNA replication *in vitro*. Regulatory role of histone like protein HU in *Escherichia coli*. *EMBO J.* **8**, 2393–405.

Mermelstein, F., Young, K., Cao, J., Inostroza, J.R., Erdjument-Bromage, H., Eagelson, K., Landsman, D. and Reinberg, D. (1996). Requirement of a corepressor for DR1-mediated repression of transcription. *Genes & Dev.* **10**, 1033–48.

Merriam, R.W. (1969). Movement of cytoplasmic proteins into nuclei induced to enlarge and initiate DNA or RNA synthesis. *J. Cell. Sci.* **5**, 333–49.

Mertz, J.E. (1982). Linear DNA does not form chromatin containing regularly spaced nucleosomes. *Mol. Cell. Biol.* **2**, 1608–18.

Messmer, S., Franke, A. and Paro, R. (1992). Analysis of the functional role of the Polycomb chromo domain in *Drosophila melanogaster. Genes Dev.* **6**, 1241–54.

Micheli, G., Luzzatto, A. R., Carri, M. T., de Capoa, A. and Pelliccia, F. (1993). Chromosome length and DNA loop size during early embryonic development of Xenopus laevis. *Chromosoma.* **102**, 478–83.

Miller, A.M. and Nasmyth, K.A. (1984). Role of DNA replication in the repression of silent mating type loci in yeast. *Nature (London)* **312**, 247–51.

Mills, A.D., Laskey, R.A., Black, P. and De Robertis, E.M. (1980). An acidic protein which assembles nucleosomes *in vitro* is the most abundant protein in *Xenopus* oocyte nuclei. *J. Mol. Biol.* **139**, 561–8.

Mills, A.D., Blow, J.J., White, J.G., Amos, W.B., Wilcock, D. and Laskey, R.A. (1989). Replication occurs at discrete foci spaced throughout nuclei replicating *in vitro. J. Cell. Sci.* **94**, 471–7.

Mirkovitch, J. and Darnell, J.E. (1991). Rapid *in vivo* footprinting technique identifies proteins bound to the TTR gene in mouse liver. *Genes Dev.* **5**, 83–93.

Mirkovitch, J., Mirault, M.E. and Laemmli, U. (1984). Organization of the higher-order chromatin loop: specific DNA attachment sites on nuclear scaffold. *Cell* **39**, 223–32.

Mirzabekov, A.D., Shick, V.V., Belyavsky, A.V., Karpov, V.L. and Bavykin, S.G. (1977). The structure of nucleosomes: the arrangement of histones in the DNA grooves and along the DNA chain. *Cold Spring Harb. Symp. Quant. Biol.* **42**, 149–55.

Mirzabekov, A.D., Bavykin, S.G., Karpov, V.L. Preobrazhenskaya, O.V., Elbradise, K.K., Tuneev, V.M., Melinkova, A.F., Goguadze, E.G., Chenchick, A.A. and Beabealashvili, R.S. (1982). Structure of nucleosomes, chromatin and RNA polymerase promoter complex as revealed by DNA protein cross-linking. *Cold Spring Harb. Symp. Quant. Biol.* **47**, 503–9.

Mirzabekov, A.D., Pruss, D.V. and Elbralidse, K.K. (1990). Chromatin super-structure-dependent cross-linking with DNA of the histone H5 residues Thr1, His25 and His62. *J. Mol. Biol.* **211**, 479–91.

Mita-Miyazawa, I., Ikegami, S. and Satoh, N. (1985). Hisospecific acetylcholinesterase development in the presumptive muscle cells isolated from 16-cell-stage ascidian embryos with respect to the number of DNA replications. *J. Embryol. Exp. Morphol.* **87**, 1–12.

Mitchell, D.L., Nguyen, T.D. and Cleaver, J.E. (1990). Nonrandom induction of pyrimidine-pyrimidone (6-4) photoproducts in ultraviolet-irradiated human chromatin. *J. Biol. Chem.* **265**, 5353–6.

Mitchell, P.J. and Tjian, R. (1989). Transcriptional regulation in mammalian cells by sequence-specific DNA binding proteins. *Science* **245**, 371–8.

Mizzen, C.A., Yang, X.J., Kobuko, T., Brownell, J.E., Bannister, A.J., Owen-Hughes, T., Workman, J., Berger, S.L., Kouzavides, T., Nakatani, Y., and Allis, C.D. (1996). The TAF$_{II}$250 subunit of TFIID has histone acetyltransferase activity. *Cell* **87**, 1261–70.

Moav, B. and Nemer, M. (1971). Histone synthesis: assignment to a special class of polyribosomes in sea urchin embryos. *Biochemistry* **10**, 881–9.

Moazed, D. and Johnson, A.D. (1996). A deubbiquitinating enzyme interacts with SIR4 and regulates silencing in *S. cerevisiae. Cell* **86**, 667–77.

Moehrle, A. and Paro, R. (1994). Spreading the silence: epigenetic transcriptional regulation during *Drosophila* development. *Dev. Genet.* **15**, 478–84.

Mohr, E., Trieschmann, L. and Grossbach, U. (1989). Histone H1 in two subspecies of *Chironomous thummi* with different genome sizes: homologous chromosome sites differ largely in their content of a specific H1 variant. *Proc. Natl Acad. Sci. USA* **86**, 9308–12.

Monneron, A. and Bernhard, W. (1969). Fine structural organization of the interphase nucleus in some mammalian cells. *J. Ultrastruct. Res.* **27**, 266–88.

Moore, G.D., Sinclair, D.A. and Grigliatti, T.A. (1983). Histone gene multiplicity and position effect variegation in *Drosophila melanogaster*. *Genetics* **105**, 327–44.

Morata, G., Botas, J., Kerridge, S. and Struhl, G. (1983). Homeotic transformations of the abdominal segments of *Drosophila* caused by breaking or deleting a central portion of the *bithorax* complex. *J. Embryol. Exp. Morphol.* **78**, 319–41.

Morcillo, G., de La Torre, C. and Gimenez-Martin, G. (1976). Nucleolar transcription during plant mitosis. *Exo. Cell Res.* **102**, 311–16.

Moreau, N., Angelier, N., Bonnanfant-Jais, M-L., Gounon, P. and Kubisz, P. (1986). Association of nucleoplasmin with transcription products as revealed by immunolocalization in the amphibian oocyte. *J. Cell. Biol.* **103**, 683–90.

Morgan, T.H. (1934). *Embryology and Genetics*. Columbia University Press, New York.

Morse, R.H. (1989). Nucleosomes inhibit both transcriptional initiation and elongation by RNA polymerase III *in vitro*. *EMBO J.* **8**, 2343–51.

Morse, R.H. (1992). Transcribed chromatin. *Trends Biochem. Sci.* **17**, 23–6.

Morse, R.H. (1993). Nucleosome disruption by transcription factor binding in yeast. *Science* **262**, 1563–6.

Morse, R.H. and Simpson, R.T. (1988). DNA in the nucleosome. *Cell* **54**, 285–8.

Morse, R.H., Roth, S.Y. and Simpson, R.T. (1992). A transcriptionally active tRNA gene interferes with nucleosome positioning *in vivo*. *Mol. Cell. Biol.* **12**, 4015–25.

Mottus, R., Reeves, R. and Grigliatti, T.A. (1980). Butyrate suppression of position effect variegation in *Drosophila melanogaster*. *Mol. Gen. Genet.* **178**, 465–9.

Muchardt, C. and Yaniv, M. (1993). A human homolog of *Saccharomyces cerevisiae* SNF2/SWI2 and *Drosophila* brm genes potentiates transcriptional activation by the glucocorticoid receptor. *EMBO J.* **12**, 4279–90.

Muchardt, C., Sardet, C., Bourachot, B., Onufryk, C. and Yaniv, M. (1995). A human protein with homology to *S. cerevisiae* SNF5 interacts with the potential helicase hbrm. *Nucleic. Acids Res.* **23**, 1127–32.

Muchardt, C., Roges, J.C., Bourarchot, B., Legouy, E., and Yaniv, M. (1996). BRG-1 proteins, components of the human SNF/SWI complex are phosphorylated and excluded from the condensed chromosomes during mitosis. *EMBO J.* **15**, 3394–402.

Mueller, R.D., Yasuda, H., Hatch, C.L., Bonner, W.M. and Bradbury, E.M. (1985). Identification of ubiquitinated histones H2A and H2B in *Physarum polycephalum*. Disappearance of these proteins at metaphase and reappearance at anaphase. *J. Biol. Chem.* **260**, 5147–53.

Muller-Storm, H., Sogo, J.M. and Schaffner, W. (1989). An enhancer stimulates transcription in *trans* when attached to the promoter via a promoter bridge. *Cell* **58**, 767–77.

Murray, A.W. and Kirschner, M.W. (1989). Cyclin synthesis drives the early embryonic cell cycle. *Nature* **339**, 275–80.

Murre, C., McCaw, P.S. and Baltimore, D. (1989). A new DNA binding and dimerization motif in immunoglobulin enhancer binding, daughterless, myoD and myc proteins. *Cell* **56**, 777–83.

Nacheva, G.A., Guschin, D.Y., Preobrazhenskaya, O.V., Karpov, V.L., Elbradise, K.K. and Mirzabekov, A.D. (1989). Change in the pattern of histone binding to DNA upon transcriptional activation. *Cell* **58**, 27–36.

Nakamura, H., Morita, T. and Sato, C. (1986). Structural organization of replicon domains during DNA synthetic phase in the mammalian nucleus. *Exp. Cell Res.* **165**, 291–7.

Nakatani, Y., Bagby, S. and Ikura, M. (1996). The histone folds in transcription factor TFIID. *J. Mol. Biol.* **271**, 6575–8.

Nan, X., Meehan, R.R. and Bird, A.P. (1993). Dissection of the methyl-CpG binding domain from the chromosomal protein MeCP2. *Nucleic. Acids Res.* **21**, 4886–92.

Nan, X., Tate, P., Li, E. and Bird, A.P. (1996). MeCP2 is a transcriptional repressor with abundant binding sites in genomic chromatin. *Mol. Cell. Biol.* **16**, 414–21.

Nan, X., Campoy, J. and Bird, A. (1997). MeCP2 is a transcriptional repressor with abundant binding sites in genomic chromatin. *Cell* **88**, 1–11.

Nan, X., Ng, H-H., Johnson, C.A.,. Laherty, C.D., Turner, B.M., Eisenman, R.N. and Bird, A. (1998). Transcriptional repression by the methyl-CpG binding protein MeCP2 involves the mSIN3-histone deacetylase repression complex. *Nature* (in press).

Neigeborn, L. and Carlson, M. (1984). Genes affecting the regulation of *SUC2* gene repressor in *saccharymyces cerevisiae*. *Genetics* **108**, 845–58.

Nelson, H.C.M., Finch, J.T., Luisi, B.F. and Klug, A. (1987). The structure of an oligo (dA).oligo (dT) tract and its biological implication. *Nature (London)* **330**, 221–6.

Nelson, P.P., Albright, S.C., Wiseman, J.M. and Garrard, W.T. (1979). Reassociation of histone H1 with nucleosomes. *J. Biol. Chem.* **254**, 11751–60.

Nelson, T., Wiegand, R. and Brutlag, D. (1981). Ribonucleic acid and other polyanions facilitate chromatin assembly *in vitro*. *Biochemistry* **20**, 2594–601.

Nenney, D.L. (1953). Nucleocytoplasmic interaction during conjugation in *Tetrahymena*. *Biol. Bull* **105**, 133–48.

Ner, S.S. and Travers, A.A. (1994). HMG-D, the *Drosophila melanogaster* homologue of HMG1 protein, is associated with early embryonic chromatin in the place of histone H1. *EMBO J.* **13**, 1817–22.

Newlon, C.S. (1988). Yeast chromosome replication and segregation. *Microb. Rev.* **52**, 568–601.

Newlon, C.S., Collins, I., Dershowitz, A., Deshpande, A.M., Greenfelder, S.A., Ong, L.Y. and Theis, J.F. (1993). Analysis of replication origin function on chromosome III of *Saccharomyces cerevisiae*. *Cold Spring Harb. Symp. Quant. Biol.* **58**, 415–23.

Newport, J. (1987). Nuclear reconstitution *in vitro*: stages of assembly around protein-free DNA. *Cell* **48**, 205–17.

Newport, J. and Spann, T. (1987). Disassembly of the nucleus in mitotic extracts: membrance vesicularization, lamin disassembly, and chromosome condensation are independent processes. *Cell* **48**, 219–30.

Newport, J., Wilson, K.L. and Dunphy, W.G. (1990). A lamin-independent pathway for nuclear envelope assembly. *J. Cell Biol.* **111**, 2247–59.

Newport, J.W. and Kirschner, M.W. (1982a). A major developmental transition in early *Xenopus* embryos. 1. Characterization and timing of cellular changes at the mid blastula stage. *Cell* **30**, 675–86.

Newport, J.W. and Kirschner, M.W. (1982b). A major developmental transition in early *Xenopus* embryos. II. Control of the onset of transcription. *Cell* **30**, 687–96.

Newrock, K.M., Alfageme, C.R., Nardi, R.V. and Cohen, L.H. (1977). Histone changes during chromatin remodeling in embryogenesis. *Cold Spring Harb. Symp. Quant. Biol.* **42**, 421–31.

Niggli, H. and Cerutti, P. (1982). Nucleosomal distribution of thymine photodimers following far and near ultraviolet irradiation. *Biochem. Biophys. Res. Commun.* **105**, 1215–23.

Nightingale, K. and Wolffe, A.P. (1995). Methylation at CpG sequences does not influence histone H1 binding to a nucleosome including a *Xenopus borealis* 5S rRNA gene. *J. Biol. Chem.* **270**, 4197–200.

Nightingale, K., Dimitrov, S., Reeves, R. and Wolffe, A.P. (1996). Evidence for a shared structural role for HMG1 and linker histones B4 and H1 in organizing chromatin. *EMBO J.* **15**, 548–61.

Nightingale, K.P., Pruss, D. and Wolffe, A.P. (1996). A single high affinity binding site for histone H1 in a nucleosome containing the *Xenopus borealis* 5S ribosomal RNA gene. *J. Biol. Chem.* **271**, 7090–4

Nikolov, D.B. and Burley, S.K. (1994). 2.1 Å resolution refined structure of a TATA box-binding protein (TBP). *Nature Struct. Biol.* **1**, 621–37.

Nissen, K.A., Lan, S.Y. and Smerdon, M.J. (1986). Stability of nuclesome placement in newly repaired regions of DNA. *J. Biol. Chem.* **261** 8585–8.

Nobile, C., Nickol, J. and Martin, R.G. (1986). Nucleosome phasing on a DNA fragment from the replication origin of simian virus 40 and rephasing upon cruciform formation of the DNA. *Mol. Cell. Biol.* **6**, 2916–22.

Noll, M. (1974a). Subunit structure of chromatin. *Nucleic. Acids Res.* **1**, 1573–8.

Noll, M. (1974b). Internal structure of the chromatin subunit. *Nature (London)* **251**, 249–51.

Noll, M. and Kornberg, R.D. (1977). Action of micrococcal nuclease on chromatin and the location of histone H1. *J. Mol. Biol.* **109**, 393–404.

Norton, V.G., Imai, B.S., Yau, P. and Bradbury, E.M. (1989). Histone acetylation reduces nucleosome core particle linking number change. *Cell* **57**, 449–57.

Norton, V.G., Marvin, K.W., Yau, P. and Bradbury, E.M. (1990). Nucleosome linking member change controlled by acetylation of histones H3 and H4. *J. Biol. Chem.* **265**, 19848–52.

Ochs, R. L., Lischwe, M. A., Spohn, W. H. and Busch, H. (1985). Fibrillarin: a new protein of the nucleolus identified by autoimmune sera. *Biol. Cell* **54**, 123–33.

O'Donohue, M.F., Duband-Goulet, I., Hamiche, A. and Prunell, A. (1994). Octamer displacement and redistribution in transcription of single nucleosome. *Nucleic. Acids Res.* **22** 937–45.

Ogryzko, V.V., Schiltz, R.L., Russanova, V., Howard, B.H. and Nakatani, Y. (1996). The transcriptional coactivators p300 and CBP are histone acetyltransferases. *Cell* **87**, 953–9.

Ohaviano, Y. and Gerace, L. (1985). Phosphorylation of the nuclear lamin during interphase and mitosis. *J. Biol. Chem.* **260**, 624–32.

Ohkuma, Y., Horikoshi, M., Roeder, R.G. and Desplan, C. (1990). Engrailed, a homeodomain protein, can repress *in vitro* transcription by competition with the TATA box binding protein transcription factor TFIID. *Proc. Natl Acad. Sci. USA* **87**, 2289–93.

Ohkuma, Y., Hoshimoto, S., Wang, C.K., Horikoshi, M. and Roeder, R.G. (1995). Analysis of the role of TFIIE in basal transcription and TFIIH-mediated carboxyl-terminal domain phosphorylation through structure-function studies of TFIIE-α. *Mol. Cell Biol.* **15**, 4856–66.

Ohlenbusch, H.H., Olivera, B.M., Tuan, D. and Davidson, N. (1967). Selective dissociation of histones from calf thymus nucleoprotein. *J. Mol. Biol.* **25**, 299–315.

Ohsumi, K., Katagiri, C. and Kishimoto, T. (1993). Chromosome condensation in *Xenopus* mitotic extracts without histone H1. *Science* **262**, 2033–5.

O'Keefe, R. T., Henderson, S. C. and Spector, D. L. (1992). Dynamic organization of DNA replication in mammalian cell nuclei: spatially and temporally defined replication of chromosome-specific alpha-satellite DNA sequences. *J. Cell Biol.* **116**, 1095–110.

Olins, A.L. and Olins, D.E. (1974). Spheroid chromatin units (V bodies). *Science* **183**, 330–2.

Oliva, R., Bazett-Jones, D.P., Locklear, L. and Dixon, G.H. (1990). Histone hyperacetylation can induce unfolding of the nucleosome core particle. *Nucleic. Acids Res.* **18**, 2739–47.

Onate, S.E., Prendergast, P., Wagner, J.P., Nissen, M., Reeves, R., Pettijohn, D.E. and Edwards, D.P. (1994). The DNA-bending protein HMG-1 enhances progesterone receptor binding to its target DNA sequences. *Mol. Cell. Biol.* **14**, 3376–91.

O'Neill, T.E., Roberge, M. and Bradbury, E.M. (1992). Nucleosome arrays inhibit both initiation and elongation of transcription by T7 RNA polymerase. *J. Mol. Biol.* **233**, 67–78.

Onnuki, Y. (1968). Structure of chromosomes: morphological studies on the spiral structure of human somatic chromosomes. *Chromosoma* **25**, 402–8.

Orlando, V. and Paro, R. (1993). Mapping Polycomb-repressed domains in the bithorax complex using *in vivo* formaldehyde cross-linked chromatin. *Cell* **75**, 1187–98.

Ostrowski, M.C., Richard-Foy, H., Wolford, R.G., Berard, D.S. and Hager, G.L. (1983). Glucocorticoid regulation of transcription at an amplified episomal promoter. *Mol. Cell Biol.* **3**, 2045–57.

Otten, A.D. and Tapscott, S.J. (1995). Triplet repeat expansion in myotinic dystrophy alters the adjacent chromatin structure. *Proc. Natl Acad. Sci. USA* **92**, 5465–9.

Otting, G., Qian, Y.Q., Billeter, M., Muller, M., Affolter, M., Gehring, W.J. and Wuthrich, K. (1990). ProteinDNA contacts in the structure of a homeo-domain DNA complex determined by nuclear magnetic resonance spectroscopy in solution. *EMBO J.* **9**, 3085–92.

Owen-Hughes, T. and Workman, J.L. (1995). Experimental analysis of chromatin function in transcriptional control. *Crit. Rev. Euk. Gene Exp.* **4**, 1–39.

Owen-Hughes, T. and Workman, J.L. (1996). Remodeling the chromatin structure of a nucleosome array by transcription factor-targeted trans-displacement of histones. *EMBO J.* **15**, 4702–12.

Owen-Hughes, T., Utley, R.T., Cote, J., Peterson, C.L. and Workman, J.L. (1996). Persistent site specific remodeling of a nucleosome array by transient action of the SWI/SNF complex. *Science* **273**. 515–16.

Palladino, F., Laroche, T., Gilson, E., Axelrod, A., Pillus, L. and Gasser, S.M. (1993). SIR3 and SIR4 proteins are required for the positioning and integrity of yeast telomeres. *Cell* **75**, 543–55.

Palmer, D.K., O'Day, K., Wener, W.H., Andrews, B.S. and Margolis, R.L. (1987). A 17 kDa centromere protein (CENP-A) copurifies with nucleosome core particles and with histones. *J. Cell Biol.* **104**, 805–15.

Palmer, D.K., O'Day, K., Trong, H.L., Charbonneau, H. and Margolis, R.L. (1991). Purification of the centromere-specific protein CENP-A and demonstration that it is a distinctive histone. *Proc. Natl Acad. Sci. USA* **88**, 3734–8.

Panning, B. and Jaenisch, R. (1996). DNA hypomethylation can activate Xist expression and silence X-linked genes. *Genes Dev.* **10**, 1991–2002.

Paranjape, S.M., Kamakaka, R.T. and Kadonaga, J.T. (1994). Role of chromatin structure in the regulation of transcription by RNA polymerase II. *Annu. Rev. Biochem.* **63**, 265–97.

Paranjape, S.M., Krumm, A. and Kadonaga, J.T. (1995). HMG17 is a chromatin specific transcriptional coactivator that increases the efficiency of transcription initiation. *Genes Dev.* **9**, 1978–91.

Pardon, J.F., Worcester, D.L., Wooley, J.C., Tatchell, K., van Holde, K.E. and Richards, B.M. (1975). Low-angle neutron scattering from chromatin subunit particles. *Nucleic. Acids Res.* **2**, 2163–75.

Pardue, M.L. and Hennig, W. (1990). Heterochomatin: junk or collectors item? *Chromosoma* **100**, 3–7.

Park, E.C. and Szostak, J.W. (1990). Point mutations in the yeast histone H4 gene prevent silencing of the silent mating type locus HML. *Mol. Cell. Biol.* **10**, 4932–4.

Parker, C.S. and Roeder, R.G. (1977). Selective and accurate transcription of the *Xenopus laevis* 5S RNA genes in isolated chromatin by purified RNA polymerase III. *Proc. Natl Acad. Sci. USA* **74**, 44–8.

Parkhurst, S.M., Harrison, D.A., Remington, M.P., Spana, C., Kelley, R.L., Coyne, R.S. and Corces, V.G. (1988). The *Drosophila su(Hw)* gene, which controls the phenotypic effect of the *gypsy* transposable element, encodes a putative DNA binding protein. *Genes Dev.* **2**, 1205–15.

Parthun, M.R., Widom, J. and Gottschling, D.E. (1996). The major cytoplasmic histone acetyltransferase in yeast: links to chromatin replication and histone metabolism. *Cell* **87**, 85–94.

Pasero, P., Braguglia, D. and Gasser, S.M. (1997). ORC-dependent and origin-specific initiation of DNA replication at defined foci in isolated yeast nuclei. *Genes Dev.* **11** 1504–18.

Patterton, D. and Wolffe, A.P. (1996). Developmental roles for chromatin and chromosomal structure. *Dev. Biol.* **173**, 2–13.

Patterton, H.G. and Simpson, R.T. (1994). Nucleosomal location of the *STE6* TATA box and Matα2p-mediated repression. *Mol. Cell Biol.* **14**, 4002–10.

Paule, M.R. (1990). In search of the single factor. *Nature (London)* **344**, 819–20.

Paull, T.T., Haykinson, M.J. and Johnson, R.C. (1993). The non specific DNA-binding and -bending proteins HMG1 and HMG2 promote the assembly of complex nucleoprotein structures. *Genes Dev.* **7**, 1521–34.

Paulson, J.R. and Laemmli, U.K. (1977). The structure of histone-depleted metaphase chromosomes. *Cell* **12**, 817–28.

Pavletich, N.P. and Pabo, C.O. (1991). Zinc fingerDNA recognition: crystal structure of a Zif 268DNA complex at 2.1 Å. *Science* **252**, 809–16.

Pavlovic, B. and Horz, W. (1988). The chromatin structure at the promoter of a glyceraldehyde phosphate dehydrogenase gene from *Saccharomyces cerevisiae* reflects its functional state. *Mol. Cell. Biol.* **8**, 5513–20.

Pavlovic, J., Banz, E. and Parish, R.W. (1989). The effects of transcription on the nucleosome structure of four *Dictyostelium* genes. *Nucleic. Acids Res.* **17**, 2315–32.

Pays, E., Donaldson, D. and Gilmour, R.S. (1979). Specificity of chromatin transcription *in vitro* anomalies due to RNA-dependent RNA synthesis. *Biochim. Biophys. Acta* **562**, 112–30.

Pazin, M.J., Kamakaka, R.T. and Kadonaga, J.T. (1994). ATP-dependent nucleosome reconfiguration and transcriptional activation from pre-assembled chromatin templates. *Science* **266**, 2007–11.

Pazin, M.J., Bhargava, P., Geiduschek, E.P. and Kadonaga, J.T. (1997). Nucleosome mobility and the maintenance of nucleosome positioning. *Science* **276**, 809–12.

Pearson, C.E. and Sinden, R.R. (1996). Alternative structures in duplex DNA formed within the trinucleotide repeats of the myotinic dystrophy and fragile X loci. *Biochemistry* **35**, 5041–53.

Pederson, D.S. and Fidrych, T. (1994). Heat shock factor can activate transcription while bound to nucleosomal DNA in *Saccharomyces cerevisiae. Mol. Cell Biol.* **14**, 189–99.

Pederson, D.S. and Morse, R.H. (1990). Effect of transcription of yeast chromatin on DNA topology *in vivo. EMBO J.* **9**, 1873–81.

Pederson, D.S., Venkatesan, M., Thoma, F. and Simpson, R.T. (1986). Isolation of an episomal yeast gene and replication origin as chromatin. *Proc. Natl Acad. Sci. USA* **83**, 7206–10.

Pehrson, J.R. (1989). Thymine dimer formation as a probe of the path of DNA in and between nucleosomes in intact chromatin. *Proc. Natl Acad. Sci. USA* **86**, 9149–53.

Pehrson, J.R. (1995). Probing the conformation of nucleosome linker DNA in situ with pyrimidine dimer formation. *J. Biol. Chem.* **270**, 22440–4.

Pehrson, J.R. and Cohen, L.H. (1992). Effects of DNA looping on pyrimidine dimer formation. *Nucleic. Acids Res.* **20**, 1321–4.

Pehrson, J.R. and Fried, V.A. (1992). Macro H2A, a core histone containing a large nonhistone region. *Science* **257**, 1396–400.

Peifer, M., Karch, F. and Bender, W. (1987). The bithorax complex: control of segment identity. *Genes Dev.* **1**, 891–8.

Pelham, H.R.B. and Brown, D.D. (1980). A specific transcription factor can bind either the 5S RNA gene or 5S RNA. *Proc. Natl Acad. Sci. USA* **77**, 4170–4.

Pennings, S., Meersseman, G. and Bradbury, E.M. (1991). Mobility of positioned nucleosomes on 5S rDNA. *J. Mol. Biol.* **220**, 101–10.

Pennings, S., Meersseman, G. and Bradbury, E.M. (1994). Linker histones H1 and H5 prevent the mobility of positioned nucleosomes. *Proc. Natl Acad. Aci. USA* **91**, 10275–9.

Pentiggia, A., Rimini, R., Harley, V.R., Goodfellow, P.N., Lovell-Badge, R. and Bianchi, M.E. (1994). Sex-reversing mutations affect the architecture of SRY-DNA complexes. *EMBO J.* **13**, 6115–24.

Perez, A., Kastner, P., Sethi, S., Lutz, Y., Reibel, C. and Chambon, P. (1993). PMLRAR homodimers: distinct DNA binding properties and heteromeric interactions with RXR. *EMBO J.* **12**, 3171–82.

Perlmann, T. and Wrange, O. (1988). Specific glucocorticoid receptor binding to DNA reconstituted in a nucleosome. *EMBO J.* **7**, 3073–83.

Perlmann, T. and Wrange, O. (1991). Inhibition of chromatin assembly in *Xenopus* oocytes correlates with derepression of the mouse mammary tumor virus promoter. *Mol. Cell Biol.* **11**, 5259–65.

Perry, C.A. and Annunziato, A.T. (1989). Influence of histone acetylation on the solubility, H1 content and DNase I sensitivity of newly assembled chromatin. *Nucleic. Acids Res.* **17**, 4275–91.

Perry, C.A., Allis, C.D. and Annunziato, A.T. (1993). Parental nucleosomes segregated to newly replicated chromatin are underacetylated relative to those assembled *de novo*. *Biochemistry* **32**, 13615–23.

Perry, M. and Chalkley, R. (1981). The effect of histone hyperacetylation on the nuclease sensitivity and the solubility of chromatin. *J. Biol. Chem.* **256**, 3313–8.

Perry, M., Thomson, G.H. and Roeder, R.G. (1986). Major transitions in histone gene expression do not occur during development in *Xenopus laevis*. *Dev. Biol.* **116**, 532–8.

Petersen, M.G. and Tjian, R. (1992). The tell tail trigger. *Nature* **358**, 620–1.

Peterson, C.L. and Herskowitz, I. (1992). Characterization of the yeast SWI1, SWI2 and SWI3 genes which encode a global activator of transcription. *Cell* **68**, 573.

Peterson, C.L. and Tamkun, J.W. (1995). The SWI/SNF complex: a chromatin remodeling machine? *TIBS* **20**, 143–6.

Peterson, C.L., Kruger, W., and Herskowitz, L. (1991). A functional interaction between the C-terminal domain of RNA polymerase II and the negative regulator SIN1. *Cell* **64**, 1135–43.

Peterson, C.L., Dingwall, A. and Scott, M.P. (1994). SWI/SNF gene products are components of a large multisubunit complex required for transcriptional enhancement. *Proc. Natl Acad. Sci. USA* **91**, 2905–8.

Pfaffle, P., Gerlach, V., Bunzel, L. and Jackson, V. (1990). *In vitro* evidence that transcription induced stress causes nucleosome dissolution and regeneration. *J. Biol. Chem.* **265**, 16830–40.

Pfeifer, K. and Tilghman, S.M. (1994). Allele-specific gene expression in mammals: the curious case of the imprinted RNAs. *Genes Dev.* **8**, 1867–74.

Philipsen, S., Talbot, O., Fraser, P. and Grosveld, F. (1990). The β-globin dominant control region: hypersensitive site. *EMBO J.* **9**, 2159–68.

Philpott, A. and Leno, G.H. (1992). Nucleoplasmin remodels sperms chromatin in *Xenopus* egg extracts. *Cell* **69**, 759–67.

Philpott, A., Leno, G.H. and Laskey, R.A. (1991). Sperm decondensation in *Xenopus* egg cytoplasm is mediated by nucleoplasmin. *Cell* **65**, 569–78.

Phi-Van, L. and Stratling, W.H. (1988). The matrix attachment regions of the chicken lysozyme gene co-map with the boundaries of the chromatin domain. *EMBO J.* **7**, 655–64.

Pieler, T., Hamm, J. and Roeder, R.G. (1987). The 5S internal control region is composed of three distinct sequence elements, organized as two functional domains with variable spacing. *Cell* **48**, 91–100.

Pina, B., Bruggemeier, U. and Beato, M. (1990a). Nucleosome positioning modulates accessibility of regulatory proteins to the mouse mammary tumor virus promoter. *Cell* **60**, 719–31.

Pina, B., Barettino, D., Truss, M. and Beato, M. (1990b). Structural features of a regulatory nucleosome. *J. Mol. Biol.* **216**, 975–90.

Pinol-Roma, S. and Dreyfuss, G. (1992). Shuttling of pre-mRNA binding proteins between nucleus and cytoplasm. *Nature* **355**, 730–2.

Pirotta, V. (1997). Chromatin-silencing mechanisms in *Drosophila* maintain patterns of gene expressin. *Trends Genet.* **13**, 314–15.

Pirotta, V. and Rastelli, L. (1994). *White* gene expression, repressive chromatin domains and homeotic gene regulation in *Drosophila*. *BioEssays* **16**, 549–56.

Platero, J.S., Hartnett, T. and Eissenberg, J.C. (1995). Functional analysis of the chromodomain of HP1. *EMBO J.* **14**, 3977–86.

Pluta, A.F., Cooke, C.A. and Earnshaw, W.C. (1990). Structure of the human centromere at metaphase. *Trends Biochem. Sci.* **15**, 181–5.

Pluta, A.F., Saitoh, N., Goldberg, I. and Earnshaw, W.C. (1992). Identification of a subdomain of CENP-B that is necessary and sufficient for localization to the human centromere. *J. Cell Biol.* **116**, 1081–93.

Pluta, A.F., Mackay, A.M., Ainsztein, A.M., Goldberg, I.G. and Earnshaw, W.C. (1995). The centromere: hub of chromosomal activities. *Science* **270**, 1591–4.

Poccia, D. (1986). Remodeling of nucleoproteins during gametogenesis, fertilization, and early development. *Int. Rev. Cytol.* **105**, 1–65.

Poccia, D.L. and Green, G.R. (1992). Packaging and unpackaging the sea urchin genome. *Trends Biochem. Sci.* **17**, 223–7.

Poccia, D., Salik, J. and Krystal, G. (1981). Transitions in histone variants of the male pronucleus following fertilization and evidence for a maternal store of cleavage stage histones in the sea urchin egg. *Dev. Biol.* **82**, 287–99.

Poccia, D., Wolff, R., Kragh, S. and Williamson, P. (1985). RNA synthesis in male pronuclei of the sea urchin. *Biochim. Biophys. Acta* **824**, 349–59.

Polach, K.J. and Widom, J. (1995). Mechanism of protein access to specific DNA sequences in chromatin: a dynamic equilibrium model for gene regulation. *J. Mol. Biol.* **254**, 130–49.

Polach, K.J. and Widom, J. (1996). A model for cooperative binding of eukaryotic regulatory proteins to nucleosomal target sites. *J. Mol. Biol.* **258**, 800–12.

Poljak, L.G. and Gralla, J.D. (1987). Competition for formation of nucleosomes on fragmented SV40 DNA: a hyperstable nucleosome forms on the termination region. *Biochemistry* **26**, 295–303.

Poljak, L.G., Seum, C., Mattioni, T. and Laemmli, U.K. (1994). SARS stimulate but do not confer position effect gene expression. *Nucleic. Acids Res.* **22**, 4386–94.

Postnikov, Y.V., Trieschmann, L., Rickers, A. and Bustin, M. (1995). Homodimers of chromosomal proteins HMG-14 and HMG-17 in nucleosome cores. *J. Mol. Biol.* **252**, 423–32.

Powers, J.A. and Eissenberg, J.C. (1993). Overlapping domains of the heterochromatin associated protein HP1 mediate nuclear localization and heterochromatin binding. *J. Cell Biol.* **120**, 2653–68.

Prescott, D.M. and Bender, M.A. (1962). Synthesis of RNA and protein during mitosis in mammalian tissue culture cells. *Exp. Cell. Res.* **26**, 260–8.

Price, D.H., Sluder, A.E. and Greenleaf, A.L. (1989). Dynamic interaction between a *Drosophila* transcription factor and RNA polymerase II. *Mol. Cell Biol.* **9**, 1465–75.

Prioleau, M.-N., Huet, J., Sentenac, A. and Méchali, M. (1994). Competition between chromatin and transcription complex assembly regulates gene expression during early development. *Cell* **77**, 439.

Prioleau, M.-N., Buckle, R.S. and Méchali, M. (1995). Programming a repressed but committed chromatin structure during early development. *EMBO J.* **14**, 5073–84.

Prior, C.P., Cantor, C.R., Johnson, E.M., Littau, V.C. and Allfrey, V.G. (1983). Reversible changes in nucleosome structure and histone H3 accessibility in transcriptionally active and inactive states of rDNA chromatin. *Cell* **34**, 1033–42.

Privé, G.G., Yanagi, K. and Dickerson, R.E. (1991). Structure of the B-DNA decamer CCAACGTTGG and comparison with isomorphous decamers CCAAGATTGG and CCAGGCCTGG. *J. Mol. Biol.* **217**, 177–91.

Protacio, R.U. and Widom, J. (1996). Nucleosome transcription studied in a real time synchronous system: test of the lexosome model and direct measurement of effects due to histone octamer. *J. Mol. Biol.* **256** 458–72.

Prunell, A. (1982). Nucleosome reconstitution on plasmid-inserted poly(dA).poly(dT). *EMBO J.* **1**, 173–9.

Pruss, D. and Bavykin, S.G. (1997). Chromatin studies by DNA–protein crosslinking. *Methods* **12**, 36–47.

Pruss, D. and Wolffe, A.P. (1993). Histone–DNA contacts in a nucleosome core containing a *Xenopus* 5S rRNA gene. *Biochemistry* **32**, 6810–14.

Pruss, D., Bushman, F.D. and Wolffe, A.P. (1994). HIV integrase directs integration to sites of severe DNA distortion within the nucleosome core. *Proc. Natl Acad. Sci. USA* **91**, 5913–17.

Pruss, D., Hayes, J.J. and Wolffe, A.P. (1995). Nucleosomal anatomy – where are the histones? *BioEssays* **17**, 161–70.

Pruss, D., Bartholomew, B., Persinger, J., Hayes, J., Arents, G., Moudrianakis, E.N. and Wolffe, A.P. (1996). An asymmetric model for the nucleosome: a binding site for linker histones inside the DNA gyres. *Science* **274**, 614–17.

Ptashne, M. (1986). Gene regulation by proteins acting nearly and at a distance. *Nature (London)* **322**, 697–701.

Puerta, C., Hernandez, F., Gutierrez, C., Pineiro, M., Lopez-Alarcon, L. and Palacian, E. (1993). Efficient transcription of a DNA template associated with histone (H3.H4)$_2$ tetramers. *J. Biochem.* **268**, 26663–7.

Pugh, B.F. and Tjian, R. (1990). Mechanism of transcriptional activation by SP1: evidence for coactivators. *Cell* **61**, 1187–97.

Puhl, H.L. and Behe, M.J. (1993). Structure of nucleosomal DNA at high salt concentration as probed by hydroxyl radical. *J. Mol. Biol.* **229**, 827–32.

Puhl, H.L. and Behe, M.J. (1995). Poly (dA).poly (dT) forms very stable nucleosomes at higher temperatures. *J. Mol. Biol.* **245**, 559–67.

Puvion-Dutilleul, F., Bachellerie, J. P. and Puvion, E. (1991). Nucleolar organization of HeLa cells as studied by in situ hybridization. *Chromosoma.* **100**, 395–409.

Puvion-Dutilleul, F., Chelbi-Alix, M. K., Koken, M., Quignon, F., Puvion, E. and de The, H. (1995). Adenovirus infection induces rearrangements in the intranuclear distribution of the nuclear body-associated PML protein. *Exp. Cell Res.* **218**, 9–16.

Qui, H., Park, E., Prakash, L. and Prakesh, S. (1993). The *Saccharomyces cerevisiae* DNA repair gene RAD25 is required for transcription by RNA polymerase II. *Genes Dev.* **7**, 2161–71.

Rabl, C. (1885). Uber zellteilung. *Morphol. Jahrbuch* **10**, 214–230.

Ram, P.T. and Schultz, R.M. (1993). Reporter gene expression in G2 of the 1 cell mouse embryo. *Dev. Biol.* **156**, 552–6.

Ramakrishnan, V., Finch, J.T., Graziano, V., Lee, P.L. and Sweet, R.M. (1993). Crystal structure of globular domain of histone H5 and its implications for nucleosome binding. *Nature* **362**, 219–23.

Ramanathan, B. and Smerdon, M.J. (1989). Enhanced DNA repair synthesis in hyperacetylated nucleosomes. *J. Biol. Chem.* **264**, 11026–34.

Ramsay, N., Felsenfeld, G., Rushton, B.M. and McGhee, J.D. (1984). A 145 base pair DNA sequence that positions itself precisely and asymmetrically on the nucleosome core. *EMBO J.* **3**, 2605–11.

Ranjan, M., Wong, J., and Shi, Y.-B. (1994). Transcriptional repression of *Xenopus* TRβ gene is mediated by a thyroid hormone response element located near the start site. *J. Biol. Chem.* **269**, 24699–705.

Raska, I., Reimer, G., Jarnik, M., Kostrouch, Z. and Raska, K.,Jr. (1989). Does the synthesis of ribosomal RNA take place within nucleolar fibrillar centers or dense fibrillar components? *Biol. Cell* **65**, 79–82.

Rasmussen, R., Benvegnu, D., O'Shea, E.K., Kim, P.S. and Albe, T. (1991). X-ray scattering indicates that the leucine zipper is a coiled coil. *Proc. Natl Acad. Sci. USA* **88**, 561–4.

Rastelli, L., Chan, C.S. and Pirotta, V. (1993). Related chromosome binding sites for zeste, suppressors of zeste and Polycomb group proteins in *Drosophila* and their dependence on enhancer of zeste function. *EMBO J.* **12**, 1513–22.

Rattner, J.B. and Lin, C.C. (1985). Radical loops and helical coils coexist in metaphase chromosomes. *Cell* **42**, 291–6.

Read, C.M., Cary, P.D., Crane-Robinson, C., Driscoll, P.C. and Norman, D.G. (1993). Solution structure of a DNA-binding domain from HMG1. *Nucleic. Acids Res.* **21**, 3427–37.

Reddy, B.A., Etkin, L.D. and Freemont, P.S. (1992). A novel zinc finger coiled-coil domain in a family of nuclear proteins. *Trends Biochem. Sci.* **17**, 344–5.

Reeves, R., Gorman, C.M. and Howard, B. (1985). Minichromosome assembly of non-integrated plasmid DNA transfected into mammalian cells. *Nucleic. Acids Res.* **13**, 3599–615.

Reifsnyder, C., Lowell, J., Clarke, A. and Pillus, L. (1996). Yeast SAS silencing genes and human genes associated with AML and HIV-1Tat interactions are homologous with acetyltransferases. *Nat. Genet.* **14**, 42–9.

Reik, A., Schutz, G. and Stewart, A.F. (1991). Glucocorticoids are required for establishment and maintenance of an alteration in chromatin structure: induction leads to a reversible disruption of nucleosomes over an enhancer. *EMBO J.* **10**, 2569–76.

Reik, W. and Surani, M.A. (1989). Cancer genetics. Genomic imprinting and embryonal tumours. *Nature* **338**, 112–13.

Reitman, M. and Felsenfeld, G. (1990). Developmental regulation of topoisomerase II sites and DNase I-hypersensitive sites in the chicken β-globin locus. *Mol. Cell Biol.* **10**, 2774–86.

Reitman, M., Lee, E., Westphal, H. and Felsenfeld, G. (1990). Site-independent expression of the chicken β^A-globin gene in transgenic mice. *Nature (London)* **348**, 749–52.

Reitman, M., Lee, E., Westphal, H. and Felsenfeld, G. (1993). An enhancer/locus control region is not sufficient to open chromatin. *Mol. Cell. Biol.* **13**, 3990–8.

Renauld, H., Aparicio, O.M., Zierath, P.D., Billington, B.L., Chhablani, S.K. and Gottschling, D.E. (1993). Silent domains are assembled continuously from the telomere and are defined by promoter distance and strength, and by SIR 3 dosage. *Genes Dev.* **7**, 1133–45.

Renauld, J.P., Rochel, N. Ruff, M., Vivat, V., Chambon, P., Gronemeyer, H. and Moras, D. (1995). Crystal structure of the RAR-γ ligand binding. *Nature* **378**, 681–9.

Rendon, M.C., Rodrigo, R.M., Goenechea, L.G., Garcia-Herdugo, V., Valdivia, M.M., Moreno, F.J. (1992). Characterization and immunolocalization of a nucleolar antigen with anti-NOR serum in HeLa cells. *Exp. Cell. Res.* **200**, 393–403.

Reuter, G. and Spierer, P. (1992). Position-effect variegation and chromatin proteins. *BioEssays* **14**, 605–12.

Reuter, G., Giarre, M., Farah, J., Gausz, J., Spierer, A. and Spierer, P. (1990). Dependence of position-effect variegation in V *Drosophila* on dose of a gene encoding an unusual zinc-finger protein. *Nature (London)* **344**, 219–23.

Reznikoff, W.S., Siegele, D.A., Cowing, D.W. and Gross, C.A. (1985). The regulation of transcription initiation in bacteria. *Ann. Rev. Genet.* **19**, 355–87.

Rhodes, D. (1985). Structural analysis of a triple complex between the histone octamer, a *Xenopus* gene for 5S RNA and transcription factor IIIA. *EMBO J.* **4**, 3473–82.

Rhodes, D. and Klug, A. (1980). Helical periodicity of DNA determined by enzyme digestion. *Nature (London)* **286**, 573–8.

Richard-Foy, H. and Hager, G.L. (1987). Sequence specific positioning of nucleosomes over the steroid-inducible MMTV promoter. *EMBO J.* **6**, 2321–8.

Richmond, T.J., Finch, J.T., Rushton, B., Rhodes, D. and Klug, A. (1984). Structure of the nucleosome core particle at 7 Å resolution. *Nature (London)* **311**, 532–7.

Richmond, T.J., Rechsteiner, T. and Luger, K. (1993). Studies of nucleosome structure. *Cold Spring Harb. Symp. Quant. Biol.* **58**, 265–72.

Ridsdale, J.A., Henzdel, M.F., Decluve, G.P. and Davie, J.R. (1990). Histone acetylation alters the capacity of the H1 histones to condense transcriptionally active/competent chromatin. *J. Biol. Chem.* **265**, 5150–6.

Riggs, A.D. and Porter, T.N. (1996). In *Epigenetic Mechanisms of Gene Regulation* eds X. Russo, R.A. Martienssen, and A.D. Riggs, Cold Spring Harbor Laboratory Press, 29–45.

Riggs, A.P. and Pfeifer, G.P. (1992). X-chromosome inactivation and cell memory. *Trends Genet.* **8**, 169–74.

Riley, D. and Weintraub, H. (1979). Conservative segregation of parental histones during replication in the presence of cycloheximide. *Proc. Natl Acad. Sci. USA* **76**, 328–32.

Ringertz, N.R. and Savage, R.E. (1976). *Cell Hybrids.* Academic Press, New York.

Ringertz, N.R., Nyman, U. and Bergman, M. (1985). DNA replication and H5 histone exchange during reactivation of chick erythrocyte nuclei in heterokaryons. *Chromosoma* **91**, 391–6.

Risley, M.S. (1983). Spermatogenic cell differentiation *in vitro*. *Gamete Res.* **4**, 331–46.

Risley, M.S. and Eckhardt, R.A. (1981). H1 histone variants in *Xenopus laevis*. *Dev. Biol.* **84**, 79–87.

Roberts, M.S., Fragoso, G. and Hagers, G.L. (1995). The MMTV LTR B nucleosome adopts multiple translational and rotational positions during *in vitro* reconstitution. *Biochemistry* **34**, 12470–80.

Roberts, S.G.E., Ha, I., Maldonado, E., Reinberg, D. and Green, M. (1993). Interaction between an acidic activator and transcription factor TFIIB is required for transcriptional activation. *Nature* **363**, 741–4.

Roberts, S.S., Miller, J., Lane, W.S., Lee, S. and Hahn, S. (1996). Cloning and functional characterization of the gene encoding the TFIIIB90 subunit of RNA polymerase III transcription factor TFIIIB. *J. Biol. Chem.* **271** 14903–9.

Robinett, C.C., Straight, A.A., Li, G., Willhelm, C., Sudlow, G., Murray, A.W. and Belmont, A.S. (1996). *In vivo* localization of DNA sequences and visualization of large-scale chromatin organization using lac operator-repressor recognition. *J. Cell Biol.* **135**, 1685–700.

Robinson, G.W. and Hallick, L.M. (1982). Mapping the *in vivo* arrangement of nucleosomes on simian virus 40 with hydroxymethyltrimethylpsoralen. *J. Virol.* **41**, 78–87.

Robl, J.M., Gilligan, B., Critser, E.S. and First, N.L. (1986). Nuclear transplantation in mouse embryos: assessment of recipient cell stage. *Biol. Reprod.* **34**, 733–9.

Rocha, E., Davie, J.R., van Holde, K.E. and Weintraub, H. (1984). Differential salt fractionation of active and inactive genomic domains in chicken erythrocyte. *J. Biol. Chem.* **259**, 8558–63.

Rodriguez-Campos, A., Shimamura, A. and Worcel, A. (1989). Assembly and properties of chromatin containing histone H1. *J. Mol. Biol.* **209**, 135–50.

Roeder, R.G. (1996). The role of general initiation factors in transcription by RNA polymerase II. *Trends Biochem. Sci.* **21**, 327–35.

Rose, S.M. and Garrard, W.T. (1984). Differentiation dependent chromatin alterations precede and accompany transcription of immunoglobulin light chain genes. *J. Biol. Chem.* **259**, 8534–44.

Roseman, R.R., Pirrotta, V. and Geyer, P.K. (1993). Insulates expression of the *Drosophila melanogaster* white gene from chromosomal position-effects. *EMBO J.* **12**, 435–42.

Roth, M.B. and Gall, J.G. (1987). Monoclonal antibodies that recognize transcription unit proteins on newt lampbrush chromosomes. *J. Cell. Biol.* **105**, 1047–54.

Roth, S.Y. and Allis, C.D. (1992). H1 phosphorylation and chromatin condensation: exceptions which define the rule? *Trends Biochem. Sci.* **17**, 93–8.

Roth, S.Y., Schulman, I.G., Richman, R., Cook, R.G. and Allis, C.D. (1988). Characterization of phosphorylation sites in histone H1 in the amitotic macronucleus of *Tetrahymena* during different physiological states. *J. Cell Biol.* **107**, 2473–82.

Roth, S.Y., Dean, A. and Simpson, R.T. (1990). Yeast α2 repressor positions nucleosomes in TRP1/ARS1 chromatin. *Mol. Cell. Biol.* **10**, 2247–60.

Roth, S.Y., Shimizu, M., Johnson, L., Grunstein, M. and Simpson, R.T. (1992). Stable nucleosome positioning and complete repression by the yeast α2 repressor are disrupted by amino-terminal mutations in histone H4. *Genes Dev.* **6**, 411–25.

Rougvie, A.E. and Lis, J.T. (1988). The RNA polymerase II molecule at the 5′ end of the uninduced hsp 70 gene of *D. melanogaster* is transcriptionally engaged. *Cell* **54**, 795–804.

Rouviere-Yaniv, J., Yaniv, M. and Germond, J.E. (1979). *E. coli* DNA binding protein HU forms nucleosomelike structures with circular double-stranded DNA. *Cell* **17**, 265–80.

Rudolph, H. and Hinnen, A. (1987). The yeast PHO5 promoter: Phosphate-control elements and sequences mediating mRNA start-site selection. *Proc. Natl Acad. Sci. USA* **84**, 1340–4.

Ruiz-Carillo, A., Wangh, L.J. and Allfrey, V.G. (1975). Processing of newly synthesized histone molecules. *Science* **190**, 117–28.

Ruiz-Carillo, A., Jorcano, J.L., Eder, G. and Lurz, R. (1979). *In vitro* core particle and nucleosome assembly at physiological ionic strength. *Proc. Natl Acad. Sci. USA* **76**, 3284–8.

Rundlett, S.E., Carmen, A.A., Kobayashi, R., Bavykin, S., Turner, B.M. and Grunstein, M. (1996). HDA1 and RPD3 are members of distinct histone deacetylase complexes that regulate silencing and transcription. *Proc. Natl Acad. Sci. USA* **93**, 14503–8.

Rungger, D., Rungger-Brandle, E., Chaponnier, C. and Gabbiani, G. (1979). Intranuclear injection of anti-actin antibodies into *Xenopus* oocytes blocks chromosome condensation. *Nature* **282**, 320–1.

Rupp, R.A.W. and Weintraub, H. (1991). Ubiquitous Myo D transcription at the mid-blastula transition precedes induction-dependent Myo D expression in presumptive mesoderm of *X. laevis*. *Cell* **65**, 927–37.

Rusconi. S. and Schaffner, W. (1981). Transformation of frog embryos with a rabbit beta globin gene. *Proc. Natl Acad. Sci. USA* **78**, 5051–5.

Ruth, J., Conesa, C., Dieci, G., Lefebvre, O., Dusterhoft, A., Ottonello, S. and Sentenac, A. (1996). A suppressor of mutation in the class III transcription system encodes a component of yeast TFIIIB. *EMBO J.* **15**, 1941–9.

Rutledge, R.G., Neelin, J.M. and Selighy, V.L. (1988). Isolation and expression of cDNA clones coding for two sequence variants of *Xenopus laevis* histone H5. *Gene* **70**, 117–26.

Ryan, T.M., Rehringer, R.R., Martin, N.C., Townes, T.M., Palmiter, R.D. and Brinster, R.L. (1989). A single erythroid-specific DNase I super-hypersensitive site activates high levels of human β-globin gene expression in transgenic mice. *Genes Dev.* **3**, 314–23.

Ryoji, M. and Worcel, A. (1984). Chromatin assembly in *Xenopus* oocytes: *in vivo* studies. *Cell* **37**, 21–32.

Saffer, L.D. and Miller, O.L. Jr (1986). Electron microscopic study of *Saccharomyces cerevisiae* rDNA chromatin replication. *Mol. Cell. Biol.* **6**, 1147–57.

Sahasrabuddhe, C.G. and van Holde, K.E. (1974). The effect of trypsin on nuclease-resistant chromatin fragments. *J. Biol. Chem.* **249**, 152–6.

Saitoh, H., Tomkiel, J., Cooke, C.A., Ratnie III, H., Maurer, M., Rothfield, N.F. and Earnshaw, W.C. (1992). CENP-C, an autoantigen in scleroderma, is a component of the human inner kinetochore plate. *Cell* **70**, 115–25.

Saitoh, H. Goldberg, I., Wood, W.R. and Earnshaw, W.C. (1994). Sc II, an abundant chromosome scaffold protein is a member of a family of putative ATPases with an unusual predicted tertiary structure. *J. Cell Biol.* **127**, 303–18.

Saitoh, N., Goldberg, I. and Earnshaw, W.C. (1995). The SMC proteins and the coming of age of the chromosome scaffold hypothesis. *BioEssays* **17**, 759–66.

Saitoh, Y. and Laemmli, U.K. (1994). Metaphase chromosome structure: bands arise from a differential folding path of the high AT-rich scaffold. *Cell* **76**, 609–22.

Salik, J., Herlands, L., Hoffman, H.P. and Poccia, D. (1981). Electrophoretic analysis of the stored histone pool in unfertilized sea urchin eggs: quantification and identification by antibody binding. *J. Cell Biol.* **90**, 385–96.

Saluz, H.P., Jiricny, J. and Jost, J.P. (1986). Genomic sequencing reveals a positive correlation between the kinetics of strand specific DNA demethylation of the overlapping stratiol/glucocorticoid receptor binding sites and the rate of avian vitellogenin mRNA synthesis. *Proc. Natl Acad. Sci. U.S.A.* **83**, 7167–71.

Samejima, I., Matsumoto, T., Nakaseko, Y., Beach, D. and Yanagida, M. (1993). Identification of seven new cut genes involved in Schizosaccharomyces pombe mitosis. *J. Cell Sci.* **105**, 135–43.

Sandaltzopoulos, R., Blank, T. and Becker, P.B. (1994). Transcriptional repression by nucleosomes but not h1 in reconstituted preblastoderm *Drosophila* chromatin. *EMBO J.* **13**, 373–9.

Sandeen, G., Wood, W.I. and Felsenfeld, G. (1980). The interaction of high mobility proteins HMG14 and 17 with nucleosomes. *Nucleic. Acids Res.* **8**, 3757–78.

Sandman, K., Krzycki, J.A., Dobrinski, B., Lurz, R. and Reeve, J.N. (1990). HMf, a DNA-binding protein isolated from the hyperthermophilic archaeon *Methanofermus fervidus,* is most closely related to histones. *Proc. Natl. Acad. Sci. USA* **92**, 1624–8.

Santisteban, M.S., Arents, G., Moudrianakis, E.N. and Smith, M.M. (1997). Histone octamer function *in vivo*: mutations in the dimer-tetramer interfaces disrupt both gene activation and repression. *EMBO J.* **16**, 2493–506.

Satchwell, S.C. and Travers, A.A. (1989). Asymmetry and polarity of nucleosomes in chicken erythrocyte chromatin. *EMBO J.* **8**, 229–38.

Satchwell, S.C., Drew, H.R. and Travers, A.A. (1986). Sequence periodicities in chicken nucleosome core DNA. *J. Mol. Biol.* **191**, 659–75.

Sauer, F. and Tjian, R. (1997). Mechanisms of transcriptional activation: differences and similarities between yeast, *Drosophila* and man. *Curr. Opin. Genet. Dev.* **7**, 176–81.

Sauer, F., Wasserman, D.A., Rubin, G.M. and Tjian, R. (1996). TAF(II)s mediate activation of transcription in the *Drosophila* embryo. *Cell* **87**, 1271–84.

Saunders, W.C., Chue, C., Goebl, M., Craig, C., Clark, R.F., Powers, J.A., Eissenberg, J.C., Elgin, S., Rothfeld, N.F. and Earnshaw, W.C. (1993). Molecular cloning of a human homolog of *Drosophila* heterochromatin protein HP1 using anti-centromere autoantibodies with anti-chromo specificity. *J. Cell Sci.* **104**, 573–85.

Scarlato, V., Arico, B., Goyard, S., Rici, S., Manetti, R., Prugnola, A., Manetti, R., Polverino-De-Laureto, P., Ullman, A. and Rappuoli, R. (1995). A novel chromatin-forming histone H1 homologue is encoded by a dispensible and growth-regulated gene in *Bordetella pertussis*. *Mol. Microbiol.* **15**, 871–81.

Schaeffer, L., Roy, R., Humbert, S., Moucollin, V., Vermenlin, W., Hoeijmakers, J.H.J., Chambon, P. and Egly, J.M. (1993). DNA repair helicase: a component of BTF2 (TFIIH) basic transcription factor. *Science* **260**, 58–63.

Schaffer, C.D., Wallrath, L.L. and Elgin, S.C.R. (1993). Regulating genes by packaging domains: bits of heterochromatin in euchromatin. *Trends Genet.* **9**, 3537.

Schaffner, G., Schirm, S., Muller-Baden, B., Weber, F. and Schaffner, W. (1988). Redundancy of information in enhancers as a principle of mammalian transcriptional control. *J. Mol. Biol.* **201**, 81–90.

Scheer, U. and Benavente, R. (1990). Functional and dynamic aspects of the mammalian nucleolus. *Bioessays* **12**, 14–21.

Scheer, U. and Rose, K. M. (1984). Localization of RNA polymerase I in interphase cells and mitotic chromosomes by light and electron microscopic immunocytochemistry. *Proc. Natl Acad. Sci. USA* **81**, 1431–5.

Schild, C., Claret, F-X., Wahli, W. and Wolffe, A.P. (1993). A nucleosome-dependent static loop potentiates estrogen-regulated transcription from the *Xenopus* vitellogenin B1 promoter *in vitro*. *EMBO J.* **12**, 423–33.

Schindelin, H., Marahiel, M.A. and Heinemann, U. (1993). Universal nucleic acid-binding domain revealed by crystal structure of the *B. subtilis* major cold-shock protein. *Nature* **364**, 164–8.

Schlissel, M.S. and Baltimore, D. (1989). Activation of immunoglobulin kappa gene rearrangement correlates with induction of kappa gene transcription. *Cell* **58**, 1001–7.

Schlissel, M.S. and Brown, D.D. (1984). The transcriptional regulation of *Xenopus* 5S RNA genes in chromatin: the roles of active stable transcription complexes and histone H1. *Cell* **37**, 903–11.

Schmid, A., Fascher, K.D. and Horz, W. (1992). Nucleosome disruption at the PHO5 promoter upon PHO5 induction occurs in the absence of DNA replication. *Cell* **71**, 853–64.

Schnos, M., Zahn, K., Inman, R.B. and Blattner, F.R. (1988). Initiation protein induced helix destabilization at the λ origin: A prepriming step in DNA replication. *Cell* **52**, 385–6.

Schnuckel, A., Wiltscheck, R., Czisch, M., Herrier, M., Willmsky, G., Graumann, P., Marahiel, M.A. and Holak, T.A. (1993). Structure in solution of the major cold shock protein from *Bacillus subtilis*. *Nature* **364**, 169–71.

Schulman, I.G., Wang, T., Stargell, L.A., Gorovsky, M. and Allis, C.D. (1991). Cell–cell interactions trigger the rapid induction of a specific high mobility group-like protein during early stages of conjugation in *Tetrahymena*. *Dev. Biol.* **143**, 248–57.

Schultz, J., and Carlson, M. (1987). Molecular analysis of SSN6 a gene functionally related to SNF1 kinase of *Saccharomyces cerevisiae*. *Mol. Cell Biol.* **7**, 3637–45.

Schultz, M.C., Hockman, D.J., Harkness, T.A., Garinther, M.I. and Altheim, B.A. (1997). Chromatin assembly in a yeast whole-cell extract. *Proc. Natl Acad. Sci. USA* **94**, 9034–9.

Schultz, R.M. (1993). Regulation of zygotic gene activation in the mouse. *BioEssays* **15**, 531–8.

Schwabe, J.W.R., Neuhans, D. and Rhodes, D. (1990). Solution structure of the DNA-binding domain of the estrogen receptor. *Nature (London)* **348**, 458–61.

Schwartz, D.C. and Cantor, C.R. (1984). Separation of yeast chromosome-sized DNAs by pulsed field gradient gel electrophoresis. *Cell* **37**, 67–75.

Schwarz, P.M. and Hansen, J.C. (1994). Formation and stability of higher order chromatin structures. *J. Biochem.* **269**, 16284–98.

Schwarz, P.M., Felthauser, A., Fletcher, T.M. and Hansen, J.C. (1996). Reversible oligonucleosome self-association: dependence on divalent cations and core histone tail domains. *Biochemistry* **35**, 4009–15.

Sealy, L., Cotten, M. and Chalkley, R. (1986). *Xenopus* nucleoplasmin: egg vs oocyte. *Biochemistry* **25**, 3064–72.

Sedat, J. and Manuelidis, L. (1978). A direct approach to the structure of mitotic chromosomes. *Cold Spring Harb. Symp. Quant. Biol.* **42**, 331–50.

Segall, J., Matsui, T. and Roeder, R.G. (1980). Multiple factors are required for the accurate transcription of purified genes by RNA polymerase III. *J. Biol. Chem.* **255**, 11986–91.

Segil, N., Guermah, M., Hoffman, A., Roeder, R.G. and Heintz, N. (1996). Mitotic regulation of TFIID: inhibition of activator-dependent transcription and changes in subcellular localization. *Genes Dev.* **10**, 2389–400.

Seidman, M.M., Levine, A.J. and Weintraub, H. (1979). The asymmetric segregation of paternal nucleosomes during chromosomal replication. *Cell* **18**, 439–49.

Senshu, T., Fukada, M. and Ohashi, M. (1978). Preferential association of newly synthesized H3 and H4 histones with newly synthesized replicated DNA. *J. Biochem. (Japan)* **84**, 985–8.

Sera, T. and Wolffe, A.P. (1998). The role of histone H1 as an architectural determinant of chromatin structure and as a specific repressor of transcription on the *Xenopus* oocyte 5S rRNA gene. *Mol. Cell Biol.* in press.

Serfling, E., Jasin, M. and Schaffner, W. (1985). Enhancers and eukaryotic gene transcription. *Trends Genet.* **1**, 224–30.

Serizawa, H., Conaway, J.W. and Conaway, R.C. (1993). Phosphorylation of C-terminal domain of RNA polymerase II is not required in basal transcription. *Nature* **363**, 371–4.

Sheehan, M.A., Mills, A.D., Sleeman, A.M., Laskey, R.A. and Blow, J.J. (1988). Steps in the assembly of replication-competent nuclei in a cell-free system from *Xenopus* eggs. *J. Cell Biol.* **106**, 1–12.

Shelby, R.D., Vafa, O. and Sullivan, K.F. (1997). Assembly of CENP-A into centromeric chromatin requires a cooperative array of nucleosomal DNA contact sites. *J. Cell Biol.* **136**, 501–13.

Shen, X. and Gorovsky, M.A. (1996). Linker histone H1 regulates specific gene expression but not global transcription *in vivo*. *Cell* **86**, 475–83.

Shen, X., Yu, L., Weir, J.W. and Gorovsky, M.A. (1995). Linker histones are not essential and affect chromatin condensation *in vivo*. *Cell* **82**, 47–56.

Sheridan, P.L., Sheline, C.T., Cannon, K., Voz, M.L., Pazin, M.J., Kadonaga, J.T. and Jones, K.A. (1995). Activation of the HIV-1 enhancer by the LEF1 HMG protein on nucleosome assembled DNA *in vitro*. *Genes Dev.* **9**, 2090–104.

Shermoen, A.W. and O'Farrell, P.H. (1991). Progression of the cell cycle through mitosis leads to abortion of nascent transcripts. *Cell* **67**, 303–10.

Shi, Y., Seto, E., Chang, L.-S. and Shenk, T. (1991). Transcriptional repression by YY1, a human GL1-Kruppel-related protein, and relief of repression by adenovirus E1A protein. *Cell* **67**, 377–88.

Shick, V.V., Belyavsky, A.V., Bavykin, S.G. and Mirzabekov, A.D. (1980). Primary organization of the nucleosome core particles: sequential arrangement of histones along DNA. *J. Mol. Biol.* **139**, 491–517.

Shick, V.V., Belyavsky, A.V. and Mirzabekov, A.D. (1985). Primary organization of nucleosomes: interaction of non-histone high mobility group proteins 14 and 17 with nucleosomes, as revealed by DNA–protein cross-linking and immuno-affinity isolation. *J. Mol. Biol.* **185**, 329–59.

Shim, E.Y., Woodcock, C. and Zaret, K.S. (1998). Nucleosome positioning by the winged helix transcription factor HNF3. *Genes Dev.* **12**, 5–10.

Shimamura, A. and Worcel, A. (1989). The assembly of regularly spaced nucleosomes in the *Xenopus* oocyte S150 extract is accompanied by deacetylation of histone H4. *J. Biol. Chem.* **264**, 14524–30.

Shimamura, A., Tremethick, D. and Worcel, A. (1988). Characterization of the repressed 5S DNA minichromosomes assembled *in vitro* with a high-speed supernatant of *Xenopus laevis* oocytes. *Mol. Cell Biol.* **8**, 4257–69.

Shimamura, A., Sapp, M., Rodriquez-Campos, A. and Worcel, A. (1989). Histone H1 represses transcription from minichromosomes assembled *in vitro*. *Mol. Cell Biol.* **9**, 5573–84.

Shimizu, M., Roth, S.Y., Szent-Gyorgi, C. and Simpson, R.T. (1991). Nucleosome are positioned with base pair precision adjacent to the α2 operator in *Saccharomyces cerevisiae*. *EMBO J.* **10**, 3033–41.

Shiokawa, K., Fu, Y., Nakakura, N., Tashiro, K., Sameshima, M. and Hosokawa, K. (1989). Effects of the injection of exogenous DNAs on gene expression in early embryos and coenocytic egg cells of *Xenopus laevis*. *Roux's Arch Dev. Biol.* **198**, 78–84.

Shrader, T.E. and Crothers, D.M. (1989). Artificial nucleosome positioning sequences. *Proc. Natl Acad. Sci. USA* **86**, 7418–22.

Shrader, T.E. and Crothers, D.M. (1990). Effects of DNA sequence and histonehistone interactions on nucleosome placement. *J. Mol. Biol.* **216**, 69–84.

Shykind, B.M., Kim, J. and Sharp, P.A. (1995). Activation of the TFIID-TFIIA complex with HMG-2. *Genes Dev.* **9**, 1354–65.

Sidik, K. and Smerdon, M.J. (1990). Nucleosome rearrangement in human cells following short patch repair of DNA damaged by bleomycin. *Biochemistry* **29**, 7501–11.

Siegfried, Z. and Cedar, H. (1997). DNA methylation: a molecular lock. *Curr. Biol.* **7**, R305–R307.

Simpson, R.T. (1978). Structure of the chromatosome, a chromatin core particle containing 160 base pairs of DNA and all the histones. *Biochemistry* **17**, 5524–31.

Simpson, R.T. (1990). Nucleosome positioning can affect the function of a cis-acting DNA element *in vivo*. *Nature (London)* **343**, 387–9.

Simpson, R.T. (1991). Nucleosome positioning: occurrence, mechanisms and functional consequences. *Progr. Nucleic. Acids Res. Mol. Biol.* **40**, 143–84.

Simpson, R.T. and Bergman, L.W. (1980). Structure of sea urchin sperm chromatin core particles. *J. Biol. Chem.* **255**, 10702–9.

Simpson, R.T. and Stafford, D.W. (1983). Structural features of a phased nucleosome core particle. *Proc. Natl Acad. Sci. USA* **80**, 51–5.

Simpson, R.T., Thoma, F. and Brubaker, J.M. (1985). Chromatin reconstituted from tandemly repeated cloned DNA fragments and core histones: a model system for study of higher order structure. *Cell* **42**, 799–808.

Simpson, R.T., Roth, S.Y., Morse, R.H., Patterton, H.G., Cooper, J.P., Murphy, M., Kladde, M.P. and Shimizu, M. (1993). Nucleosome positioning and transcription. *Cold Spring Harb. Symp. Quant. Biol.* **58**, 237–45.

Shrivastava, A. and Calame, K. (1994). An analysis of genes regulated by the multi-functional transcriptional regulatory Yin yang 1. *Nucleic. Acids Res.* **22**, 5151–5.

Sinden, R.R. (1994). *DNA Structure and Function*. Academic Press, San Diego.

Sinha, S., Maity, S.N., Lu, J. and de Crombrugghe, B. (1995). Recombinant rate CBF-C, the third subunit of CBF/NFY, allows formation of a protein–DNA complex with CBF-A and CBF-B and with yeast HAP2 and HAP3. *Proc. Natl Acad. Sci. USA* **92**, 1624–8.

Singer, D.S. and Singer, M.F. (1976). Studies on the interaction of histone H1 with superhelical DNA: characterization of the recognition and binding regions of H1 histone. *Nucleic. Acids Res.* **3**, 2531–47.

Singh, J. and Rao, M.R.S. (1987). Interaction of rat testis protein, TP with nucleic acids *in vitro*. *J. Biol. Chem.* **262**, 734–40.

Singh, P.B. (1994). Molecular mechanisms of cellular determination: their relation to chromatin structure and paternal imprinting. *J. Cell Sci.* **107**, 2653–68.

Sisodia, S. S., Sollner-Webb, B. and Cleveland, D. W. (1987). Specificity of RNA maturation pathways: RNAs transcribed by RNA polymerase III are not substrates for splicing or polyadenylation. *Mol. Cell Biol.* **7**, 3602–12.

Sivolob, A.V. and Khrapunov, S.N. (1995). Translational positioning of nucleosomes on DNA: the role of sequence-dependent isotropic DNA bending stiffness. *J. Mol. Biol.* **247**, 918–31.

Smale, S.T., Schmidt, M.C., Berk, A.J. and Baltimore, D. (1990). Transcriptional activation by SP1 as directed through TATA or initiator: specific requirement for mammalian transcription factor IID. *Proc. Natl Acad. Sci. USA* **87**, 4509–13.

Smerdon, M.J. (1986). Completion of excision repair in human cells. Relationship between ligation and nucleosome formation. *J. Biol. Chem.* **261**, 244–52.

Smerdon, M.J. and Lieberman, M.W. (1978). Nucleosome rearrangement in human chromatin during UV-induced DNA repair synthesis. *Proc. Natl Acad. Sci. USA* **75**, 4238–41.

Smerdon, M.J. and Thoma, F. (1990). Site-specific DNA repair at the nucleosome level in a yeast minichromosome. *Cell* **61**, 675–84.

Smerdon, M.J., Lan, S.Y., Calza, R.E. and Reeves, R. (1982). Sodium butyrate stimulates DNA repair in UV-irradiated normal and xeroderma pigmentosum human fibroblasts. *J. Biol. Chem.* **257**, 13441–7.

Smerdon, M.J., Bedoyan, J. and Thoma, F. (1990). DNA repair in a small yeast plasmid folded into chromatin. *Nucleic. Acids Res.* **18**, 2045–51.

Smith, B.L. and Macleod, M.C. (1993). Covalent binding of the carcinogen benzo(a)pyrene diol epoxide to *Xenopus laevis* 5S DNA reconstituted into nucleosomes. *J. Biol. Chem.* **268**, 20620–9.

Smith, B.L., Bauer, G.B. and Povirk, L.F. (1994). DNA damage induced by bleomycin, neocarzinostatin, and melphalan in a precisely positionined nucleosome. Asymmetry in protection at the periphery of nucleosome bound DNA. *J. Biol. Chem.* **269**, 30587–94.

Smith, J.S. and Boeke, J.D. (1997). An unusual form of transcriptional silencing in yeast ribosomal DNA. *Genes Dev.* **11**, 241–54.

Smith, M.M., Yang, P., Santisteban, S., Boone, P.W., Goldstein, A.T. annd Megee, P.C. (1996). A novel histone H4 mutant defective in nuclear division and mitotic chromosome transmission. *Mol. Cell Biol.* **16**, 1017–26.

Smith, P.A., Jackson, V. and Chalkley, R. (1984). Two-stage maturation process for newly replicated chromatin. *Biochemistry* **23**, 1576–81.

Smith, R.C., Dworkin-Rastl, E. and Dworkin, M.D. (1988). Expression of a histone H1-like protein is restricted to early *Xenopus* development. *Genes Dev.* **2**, 1284–95.

Smith, S. and Stillman, B. (1989). Purification and characterization of CAF-1 a human cell factor required for chromatin assembly during DNA replication *in vitro*. *Cell* **58**, 15–25.

Smith, S. and Stillman, B. (1991a). Stepwise assembly of chromatin during DNA replication *in vitro*. *EMBO J.* **10**, 971–80.

Smith, S. and Stillman, B. (1991b). Immunological characterization of chromatin assembly factor 1, a human cell factor required for chromatin assembly during DNA replication *in vitro*. *J. Biol. Chem.* **266**, 12041–7.

Sobczak-Thepot, J., Harper, F., Florentin, Y., Zindy, F., Brechot, C. and Puvion, E. (1993). Localization of cyclin A at the sites of cellular DNA replication. *Exp. Cell Res.* **206**, 43–8.

Sogo, J.M., Stahl, H., Koller, Th. and Knippers, R. (1986). Structure of replicating SV40 minichromosomes: The replication fork, core histone segregation and terminal structures. *J. Mol. Biol.* **189**, 189–204.

Solomon, M.J. and Varshavsky, A. (1987). A nuclease-hypersensitive region forms de novo after chromosome replication. *Mol. Cell. Biol.* **7**, 3822–5.

Solomon, M.J., Strauss, F. and Varshavsky, A. (1986). A mammalian high mobility group protein recognizes any stretch of six A-T base pairs in duplex DNA. *Proc. Natl Acad. Sci. USA* **83**, 1276–80.

Solomon, M.J., Larsen, P.L. and Varshavsky, A. (1988). Mapping protein–DNA interactions *in vivo* with formaldehyde: evidence that histone H4 is retained on a highly transcribed gene. *Cell* **53**, 937–47.

Sopta, M., Burton, Z.F. and Greenblatt, J. (1989). Structure and associated DNA helicase activity of a general transcription intiation factor that binds to RNA polymerase II. *Nature (London)* **341**, 410–14.

Spector, D. L. (1990). Higher order nuclear organization: three-dimensional distribution of small nuclear ribonucleoprotein particles [published erratum appears in *Proc. Natl Acad. Sci. USA* 1990 87, 2384]. *Proc. Natl Acad. Sci. USA* **87**, 147–51.

Spector, D. L. (1993). Macromolecular domains within the cell nucleus. *Annu. Rev. Cell Biol.* **9**, 265–315.

Spencer, C.A. and Groudine, M. (1990). Transcription elongation and eukaryotic gene regulation. *Oncogene* **5** 777–85.

Spradling, A.C. and Karpen, G.H. (1990). Sixty years of mystery. *Genetics* **126**, 779–84.

Srikantha, T., Landsman, D. and Bustin, M. (1988). Cloning of the chicken chromosomal protein HMG-14 cDNA reveals a unique protein with a conserved DNA binding domain. *J. Biol. Chem.* **263**, 13500–3.

Stadler, M., Chelbi-Alix, M. K., Koken, M. H., Venturini, L., Lee, C., Saib, A., Quignon, F., Pelicano, L., Guillemin, M. C. and Schindler, C. (1995). Transcriptional induction of the PML growth suppressor gene by interferons is mediated through an ISRE and a GAS element. *Oncogene.* **11**, 2565–73.

Stargell, L.A., Bowen, J., Dadd, C.A., Dedon, P.A., Davis, M., Cook, R.G., Allis, C.D. and Gorovsky, M.A. (1993). Temporal and spatial association of histone H2A variant hv1 with transcriptionally competent chromatin during nuclear development in *Tetrahymena thermophila*. *Genes Dev.* **7**, 2641–51.

Stavenhagen, J.B. and Zakian, V.A. (1994). Internal tracts of telomeric DNA act as silencers in *Saccharomyces cerevisiae*. *Genes Dev.* **8**, 1411–22.

Staynov, D.Z. and Crane-Robinson, C. (1988). Footprinting of linker histones H5 and H1 on the nucleosome. *EMBO J.* **7**, 3685–91.

Stedman, E. and Stedman, E. (1947). The chemical nature and functions of components of cell nuclei not histone but protein. *Cold Spring Harb. Symp.* **12**, 224–36.

Steger, D.J. and Workman, J.L. (1997). Stable co-occupancy of transcription factors and histones at the HIV-1 enhanceer. *EMBO J.* **16**, 2463–72.

Stein, A. and Bina, M. (1984). A model chromatin assembly system: factors affecting nucleosome spacing. *J. Mol. Biol.* **178**, 341–63.

Stein, A., Whitlock, J.P. and Bina, M. (1979). Acidic polypeptides can assemble both histones and chromatin *in vitro* at physiological ionic strength. *Proc. Natl Acad. Sci. USA* **76**, 5000–4.

Steinbach, O.C., Wolffe, A.P. and Rupp, R. (1997). Accumulation of somatic linker histones causes loss of mesodermal competence in *Xenopus*. *Nature* **389**, 395–9.

Stelzer, G.A., Goppelt, A., Lottspeich, F. and Meisterernst, M. (1994). Repression of basal transcription by HMG2 is counteracted by TFIIH associated factors in an ATP dependent process. *Mol. Cell. Biol.* **14**, 4712–21.

Stern, M.J., Jensen, R.E. and Herskowitz, I. (1984). Five SWI genes are required for expression of the HO gene in yeast. *J. Mol. Biol.* **178**, 853–68.

Sterner, R., Boffa, L., Chen, T.A. and Allfrey, V.G. (1987). Cell cycle-dependent changes in conformation and composition of nucleosomes containing human histone gene sequences. *Nucleic. Acids Res.* **15**, 4375–91.

Stick, R. and Hansen, P. (1985). Changes in the nuclear lamina composition during early development of *Xenopus laevis*. *Cell* **41**, 191–200.

Stief, A., Winter, D.M., Stratting, W.E.H. and Sippel, A.E. (1989). A nuclear DNA attachement element mediates elevated and position independent gene activity. *Nature (London)* **341**, 343–5.

Stillman, B. (1986). Chromatin assembly during SV40 DNA replication *in vitro*. *Cell* **45**, 555–65.

Stillman, B. (1989). Initiation of eukaryotic DNA replication *in vitro*. *Annu. Rev. Cell. Biol.* **5**, 197–245.

Stoler, S., Keith, K.C., Curnick, K.E. and Fitzgerald-Hayes, M. (1995). A mutation in CSE4, an essential gene encoding a novel chromatin associated protein in yeast, causes chromatin non disjunction and cell cycle arrest at mitosis. *Genes Dev.* **9**, 573–86.

Straka, C. and Horz, W. (1991). A functional role for nucleosomes in the repression of a yeast promoter. *EMBO J.* **10**, 361–8.

Strick, R. and Laemmli, U.K. (1995). SARs are cis DNA elements of chromosome dynamics: synthesis of a SAR repressor protein. *Cell* **83**, 1137–48.

Strickland, W.N., Strickland, M., Brandt, W.F., Von Holt, C., Lehmann, A. and Wittmann-Liebold, B. (1980). The primary structure of histone H1 from sperm of the sea urchin *Parechinus angulosus*. *Eur. J. Biochem.* **104**, 567–78.

Stringer, K.F., Ingles, C.J. and Greenblatt, J. (1990). Direct and selective binding of an acidic activation domain to the TATA-box factor TFIID. *Nature (London)* **345**, 783–6.

Strouboulis, J. and Wolffe, A.P. (1996). The functional compartmentalization of the nucleus. *J. Cell Sci.* **109**, 1991–2000.

Strunkel, W., Kober, I. and Seifart, K.H. (1997). A nucleosome positioned in the distal promoter region activates transcription of the human U6 gene. *Mol. Cell Biol.* **17**, 4397–405.

Strunnikov, A.V., Larionov, V.L. and Koshland, D. (1993). SMC1: an essential yeast gene encoding a putative head-rod-tail protein is required for nuclear division and defines a new ubiquitous protein family. *J. Cell Biol.* **123**, 1635–48.

Strunnikov, A.V., Hogan, E. and Koshland, D. (1995). SMC2, a *Saccharomyces cerevisiae* gene essential for chromosome segregation and condensation defines a subgroup within the SMC-family. *Genes Dev.* **9**, 587–99.

Studitsky, V.M., Clark, D.J. and Felsenfeld, G. (1994). A histone octamer can step around a transcribing polymerase without leaving the template. *Cell* **76**, 371–82.

Studitsky, V.M., Clark, D.J. and Felsenfeld, G. (1995). Overcoming a nucleosome barrier to transcription. *Cell* **83** 19–27.

Studitsky, V.M., Kassavetis, G.A., Geiduschek, E.P. and Felsenfeld, G. (1997). Mechanism of transcription through the nucleosome by eukaryotic RNA polymerase. *Science* **278**, 1960–1963.

Su, W., Jackson, S., Tjian, R. and Echols, H. (1991). DNA looping between sites for transcriptional activation: self association of DNA-bound Sp1. *Genes Dev.* **5**, 820–6.

Suau, P., Bradbury, E.M. and Baldwin, J.P. (1979). Higher-order structures of chromatin in solution. *Eur. J. Biochem.* **97**, 593–602.

Sugita, K., Koizumi, K. and Yoshida, M. (1992). Morphological reversion of sis-transformed NIH3T3 cells by trichostatin A. *Cancer Res.* **52**, 168–72.

Sullivan, C.H. and Grainger, R.M. (1986). γ-Crystallin genes become hypomethylated in postmitotic lens cells during chicken development. *Proc. Natl Acad. Sci. USA* **83**, 329–33.

Sullivan, K.F., Hechenberger, M. and Masri, K. (1994). Human CENP-A contains a histone H3 related histone fold domain that is required for targeting to the centromer. *J. Cell Biol.* **127**, 581–92.

Sun, J.-M., Wiaderkiewicz, R. and Ruiz-Carrillo, A. (1989). Histone H5 in the control of DNA synthesis and cell proliferation. *Science* **245**, 68–71.

Sung, M.T. and Dixon, G.H. (1970). Modification of histones during spermiogenesis in trout: a molecular mechanism for altering histone binding to DNA. *Proc. Natl Acad. Sci. USA* **67**, 1616–23.

Suquet, C. and Smerdon, M.J. (1995). Repair of UV induced (6-4) photo-products in nucleosome core DNA. *J. Biol. Chem.* **270**, 16507–9.

Sutherland, G.R. and Richards, R.I. (1995). Simple tandem DNA repeats and human genetic disease. *Proc. Natl Acad. Sci. USA* **92**, 3636–41.

Svaren, J. and Chalkley, R. (1990). The structure and assembly of active chromatin. *Trends Genet.* **6**, 52–6.

Svaren, J. and Horz, W. (1993). Histones, nucleosomes and transcription. *Curr. Opin. Genet. Dev.* **3**, 219–25.

Svaren, J. and Horz, W. (1995). Interplay between nucleosomes and transcription factors at the yeast PHO5 promoter. *Semin. Cell Biol.* **6**, 177–83.

Svaren, J. and Horz, W. (1997). Transcription factor vs nucleosomes: regulation of the PHO5 promoter in yeast. *Trends Biochem Sci.* **22**, 93–7.

Svaren, J., Schmitz, J. and Horz, W. (1994a). The transactivation domain of Pho4 is required for nucleosome disruption at the PHO5 promoter. *EMBO J.* **13**, 4856–62.

Svaren, J., Klebanow, E., Sealy, L. and Chalkley, R. (1994b). Analysis of the competition between nucleosome formation and transcription factor binding. *J. Biol. Chem.* **269**, 9335–44.

Svejstrup, J.Q., Vichi, P. and Egly, J.M. (1996). The multiple roles of transcription/repair factor TFIIH. *Trends Biochem. Sci.* **21**, 346–50.

Sweet, M.T., Jones, K. and Allis, C.D. (1996). Phosphorylation of linker histone in associated with transcriptional activation in a normally silent nucleus. *J. Cell Biol.* **135**, 1219–28.

Szent-Gyorgi, C., Finkelstein, D.B. and Garrard, W.T. (1987). Sharp boundaries demarcate the chromatin structure of a yeast heat-shock gene. *J. Mol. Biol.* **193**, 71–80.

Szyf, M. (1996). The DNA methylation machinery as a target for anticancer therapy. *Pharmacol. Ther.* **70**, 1–37.

Tafuri, S.R. and Wolffe, A.P. (1990). *Xenopus* Y-box transcription factors: Molecular cloning, functional analysis and developmental regulation. *Proc. Natl Acad. Sci. USA* **87**, 9028–32.

Tafuri, S.R. and Wolffe, A.P. (1993). Selective recruitment of masked maternal mRNA from messenger ribonucleoprotein particles containing FRGY2 (mRNP4). *J. Biol. Chem.* **257**, 24255–61.

Takagi, N. (1974). Differentiation of X chromosomes in early female mouse embryos. *Exp. Cell. Res.* **86**, 127–35.

Takata, C., Albright, J.F. and Yomade, T. (1964). Lens antigens in a lens regenerating system studied by the immunofluorescent technique. *Dev. Biol.* **9**, 385–97.

Takeichi, T., Satoh, S., Tashiro, K. and Shiokawa, K. (1985). Temporal control of rRNA synthesis in cleavage-arrested embryos of *Xenopus laevis*. *Dev. Biol.* **112**, 443–50.

Talbot, D. and Grosveld, F. (1991). The 5′ HS 2 of the globin locus control region enhances transcription through the interaction of a multimeric complex binding at two functionally distinct NF-E2 binding sites. *EMBO J.* **10**, 1391–8.

Talbot, D., Philipsen, S., Fraser, P. and Grosveld, F. (1990). Detailed analysis of the site 3 region of the human β-globin dominant control region. *EMBO J.* **9**, 2169–78.

Tamkun, J.W., Deuring, R., Scott, M.P., Kissinger, M., Patlatucci, A.M., Kaufman, T.C. and Kennison, J.A. (1992). Brahma: a regulator of *Drosophila* homeotic genes structurally related to the yeast transcriptional activator SNF2/SWI2. *Cell* **68**, 561–72.

Tan, S., Aso, T., Conaway, R.C. and Conaway, J.W. (1994). Roles for both the RAP30 and RAP74 subunits of transcription factor IIF in transcription initiation and elongation by RNA polymerase II. *J. Biol. Chem.* **269** 25684–91.

Tanese, N., Pugh, B.F. and Tjian, R. (1991). Coactivators for a proline-rich activator purified from the multisubunit human TFIID complex. *Genes Dev.* **12**, 2212–24.

Taniura, H., Glass, C. and Gerace, L. (1995). A chromatin binding site in the tail domain of nuclear lamins that interacts with the core histones. *J. Cell Biol.* **131**, 33–44.

Tate, P., Skarnes, W. and Bird, A. (1996). The methyl-CpG binding protein MeCP2 is essential for embryonic development in the mouse. *Nat. Genet.* **12**, 205–8.

Tate, P.H. and Bird, A.P. (1993). Effects of DNA methylation on DNA-binding proteins and gene expression. *Curr. Opin. Genet. Dev.* **3**, 226–31.

Taunton, J., Hassig, C.A. and Schreiber, S.L. (1996). A mammalian histone deacetylase related to a yeast transcriptional regulator Rpd3. *Science* **272**, 408–11.

Taylor, I.C.A., Workman, J.L., Schmetz, T-J. and Kingston, R.E. (1991). Facilitated binding of GAL4 and heat shock factor to nucleosomal templates: differential function of DNA-binding domains. *Genes Dev.* **5**, 1285–98.

Tazi, J. and Bird, A. (1990). Alternative chromatin structure at CpG islands. *Cell* **60**, 902–20.

Telford, N.A., Watson, A.J. and Schultz, G.A. (1990). Transition from maternal to embryonic control in early mammalian development: a comparison of several species. *Mol. Reprod. Devel.* **26**, 90–100.

Terris, B., Baldin, V., Dubois, S., Degott, C., Flejou, J. F., Henin, D. and Dejean, A. (1995). PML nuclear bodies are general targets for inflammation and cell proliferation. *Cancer Res.* **55**, 1590–7.

Thanos, D. and Maniatis, T. (1992). The high mobility group protein HMGI (Y) is required for NF-κB-dependent virus induction of the human IFN-β gene. *Cell* **71**, 777–89.

Theulaz, I., Hipskind, R., TenHeggeler-Bordier., B., Green, S., Kumar, V., Chambon, P. and Wahli, W. (1988). Expression of human estrogen receptor mutants in *Xenopus* oocytes: correlation between transcriptional activity and ability to form protein–DNA complexes. *EMBO J.* **7**, 1653–60.

Thiry, M. (1992a). Ultrastructural detection of DNA within the nucleolus by sensitive molecular immunocytochemistry. *Exp. Cell Res.* **200**, 135–44.

Thiry, M. (1992b). New data concerning the functional organization of the mammalian cell nucleolus: detection of RNA and rRNA by in situ molecular immunocytochemistry. *Nucleic. Acids Res.* **20**, 6195–200.

Thiry, M. and Goessens, G. (1992). Location of DNA within the nucleolus of rat oocytes during the early stages of follicular growth. *Int. J. Dev. Biol.* **36**, 139–42.

Thoma, F. (1986). Protein–DNA interactions and nuclease sensitive regions determine nucleosome positions on yeast plasmid chromatin. *J. Mol. Biol.* **190**, 177–90.

Thoma, F. (1991). Structural changes in nucleosomes during transcription: strip, split or flip? *Trends Genet.* **7**, 175–7.

Thoma, F. and Simpson, R.T. (1985). Local protein–DNA interactions may determine nucleosome positions on yeast plasmids. *Nature (London)* **315**, 250–3.

Thoma, F. and Zatchej, M. (1988). Chromatin folding modulates nucleosome positioning in yeast minichromosomes. *Cell* **55**, 945–53.

Thoma, F., Koller, T. and Klug, A. (1979). Involvement of histone H1 in the organization of the nucleosome and the salt-dependent superstructures of chromatin. *J. Cell Biol.* **83**, 402–27.

Thoma, F., Bergman, L. W. and Simpson, R.T. (1984). Nuclease digestion of circular TRP1ARS1 chromatin reveals positioned nucleosomes separated by nuclease sensitive regions. *J. Mol. Biol.* **177**, 715–33.

Thomas, G.H. and Elgin, S.C.R. (1988). Protein/DNA architecture of the DNase I hypersensitive region of the *Drosophila* hsp26 promoter. *EMBO J.* **7**, 2191–201.

Thomas, G.P. and Mathews, M.B. (1980). DNA replication and the early to late transition in adenovirus infection. *Cell* **22**, 523–33.

Thomas, J.O. (1989). Chemical radiolabeling of lysines that interact strongly with DNA in chromatin. *Methods Enzymol.* **170**, 369–85.

Thomas, J.O., Rees, C. and Finch, J.T. (1992). Cooperative binding of the globular domains of histones H1 and H5 to DNA. *Nucleic. Acids Res.* **20**, 187–94.

Thompson, C.M., Koleske, A.J., Chao, D.M. and Young, R.A. (1993). A multisubunit complex associated with the RNA polymease II CTD and TATA binding protein in yeast. *Cell* **73** 1361–75.

Thompson, E.M., Christians, E., Stinnakre, M.-G. and Renard, J.-P. (1994b). Scaffold attachment regions stimulate HSP70.1 expression in mouse preimplantation embryos but not in differentiated tissues. *Mol. Cell. Biol.* **14**, 4694–703.

Thompson, E.M., Legony, E., Christians, E. and Renard, J.-P. (1995). Progressive maturation of chromatin structure regulates HSP70.1 gene expression in the preimplantation mouse embryo. *Development* **121**, 3425–7.

Thompson, J.S., Ling, X. and Grunstein, M. (1994a). Histone h3 amino terminus is required for telomeric and silent mating locus repression in yeast. *Nature* **369**, 245–7.

Thon, G. and Klar, A.J.S. (1992). The clr1 locus regulates the expression of the cryptic mating type loci in fission yeast. *Genetics* **131**, 287–96.

Thornton, C.A., Wymer, J.P., Simmons, Z., McClain, C. and Moxley, R.T. (1997). Expansion of the myotonic dystrophy CTG repeat reduces expression of the flanking DMAHP gene. *Nature Genet.* **16**, 407–9.

Thrall, B.D., Mann, D.B., Smerdon, M.J. and Springer, D.L. (1994). Nucleosome structure modulates benzo(a)pyrenediol epoxide adduct formation. *Biochemistry* **33**, 2210–16.

Tjian, R. and Maniatis, T. (1994). Transcriptional activation: a complex puzzle with few easy pieces. *Cell* **77**, 5–8.

Tomaszewski, R. and Jerzmanowski, A. (1997). The AT-flanks of the oocyte-type 5S RNA gene of *Xenopus laevis* act as a strong signal for histone H1-mediated chromatin reorganization *in vitro*. *Nucleic. Acids Res.* **25**, 458–65.

Tomkiel, J., Cooke, C.A., Saitoh, H., Bernat, R.L. and Earnshaw, W.C. (1994). CENP-C is required for maintaining proper kinetochore size and for a timely transition to anaphase. *J. Cell Biol.* **125**, 531–45.

Tommerup, H., Dousmanis, A. and de Lange, T. (1994). Unusual chromatin in human telomeres. *Mol. Cell. Biol.* **14**, 5777–89.

Travers, A.A. (1989). DNA conformation and protein binding. *Annu. Rev. Biochem.* **58**, 427–52.

Travers, A.A. (1994). Chromatin structure and dynamics. *BioEssays* **16**, 657–62.

Travers, A.A. and Klug, A. (1987). The bending of DNA in nucleosomes and its wider implications. *Phil. Trans. R. Soc. London B* **317**, 537–61.

Travers, A.A., Ner, S.S. and Churchill, M.E.A. (1994). DNA chaperones: a solution to a persistence problem. *Cell* **77**, 167–9.

Tremethick, D.J. (1994). High mobility group proteins 14 and 17 can space nucleosomal particles deficient in histones H2A and H2B creating a template that is transcriptionally active. *J. Biol. Chem.* **269**, 28436–42.

Tremethick, D.J. and Drew, H.R. (1993). High mobility group proteins 14 and 17 can space nucleosomes *in vitro*. *J. Biol. Chem.* **268**, 11389–93.

Tremethick, D.J. and Frommer, M. (1992). Partial purification, from *Xenopus laevis* oocytes of an ATP-dependent activity required for nucleosome spacing *in vitro*. *J. Biol. Chem.* **267**, 15041–8.

Tremethick, D.J. and Molloy, P.L. (1986). High mobility group proteins 1 and 2 stimulate transcription *in vitro* by RNA polymerases II and III. *J. Biol. Chem.* **261**, 6986.

Tremethick, D., Zucker, D. and Worcel, A. (1990). The transcription complex of the 5S RNA gene, but not the transcriptional factor TFIIIA alone, prevents nucleosomal repression of transcription. *J. Biol. Chem.* **265**, 5014–23.

Trieschmann, L., Alfonso, P.J., Crippa, M.P., Wolffe, A.P. and Bustin, M. (1995a). Incorporation of chromosomal proteins HMG-14/-17 into nascent nucleosomes induces an extended chromatin conformation and enhances the utilization of active transcription complexes. *EMBO J.* **14**, 1478–89.

Trieschmann, L., Postnikov, Y.V., Rickers, A. and Bustin, M. (1995). Modular structure of chromosomal proteins HMG-14 and HMG-17: definition of a transcriptional enhancement domain distinct from the nucleosomal binding domain. *Mol. Cell Biol.* **15**, 6663–9.

Truss, M., Bartsch, J., Schelbert, A., Haché, R.J.G. and Beato, M. (1995). Hormone induces binding of receptors and transcription factors to a rearranged nucleosome on the MMTV promoter *in vivo*. *EMBO J.* **14**, 1737–51.

Tsanev, R. and Sendov, B. (1971). Possible molecular mechanism for cell differentiation in multicellular organisms. *J. Theor. Biol.* **30**, 337–93.

Tsukiyama, T., Becker, P.B. and Wu, C. (1994). ATP-dependent nucleosome disruption at a heat-shock promoter mediated by binding of GAGA transcription factor. *Nature* **367**, 525–32.

Tsukiyama, T. and Wu, C. (1995). Purification and properties of an ATP dependent nucleosome remodeling factor. *Cell* **83**, 1011–20.

Tsukiyama, T., Daniel, C., Tamkun, J. and Wu, C. (1995). ISWI, a member of the SWI2/SNF2 ATPase family, encodes the 140 kDa subunit of the nucleosme remodelling factor. *Cell* **83**, 1021–6.

Tuan, D., Solomon, W., Li, Q. and London, I.M. (1985). The '*β*-like-globin' gene domain in human erythroid cells. *Proc. Natl Acad. Sci. USA* **82**, 6384–8.

Tullius, T.D. and Dombroski, B.A. (1985). Iron (II) EDTA used to measure the helical twist along any DNA molecule. *Science* **230**, 679–81.

Turner, B.M. (1991). Histone acetylation and control of gene expression. *J. Cell. Sci.* **99**, 13–20.

Turner, B.M., Franchi, L. and Wallace, H. (1990). Islands of acetylated histone H4 in polytene chromosomes and their relationship to chromatin packaging and transcriptional activity. *J. Cell Sci.* **96**, 335–46.

Turner, B.M., Birley, A.J. and Lavender, J. (1992). Histone H4 isoforms acetylated at specific lysine residues define individual chromosomes and chromatin domains in *Drosophila* polytene nuclei. *Cell* **69**, 375–84.

Turner, B.M. (1993). Decoding the nucleosome. *Cell* **75**, 5–8.

Tyler, J.K., Bulger, M., Kamakaka, R.T., Kobayashi, R. and Kadonaga, J.T. (1996). The p55 subunit of *Drosophila* chromatin assembly factor 1 is homologous to a histone deacetylase-associated protein. *Mol. Cell Biol.* **16**, 6149–59.

Tzamarias, D. and Struhl, K. (1994). Functional dissection of the yeast Cyc8-Tup1 transcriptional co-repressor complex. *Nature* **369**, 758–61.

Uemura, T. and Yanagida, M. (1986). Mitotic spindle pulls but fails to separate chromosomes in type II DNA topoisomerase mutants: uncoordinated mitosis. *EMBO J.* **5**, 1003–10.

Ulitzer, N. and Gruenbaum, Y. (1989). Nuclear envelope assembly around sperm chromatin in cell-free preparations from *Drosophila* embryos. *FEBS Lett.* **259**, 113–16.

Ura, K., Wolffe, A.P. and Hayes, J.J. (1994). Core histone acetylation does not block linker histone binding to a nucleosome including a *Xenopus borealis* 5S rRNA gene. *J. Biol. Chem.* **269**, 27171–4.

Ura, K., Hayes, J.J. and Wolffe, A.P. (1995). A positive role for nucleosome mobility in the transcriptional activity of chromatin templates: restriction by linker histones. *EMBO J.* **14**, 3752–65.

Ura, K., Nightingale, K., and Wolffe, A.P. (1996). Differential association of HMG1 and linker histones B4 and H1 with dinucleosomal DNA: structural transitions and transcriptional repression. *EMBO J.* **15**, 4959–69.

Ura, K., Kurumizaka, H., Dimitrov, S., Almouzni, G., and Wolffe, A.P. (1997). Histone acetylation: influence on transcription by RNA polymerase, nucleosome mobility and positioning, and linker histone dependent transcriptional repression. *EMBO J.* **16**, 2096–107.

Usachenko, S.I., Bavykin, S.G., Gavin, I.M. and Bradbury, E.M. (1994). Rearrangement of the histone H2A C-terminal domain in the nucleosome. *Proc. Natl Acad. Sci. USA* **91**, 6845–9.

Usachenko, S.I., Gavin, I.M. and Bavykin, S.G. (1996). Alterations in nucleosome core structure in linker histone-depleted chromatin. *J. Biol. Chem.* **271** 3831–6.

Van Daal, A. and Elgin, S.C.R. (1992). A histone variant, H2AvD, is essential in *Drosophila melanogaster. Mol. Biol. Cell.* **3**, 593–602.

Van Daal, A., White, E.M., Gorovsky, M.A. and Elgin, S.C.R. (1988). *Drosophila* has a single copy of the gene encoding a highly conserved histone H2A variant of the H2A.F/Z type. *Nucleic. Acids Res.* **16**, 7487–97.

van Driel, R., Wansink, D. G., van Steensel, B., Grande, M. A., Schul, W. and de Jong, L. (1995). Nuclear domains and the nuclear matrix. *Int. Rev. Cytol.* **162A**, 151–89.

van Dyke, M.W., Roeder, R.G. and Sawadogo, M. (1988). Physical analysis of transcription preinitiation complex assembly on a class II gene promoter. *Science* **241**, 1335–8.

Van Holde, K.E. (1988). *Chromatin.* Springer-Verlag, New York.

van Holde, K. (1993). The omnipotent nucleosome. *Nature* **362**, 111–12.

van Holde, K.E. and Zlatanova, J. (1995). Chromatin higher order structure: chasing a mirage? *J. Biol. Chem.* **270**, 8373–6.

van Holde, K. and Zlatanova, J. (1996). What determines the folding of the chromatin fiber. *Proc. Natl Acad. Sci. USA* **93**, 10548–55.

van Lint, C., Emillani, S., Ott, M. and Verdin, E. (1996). Transcriptional activation and chromatin remodeling of the HIV-1 promoter in response to histone acetylation. *EMBO J.* **15**, 1112–20.

Van Steensel, B., van Binnendijk, E.P., Hornsby, C.D., van der Voort, H.T.M., Krozowski, Z.S., de Kloet, E.R. and van Driel, R. (1996). Partial colocalization of glucocorticoid and mineralocorticoid receptors in discrete compartments of rat hippocamus neurons. *J. Cell. Sci.* **109**, 787–92.

Varga-Weisz, P., Zlatanova, J., Leuba, S.H., Schroth, G.P. and van Holde, K. (1994). Binding of histones H1 and H5 and their globular domains to four way junction DNA. *Proc. Natl Acad. Sci. USA* **91**, 3525–9.

Varga-Weisz, P.D., Blank, T.A. and Becker, P.B. (1995). Energy-dependent chromatin accessibility and nucleosome mobility in a cell-free system. *EMBO J.* **14**, 2209–16.

Varga-Weisz, P.D., Wilm, M., Boute, E., Dumas, K., Mann, M. and Becker, P.B. (1997). Chromatin remodeling factor CHRAC contans the ATPses ISWI and topoisomerase II. *Nature* **388**, 598–602.

Varshavsky, A.J., Bakayev, V.V. and Georgiev, G.P. (1976). Heterogeneity of chromatin subunits *in vitro* and location of histone H1. *Nucleic. Acids Res.* **3**, 477–92.

Varshavsky, A.J., Sundin, O.H. and Bohn, M.J. (1978). SV40 viral minichromosome: preferential exposure of the origin of replication as probed by restriction endonucleases. *Nucleic. Acids Res.* **5**, 3469–77.

Venter, U., Svaren, J., Schmitz, J., Schmid, A. and Horz, W. (1994). A nucleosome precludes binding of the transcription factor Pho4 *in vivo* to a critical target site in the PHO5 activation. *J. Mol. Biol.* **231**, 658–67.

Verdin, E., Paras Jr, P. and Van Lint, C. (1993). Chromatin disruption in the promoter of human immunodeficiency virus type 1 during transcriptional activation. *EMBO J.* **12**, 3249–59.

Vernet, G., Sala-Rovira, M., Maeder, M., Jacques, F. and Herzog, M. (1990). Basic nuclear proteins of the histone less eukaryote *Gypthecodinium cohnii* (Pyrrhophyta): two dimensional electrophoresis and DNA binding properties. *Biochim. Biophys. Acta* **1048**, 281–9.

Verreault, A., Kaufman, P.D., Kobayashi, R. and Stillman, B. (1996). Nucleosome assembly by a complex of CAF-1 and acetylated histones H3/H4. *Cell* **87**, 95–104.

Verrijzer, C.P. and Tjian, R. (1996). TAFs mediate transcriptional activation and promoter selectivity. *Trends Biochem. Sci.* **21**, 338–42.

Verrijzer, C.P., Chen, J.L., Yokomori, K. and Tjian, R. (1995). Binding of TAFs to core elements directs promoter selectivity by RNA polymerase II. *Cell* **81**, 1115–25.

Vettesse-Dadey, M., Walter, P., Chen, H., Juan, L-J. and Workman, J.L. (1994). Role of the histone amino termini in facilitated binding of a transcription factor, GAL4-AH to nucleosome cores. *Mol. Cell. Biol.* **14**, 970–81.

Vettesse-Dadey, M., Grant, P.A., Hebbes, R.T., Crane-Robinson, C., Allis, C.D., and Workman, J.L. (1996). Acetylation of histone H4 plays a primary role in enhancing transcription factor binding to nucleosomal DNA *in vitro*. *EMBO J.* **15**, 2508–18.

Vidal, M., Strich, R., Esposito, R.E., and Gaber, R.F. (1991). RPD1 (SIN3/UME4) is required for maximal activation and repression of diverse yeast genes. *Mol. Cell. Biol.* **11**, 6306–16.

Vidal, M., and Gaber, R.F. (1991). RPD3 encodes a second factor required to achieve maximum positive and negative transcriptional states in Saccharomyces cerevisiae. *Mol. Cell. Biol.* **11**, 6317–27.

Visa, N., Alzhanova-Ericsson, A. T., Sun, X., Kiseleva, E., Bjorkroth, B., Wurtz, T. and Daneholt, B. (1996). A pre-mRNA-binding protein accompanies the RNA from the gene through the nuclear pores and into polysomes. *Cell* **84**, 253–64.

Visa, N., Puvion-Dutilleul, F., Harper, F., Bachellerie, J. P. and Puvion, E. (1993a). Intranuclear distribution of poly(A) RNA determined by electron microscope in situ hybridization. *Exp. Cell Res.* **208**, 19–34.

Visa, N., Puvion-Dutilleul, F., Bachellerie, J. P. and Puvion, E. (1993b). Intranuclear distribution of U1 and U2 snRNAs visualized by high resolution in situ hybridization: revelation of a novel compartment containing U1 but not U2 snRNA in HeLa cells. *Eur. J. Cell Biol.* **60**, 308–21.

Voeller, B.R. (1968). *The Chromosome Theory of Inheritance: Classic Papers in Development and Hereditary.* Appleton, New York.

Wabl, M.R. and Burrows, P.D. (1984). Expression of immunoglobulin heavy chain at a high level in the absence of a proposed immunoglobulin enhancer in cis. *Proc. Natl Acad. Sci. USA.* **81**, 2452–5.

Wade, P.A., Pruss, D. and Wolffe, A.P. (1997). Histone acetylation: chromatin in action. *Trends Biochem. Sci.* **22**, 128–32.

Waga, S., Mizuno, S. and Yoshida, M. (1989). Non histone proteins HMG1 and HMG2 suppress the nucleosome assembly at physiological ionic strength. *Biochim. Biophys. Acta* **1007**, 209.

Wagner, R.L., Apriletti, J.W., McGrath, M.E., West, B.L., Baxter, J.D. and Fletterick, R.J. (1995). A structural role for hormone in the thyroid hormone receptor. *Nature* **378**, 690–7.

Wahi, M., and Johnson, A.D. (1995). Identification of genes required for α2 repression in *Sacchromyces cerevisiae. Genetics* **140**, 79–90.

Wakefield, L. and Gurdon, J.B. (1983). Cytoplasmic regulation of 5S RNA genes is nuclear-transplant embryos. *EMBO J.* **2**, 1613–19.

Walker, J., Chen, T.A., Sterner, R., Berger, M., Winston, F. and Allfrey, V.G. (1990). Affinity chromatography of mammalian and yeast nucleosomes: two modes of binding of transcriptionally active mammalian nucleosomes to organomercurial columns and contrasting behaviour of the active nucleosomes of yeast. *J. Biol. Chem.* **265**, 5736–46.

Walker, S.S., Shen, W.C., Reese, J.C. Apone, L.M. and Green M.R. (1997). Yeast TAF(II)145 required for transcription of G1/S cyclin genes and regulated by the cellular growth state. *Cell* **90**, 607–14.

Wall, G., Varga-Weisz, P.D., Sandaltzopoulos, R. and Becker, P.B. (1995). Chromatin remodeling by GAGA factor and heat shock factor at the hypersensitive *Drosophila* hsp26 promoter *in vitro. EMBO J.* **14**, 1727–36.

Wallrath, L.L., Lu, Q., Granok, H. and Elgin, S.C.R. (1994). Architectural variations of inducible eukaryotic promoters: present and remodeling chromatin structures. *BioEssays* **16**, 165–70.

Walter, P.P., Owen-Hughes, T.A., Cote, J. and Workman, J.L. (1995). Stimulation of transcription factor binding and histone displacement by nucelosome assembly protein 1 and nucleoplasmin requires disruption of the histone octamer. *Mol. Cell Biol.* **15**, 6178–87.

Walters, M.C., Magis, W., Fiering, S., Eidemiller, J., Scalzo, D., Groudine, M. and Martin, D.I.K. (1996). Transcription enhancers act in *cis* to suppress position-effect variegation. *Genes Dev.* **10** 185–95.

Wan, J.S., Mann, R.K. and Grunstein, M. (1995). Yeast histone H3 and H4 N-termini function through different GAL1 regulatory elements to repress and activate transcription. *Proc. Natl Acad. Sci. USA* **92**, 5664–8.

Wang, H., and Stillman, D.J. (1990). In vitro regulation of a SIN3-dependent DNA-binding activity by stimulatory and inhibitory factors. *Proc. Natl Acad. Sci. USA* **87**, 9761–5.

Wang, H. and Stillman, D.J. (1993). Transcriptional repression in *Saccharomyces cerevisiae* by a SIN2-LexA fusion protein. *Mol. Cell Biol.* **13**, 1805–14.

Wang, H., Clark, I., Nicholon, P.R., Herskowitz, I., and Stillman, D.J. (1990). The *Saccharomyces cerevisiae* SIN3 gene a negative regulator of HO contains four paired amphipathic helix motifs. *Mol. Cell. Biol.* **10**, 5927–36.

Wang, J., Cao, L. G., Wang, Y. L. and Pederson, T. (1991). Localization of pre-messenger RNA at discrete nuclear sites. *Proc. Natl Acad. Sci. USA* **88**, 7391–5.

Wang, T. and Allis, C.D. (1992). Replication-dependent and independent regulation of HMG expression during the cell cycle and conjugation in *Tetrahymena*. *Nucleic. Acids Res.* **20**, 6525–33.

Wang, W., Côté, J., Xue, Y., Zhou, S., Khavara, P.A., Biggar, S.R., Muchardt, C., Kalpana, G.V., Goff, S.P., Yaniv, M., Workman, J.L., and Crabtree, G.R. (1996a). Purification and biochemical heterogeneity of the mammalian SWI/SNF complex. *EMBO J.* **15**, 5370–82.

Wang, W., Xue, Y., Zhou, S., Kuo, A., Cairns, B.R., and Crabtree, G.R. (1996b). Diversity and specialization of mammalian SWI/SNF complexes. *Genes & Dev.* **10**, 2117–30.

Wang, X.F. and Calame, K. (1986). SV40 enhancer-binding factors are required at the establishment but not the maintenance step of enhancer-dependent transcriptional activation. *Cell* **47**, 241–7.

Wang, Y.-H., Amirhaeri, S., Kang, S., Wells, R.D. and Griffith, J.D. (1994). Preferential nucleosome assembly at DNA triplet repeats from the myotonic dystrophy gene. *Science* **265**, 1709–12.

Wang, Y.-H. and Griffith, J. (1995). Expanded CTG triplet blocks from the myotonic dystrophy gene create the strongest known natural nucleosome positioning elements. *Genomics* **25**, 570–3.

Wang, Y.W. and Griffith, J. (1996). Methylation of expanded CCG triplet repeat DNA from fragile X syndrome patients enhances nucleosome exclusion. *J. Biol. Chem.* **271**, 22937–40.

Wang, Z.Q., Auer, B., Stingl, L., Berghammer, H., Haidacher, D., Schweiger, M. and Wagner, E.F. (1995). Mice lacking ADPRT and poly(ADP-ribosyl)ation develop normally but are susceptible to skin disease. *Genes Dev.* **9**, 509–20.

Wansink, D. G., Manders, E. E., van der Kraan, I., Aten, J. A., van Driel, R. and de Jong, L. (1994). RNA polymerase II transcription is concentrated outside replication domains throughout S-phase. *J. Cell Sci.* **107**, 1449–56.

Wansink, D. G., Schul, W., van der Kraan, I., van Steensel, B., van Driel, R. and de Jong, L. (1993). Fluorescent labeling of nascent RNA reveals transcription by RNA polymerase II in domains scattered throughout the nucleus. *J. Cell Biol.* **122**, 283–93.

Warburton, P.E. and Earnshaw, W.C. (1997). Untangling the role of DNA topoisomerase II in mitotic chromosome structure and function. *BioEssays* **19**, 97–9.

Warrell, R. P.,Jr., de The, H., Wang, Z. Y. and Degos, L. (1993). Acute promyelocytic leukemia [see comments]. *N. Engl. J. Med.* **329**, 177–89.

Wasylyk, B. and Chambon, P. (1979). Transcription by eukaryotic RNA polymerases A and B of chromatin assembled *in vitro*. *Eur. J. Biochem.* **98**, 317–27.

Waterborg, J.H. and Matthews, H.R. (1982). Control of histone acetylation. Cell cycle dependence of deacetylase activity in *Physarum* nuclei. *Exp. Cell Res.* **138**, 462–6.

Watson, J.D. and Crick, F.H.C. (1953). A structure for deoxyribosenucleic acids. *Nature (London)* **171**, 737–8.

Wechser, M.A., Kladde, M.P., Alfieri, J.A., and Peterson, C.L. (1997). Effects of Sin versions of histone H4 on yeast chromatin structure and function. *EMBO J.* **16**, 2086–95.

Wechsler, D.S., Papoulas, O., Dang, C.V. and Kingston, R.E. (1994). Differential binding of c-Myc and Max to nucleosomal DNA. *Mol. Cell. Biol.* **14**, 4097–107.

Weih, F., Nitsch, D., Reik, A., Schutz, G. and Becker, P.B. (1991). Analysis of CpG methylation and genomic footprinting at the tyrosine amino transferase gene: DNA methylation alone is not sufficient to prevent protein binding *in vivo*. *EMBO J.* **10**, 2559–67.

Weil, P.A., Luse, D.S., Segall, J. and Roeder, R.G. (1979). Selective and accurate initiation of transcription at the Ad2 major late promoter in a soluble system dependent on purified RNA polymerase II and DNA. *Cell* **18**, 469–84.

Weintraub, H. (1978). The nucleosome repeat length increases during electrophoresis in the chick. *Nucleic. Acids Res.* **5**, 1179–88.

Weintraub, H. (1984). Histone H1-dependent chromatin superstructures and the suppression of gene activity. *Cell* **38**, 17–27.

Weintraub, H. (1985). Assembly and propagation of repressed and derepressed chromosomal states. *Cell* **42**, 705–11.

Weintraub, H. (1988). Formation of stable transcription complexes as assayed by analysis of individual templates. *Proc. Natl Acad. Sci. USA* **85**, 5819–23.

Weintraub, H. and Groudine, M. (1976). Chromosomal subunits in active genes have an altered conformation. *Science* **193**, 848–56.

Weintraub, H., Worcel, A. and Alberts, B. (1976). A model for chromatin based upon two symmetrically paired half nucleosomes. *Cell* **9**, 409–17.

Weintraub, H., Beug, H., Groudine, M. and Graf, T. (1982). Temperature sensitive changes in the structure of globin chromatin in lines of red cell precursors transformed by ts-AEV. *Cell* **28**, 931–40.

Weir, H.M., Kraulis, P.J., Hill, C.S., Raine, A.R.C., Laue, E.D. and Thomas, J.O. (1993). Structure of the HMG box motif in the B domain of HMG1. *EMBO J.* **12**, 1311–19.

Weis, K., Rambaud, S., Lavau, C., Jansen, J., Carvalho, T., Carmo-Fonseca, M., Lamond, A. and Dejean, A. (1994). Retinoic acid regulates aberrant nuclear localization of PML-RAR alpha in acute promyelocytic leukemia cells. *Cell* **76**, 345–56.

Weisbrod, S., Wickens, M.P., Whytock, S. and Gurdon, J.B. (1982). Active chromatin of oocytes injected with somatic cell nuclei or cloned DNA. *Dev. Biol.* **94**, 216–29.

Weiss, E., Ghose, D., Schultz, P. and Oudet, P. (1985). Tr antigen is the only detectable protein on the nucleosome-free origin region of isolated simian virus 40 minichromosomes. *Chromosoma (Berl)* **92**, 391–400.

Wells, D. and Brown, D. (1991). Histone and histone gene compilation and alignment update. *Nucleic Acids Res.* **19**, 2173–88.

Wenkert, D. and Allis, C.D. (1984). Timing of the appearance of the macronuclear-specific histone variant hv1 and gene expression in developing new macronuclei of *Tetrahymena thermophilia*. *J. Cell Biol.* **98**, 2107–17.

Werner, M.H., Huth, J.R., Gronenborn, A.M. and Clore, G.M. (1995). Molecular basis of human 46X, Y sex reversal revealed from the three-dimensional solution structure of the human SRY-DNA complex. *Cell* **81**, 705–14.

West, M.H.P. and Bonner, W.M. (1980). Histone H2B can be modified by the attachment of ubiquitin. *Nucleic Acids Res.* **8**, 4671–80.

Westerman, R. and Grossbach, U. (1984). Localization of nuclear proteins related to high mobility group protein 14 (HMG14) in polytene chromosomes. *Chromosoma* **90**, 355–65.

White, E., Shapiro, D.L., Allis, C.D. and Gorovsky, M.A. (1988a). Sequence and properties of the message encoding *Tetrahymena* hv1, a highly conserved histone H2A variant that is associated with active genes. *Nucleic. Acids Res.* **16**, 179–98.

White, J.H., Cozzarelli, N.R. and Bauer, W.R. (1988b). Helical repeat and linking number of surface wrapped DNA. *Science* **241**, 323–7.

White, M.J.D. (1973). *Animal Cytology and Evolution*. Cambridge University Press, Cambridge, pp. 1–58.

White, R.J. (1994). *RNA Polymerase III Transcription*. R.G. Landes, Austin, TX.

Widlund, H.R., Cao, H., Simonsson, S., Magnusson, E., Simonsson, T., Nielsen, P.E., Kahn, J.D., Crothers, D.M. and Kubista, M. (1997). Identification and characterization of genomic nucleosome-positioning sequence. *J. Mol. Biol.* **267**, 807–17.

Widom, J. and Klug, A. (1985). Structure of the 300 Å chromatin filament: X-ray diffraction from orientated samples. *Cell* **43**, 207–13.

Wiekowski, M., Miranda, M. and De Pamphilis, M.L. (1991). Regulation of gene expression in preimplantation mouse embryos: effects of zygotic gene expression and the first mitosis on promoter and enhancer activities. *Dev. Biol.* **147**, 403–14.

Wiekowski, M., Miranda, M. and De Pamphilis, M.L. (1993). Requirements for promoter activity in mouse oocytes and embryos distinguish paternal pronuclei from maternal and zygotic nuclei. *Dev. Biol.* **159**, 366–78.

Wijgerde, M., Grosveld, F. and Fraser, P. (1995). Transcription complex stability and chromatin dynamics *in vivo*. *Nature* **377**, 209–13.

Wildeman, A.G., Zenke, M., Schatz, C., Wintzerith, M., Grundstrom, T.T., Matthes, H., Takahaski, K. and Chambon, P. (1986). Specific protein binding to the simian virus 40 enhancer *in vitro*. *Mol. Cell. Biol.* **6**, 2098–105.

Willard, H.F., Brown, C.J., Carrel, L., Hendrich, B. and Miller, A.P. (1993). Epigenetic and chromosomal control of gene expression: molecular and genetic analysis of X chromosome inactivation. *Cold Spring Harb. Symp. Quant. Biol.* **58**, 315–25.

Williams, F.E. and Trumbly, R.J. (1990). Characterization of TUP1 a mediator of glucose repression in *Saccharomyces cerevisiae*. *Mol. Cell Biol.* **10**, 6500–11.

Williams, S.P. and Langmore, J.P. (1991). Small angle x-ray scattering of chromatin. *Biophys. J.* **69**, 606–18.

Williamson, P. and Felsenfeld, G. (1978). Transcription of histone-covered T7 DNA by *Escherichia coli* RNA polymerase. *Biochemistry* **17**, 5695–705.

Williamson, R. (1970). Properties of rapidly labelled deoxyribonucleic acid fragments isolated from the cytoplasm of primary cultures of embryonic mouse liver cells. *J. Mol. Biol.* **51**, 157–68.

Wilmut, I., Schnieke, A.E., McWhir, J., Kind, A.J. and Campbell, K.H. (1997). Viable offspring derived from fetal and adult mammalian cells. *Nature* **385**, 810–13.

Wilson, C.J., Chao, D.M., Imbalzano, A.N., Schnitzer, G.R., Kingston, R. and Young, R.A. (1996). RNA polymerase II holoenzyme contains SWI/SNF regulators involved in chromatin remodelling. *Cell* **84**, 235–44.

Wilson, E.B. (1925). *The Cell in Development and Heredity*. Macmillan, New York.

Wilt, F. (1963). The synthesis of ribonucleic acid in sea urchin embryos. *Biochem. Biophys. Res. Commun.* **11** 447–57.

Wilt, F. (1964). Ribonucleic acid synthesis during sea urchin embryogenesis. *Dev. Biol.* **9**, 299–315.

Wingender, E., Jahn, D. and Seifart, K.H. (1986). Association of RNA polymerase III with transcription factors in the absence of DNA. *J. Biol. Chem.* **261**, 1409–13.

Winston, F. and Carlson, M. (1992). Yeast SNF/SWI transcriptional activators and the SPT/SIN chromatin connection. *Trends Genet.* **8**, 387–91.

Wisniewski, J.R. and Schulze, E. (1994). High affinity interaction of dipteran high mobility group (HMG) proteins 1 with DNA is modulated by COOH-terminal regions flanking the HMG box domain. *J. Biol. Chem.* **269**, 10713–19.

Wolffe, A.P. (1988). Transcription fraction TFIIIC can regulate differential *Xenopus* 5S RNA gene transcription *in vitro*. *EMBO J.* **7**, 1071–9.

Wolffe, A.P. (1989a). Transcriptional activation of *Xenopus* class III genes in chromatin isolated from sperm and somatic nuclei. *Nucleic. Acid Res.* **17**, 767–80.

Wolffe, A.P. (1989b). Dominant and specific repression of *Xenopus* oocyte 5S RNA genes and satellite I DNA by histone H1. *EMBO J.* **8**, 527–37.

Wolffe, A.P. (1990a). Transcription complexes. *Prog. Clin. Biol. Res.* **322**, 171–86.

Wolffe, A.P. (1990b). New approaches to chromatin function. *New Biol.* **2**, 211–18.

Wolffe, A.P. (1991a). Developmental regulation of chromatin structure and function. *Trends Cell Biol.* **1**, 61–6.

Wolffe, A.P. (1991b). RNA polymerase III transcription. *Curr. Opin. Cell Biol.* **3**, 461–6.

Wolffe, A.P. (1991c). Implications of DNA replication for eukaryotic gene expression. *J. Cell Sci.* **99**, 201–6.

Wolffe, A.P. (1993). Replication timing and *Xenopus* 5S RNA gene transcription *in vitro*. *Dev. Biol.* **157**, 224–31.

Wolffe, A.P. (1994a). Insulating chromatin. *Curr. Biol.* **4**, 85–7.

Wolffe, A.P. (1994b). Switched-on chromatin. *Curr. Biol.* **4**, 525–7.

Wolffe, A.P. (1994c). Nucleosome positioning and modification: chromatin structures that potentiate transcription. *Trends Biochem. Sci.* **19**, 240–4.

Wolffe, A.P. (1994d). The role of transcription factors, chromatin structure and DNA replication in 5S RNA gene regulation. *J. Cell. Sci.* **107**, 2055–63.

Wolffe, A.P. (1996). Histone deacetylase: a regulatory of transcription. *Science* **272**, 371–2.

Wolffe, A.P. (1997). Sinful repression. *Nature* **387**, 16–17.

Wolffe, A.P. and Brown, D.D. (1986). DNA replication *in vitro* erases a *Xenopus* 5S RNA gene transcription complex. *Cell* **47**, 217–27.

Wolffe, A.P. and Brown, D.D. (1987). Differential 5S RNA gene expression *in vitro*. *Cell* **51**, 733–40.

Wolffe, A.P. and Brown, D.D. (1988). Developmental regulation of two 5S ribosomal RNA genes. *Science* **241**, 1626–32.

Wolffe, A.P. and Drew, H.R. (1989). Initiation of transcription on nucleosomal templates. *Proc. Natl Acad. Sci. USA* **86**, 9817–21.

Wolffe, A.P. and Morse, R.H. (1990). The transcription complex of the *Xenopus* somatic 5S RNA gene. *J. Biol. Chem.* **265**, 4592–9.

Wolffe, A.P. and Pruss, D. (1996a). Deviant nucleosomes: the functional specialization of chromatin. *Trends Genet.* **12**, 58–62.

Wolffe, A.P. and Pruss, D. (1996b). Targeting chromatin disruption: transcription regulators that acetylate histones. *Cell* **84**, 817–19.

Wolffe, A.P. and Schild, C. (1991). Chromatin assembly. *Methods Cell Biol.* **36**, 541–59.

Wolffe, A.P., Jordan, E. and Brown, D.D. (1986). A bacteriophage RNA polymerase transcribes through a *Xenopus* 5S RNA gene transcription complex without disrupting it. *Cell* **44**, 381–9.

Wolffe, A.P., Andrews, M.T., Crawford, E.T., Losa, R.M. and Brown, D.D. (1987). Negative supercoiling is not required for 5S RNA transcription *in vitro*. *Cell* **49**, 301–3.

Wolffe, A.P., Khochbin, S. and Dimitrov, S. (1997a). What do linker histones do in chromatin? *BioEssays* **19**, 249–55.

Wolffe, A.P., Wong, J. and Pruss, D. (1997b). Activators and repressors: making use of chromatin to regulate transcription. *Genes to Cells* **2**, 291–302.

Wong, E.H. and Tjian, R. (1994). Promoter selective transcriptional defect in cell cycle mutant ts13 rescued by hTAFII250. *Science* **263**, 811–14.

Wong, J., Shi, Y.-B. and Wolffe, A.P. (1995). A role for nucleosome assembly in both silencing and activation of the *Xenopus* TRβA gene by the thyroid hormone receptor. *Genes Dev.* **9**, 2696–711.

Wong, J., Shi, Y.-B. and Wolffe, A.P. (1997a). Determinants of chromatin disruption and transcriptional regulation instigated by the thyroid hormone receptor: hormone regulated chromatin disruption is not sufficient for transcriptional activation. *EMBO J.* **16**, 3158–71.

Wong, J., Li, Q., Levin, B.-Z., Shi, Y.-B. and Wolffe, A.P. (1997b). Structural and functional features of a specific nucleosome containing a recognition element for the thyroid hormone receptor. *EMBO J.* **16**, 7130–45.

Wong, J., Patterton, D., Imhof, A., Guschin, D., Shi, Y.-B. and Wolffe, A.P. (1998). Distinct requirements for chromatin assembly in transcriptional repression by thyroid hormone receptor and histone deacetylase. *EMBO J.* **17**, 520–534.

Wood, W.I. and Felsenfeld, G. (1982). Chromatin structure of the chicken β-globin gene region: sensitivity to DNaseI, micrococcal nuclease, and DNaseII. *J. Biol. Chem.* **257**, 7730–6.

Woodcock, C.L.F. (1973). Ultrastructure of inactive chromatin. *J. Cell. Biol.* **59**, 368a.

Woodcock, C.L. (1994). Chromatin fibers observed *in situ* in frozen hydrated sections. Native fiber diameter is not correlated with nucleosome repeat length. *J. Cell Biol.* **125**, 11–19.

Woodcock, C.L., Frado, L.L. and Rattner, J.B. (1984). The higher-order structure of chromatin: evidence for a helical ribbon arrangement. *J. Cell. Biol.* **99**, 42–52.

Woodcock, C.L., Grigoryer, S.A., Horowitz, R.A. and Whitaker, N. (1993). A folding model for chromatin that incorporates linker DNA variability produces fibers that mimic the native structures. *Proc. Natl Acad. Sci. USA* **90**, 9021–25.

Woodland, H.R. and Adamson, E.D. (1977). The synthesis and storage of histones during the oogenesis of *Xenopus laevis*. *Dev. Biol.* **57**, 118–35.

Woodland, H.R., Flynn, J.M. and Wyllie, A.J. (1979). Utilization of stored mRNA in *Xenopus* embryos and its replacement by newly synthesized transcripts: histone H1 synthesis using interspecies hybrids. *Cell* **18**, 165–71.

Worcel, A. (1978). Molecular architecture of the chromatin fibre. *Cold Spring Harb. Symp. Quant. Biol.* **42**, 313–24.

Worcel, A. and Burgi, E. (1972). On the structure of the folded chromosome of *E. coli*. *J. Mol. Biol.* **71**, 127–48.

Worcel, A., Han, S. and Wong, M.L. (1978). Assembly of newly replicated chromatin. *Cell* **15**, 969–77.

Workman, J.L. and Kingston, R.E. (1992). Nucleosome core displacement *in vitro* via a metastable transcription factor/nucleosome complex. *Science* **258**, 1780–4.

Workman, J.L. and Roeder, R.G. (1987). Binding of transcription factor TFIID to the major late promoter during *in vitro* nucleosome assembly potentiates subsequent initiation by RNA polymerase II. *Cell* **51**, 613–22.

Workman, J.L., Abmayr, S.M., Cromlish, W.A. and Roeder, R.G. (1988). Transcriptional regulation of the immediate early protein of pseudorabies virus during *in vitro* nucleosome assembly. *Cell* **55**, 211–19.

Workman, J.L., Roeder, R.G. and Kingston, R.E. (1990). An upstream transcription factor, USF (MLTF), facilitates the formation of preinitiation complexes. *EMBO J.* **9**, 1299–308.

Workman, J.L., Taylor, I.C.A. and Kingston, R.E. (1991). Activation domains of stably bound GAL4 derivatives alleviate repression of promoters by nucleosomes. *Cell* **64**, 533–44.

Wormington, W.M., and Brown, D.D. (1983). Onset of 5S RNA gene regulation during *Xenopus* embryogenesis. *Dev. Biol.* **99**, 248–57.

Wormington, W.M., Schlissel, M. and Brown, D.D. (1982). Developmental regulation of *Xenopus* 5S RNA genes. *Cold Spring Harb. Symp. Quant. Biol.* **47**, 879–84.

Worrad, D.M., Turner, B.M. and Schultz, R.M. (1995). Temporally restricted spatial localization of acetylated isoforms of histone H4 and RNA polymerase II in the 2-cell mouse embryo. *Development* **121**, 2949–59.

Wu, C. (1984). Two protein-binding sites in chromatin implicated in the activation of heat shock genes. *Nature* **309**, 229–34.

Wu, C. and Gilbert, W. (1981). Tissue-specific exposure of chromatin structure at the 5′ terminus of the rat prepro insulin II gene. *Proc. Natl Acad. Sci. USA* **78**, 1577–80.

Wu, C., Binham, P.M., Livak, K.J., Holmgren, R. and Elgin, S.C.R. (1979). The chromatin structure of specific genes: evidence for higher order domains of defined DNA sequence. *Cell* **16**, 797–806.

Wu, G.J. (1978). Adenovirus DNA-directed transcription of 5S RNA *in vitro*. *Proc. Natl Acad. Sci. USA* **75**, 2175–9.

Wu, M., Allis, C.D., Richman, R., Cook, R.G. and Govorsky, M.A. (1986). An intervening sequence in an unusual histone H1 gene of *Tetrahymena thermophila*. *Proc. Natl Acad. Sci. USA* **83**, 8674–8.

Wu, M., Allis, C.D. and Gorovsky, M.A. (1988). Cell cycle regulation as a mechanism for targeting proteins to specific DNA sequences in *Tetrahymena thermophila*. *Proc. Natl Acad. Sci. USA* **85**, 2205–9.

Wu, M., Allis, C.D., Sweet, M.T., Cook, R.G. and Gorovsky, M.A. (1994). Four distinct and unusual linker histones in a mitotically dividing nucleus are derived from a 71 Kd polyprotein lack p34 cdc2 sites and contain protien A kinase sites. *Mol. Cell. Biol.* **14**, 10–20.

Wuarin, J. and Schibler, U. (1994). Physical isolation of nascent RNA chains transcribed by RNA polymerase II: evidence for cotranscriptional splicing. *Mol. Cell Biol.* **14**, 7219–25.

Wyllie, A.H., Gurdon, J.B. and Price, J. (1977). Nuclear localization of an oocyte component required for the stability of injected DNA. *Nature (London)* **268**, 150–2.

Wyllie, A.H., Laskey, R.A., Finch, J. and Gurdon, J.B. (1978). Selective DNA conservation and chromatin assembly after injection of SV40 DNA into *Xenopus* oocytes. *Dev. Biol.* **64**, 178–88.

Xie, X., Kokubo, T., Cohen, S.L., Mirza, U.A., Hoffmann, A., Chait, B.T., Roeder, R.G., Nakatani, Y. and Burley, S.K. (1996). Structural similarity between TAFs and the heterotetrameric core of the histone octamer. *Nature* **380**, 316–22.

Xing, Y. and Lawrence, J.B. (1993). Nuclear RNA tracks: structural basis for transcription and splicing? *Trends in Cell Biol.* **3**, 346–53.

Xing, Y., Johnson, C. V., Dobner, P. R. and Lawrence, J. B. (1993). Higher level organization of individual gene transcription and RNA splicing [see comments]. *Science* **259**, 1326–30.

Xing, Y., Johnson, C. V., Moen, P. T., Jr., McNeil, J. A. and Lawrence, J. (1995). Nonrandom gene organization: structural arrangements of specific pre-mRNA transcription and splicing with SC-35 domains. *J. Cell Biol.* **131**, 1635–47.

Xu, M., Barnard, M.B., Rose, S.M., Cockerill, P.N., Huang, S-Y. and Garrard, W.T. (1986). Transcription termination and chromatin structure of the active immunoglobulin K gene locus. *J. Biol. Chem.* **261**, 3838–45.

Yamamoto, K.R. (1985). Steroid receptor regulated transcription of specific genes and gene networks. *Annu. Rev. Genet.* **19**, 209–52.

Yanagi, K., Privé, G.G. and Dickerson, R.E. (1991). Analysis of local helix geometry in three B-DNA decamers and eight dodecamers. *J. Mol. Biol.* **217**, 201–14.

Yang, C.C. and Nash, H.A. (1989). The interaction of *E. coli* IHF protein with its specific binding sites. *Cell* **57**, 869–80.

Yang, G., Leuba, S.H., Bustamente, C., Zlatanova, J. and van Holde, K. (1994). Role of linker histones in extended fibre structure. *Nature Struct. Biol.* **1**, 761–3.

Yang, W.M., Inouye, C., Zeng, Y., Bearss, D., and Soto, E. (1996a). Transcriptional repression by YY1 is mediated by interaction with a mammalian homolog of the yeast global regulator RPD3. *Proc. Natl Acad. Sci. USA* **93**, 12845–50.

Yang, X.-J., Ogryzko, V.V., Nishikawa, J.-I., Howard, B. and Nakatani, Y. (1996b). A p300/CBP-associated factor that competes with the adenoviral E1A oncoprotein. *Nature* **382**, 319–24.

Yao, J., Lowary, P.T. and Widom, J. (1990). Direct detection of linker DNA bending in defined-length oligomers of chromatin. *Proc. Natl Acad. Sci. USA* **87**, 7603–7.

Yao, J., Lowary, P.T. and Widom, J. (1991). Linker DNA bending by the core histones of chromatin. *Biochemistry* **30**, 8408–14.

Ye, Q. and Worman, H.J. (1994). Primary structure analysis and lamin B and DNA binding of human LBR, an integral protein of the nuclear envelope inner membrane. *J. Biol. Chem.* **269**, 11306–11.

Ye, Q. and Worman, H.J. (1996). Interaction of an integral protein of the nuclear envelop inner membrane and human chromodomain proteins homolgous to *Drosophila* HP1. *J. Biol. Chem.* **271**, 14653–6.

Ye, Q., Callebraut, I., Pezhman, A., Courvalin, J.C. and Worman, H.J. (1997). Domain-specific interactions of human HP1-type chromodomain proteins and inner nuclear membrane protein LBR. *J. Biol. Chem.* **272**, 14983–9.

Yen, T.J., Compton, D.A., Wise, D., Zinkowski, R.P., Bunkley, B.R., Earnshaw, W.C. and Cleveland, D.W. (1991). CENP-E, a human centromere-associated protein required for progression from metaphase to anaphase. *EMBO J.* **10**, 1245–54.

Yen, T.J., Li, G., Schaar, B.T., Szilak, I. and Cleveland, D.W. (1992). CENP-E is a putative kinetochore metor that accumulates just before mitosis. *Nature* **359**, 536–9.

Yoda, K., Kitagawa, K., Matsumoto, H., Muro, Y. and Okazaki, T. (1992). A human centromere protein, CENP-B, has a DNA binding domain containing four potential alpha-helices at the NH_2-terminus, which is separable from dimerizing activity. *J. Cell Biol.* **119**, 1413–27.

Yoda, K., Nakamura, T., Matsumoto, M., Suzuki, N., Kitagawa, K., Nakano, M., Shinjo, A. and Okazaki, T. (1996). Centromere protein B of African Green Monkey cells: gene structure, cellular expression and centromeric localization. *Mol. Cell Biol.* **16** 5169–77.

Yoder, J.A., Walsh, C.P. and Bestor, T.H. (1997). Cytosine methylation and the ecology of intragenomic parasites. *Trends Genet.* **13**, 335–40.

Yoshida, I., Kashio, N. and Takagi, N. (1993). Cell fusion-induced quick change in replication time of the inactive mouse X chromosome: an implication for the maintenance mechanism of late replication. *EMBO J.* **12**, 4397–405.

Yoshida, M., Kijima, M., Akita, M., and Beppu, T. (1990). Potent and specific inhibition of mammalian histone deacetylase both *in vivo* and *in vitro* by Trichostatin A. *J. Biol. Chem.* **265**, 17174–9.

Yoshida, M., Nomura, S. and Beppu, T. (1987). Effects of trichostatins on differentiation of murine erythroleukemia cells. *Cancer Res.* **47**, 3688–91.

Yoshida, M., Horinouoshi, S. and Beppu, T. (1995). Trichostatin A and trapoxin: novel chemical probes for the role of histone acetylation in chromatin structure and function. *BioEssays* **17**, 423–30.

Yoshinaga, S.K., Peterson, S.L., Herskowitz, I. and Yamamoto, K.R. (1992). Roles of SWI1, SWI2 and SWI3 proteins for transcriptional enhancement by steroid receptors. *Science* **258**, 1598–604.

Young, R.A. (1991). RNA polymerase II. *Annu. Rev. Biochem.* **60**, 689–715.

Zachar, Z., Kramer, J., Mims, I. P. and Bingham, P. M. (1993). Evidence for channeled diffusion of pre-mRNAs during nuclear RNA transport in metazoans. *J. Cell Biol.* **121**, 729–42.

Zakian, V.A. (1989). Structure and function of telomeres. *Ann. Rev. Genet.* **23**, 579–604.

Zaret, K.S. and Yamamoto, K.R. (1984). Reversible and persistent changes in chromatin structure accompany activation of a glucocorticoid dependent enhancer element. *Cell* **38**, 29–38.

Zawel, L. and Reinberg, D. (1992). Advances in RNA polymerase II transcription. *Curr. Opin. Cell Biol.* **4**, 488–95.

Zawel, L. and Reinberg, D. (1993). Initiation of transcription by RNA polymerase II. A multistep process. *Prog. Nucleic. Acids Res. Mol. Biol.* **44**, 67–108.

Zenke, M., Grundstrom, T., Matthes, H., Wintzerith, M., Schatz, C., Wildeman, A. and Chambon, P. (1986). Multiple sequence motifs are involved in SV40 enhancer function. *EMBO J.* **5**, 387–97.

Zentgraf, H. and Franke, W.W. (1984). Differences of supra nucleosomal organization in different kinds of chromatin: cell type-specific globular subunits containing different numbers of nucleosomes. *J. Cell Biol.* **99**, 272–86.

Zervos, A.S., Gyuris, J. and Brent, R. (1993). Mxi1, a protein that specifically interacts with Max to bind Myc-Max recognition sites. *Cell* **72**, 223–32.

Zhang, G., Taneja, K. L., Singer, R. H. and Green, M. R. (1994). Localization of pre-mRNA splicing in mammalian nuclei [see comments]. *Nature* **372**, 809–12.

Zhao, K., Kas, E., Gonzalez, E. and Laemmli, U.K. (1993). SAR-dependent mobilization of histone H1 by HMG-I/Y *in vitro*: HMG-I/Y is enriched in H1-depleted chromatin. *EMBO J.* **12**, 3237–47.

Zhong, Z., Shine, L., Kaplan, S. and de Lange, T. (1992). A mammalian factor that binds telomeric TTAGGG repeats *in vitro*. *Mol. Cell. Biol.* **12**, 4834–43.

Zink, B. and Paro, R. (1989). *In vivo* binding pattern of a *trans* regulator of homeotic genes in *Drosophila melanogaster*. *Nature (London)* **337**, 468–71.

Zink, D. and Paro, R. (1995). *Drosophila* polycomb-group regulated chromatin inhibits the accessibility of a trans-activator to its target DNA. *EMBO J.* **14**, 5660–71.

Zirbel, R. M., Mathieu, U. R., Kurz, A., Cremer, T. and Lichter, P. (1993). Evidence for a nuclear compartment of transcription and splicing located at chromosome domain boundaries. *Chromosome Res.* **1**, 93–106.

Zlatanova, J. and van Holde, K. (1996). The linker histones and chromatin structure: new twists. *Prog. Nucleic. Acids Res. Mol. Biol.* **52**, 217–59.

Zweidler, A. (1980). Nonallelic histone variants in development and differentiation. *Dev. Biochem.* **15**, 47–56.

Zweidler, A. (1992). Role of individual histone tyrosines in the formation of the nucleosome complex. *Biochemistry* **31**, 9205–11.

Index